Environmental Science and Engineering

The ultimate goal of this series is to contribute to the protection of our environment, which calls for both profound research and the ongoing development of solutions and measurements by experts in the field. Accordingly, the series promotes not only a deeper understanding of environmental processes and the evaluation of management strategies, but also design and technology aimed at improving environmental quality. Books focusing on the former are published in the subseries Environmental Science, those focusing on the latter in the subseries Environmental Engineering.

Partha Pratim Adhikary · Pravat Kumar Shit · Jayasree Laha
Editors

Soil, Water Pollution and Mitigation Strategies

A Spatial Approach

Springer

Editors
Partha Pratim Adhikary
Remote Sensing Application
ICAR National Bureau of Soil Survey
and Land Use Planning
Nagpur, Maharashtra, India

Pravat Kumar Shit
Department of Geography
Raja N. L. Khan Women's College
(Autonomous)
Midnapore, West Bengal, India

Jayasree Laha
Principal
Raja N. L. Khan Women's College
(Autonomous)
Midnapore, West Bengal, India

ISSN 1863-5520 ISSN 1863-5539 (electronic)
Environmental Science and Engineering
ISBN 978-3-031-63295-2 ISBN 978-3-031-63296-9 (eBook)
https://doi.org/10.1007/978-3-031-63296-9

This Springer imprint is published by the registered company Springer Nature Switzerland AG
The registered company address is: Gewerbestrasse 11, 6330 Cham, Switzerland

Dedicated to
Young Scholars in the field of Soil Science and Environment

Preface

This book demonstrates the measurement, monitoring, mapping, and modeling of soil and water pollution, as well as strategies for mitigation. It delves into cutting-edge techniques that utilize modern geospatial methods, with a particular focus on the latest trends in data mining and robust modeling. In today's world, soil and water resources have become crucial components of the environment, supporting the foundations of agricultural ecosystems. Unfortunately, human activities in recent times have resulted in uncontrolled modifications to agricultural practices, primarily due to the excessive use of chemical fertilizers and pesticides. In light of this, spatial modeling, prediction, efficient utilization, risk assessment, protection, and management of soil and water resources within the agricultural system are paramount in achieving sustainable development goals and ensuring the promotion of economic, social, and environmental sustainability in the future. The present book aims to provide a quantitative approach and risk zone assessment for the proper mitigation of soil and water resources using geospatial technology. We express our heartfelt gratitude to all the authors who diligently completed their contributions on short notice and played a vital role in creating this enlightening and valuable publication. We firmly believe that this book will prove to be a valuable resource for individuals working in the fields of soil science, agriculture science, geography, hydro-ecology, environmental science, and others involved in soil-water resource management, including research scholars, environmentalists, and policymakers. We would also like to extend our sincere appreciation to Springer publishing house for collaborating with us on this timely publication.

Nagpur, India — Partha Pratim Adhikary
Midnapore, India — Pravat Kumar Shit
Midnapore, India — Jayasree Laha

Acknowledgments

The preparation of this book has been guided by numerous subject experts. We express our gratitude to these experts for generously devoting their time to evaluating the chapters included in this publication. Additionally, we extend our thanks to the anonymous reviewers for their constructive comments, which greatly contributed to the significant improvement in the quality of this book. As this project has taken a considerable amount of time, we would like to express our appreciation to our family and friends for their unwavering support. Their encouragement has been invaluable throughout the process. Furthermore, we are indebted to our students for their constant inspiration, our teachers for imparting their knowledge, our colleagues and collaborators for their enthusiasm, and our family for their unwavering support. Finally, we would like to acknowledge our publisher, Springer, and its publishing editor for their continuous support in bringing this book to fruition.

Contents

Editors and Contributors

About the Editors

Partha Pratim Adhikary (Ph.D.) is Principal Scientist at ICAR-National Bureau of Soil Survey and Land Use Planning, Nagpur, India. He obtained his Ph.D. in Agricultural Physics from ICAR-Indian Agricultural Research Institute, New Delhi, India. His research interests include solute transport, soil and water conservation and management, pedotransfer functions, and geospatial modeling of natural resources. Dr. Adhikary has published more than 80 research papers in peer-reviewed journals and 10 books. His other publications include book chapters, popular articles, technology brochures, technical bulletins, and scientific reports. He is the chief editor of *Agropedology and Environment* and the associate editor of *Indian Journal of Soil Conservation*. He is also the guest editor of *Environmental Science and Pollution Research* and *Applied Water Science*. Currently, he is the editor of Springer-Nature book series *GIScience and Geo-environmental Modelling*.

Pravat Kumar Shit (Ph.D.) is Assistant Professor at the PG Department of Geography, Raja N. L. Khan Women's College (Autonomous), West Bengal, India. He received his M.Sc. and Ph.D. degrees in Geography from Vidyasagar University and PG Diploma in Remote Sensing and GIS from Sambalpur University. His research interests include applied geomorphology, soil erosion, groundwater, forest resources, wetland ecosystem, environmental contaminants and pollution, and natural resources mapping and modeling. He has published twenty-six books (Springer-18, Elsevier-04, CRC Press-01, Rawat Publication-01, Other-02) and more than 100 papers in peer-reviewed journals and 90 book chapters. He is also the guest editor of *Environmental Science and Pollution Research* and *Applied Water Science*. He is currently the editor of the *GIScience and Geo-environmental Modelling* (GGM) Book Series, Springer-Nature.

Jayasree Laha (Ph.D.) is an esteemed academic and renowned scientist who currently holds the position of Principal at Raja N. L. Khan Women's College (Autonomous) in West Bengal, India. With a strong passion for both research and teaching, Dr. Laha has made significant contributions to the fields of nanotechnology, heavy metals pollution, and environmental contaminants. Dr. Laha's academic journey commenced at the prestigious Indian Institute of Technology (IIT-Kharagpur) in Kharagpur, India, where she pursued her M.Sc. and Ph.D. degrees. It was during her time at IIT-Kharagpur that Dr. Laha developed a keen interest in the fascinating overlap between chemistry and environmental science. She focused her studies on investigating the impact of nanoparticles and heavy metals on the environment, recognizing the importance of understanding their effects and finding sustainable solutions. Throughout her illustrious career, Dr. Laha has consistently demonstrated a deep commitment to scientific research, contributing immensely to her field of expertise. Her groundbreaking work has been published in over 20 research papers in prestigious peer-reviewed journals, highlighting her expertise and innovative research methodologies. Her contributions have earned widespread recognition and admiration from fellow researchers in the field. Recognizing the importance of disseminating knowledge and sharing her expertise, Dr. Laha has also edited three comprehensive books. These books provide comprehensive insights into the topics of nanoparticles, heavy metals pollution, environmental contaminants and forestry, making them valuable resources for students, researchers, and professionals in the field. As the Principal of Raja N. L. Khan Women's College (Autonomous), Dr. Laha continues to inspire and guide young minds in the pursuit of scientific excellence. She is known for her dynamic leadership, fostering an environment of innovation and academic growth within the institution.

Contributors

Partha Pratim Adhikary ICAR-Indian Institute of Water Management, Chandrasekharpur, Bhubaneswar, Odisha, India;
ICAR-National Bureau of Soil Survey & Land Use Planning, Nagpur, Maharashtra, India

Khalil Ahmad Department of Chemistry, University of Management and Technology (UMT), Sialkot, Punjab, Pakistan;
Department of Chemistry, Emerson University Multan, Multan, Pakistan

R. H. Ankhila ICAR-Indian Institute of Water Management, Chandrasekharpur, Bhubaneswar, Odisha, India

Ifzan Arshad Department of Chemistry, University of Management and Technology (UMT), Sialkot, Punjab, Pakistan

Muhammad Ashfaq Institute of Chemistry, Baghdad Ul Jadeed Campus, Islamia University of Bahawalpur, Bahawalpur, Punjab, Pakistan

Sumana Balo Soil Science and Agricultural Chemistry, School of Agriculture, Gandhi Institute of Education and Technology University (GIETU), Gunupur, Rayagada, Odisha, India

Bhaskar Behera Department of Biosciences and Biotechnology, Fakir Mohan University, Balasore, Odisha, India

Jiban Kumar Behera Department of Zoology, Fakir Mohan University, Balasore, Odisha, India

Manojit Bhattacharya Department of Zoology, Fakir Mohan University, Balasore, Odisha, India

M. Chandrakala ICAR—National Bureau of Soil Survey and Land Use Planning, Regional Centre, Bangalore, Karnataka, India

Sabyasachi Chatterjee PG Department of Botany, Ramananda College, Bishnupur, Bankura, West Bengal, India

Ashok Kumar Chaubey Nematology Laboratory, Department of Zoology, Chaudhary Charan Singh University, Meerut, India

Asim Chandra Dash College of Agriculture, Odisha University of Agriculture and Technology, Bhubaneswar, Odisha, India

Ch. Jyotiprava Dash ICAR-Indian Institute of Soil and Water Conservation, Research Centre, Sunabeda, Koraput, Odisha, India

Priyanka Devi Department of Agronomy, School of Agriculture, Lovely Professional University, Phagwara, Punjab, India

Subhadip Dhabal Department of Microbiology, Bankura Sammilani College, Bankura, West Bengal, India

Gopal Dutta Department of Agricultural Meteorology, Bidhan Chandra Krishi Viswavidyalaya, Haringhata, West Bengal, India

Subrata Dutta Department of Plant Pathology, Bidhan Chandra Krishi Viswavidyalaya, Mohanpur, West Bengal, India

Susanta Dutta Agronomy, Horticulture College, Birsa Agricultural University, Khuntpani, Ranchi, Jharkhand, India

Garima Department of Environmental Science, Kanya Gurukula Campus (Haridwar) of Gurukula Kangri (Deemed to Be University), Haridwar, Uttarakhand, India

Anadi Gayen Rajiv Gandhi National Ground Water Training and Research Institute, Central Ground Water Board, Ministry of Jal Shakti, Government of India, Naya Raipur, Chhattisgarh, India

Tehreema Iftikhar Department of Botany, GC University Lahore, Lahore, Pakistan

Sk Saruk Islam Department of Zoology, Raja Narendra Lal Khan Women's College (Autonomous), Midnapore, West Bengal, India

Anway Kumar Jena Department of Zoology, Fakir Mohan University, Balasore, Odisha, India

V. S. Joji Department of Marine Geology & Geophysics, School of Marine Sciences, Cochin University of Science and Technology, Kochi, Kerala, India

Namita Joshi Department of Environmental Science, Kanya Gurukula Campus (Haridwar) of Gurukula Kangri (Deemed to Be University), Haridwar, Uttarakhand, India

Suchetana Karmakar Department of Zoology, Raja Narendra Lal Khan Women's College (Autonomous), Midnapore, West Bengal, India

Adyant Kumar Department of Agronomy, Dr. Kalam Agricultural College, Bihar Agricultural University, Kishanganj, Bihar, India

Hemant Kumar Sam Higginbottom University of Agriculture Technology and Sciences (SHUATS), Prayagraj, Uttar Pradesh, India

Prasann Kumar Department of Agronomy, School of Agriculture, Lovely Professional University, Phagwara, Punjab, India

Randhir Kumar ICAR-Indian Institute of Soil and Water Conservation, Research Centre, Sunabeda Koraput, Odisha, India

Pratibha Kumari Krishi Vigyan Kendra Aurangabad, Bihar Agricultural University, Sirish Agricultural Farm, Aurangabad, Bihar, India

Ragini Kumari Sam Higginbottom University of Agriculture Technology and Sciences (SHUATS), Prayagraj, Uttar Pradesh, India;
Lalit Narayan Mithila University (LNMU), Darbhanga-Bihar, India

Reena Kumari Department of Agricultural Engineering, N. M. College of Agriculture, Navsari Agricultural University, Navsari, Gujarat, India

Savita Kumari Sam Higginbottom University of Agriculture Technology and Sciences (SHUATS), Prayagraj, Uttar Pradesh, India;
Mahatma Gandhi Central University (MGCU), Motihari-Bihar, India

Jayasree Laha Principal, Raja N.L. Khan Women's College (Autonomous), Midnapore, West Bengal, India

Jyotirmayee Lenka ICAR-Indian Institute of Soil and Water Conservation, Research Centre, Sunabeda Koraput, Odisha, India

N. Maddileti ICAR—National Bureau of Soil Survey and Land Use Planning, Regional Centre, Bangalore, Karnataka, India

Sumana Mahato Department of Zoology, Raja Narendra Lal Khan Women's College (Autonomous), Midnapore, West Bengal, India

Saheli Maiti Department of Zoology, Raja Narendra Lal Khan Women's College (Autonomous), Midnapore, West Bengal, India

Hammad Majeed Department of Chemistry, University of Management and Technology (UMT), Sialkot, Punjab, Pakistan

Suman Mallick Department of Microbiology, Bankura Sammilani College, Bankura, West Bengal, India

Anik Mandal Department of Microbiology, Bankura Sammilani College, Bankura, West Bengal, India

Animesh Mandal Dakshin Dinajpur KVK, Uttar Banga Krishi Viswavidyalaya, Cooch Behar, West Bengal, India

Asish Mandal PG Department of Botany, Ramananda College, Bishnupur, Bankura, West Bengal, India

S. P. Maske ICAR—National Bureau of Soil Survey and Land Use Planning, Regional Centre, Bangalore, Karnataka, India

Sujoy Midya Department of Zoology, Raja Narendra Lal Khan Women's College (Autonomous), Midnapore, West Bengal, India

Pabitra Mishra Department of Zoology, Fakir Mohan University, Balasore, Odisha, India

Pritish Mitra PG Department of Botany, Ramananda College, Bishnupur, Bankura, West Bengal, India

Malay Kanti Modak Department of Seed Science and Technology, Institute of Agricultural Science, University of Calcutta, Kolkatta, West Bengal, India

Michael E. Omeka Department of Geology, University of Calabar, Calabar, Cross River State, Nigeria

Bappa Paramanik Soil Science, Dakshin Dinajpur KVK, Uttar Banga Krishi Viswavidyalaya, Cooch Behar, West Bengal, India

Neha Pareek Sam Higginbottom University of Agriculture Technology and Sciences (SHUATS), Prayagraj, Uttar Pradesh, India;
Bihar Agricultural University (BAU), Sabour, Bhagalpur-Bihar, India

Priyanka Patra Department of Microbiology, Bankura Sammilani College, Bankura, West Bengal, India

A. Pavithra Department of Soil Science and Agricultural Chemistry, Palar Agricultural College, Affiliated to Tamil Nadu Agricultural College, Vellore, Tamilnadu, India

S. Praveen Kumar Department of Soil Science and Agricultural Chemistry, Faculty of Agriculture, Annamalai University, Chidambaram, Tamilnadu, India

Khizar Qureshi Department of Chemistry, University of Management and Technology (UMT), Sialkot, Punjab, Pakistan

P. Ramamoorthy Department of Soil Science and Agricultural Chemistry, Palar Agricultural College, Affiliated to Tamil Nadu Agricultural College, Vellore, Tamilnadu, India

V. Ramamurthy ICAR—National Bureau of Soil Survey and Land Use Planning, Regional Centre, Bangalore, Karnataka, India

R. Ray Department of Soil and Water Conservation, Faculty of Agriculture, Bidhan Chandra Krishi Viswavidyalaya, Mohanpur, Nadia, West Bengal, India

Basit Raza ICAR-Indian Institute of Soil and Water Conservation, Research Centre, Sunabeda Koraput, Odisha, India

Pravat Kumar Roul Odisha University of Agriculture and Technology, Bhubaneswar, Odisha, India

Shyamashree Roy Agronomy, Regional Research Station (OAZ), Uttar Banga Krishi Viswavidyalaya, Majhian, Dakshin Dinajpur, West Bengal, India

Sk Md Abu Imam Saadi Department of Biological Sciences, Aliah University, Kolkata, West Bengal, India

Gyanaranjan Sahoo Krishi Vigyan Kendra, Odisha University of Agriculture and Technology, Nayagarh, Odisha, India

Prativa Sahu ICAR-Indian Institute of Water Management, Chandrasekharpur, Bhubaneswar, Odisha, India

S. Sathiyamurthi Department of Soil Science and Agricultural Chemistry, Faculty of Agriculture, Annamalai University, Chidambaram, Tamilnadu, India; Horticultural Research Station,, Tamil Nadu Agricultural University, Thadiyankudisai, Perum, Thadiyankudisai, Perumbarai, Tamil Nadu, India

Krishnendu Sen Department of Plant Pathology, Bidhan Chandra Krishi Viswavidyalaya, Mohanpur, West Bengal, India;
Department of Microbiology, Vidyasagar University, Midnapore, West Bengal, India

Habib Ur Rahman Shah Institute of Chemistry, Baghdad Ul Jadeed Campus, Islamia University of Bahawalpur, Bahawalpur, Punjab, Pakistan

Babloo Sharma Krishi Vigyan Kendra Jaisalmer-1, Swami Keshwanand Rajasthan Agricultural University, Jaisalmer, Rajasthan, India

Himani Sharma Nematology Laboratory, Department of Zoology, Chaudhary Charan Singh University, Meerut, India

Pravat Kumar Shit Department of Geography, Raja N.L. Khan Women's College (Autonomous), Midnapore, West Bengal, India

Joginder Singh Department of Botany, Nagaland University, Hqrs. Lumami Zunheboto, Nagaland, India

Puja Singh Soil Science and Agricultural Chemistry, Horticulture College, Birsa Agricultural University, Khuntpani, Ranchi, Jharkhand, India

M. Sivasakthi Department of Soil Science and Agricultural Chemistry, Faculty of Agriculture, Annamalai University, Chidambaram, Tamilnadu, India

R. Srinivasan ICAR—National Bureau of Soil Survey and Land Use Planning, Regional Centre, Bangalore, Karnataka, India

Tabinda Department of Chemistry, University of Management and Technology (UMT), Sialkot, Punjab, Pakistan

Sandeep Kumar Tripathi Soil and Land Use Survey of India, Hyderabad, India; Department of Soil and Water Conservation, Faculty of Agriculture, Bidhan Chandra Krishi Viswavidyalaya, Mohanpur, Nadia, West Bengal, India

Prachi Yadav ICAR-Indian Institute of Soil and Water Conservation, Research Centre, Sunabeda Koraput, Odisha, India

Muhammad Zubair University of Agriculture Faisalabad, Faisalabad, Punjab, Pakistan

Chapter 1
Introduction to Soil, Water Pollution and Mitigation Strategies: A Spatial Approach

Partha Pratim Adhikary, Pravat Kumar Shit, and Jayasree Laha

Keywords Spatial modelling · Soil pollution · Water pollution · Agriculture practices · Geospatial technology

1.1 Introduction

At present, soil and water resources are essential elements of the environment, supporting the agrarian ecosystem's foundation. Agricultural techniques have changed as a result of human activity, particularly the unrestricted application of chemical fertilizers and pesticides. To effectively use geospatial technology for impact mitigation, it is crucial to identify and evaluate the risk zones related to soil and water resources (Shit et al. 2021a, b, c).

Achieving sustainable development goals depends in large part on spatial modeling, prediction, effective use, risk assessment, protection, and management of soil and water resources within the agricultural system. Additionally, by taking these steps, future economic, social, and environmental sustainability is maintained (Shit et al. 2022a, b).

P. P. Adhikary (✉)
ICAR-National Bureau of Soil Survey & Land Use Planning, Amravati Road, Nagpur, Maharashtra 440033, India
e-mail: Partha.Adhikary@icar.gov.in; ppadhikary@gmail.com

P. K. Shit
Department of Geography, Raja N.L. Khan Women's College (Autonomous), Midnapore, West Bengal, India

J. Laha
Principal, Raja N.L. Khan Women's College (Autonomous), Midnapore, West Bengal, India

P. P. Adhikary et al. (eds.), *Soil, Water Pollution and Mitigation Strategies*, Environmental Science and Engineering, https://doi.org/10.1007/978-3-031-63296-9_1

By upsetting the balance of ecosystems and resulting in the extinction of predator or competitor species that control their abundance, soil pollution significantly contributes to the establishment of new diseases and pests. It also helps to propagate genes and germs that are resistant to antibiotics, which makes it harder for people to fight against diseases. Additionally, over time, pollution causes soil quality to deteriorate, making it more difficult to cultivate crops. Forty percent of the world's population, or at least 3.2 billion people, are currently impacted by land and soil degradation.

The Food and Agriculture Organization (FAO) suggests that national governments develop rules on soil contamination and restrict the buildup of contaminants over set levels in order to address these issues. This assures the protection of food safety, environmental health, and human health and well-being. The United Nations Decade on Ecosystem Restoration (2021–2030) is focused on restoring ecosystems, and one of the main causes of ecosystem degradation is contaminated soil. This international program, led by UNEP, FAO, and partners, aims to scale up the restoration of terrestrial, coastal, and marine ecosystems over the following ten years while promoting sustainable soil management techniques.

Another major issue is water pollution, which is mostly brought on by agricultural practices. Agrochemicals, organic waste, drug remnants, sediments, and salty drainage are all heavily discharged into water bodies by agriculture, which accounts for 70% of all water withdrawals worldwide. For controlling agricultural practices to lessen these detrimental effects on water resources, proper diagnosis, prediction, and monitoring are essential (Adhikary et al. 2021; Shit et al. 2022b).

In a recent report, the FAO and the Water, Land, and Ecosystems (WLE) program highlight the rising global concern over agricultural water contamination. It hinders economic expansion and jeopardizes the wellbeing of billions of people. The rising demand for food with a high environmental impact, such as meat from industrial farms, adds to the unsustainable intensification of agriculture and the deterioration of water quality.

With the worldwide pesticide business currently worth over USD 35 billion yearly, the expansion of crop production has mainly relied on the intense use of inputs like pesticides and artificial fertilizers. Pesticide use intensity has increased by double digits in some nations. In many high-income countries and growing economies, agricultural pollution has surpassed contamination from cities and industries as the primary cause of the degradation of inland and coastal waters.

In recent decades, the area used for irrigation has more than doubled, causing agricultural pollution to enter waterways. The amount of area used for livestock production, which today makes up 30% of all land on the earth and 70% of all agricultural land, has also greatly increased. Since the 1980s, aquaculture has increased over 20-fold, especially in Asia. As a result, fish waste, uneaten feed, and the use of antibiotics and other chemicals have all contributed to a decline in water quality (Shit et al. 2022b). The most prevalent chemical pollutant in groundwater aquifers around the world is nitrate, which is a result of agricultural pollution, specifically nitrate contamination. Agricultural pollution also has an effect on aquatic ecosystems, such as eutrophication, which is brought on by nutrient buildup in lakes (Shit et al. 2022a) and coastal waterways and has a detrimental effect on fisheries and biodiversity. Eutrophication has been found to be occurring in several coastal locations.

In addition, a sizeable amount of the food produced is lost in the food supply chain, wasting 24% of the freshwater resources used to grow food crops, 23% of the world's cropland, and 23% of the fertilizer used globally.

With agricultural contamination affecting 38% of water bodies in the European Union, the effects of these problems are apparent. Agriculture is the third biggest source of pollution in American lakes, and it is the largest cause of pollution in rivers, streams, and wetlands. Nitrogen pollution of groundwater and surface waters in China is largely caused by agriculture.

Aquatic ecosystems, human health, and productive activities are all at danger due to this pollution. For instance, "blue baby syndrome," a potentially fatal condition in neonates, can be brought on by high nitrate levels in water. The environmental and social costs of agricultural water contamination in OECD (Organisation for Economic Co-operation and Development) nations alone are projected to be in the billions of dollars yearly (Shit et al. 2021c).

1.2 Individual Chapters

There are twenty chapters to address the soil and water contaminants. The brief description of the individual chapters is highlighted below.

1.2.1. The analysis of regional variations in soil properties and the evaluation of the effects of climate change on various agricultural systems are the main topics of Chap. 2. It offers advice on how to manage land use to slow climate change and create coping mechanisms. A comprehensive soil survey at the farm scale was conducted as part of the assessment employing a four-tiered technique that included ground truth observation, laboratory analysis, and GIS mapping. The produced maps assisted in determining which crops were suitable for Inagalur Panchayat's arid region's dry environment and sustainable crop yield.

1.2.2. In Chap. 3, it is discussed how important it is to execute the UN Sustainable Development Goals across productive landscapes all over the world, and how agroforestry and sustainable landscape management are key tactics for doing so. Agroforestry is an agro-ecological method that includes the development of timber products, integrating livestock, and working with trees and forests of all sizes. It goes beyond simply integrating trees into agriculture. In order to decrease wildfires and boost the density of forest cover in cultivated soils, silvi-pasture, which mixes forestry with pasture, should be given priority as a climate change mitigation strategy. 13% of CO_2 emissions are produced by the sectors of agriculture, forestry, and land use, which significantly contribute to climate change. Agroforestry practices can diversify food systems, boost economic output, maximize the use of natural resources, and benefit the environment. Advocating for agroforestry at the landscape scale, highlighting the numerous advantages of agroforestry for the economy, ecology, society, and culture, fostering equitable landscape governance, and assisting in the design and analysis of agroforestry systems are all components of effective sustainable landscape management through agroforestry. The connection between forest trees and agricultural crops has a huge range of possible advantages. Numerous benefits of agroforestry are compatible with sustainable production methods and lifestyles.

1.2.3. In Chap. 4, it is discussed how the rapid rise in global population has increased the demand on land resources. Due to agricultural intensification and farm consolidation, this has caused a reduction in agro-diversity and the standardization of farming practices. Diverse agroforestry systems have developed in response, offering resilience in the face of water constraint while boosting output and profitability. Due to its ability to sustainably enhance rural livelihoods, protect ecosystems, and guarantee food security, agroforestry is seen as a key instrument and the future of land management worldwide. This encourages sustainable manufacturing and production diversity, which has positive effects on the economy, the environment, and society. Tree growing with agricultural crops has long been a trend among traditional farmers worldwide. Droughts, floods, and insect infestations are made worse by climate change, emphasizing the importance of effective farmer responses and adaptation methods. Agroforestry has the potential to help agricultural landscapes adapt to climate change and grow. While minimizing the detrimental effects on natural forests, it offers a number of advantages, including as carbon sequestration, wood energy, improved local climatic conditions, and increased soil fertility. Agroforestry can provide ecologically friendly carbon sequestration solutions with multifunctional advantages by bridging the gap between adaptation and mitigation techniques, supporting both local adaptation and global efforts to lower greenhouse gas concentrations. It is crucial to place an emphasis on agroforestry and land diversification in order to achieve sustainable food security and solve the problems associated with climate change.

1.2.4. The emphasis in Chap. 5 is on the necessity of changing irrigation and nutrient management strategies in order to increase crop output and maximize input usage. In order to increase rhizome yield, soil fertility, and nutrient use efficiency, this chapter sought to identify the best irrigation strategy and nutrient supplement sources. The study included three different nutrient source combinations and four different moisture regime levels. Results showed that under the irrigation water to cumulative pan evaporation (CPE) ratio of 0.9, turmeric rhizome output was significantly increased. Additionally, the combined nutrient sources—more specifically, the recommended dose of fertilizer (RDF) of 50% plus 50% of organic sources [25% farm yard manure (FYM) + 25% vermicompost]—significantly increased rhizome output. According to post-harvest soil analysis, irrigation at 0.9 IW/CPE and 50% RDF + 50% organic source considerably increased the amount of main nutrients, electrical conductivity, and organic carbon. When compared to alternative irrigation regimes and nutrient sources, rhizome nutritional content and uptake were found to be higher under irrigation at 0.9 IW:CPE, integrated nutrient sources, and their interaction. In comparison to other irrigation techniques and nutrient sources, the interaction of irrigation at 0.9 IW:CPE and integrated nutrient sources (50% inorganic fertilizer + 50% organic manure) considerably increased the partial factor productivity (PFP) of applied nutrients. This study's findings suggest that, due to improved soil fertility, integrating integrated nutrient management (INM) and irrigation with crop water requirements can have positive effects on rhizome output and plant nutrition.

1.2.5. The exponential expansion in global population has resulted in a greater need for food and energy, which has led to a more intensive use of agricultural land, as discussed in Chap. 6. However, agricultural production in the twenty-first century faces a number of difficulties, such as climatic changes, environmental deterioration, urbanization, and problems with soil fertility. In order to increase crop yields, conventional farming methods frequently and excessively use agrochemicals, such as pesticides and synthetic fertilizers, which has a negative impact on the environment. The global scientific community is working to boost crop yield while maintaining the environment's quality and soil fertility. Future crop production methods should focus on sustainable practices, like using Plant Growth-Promoting Rhizobacteria (PGPR), a type of rhizobacterium. The physiology and development of plants are significantly influenced by the rhizospheric bacterial populations. The intricate network that links soil microbes and plants is known as the rhizosphere, or the root zone of plants. The rhizospheric microbiome is chemo-attracted by plant root exudates. Amino compounds, organic acids, phenolic compounds, and hormones are only a few of the substances generated by plant roots that can change the physicochemical characteristics of soils and the microbial makeup of the rhizosphere, or PGPR. Rhizobacteria that encourage plant growth have the power to reduce plant stress and boost agricultural production by 10–30%. This chapter offers a thorough summary of the chemical conversations that take place during multitrophic interactions between rhizobacteria and plants, as well as their effects on promoting plant development and reducing stress.

1.2.6. The present understanding of microbe-plant interactions is examined in Chap. 7, where it is shown that under salinity stress conditions, advantageous microorganisms encourage plant growth and improve stress resistance. Siderophores, phytohormones, and exopolysaccharides are all produced by microbes and help with soil structure, plant growth, and iron uptake, respectively. Additionally, bacteria have the ability to control gene expression, ion channels, and the production of osmoprotectants (OPs), which help plants resist salt stress. A vital group of helpful microorganisms known to promote plant development, nitrogen uptake, and stress tolerance is known as plant growth-promoting rhizobacteria (PGPRs). Utilizing beneficial microbiomes in agriculture has the potential to reduce the negative effects of salt stress on crop output and advance sustainable agriculture. The possible effects of microbe-plant interactions on the development of sustainable agriculture methods are discussed in this chapter.

1.2.7. The use of PGPR in place of traditional inorganic fertilizers has the potential to improve plant growth and total production, as is emphasized in Chap. 8. Implementing this strategy widely in field settings, though, is still a long way off. It requires the creation of an acceptable infrastructure, which includes well outfitted laboratories, qualified employees, and carefully monitored field studies with all necessary resources. Before releasing any PGPR formulation on the market, extensive background research is necessary to evaluate market dynamics, demand, commercial production viability, safety issues, and potential risks.

1.2.8. The environmental pollution brought on by industrial, agricultural, and urban activities poses a serious risk to both human health and the sustainability of the ecosystem, and is the subject of Chap. 9's discussion of bioremediation technology. Plant growth-promoting rhizobacteria (PGPRs) have become important players in the bioremediation process, a popular and economical method for reducing the harmful impacts of contaminants. However, there are a number of factors to take into account when using PGPR-based bioremediation, including choosing the right microbial strains, comprehending the types and sources of pollutants, and knowing how PGPRs improve remediation effectiveness. Addressing potential phytotoxic effects, biohazards, and assuring soil nutrient stability are among the difficulties. Although PGPR-based bioremediation offers potential advantages for soil health and environmental sustainability, there are still obstacles to be solved and more research is required before it can be widely used.

1.2.9. The variety and ecology of zooplankton and bacteria associated with plankton in freshwater wetlands were examined in Chap. 10. Four orders and two phyla made up the zooplankton, and their patterns of distribution across the study locations revealed a non-uniform distribution. Nine local isolates were chosen for additional research after sixty bacterial isolates from plankton appendages were screened. In vitro tests showed that the bacterial strains could tolerate a wide range of pH, temperature, and salt. Two isolates (SSI1 and SSI5) had the most notable effects on plant development out of the activities that was seen to promote plant growth. Among the studied strains, SSI5 had the greatest vigor index and antifungal activity. The relationship between plankton and bacteria as well as the potential for plankton-associated bacteria to support plant growth are clarified by these studies.

1.2.10. Nematodes, despite being diverse organisms, are frequently disregarded, according to Chap. 11. In ecosystems, they form different communities with a range of traits and eating behaviors. They are helpful markers for determining the health of the soil because of their qualities. Based on nematode communities, indicators such as the maturity index, enrichment index, and structure index are generated to assess soil quality. In order to measure soil quality, researchers have created models that concentrate on the nematode community, such as the soil food web model and the fertilizer usage efficiency model. The Caenorhabditis elegans model has also attracted interest for evaluating the effects of particular metals in a region. These nematode-based indexes offer information on the characteristics of soil and its general condition.

1.2.11. The use of bioremediation techniques to reduce soil pollution from heavy metals is discussed in Chap. 12. Widespread heavy metal pollution is a result of human activity and is a serious environmental issue. Heavy metals in soil have significant negative effects on a wide range of organisms, including bacteria, fungus, protozoans, and plants. These metals affect nutrient mobility in addition to changing the physiochemical properties of the soil. Heavy metal contamination affects soil-dwelling species at the cellular, individual, and population levels. In addition, heavy metals can bioaccumulate in plants, infiltrate the food chain, and ultimately make their way to people. A naturally stable and economical strategy to reduce heavy metal contamination is bioremediation, which includes techniques like phytoremediation, mycoremediation, and vermiremediation. Vermiremediation stands out as the most efficient, environmentally benign, and cost-effective technique for eliminating heavy metals from soil among these.

1.2.12. The use of hyperspectral imaging to address the issue of low agricultural yield is covered in Chap. 13. In order to meet the demands of an increasing human population, crop yield must be increased. Plant yield greatly depends on the amount of nutrients present in the soil. Although time-consuming and difficult, the current methods for determining soil nutrient concentration through field sampling and laboratory analysis. Additionally, the regional diversity of soil nutrient levels is not captured by these methods. A potential method to get around these restrictions is hyperspectral imaging, or HRS. HRS can identify and measure soil nutrients based on their absorption and reflection characteristics by recording and processing the spectral signatures of objects or materials. This technology is practical for remote places since it provides real-time and online monitoring capabilities.

1.2.13. In Chap. 14, it is discussed how high soil concentrations of harmful and dangerous compounds pose multiple health risks and are a major cause of global concern about soil pollution. The over use of fertilizers in agriculture has also seriously harmed ecosystems and human health. In order to achieve food self-sufficiency, fertilizers were crucial, helping to raise food output by 55%. The total nutritional intake (NPK) in India was 25.58 MT in 2018. Large quantities of reactive nitrogen and phosphorus are found in the air, soil, and water due to the disruption of the nutrient cycle in soil caused by large fertilizer inputs. These unutilized nutrients behave like toxins, harming the environment and bioaccumulating in the food chain. Farmers all across the world have started using controlled fertilizers, including slow-release, controlled-release, and nano-coated fertilizers, to help offset these problems.

These actions prevent soil contamination and slow down nutrient leaching, allowing fertilizer availability in the soil to last longer. Continuous efforts are being made to encourage sustainable agriculture practices while minimizing fertilizer output and usage. In order to achieve a less polluting economy and healthier soil conditions across the country, expenditures are also required to educate rural communities about the negative impacts of fertilizer misuse.

1.2.14. The greatest carbon store on the surface of the Earth, soil holds twice as much carbon as the atmosphere and two to three times more than all living things put together, according to Chap. 15. It is a complicated natural phenomenon made up of solids, liquids, and gases that provides a habitat and food for living things. Soil is capable of filtering water, taking part in garbage recycling, and even serving as a building material. But since soil nutrition is so important for holding onto water and minerals for plant growth as well as giving plants structural support, that is where author's main attention is focused in this article. Different soil types in the state of Odisha are influenced by a variety of elements, including the tropical climate, deltaic and coastal regions, and geological processes. Red, yellow, black, laterite, deltaic alluvial, coastal saline, brown forest, and mixed red and black soils are some of the prominent soil types in Odisha. This chapter discusses the availability of other geomaterials, addresses natural hazards, outlines the various uses of soils, highlights state-built structures, and presents a case study based on soil data gathered from various districts. It also gives a brief overview of soil nutrition in Odisha. This chapter provides a thorough overview of soil functionality and conservation initiatives in Odisha while focusing on soil health and its practical implications. In-depth discussion is given on scientific developments, present understanding, and potential directions for soil nutrient management strategies in the agricultural industry. In addition, this chapter examines the biological, chemical, and physical features of soil that are essential for long-term, sustainable agricultural productivity with little harm to the environment.

1.2.15. Chapter 16 deals with how machine learning (ML) and artificial intelligence is being used more and more for monitoring and predicting water quality. This is because water quality statistics are complicated and subject to seasonal and anthropogenic fluctuations. A systematic literature review (SLR) was done in this work to evaluate how ML models have been used to monitor and predict water quality in Nigeria over the past 20 years (from 2003 to 2023). For data extraction and visualization, the review used bibliometric coupling and meta-analysis techniques. The examination of a few studies showed that most predictions still rely on artificial neural networks (ANN), with little global exploration of hybrid machine learning (ML) models for water quality prediction. The Wavelet-Adaptive Neural Fuzzy Interference System (W-ANFIS) hybrid model and the Adaptive Neuro-Fuzzy Inference System (ANFIS) were determined to be the most accurate ANN algorithms for prediction. Temperature, dissolved oxygen (DO), pH, conductivity (EC), and total dissolved solids (TDS) were among the parameters that were commonly forecast. China and the US had the most papers on the use of ML to predict water quality, whereas Nigeria lagged behind. Given the sources of contaminants in Nigeria's aquatic systems, more attention must be paid to machine learning when predicting the country's future water quality, especially with relation to heavy metal contamination brought on by mining operations.

1.2.16. The establishment of government strategies to alleviate the water problem and ensure sustainable water resource management is discussed in Chap. 17. The Central Plains, the Bastar Plateau, and the Northern Hill Zone of Chhattisgarh are the three ecological zones, and the agriculture practices in bunded and unbunded fields present difficulties in different soil types. To take into consideration changes in topology, terrain characteristics, soil texture, and moisture retention, location-specific treatments are required. In all agro-climatic zones, deforestation has caused siltation and decreased the water recharge capacity of reservoirs, tanks, and water collecting structures. In order to strengthen water resources and encourage socioeconomic growth in the various agro-climatic zones of Chhattisgarh, it is advised that micro-irrigation projects, minor irrigation projects, dry farming, modernization of dams and canals, watershed management schemes, and precision agriculture be implemented. Rural development depends on groundwater conservation, the joint use of surface and groundwater, and the use of small-scale water technology. Priority should be given to farmers with restricted access to canal irrigation and drinking water when allocating subsidised irrigation water. Due to subsurface heterogeneity, groundwater extraction should be combined with management of surface water resources and precipitation. To address water issues, dry farming practices should be used in conjunction with proper maintenance of water conservation structures, crop diversification, water harvesting, managed aquifer recharge, systematic groundwater monitoring, community involvement, and farmer sensitization to modern agricultural technologies. In order to comprehend the water issues and management aspects in Chhattisgarh, a SWOT analysis and Fishbone analysis have been performed. These analyses highlight strengths such as substantial unconfined aquifers along major river courses, weaknesses related to diverse topography, and threats posed by a declining post-monsoon dynamic water table. The Fishbone research pinpoints the reasons for the decline in the water table, the decline in water quality, problems with the chemistry of the soil, and the absence of site-specific water interventions.

1.2.17. The topic related to Arsenic pollution is covered in Chap. 18. Because it is more poisonous, inorganic arsenic is dangerous for our health and can cause cancer, skin lesions, cardiovascular illness, and neurodevelopmental issues. The Toxicity Characteristic Leaching Procedure (TCLP), which helps regulatory authorities classify hazardous waste, is used to measure the content of arsenic in waste materials and determine the risk of exposure.

Carbon nanotubes (CNTs), graphene oxide (GO), metal–organic frameworks (MOFs), covalent organic frameworks (COFs), and their hybrids are among the cutting-edge materials that have the potential to reduce arsenic pollution. These substances have special qualities that make removal from water and soil efficient, such as high surface area, reactivity, and selectivity towards arsenic. Research is still being done to improve these materials and provide useful applications for tackling arsenic contamination in the real world.

1.2.18. In the Guntur area of Andhra Pradesh, India, Chap. 19 examined the geographical variance of nitrate in groundwater and its effects on human health. The study found that during the pre-monsoon season, groundwater levels varied from 0.2 to 39.5 m, while during the post-monsoon season, it varied from 0.2 to 22.4 m. Nitrate levels in groundwater samples ranged from 1.2 to 665.0 mg l^{-1}, with a mean value of 113.4 mg l^{-1}. 60% of groundwater samples had nitrate levels that were over the permissible limit of 45 mg l^{-1}. A health risk evaluation utilizing the hazard quotient (HQn) showed that 62.7% of samples for new-borns were in the risk range, while 49.3% of samples for males surpassed the HQn value of 1. To lessen the negative health impacts of nitrate-contaminated drinking water in the study area, it is essential to put in place the necessary safety measures, such as routine water supply inspections, health monitoring, and evaluation of subterranean pipelines.

1.2.19. The problems associated with water contamination in India are covered in Chap. 20. Due to its growing economy, India must confront substantial ecological issues, such as soil, water, and air pollution, as well as a wider range of environmental problems. Among these issues, water pollution is a serious issue in India since metropolitan areas generate over 50 billion gallons of sewage each year from home and industrial sources. To lessen river water pollution, initiatives like the Ganga action plan and the Jamuna action plan have been put in place. Indicators like concentration, load, or frequency of occurrence, coupled with elements like pH, dissolved oxygen (DO), biochemical oxygen demand (BOD), and solids content, are frequently used to gauge the quality and level of pollution in water. There have been studies on water pollution in India, but it is uncommon to find quantitative analyses that take into consideration how it affects every area of the economy. Water contamination, which has a negative impact on both human health and the environment, is primarily caused by chemical and metal discharge from factories and commercial organizations. The main cause of water source contamination worldwide is human activity. Programs for environmental education for the general public and schoolchildren are essential in resolving this issue. The status of numerous water contaminants is discussed in this chapter along with how they affect both human health and agricultural productivity. A thorough water pollution coefficient matrix is also built. The study examines the direct and indirect water pollution in India brought on by various businesses and activities, revealing differences in pollution levels between different economic sectors. The essay offers policy suggestions for dealing with water contamination as well.

1.2.20. In the last chapter, it is discussed how there is strong evidence that crop yields are negatively impacted by climate change, endangering the safety of the world's food supply. Agricultural systems must adapt to climate change while reducing ecological consequences in order to fulfil the rising demands of a changing global population and assure sustainable food production. Crop simulation models have become essential tools for determining appropriate adaptation methods in agricultural risk management and monitoring the effects of climate change on crop output. Crop simulation models address food safety as a key issue in guaranteeing a steady supply of food. To help mitigate greenhouse gas emissions, innovative agricultural methods and crops can be created that increase soil organic matter, lower net emissions of greenhouse gases, and make it easier to make biofuels. Enhancing crop yields, lowering production risks, and promoting sustainable agricultural development all depend on the implementation of mitigation and adaptation measures. Multimodal strategies that combine agricultural simulation with plant physiology, biochemistry, and gene-based modelling should be used to reduce prediction uncertainties. During the creation of new models, it is crucial to get top-notch field data for model testing. It is essential to regularly analyze agricultural output to guarantee that future generations will have access to food. A possible strategy for reaching this aim is to create a repository of information and resources to improve farming methods and reduce the effects of climate change.

1.3 Concluding Remarks

Soil, Water Pollution and Mitigation Strategies: A Spatial Approach offers a concise overview of the geospatial assessment and fate of environmental contaminants. It covers all the essential topics necessary for understanding and predicting the distribution of pollutants in soil, groundwater, and surface water. The book provides a solid foundation for effective soil and water pollution control and integrated catchment management and sustainable agriculture practices. This comprehensive reference book has been thoroughly updated and expanded, featuring numerous case studies, examples, and exercises. It is an invaluable resource for undergraduate and graduate students in Earth and Environmental Sciences, providing all the necessary material for studying soil and water contamination. Additionally, it serves as a valuable source of information for professionals in the field.

References

Adhikary PP, Shit PK, Santra P, Bhunia GS, Tiwari AK, Chaudhary BS (eds) (2021) Geostatistics and geospatial technologies for groundwater resources in India. Springer Cham, 598 pp. https://doi.org/10.1007/978-3-030-62397-5

Shit PK, Bhunia GS, Adhikary PP, Dash Ch. Jyotiprava (eds) (2021a) Groundwater and society: applications of geospatial technology. Springer, 534 pp. https://doi.org/10.1007/978-3-030-64136-8

Shit PK, Das P, Bhunia GS, Dutta D (eds) (2021b) Mapping, monitoring, and modeling land and water resources: advanced techniques using open source software. CRC Press, 348 pp. https://doi.org/10.1201/9781003181293

Shit PK, Adhikary PP, Sengupta D (eds) (2021c) Spatial modeling and assessment of environmental contaminants: risk assessment and remediation. Springer Cham, 720 pp. https://doi.org/10.1007/978-3-030-63422-3

Shit PK, Adhikary PP, Bhunia GS, Sengupta D (eds) (2022a) Soil health and environmental sustainability: application of geospatial technology. Springer Cham, 729 pp. https://doi.org/10.1007/978-3-031-09270-1

Shit PK, Bhunia GS, Adhikary PP (eds) (2022b) Case studies in geospatial applications to groundwater resources. Elsevier, 432 pp

Chapter 2
GIS-Based Major Crop Suitability Evaluation for Soil Health Maintenance and Climate Resilience in Arid Region of Andhra Pradesh, India

R. Srinivasan, M. Chandrakala, N. Maddileti, S. P. Maske, and V. Ramamurthy

Keywords Climate change · Crop suitability · Drought · Inagalur panchayet · Soil resources

2.1 Introduction

One of the biggest issues facing humanity today is climate change, which has a significant impact on the safety of the world's food supply. Particularly in semi-arid tropical regions, its effects have already been felt strongly on water supplies, the length of the growing season, and food security (IPCC 2007). According to climate model predictions based on the Inter-Governmental Panel on Climate Change (IPCC) in India, pre-industrial temperatures are anticipated to have increased by 1.7–2.0 °C by the 2030s and 3.3–4.8 °C by the 2080s. (Chaturvedi et al. 2012). The northern hemisphere's temperate zones experienced a rise in average rainfall over the past century, as well as an increase in the frequency of very heavy rainfall events. Contrarily, rainfall in the subtropics dropped, and droughts in Asia and Africa intensified and increased in frequency (IPCC 2007).

Climate change will affect agriculture due of differences in temperature, precipitation patterns, and carbon dioxide (CO_2) concentrations, particularly in rainfed regions. Crop productivity is expected to fluctuate as a result of these climatic changes, extreme weather events, and changing pest and disease conditions. Excess

R. Srinivasan (✉) · M. Chandrakala · N. Maddileti · S. P. Maske · V. Ramamurthy
ICAR—National Bureau of Soil Survey and Land Use Planning, Regional Centre, Bangalore, Karnataka 560024, India
e-mail: srinivasan.surya@gmail.com

P. P. Adhikary et al. (eds.), *Soil, Water Pollution and Mitigation Strategies*, Environmental Science and Engineering, https://doi.org/10.1007/978-3-031-63296-9_2

CO_2 concentrations can enhance plant growth and increase water consumption efficiency, according to controlled research. Plant development benefits from higher atmospheric carbon dioxide concentrations (Walthall et al. 2012). The ways in which climate change affects agriculture will differ based on where you live. It is challenging to forecast how climate change may affect agriculture because of the wide variety of consequences that are expected to occur (Roudier et al. 2011; Knox et al. 2012). Major changes in production procedures might be necessary. To reduce effects on productivity, adaptation techniques may be needed. These include producing pest, heat, and drought resistant cultivars, diversifying crop rotations, incorporating animals into agricultural production systems, and using nutrient-efficient systems. In order to create potential adaptation strategies, it is crucial to identify the location-specific effects of climate change on crop productivity and water supplies (Godfray et al. 2010).

Land resources inventory is essential for understanding the soil factors and processes responsible for the types of degradation in the arid region of Andhra Pradesh. A farm-scale approach is evolving rational, site-specific soil characterization, developing a database for each parcel of land, and mapping soil potential and problems (Srinivasan et al. 2021). Assessment of different soil constraints and development of various crop suitability maps under GIS helps to land restoration and mitigate climate change in the Inagalur Panchayat. It helps adopt a sustainable land-use system and fertility restoration toward crop productivity and ecosystem conservation (Keesstra et al. 2018). Some of the crucial soil factors, such as genesis, physiography, climate, vegetation, depth, colour, and age, among others, may be to blame for the low output of agricultural and horticultural products in arid soils. Knowing the properties of the soil will aid with accurate planning, effective land use planning, and knowing any changes that may have occurred during development.

When resources are scarce, land evaluation is a crucial phase in the land use planning process. A land evaluation analysis analyses if the features of the land effectively satisfy the requirements of land use. Therefore, the goal of evaluating the land's suitability potential is to foresee its inherent ability to sustain a certain land use for an extended length of time without degrading. Therefore, a land suitability assessment is a land evaluation that is typically done to establish a specific land use for a given site and discover characteristics that limit the production of a particular crop (Mu 2006; Abdel Rahman et al. 2016). Assessing the suitability of a piece of land depends on its capabilities as well as other elements including its quality, closeness to various access points, its ownership, consumer demand, and economic values (EI Baroudy 2016). Land suitability assessment is a planning approach to avoid environmental conflicts by the segregation of competing land uses (FAO 1976, 1991; Al-Mashreki et al. 2011). Land suitability evaluation can be either qualitative or quantitative. In order to evaluate land potential on a large scale, a qualitative technique is applied, and the results are provided in qualitative terms. FAO guidelines on land evaluation systems (FAO 1976, 1985) and physical land evaluation methods (Sys et al. 1991) were widely used for land suitability assessment.

Remote sensing and GIS frameworks offer scientific insight into real-world application systems with sustainable solutions for the new reality of climate change, and they greatly contribute to resilience to reduce future risk (Srinivasan et al. 2021). A geographical information system (GIS) has been used to evaluate land using map analysis techniques (Bandyopadhyay et al. 2009; Bobade et al. 2010). Remote sensing and GIS hold great promises for improving the convenience and accuracy of spatial data, more productive analysis and improved data access. These technologies have been used to assess the criteria required to define the suitability of land (Mishelia and Zirra 2015).

The purpose of this research was to develop a GIS-based technique for determining a land use's viability, which will assist planners and land managers in identifying areas with physical constraints for a range of designated land uses. Additionally, to assist in determining the management needs necessary to guarantee that a specific land use in a dry region of Andhra Pradesh may continue without seriously degrading the land quality either on- or off-site.

The study was conducted with the following objectives to provide a database at the farm level for the arid region of Andhra Pradesh for climate change mitigation.

1. Characterization and grouping of soils into different soil series in Inagalur Panchayat.
2. Generation of GIS-based thematic maps based on soil-site characteristics.
3. Identification of different soil and climatic problems and identification of suitable crops for land restoration.

The rationale of the study

Identification of the soil potential of each farm and assessing the climatic adaptability crop plan or alternate land use plan in Inagalur Panchayats for site-specific soil and water conservation plan implementation and create ecologically sustainable cropping system. Soil depth, light and heavy texture, coarse gravels and poor fertility are major soil constraints in Inagalur Panchayat and along with regular drought make crop loss and economic poor farming and farmers. Evaluation of different crops' suitability on a farm scale helps crop diversification and managing soil quality for climate change mitigation.

2.2 Materials and Methods

2.2.1 Location of the Study Area

Inagalur Panchayat lies within the arid Southern Deccan Plateau in Andhra Pradesh, covering 2938 ha and is situated between 14°00 to 14°50 N latitude and 78°00 to 78°30 E longitude (Fig. 2.1). It's part of Agro-Ecological Region 3, marked by hot arid conditions and deep loamy and clayey mixed red and black soils. Rainfall scarcity is severe, with drought occurring two out of three years due to its meager average annual rainfall of 574 mm. Rain mainly falls during the southwest monsoon (320 mm) from

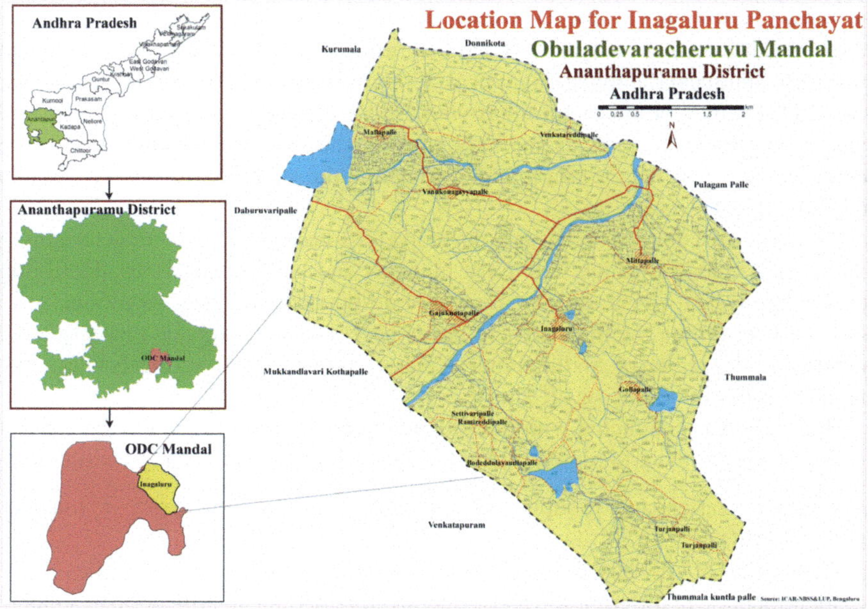

Fig. 2.1 Location map of the study area

June to September and the northeast monsoon (190 mm) from October to December. The area's temperature ranges between 17–19 °C and 30–40 °C. Geologically, granite gneiss and alluvium are prevalent, with elevations varying from 495 to 568 m above sea level, showcasing various landforms. Agriculture, horticulture, plantation, fallow, open scrub, and forestry are the dominant land uses in this region.

2.2.2 Geology

The predominant geological composition comprises significant rock formations originating from granite gneiss and alluvium parent materials (Fig. 2.2). Granite gneisses within the area exhibit hues ranging from pink to gray, characterized by a coarse to medium-grained texture. These gneisses primarily consist of quartz, feldspar, biotite, and hornblende. The gray granite gneisses undergo extensive weathering, displaying a high degree of fracturing and fissuring, particularly up to a depth of approximately 10 m.

The soil thickness of the alluvium is generally limited to less than a meter, except in lowlands where it can be considerably deeper. Soils found in lowlands have been transported and exhibit dark colors, originally formed at higher elevations and subsequently deposited in river sediment. This phenomenon underscores the dynamic

Fig. 2.2 Major types of geology Granite rock in the study area

geological processes at play in the region, shaping the characteristics of the soil profile and composition.

2.2.3 Landforms

The terrain, shaped by Granite gneiss and Alluvial materials, features distinct landforms: mounds/ridges, summits, side slopes, gently sloping uplands, and level plains. Mounds and ridges are often marked by exposed rock outcrops in this area (Fig. 2.3).

Upland Midland Lowland

Fig. 2.3 Major types of landforms in the study area

Groundnut Bajra Mango

Fig. 2.4 Major types of landuse in the study area

2.2.4 Land Use

Agriculture is the main component of the livelihood of the people. The cultivation of main crops i.e., Groundnut (*Arachis hypogaea*), Pearl millet (*Pennisetum glaucum*), and Pulses (*Cajanus cajan*) under rainfed conditions and Rice (*Oryza sativa*), Maize (*Zea mays*), and different vegetables (Tomato, Brinjal and Okra, etc.) under irrigated condition. Mango (*Mangifera indica*) is a major perennial horticultural crop extensively cultivated (Fig. 2.4).

2.2.5 Field Studies

A comprehensive soil survey was conducted in Inagalur Panchayat, utilizing village cadastral maps and satellite imagery from IRS (IRS LISS IV and Cartosat-1) at a scale of 1:10,000. False color composites of IRS imagery were analyzed for physiography, forming the basis for mapping soils. The survey involved studying soils along multiple transects, leading to the creation of a soil map featuring phases of soil series as mapping units. Morphological characteristics of the soil were investigated through vertical cuts of soil profiles, excavated up to 1.5 m or until reaching the rock or hard substratum/parent materials (Soil Survey Staff 2003). At each sampling site, latitude, longitude, and elevation data were meticulously recorded using a handheld global positioning system (GPS). One hundred and fifteen (115) soil profiles were examined following standard methods (Soil Survey Staff 1999). The Inagalur Panchayat was categorized into Eleven (11) soil series, representing homogeneous units with similar responses to land management, and further delineated into 50 mapping units (Fig. 2.5). Soil samples collected from different horizons were air-dried, processed, and sieved using a 2 mm sieve. These samples were then used for determining major soil properties. Standard procedures outlined in Jackson (1973) and Page et al. (1982) were employed to analyze various soil parameters, contributing to a comprehensive understanding of the soil composition and characteristics in Inagalur Panchayat.

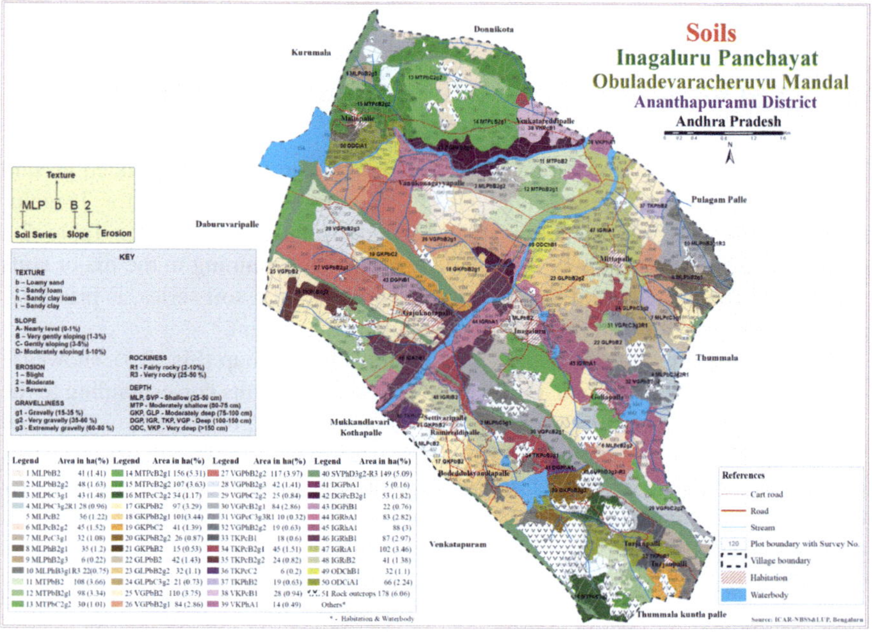

Fig. 2.5 Soil series with mapping units of Inagaluru Panchayat

2.2.6 Thematic Map Preparation

The thematic maps for slope, soil depth, erosion and major soil properties viz., texture, gravelliness, pH, EC, AWC and OC were prepared under the GIS environment. The prepared thematic maps were used to identify theconstraints and potentials of the soils and which helps to evaluate the major crop suitability inInagalur Panchayat.

2.2.7 Land Evaluation

The assessment for crop suitability followed FAO (1983), Sys et al. (1993), and NBSS and LUP (1994) guidelines. Soil suitability classes were determined by aligning soil characteristics with crop requirements, categorizing sites into highly suitable (S1), moderately suitable (S2), marginally suitable (S3), currently not suitable (N1), and permanently not suitable (N2) based on limitations in number and intensity.

2.3 Results and Discussion

2.3.1 Soil Resources

The finalization of the soil map involved a meticulous process that included ground truth checking and laboratory analysis. In creating the soil map for Inagalur Panchayat, all observations were amalgamated using GIS software. Table 2.1 provides a detailed overview of the soil characteristics pertaining to the major soils in Inagalur Panchayat. A soil phase, a subdivision within a soil series, is primarily based on surface features influencing its use and management.

The soil mapping units, represented by symbols on the map (Fig. 2.5), showcase the geographic distribution and area extent of 50 mapping units corresponding to the 11 soil series present in the Panchayat. Each mapping unit, or soil phase, depicted on the map exhibits similar soil and site characteristics. Simply put, all the farms or survey numbers encompassed within a specific phase share analogous management needs and should be treated accordingly. This approach ensures a nuanced understanding of the diverse soil variations within Inagalur Panchayat and facilitates tailored management strategies for each delineated mapping unit.

Table 2.1 Soil series and areal cover in Inagaluru Panchayat

Soil series	Area (ha)	% of TGA
Settivaripalle (SVP)	149	5.09
Mallapalle (MLP)	337	11.46
Mittapalle (MTP)	533	18.12
Gajukuntapalle (GKP)	279	9.51
Gollapalle (GLP)	96	3.25
Venukanagayyapalle (VGP)	489	16.65
Tummalakuntlapalle (TKP)	111	3.77
Venkatapuram (VKP)	42	1.42
Diguvapalle (DGP)	80	2.74
Inagalur (IGR)	400	13.62
Obaladevaracheravu (ODC)	98	3.34
Rock outcrops and others	178	6.06
Habitation and Water bodies	146	4.97
Total	2938	100

2.3.2 *Site Characteristics*

2.3.2.1 Slope

The land slope is a key factor impacting soil formation, water availability, and nutrient distribution, holding a pivotal role in natural soil structures. It's determined by its angle, shape, and length, significantly affecting soil formation, land utilization, and development. Both length and angle impact runoff, infiltration, erosion, and deposition rates. To represent the diverse terrain, soil map units were sorted into specific slope categories, generating a slope map that illustrates the distribution of these classes across the Panchayat (Fig. 2.6).

Around 358 ha (12.17%) are categorized as nearly level lands (0–1% slope), found in the northern, southern, and eastern regions of the Panchayat. The largest area, approximately 1838 ha (62.53%), constitutes very gently sloping lands (1–3% slope) spread throughout the Panchayat. Gently sloping lands cover roughly 270 ha (9.18%), while moderately gently sloping lands encompass about 149 ha (5.09%).

It's noteworthy that the entire cultivated area in the Panchayat showcases soils with favorable slope characteristics. These areas allow for diverse cultivation of both annual and perennial crops suited to the local climate without requiring extensive soil or water conservation efforts. Understanding these slope variations enriches our

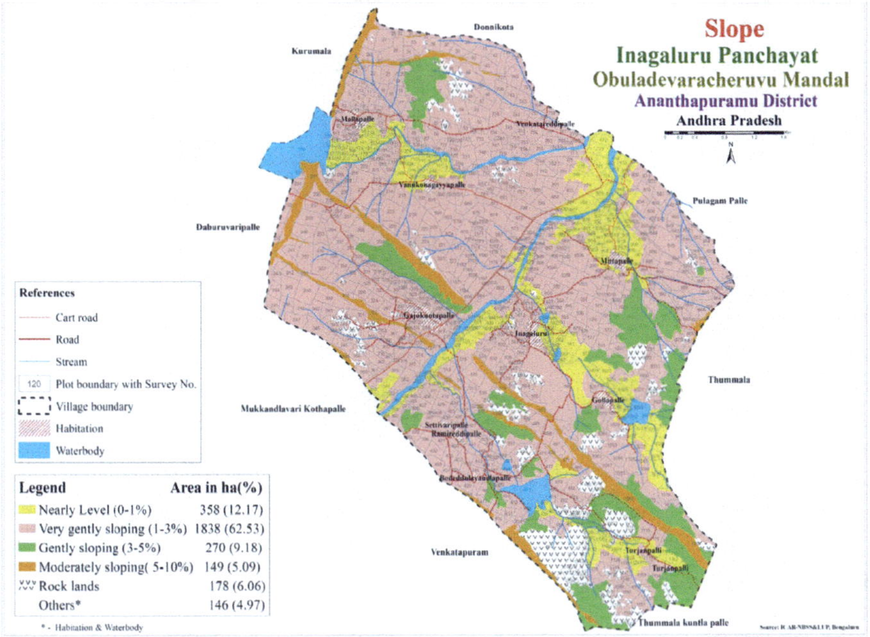

Fig. 2.6 Soil slope map of Inagaalur panchayat

grasp of the Panchayat's landscape, aiding in implementing suitable agricultural and land management practices.

2.3.2.2 Soil Depth

Soil depth is a pivotal factor used to categorize soils into distinct series, defining the measure of soil above the parent material or rock. It significantly influences root depth and water-holding capacity. Figure 2.7 showcases the distribution of diverse soil depth classes in the area. Soil identification involves depth classes like shallow (25–50 cm), moderately shallow (50–75 cm), moderately deep (75–100 cm), deep (100–150 cm), and very deep (> 150 cm).

The most productive lands, encompassing 1081 ha (36%) and 140 ha (4.77%), are classified as deep and very deep, respectively. These areas are conducive for cultivating all climatically adapted long-duration crops due to their substantial soil depth, providing optimal conditions for root development and water retention. This categorization based on soil depth enhances our understanding of the soil profile variations, aiding in the selection of suitable crops and guiding agricultural practices in the region.

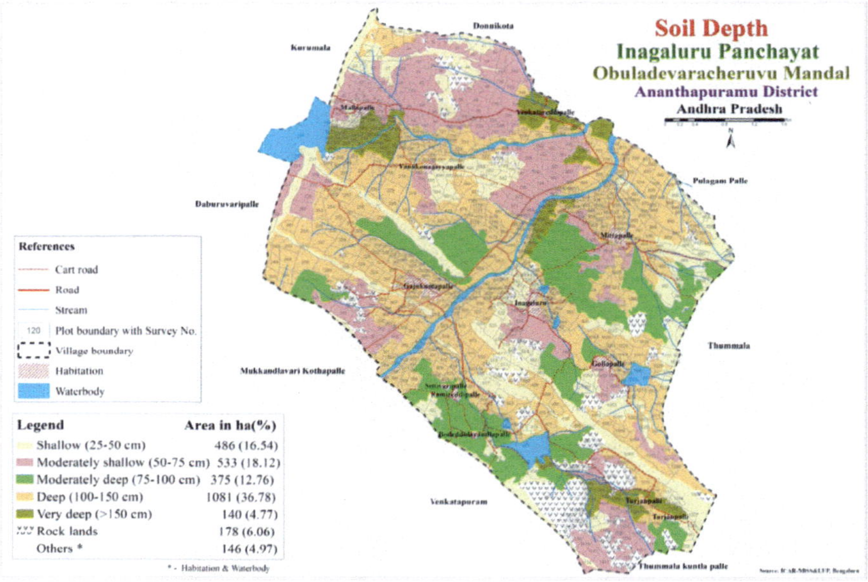

Fig. 2.7 Soil depth map of Inagalur panchayat

2.3.2.3 Soil Erosion

Soil erosion refers to the gradual erosion of the Earth's surface due to water, wind, and ice, involving the detachment and movement of soil particles through rainfall impact. Human activities like burning, overgrazing, extensive deforestation, and tillage often accelerate this process. Erosion classes assess the present erosion status, determined through field observations like rills, gullies, or a surface covered with gravel.

Three erosion classes—slight erosion (e1), moderate erosion (e2), and severe erosion (e3)—are identified based on field assessments. This led to the creation of a soil erosion map (Fig. 2.8) depicting their extent and distribution within the Panchayat. Slightly eroded soils (e1 class) cover around 545 ha (18.53%), spread across the central, western, northern, and southeastern parts. Moderately eroded soils (e2 class) encompass a larger area, about 1764 ha (60.04%), distributed across the western, northern, eastern, and southern regions. Severe erosion is observed in approximately 306 ha (10.44%) within the Panchayat. Implementing soil and water conservation, alongside other land development measures, becomes crucial in these severely eroded areas to mitigate erosion's impact and promote sustainable land use practices.

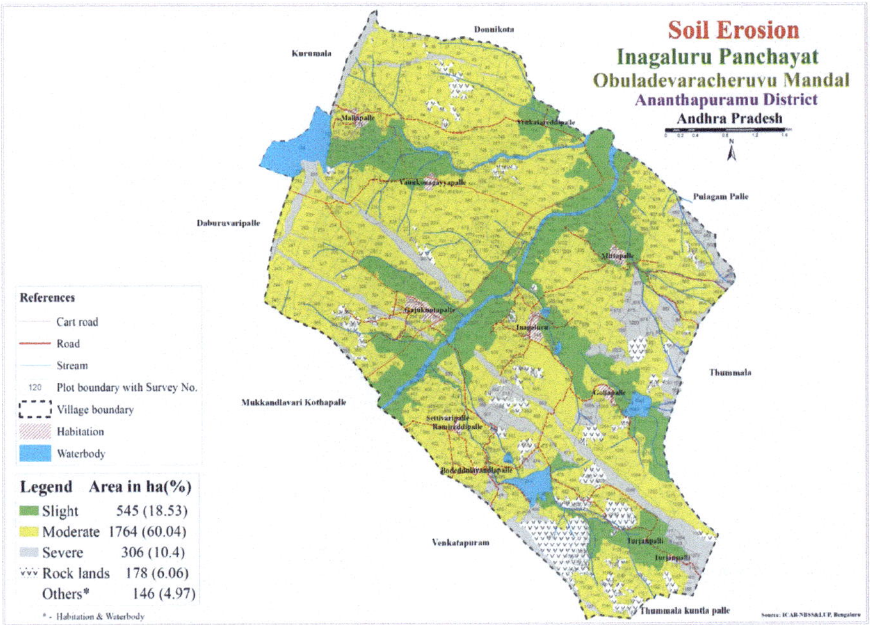

Fig. 2.8 Soil erosion map of Inagalur panchayat

2.3.3 *Soil Characteristics*

2.3.3.1 Soil Texture

Soil texture refers to the relative composition of primary particles, such as sand, silt, and clay, within the soil. Soils are classified into various textural classes, including loamy sand, sandy loam, sandy clay loam, sandy clay, and clay. The texture of the surface or plough layer plays a crucial role in influencing crop growth and yield, particularly for shallow-rooted crops with fibrous roots. While the thickness of the surface layer varies from place to place, typically ranging from 10 to 25 cm, it serves as a significant parameter for assessing soil suitability to produce specific crops. The surface texture significantly alters the infiltration rate, available water capacity, aeration, and soil fertility.

Loamy sand, covering approximately 1199 ha (40%), dominates the surface in the western, eastern, southern, and central parts of the Panchayat. An additional area of about 695 ha (23.64%) features soils that are sandy loam at the surface, occurring in the central, western, northern, northeastern, eastern, and southern parts. The entire area displays moderately productive lands concerning surface soil texture, with 40% featuring loamy sand soils (Fig. 2.9). While these soils exhibit low potential for soil–water retention and nutrient availability, they show somewhat excessive drainage without issues related to infiltration or workability.

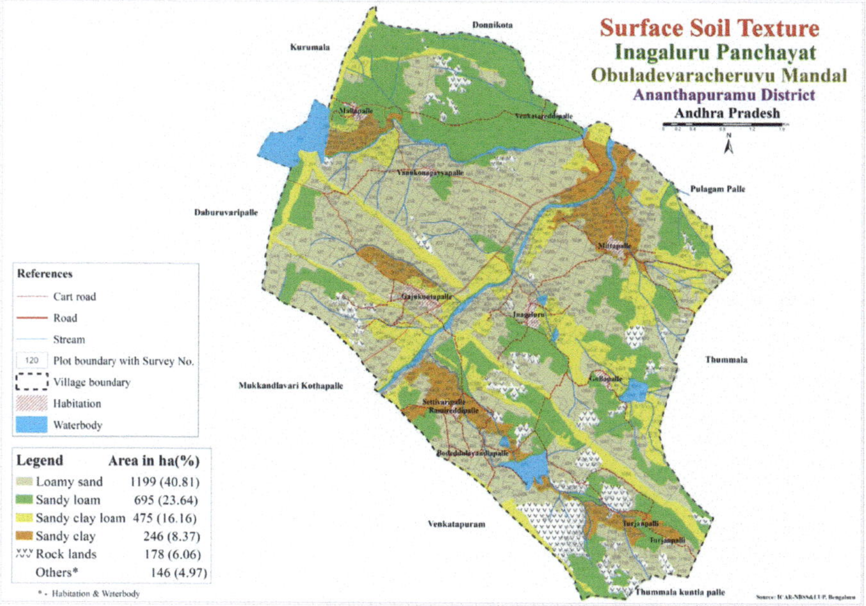

Fig. 2.9 Soil texture map of Inagalur Panchayat

Another productive land category includes sandy clay soils, covering 8.37% of the area. These soils also demonstrate moderate potential for soil–water retention and nutrient availability, without drainage or other physical problems. Understanding soil texture provides crucial insights into agricultural potential and challenges in different areas within the Panchayat.

2.3.3.2 Soil Gravelliness

Gravel consists of coarse fragments ranging from 2 mm to 7.5 cm, while stones are characterized by diameters between 7.5 and 25 cm. Their presence in soil significantly reduces the soil volume responsible for moisture and nutrient retention, affecting drainage, infiltration, and runoff. Moreover, gravel and stones hinder plant growth by impeding root development, seedling emergence, and farm mechanization. Figure 2.10 illustrates the distribution and extent of these components across the Panchayat.

The largest expanse, covering about 1100 ha (37%), consists of non-gravelly soils (< 15%), predominantly located in the northern, northeastern, eastern, and southern parts of the Panchayat. Another area, totaling about 754 ha (26%), features gravelly soils (15–35%) and is distributed across the central, western, northern, eastern, and southern regions. A third area, encompassing about 704 ha (24%), is identified as very gravelly and is deemed unsuitable for potential crops. These very gravelly soils,

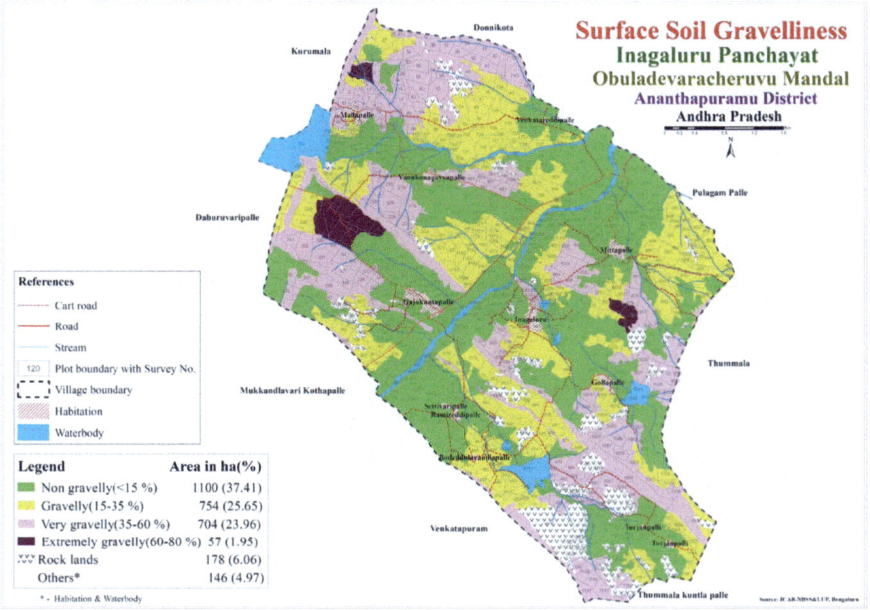

Fig. 2.10 Soil gravelliness map of Inagalur panchayat

characterized by less than 15% gravel content, do have the potential to support the growth of both annual and perennial crops. The distribution of gravel and stones in the soil landscape highlights the need for strategic land management practices in specific areas to optimize agricultural productivity.

2.3.3.3 Available Water Content (AWC)

The estimation of Soil Available Water Content (AWC) hinges on the soil's capacity to retain water across specified tensions and depths. It's determined within 100 cm or the full solum for shallower soils, accounting for factors like texture, mineralogy, depth, and gravel content. This evaluation categorizes AWC into three classes: very low (< 50 mm/m), low (50–100 mm/m), and medium (100–150 mm/m). These classifications formed the basis for creating an AWC map (Fig. 2.11), illustrating the varying water retention capacities across the surveyed area.

Approximately 1379 ha (46.91%) across the Panchayat exhibit very low AWC (< 50 mm/m), spread throughout the area. Low AWC (50–100 mm/m) covers about 696 ha (23.67%), concentrated in the western and eastern regions. Another area of around 540 ha (18.39%) displays medium AWC (101–150 mm/m), primarily in the northern and central parts.

Within the Panchayat, around 1379 ha (46.91%) face challenges due to low available water capacity, suitable mostly for short-duration crops with high risk of crop

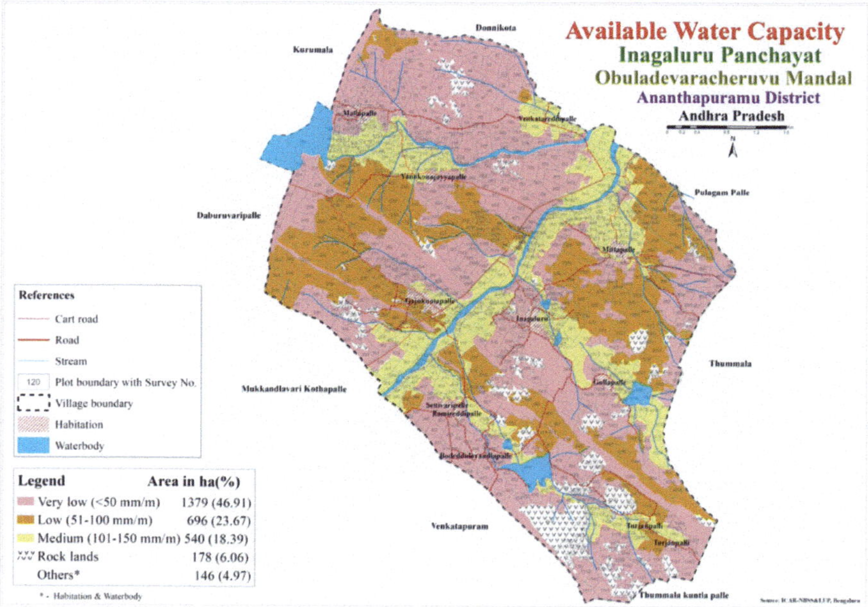

Fig. 2.11 Soil available water capacity map of Inagalur panchayat

failure. These regions might be more suitable for alternative land uses. Conversely, approximately 540 ha (18.39%) feature high potential in available water capacity (101–150 mm/m), allowing successful cultivation of all climatically adapted long-duration crops. This classification aids in identifying areas with different water-holding capacities, enabling tailored agricultural practices and land management strategies.

2.3.3.4 pH

The soil's reaction, or the acidity and alkalinity of the soil environment, is a critical parameter that significantly influences crop production. Given the marked impact of soil reaction on plant growth, its importance has been widely acknowledged over an extended period. Soil is typically categorized as acidic, neutral, or alkaline based on its pH level. Most nutrients are readily available to plants at a neutral pH, whereas some nutrients become fixed and unavailable in either acid or alkaline soils.

The soil analysis conducted in Inagalur Panchayat for soil reaction (pH) revealed that the largest area, approximately 923 ha (31.36%), exhibits a neutral soil reaction (pH 6.5–7.3). This area is predominantly distributed in the southern part of the Panchayat. About 795 ha (27.05%) is characterized as strongly acidic to slightly acidic (pH 5.0–6.5) in soil reaction, prevalent in the central, western, and northern parts of the Panchayat. Additionally, an area of about 900 ha (30.58%) is slightly alkaline to strongly alkaline (pH 7.3–9.0) in soil reaction and is found in the central, western, northern, eastern, and southern parts of the Panchayat (Fig. 2.12). This classification based on soil reaction aids in understanding the pH variations across the Panchayat, providing valuable insights for tailored agricultural practices and nutrient management strategies to optimize crop productivity.

2.3.3.5 EC

Soil electrical conductivity (EC) measures soil salt content, serving as an indirect indicator of nutrient availability. High EC (EC > 4 dS m^{-1}) signifies soil salinity issues, hindering crop growth and microbial activity. Elevated sodium levels in high EC soils often lead to poor structure, drainage, and sodium toxicity to plants. In Inagalur Panchayat, soils are classified as non-saline (< 2 dS m^{-1}), indicating no salinity-related issues.

2.3.3.6 Organic Carbon

Organic matter serves as a vital reservoir of essential soil nutrients crucial for plant growth, significantly impacting soil productivity. In Inagalur Panchayat, the soil's organic carbon content, indicative of available nitrogen, shows medium levels (0.5–0.75%) across approximately 923 ha (31.36%), spanning all areas. Conversely, the

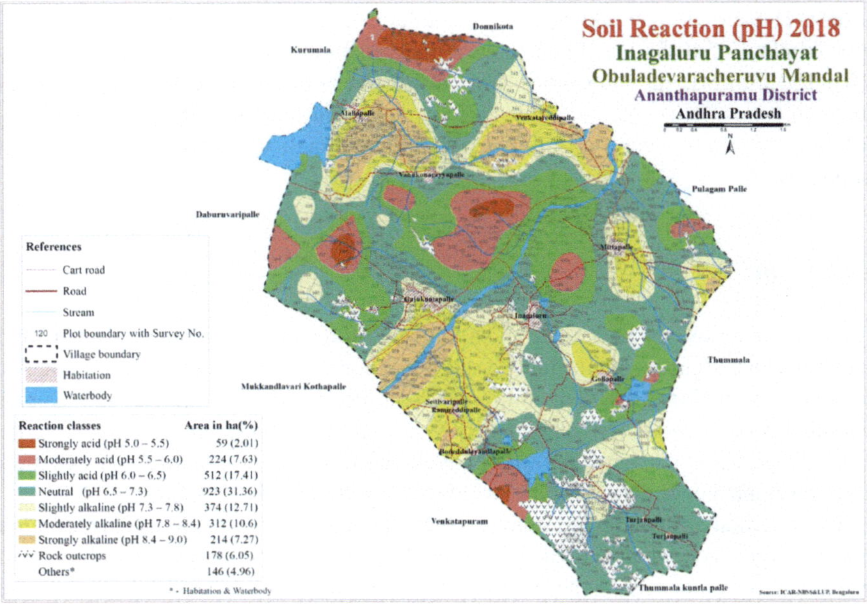

Fig. 2.12 Soil reaction (pH) map of Inagalur panchayat

largest area, about 1212 ha (41.2%), exhibits low organic carbon levels (< 0.5%) and is prevalent in the central, western, northern, eastern, and southern parts (Fig. 2.13). Additionally, around 483 ha (16.43%) showcase high organic carbon levels (> 0.5%), spread throughout the Panchayat. This classification of organic carbon content offers valuable insights into soil nutrient status, facilitating targeted soil management strategies to improve overall agricultural productivity.

2.3.4 Alternate Land Use Plan for Climate Change Resilience

Climate change'simpact on natural resources directly affects crop production and livelihood. Assessment of land resources viz, soil, water and vegetation, etc. at farm scale is highly useful and helps to resource management. The adoption of different land use systems is a spatial perspective for climate change and drought mitigation. However, farm scale assessment of soil variability and water quality could improve climate change resilience. Consideration of climate change in an arid part of Andhra Pradesh in case of frequent drought and crop loss. Allocating different land use systems could be useful for long-term adaptation and sustainability in the arid region. Hence, adopting an appropriate land use system and water and nutrients management helps restore crop productivity (Lei et al. 2016). The land resources are suitable for growing various crops like agriculture, horticultural and plantation

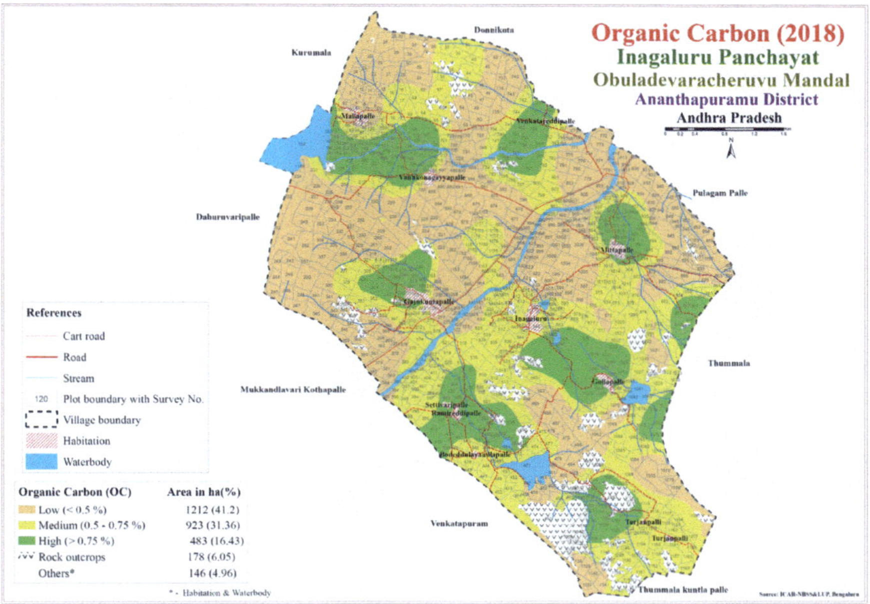

Fig. 2.13 Soil organic carbon (OC) map of Inagalur panchayat

making restoration for climate resilience. Assessing the extent and degree of suitability of the land resources in the Inagalur panchayat for various crop cultivation and adopting suitable interventions for rectifying soil and climatic problems and ways of land restoration towards sustainable crop productivity (Fig. 2.14).

Severe erosion *Alternate land use system* *Sheep Grazing*

Fig. 2.14 Soil problems and alternate landuse through restoration in the Inagalur panchayat

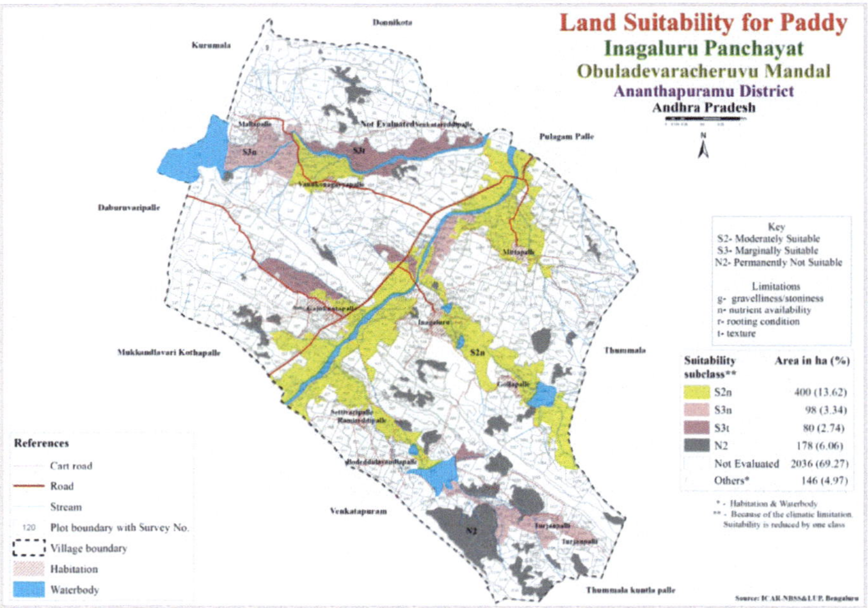

Fig. 2.15 Land suitability map of paddy in Inagalur panchayat

2.3.5 *Crop Suitability Assessment*

2.3.5.1 Paddy (*Oryza sativa*)

Since paddy is one of the main staple food crops in the area, it is quite important. A map of land suitable for paddy was created using soil variability as a foundation. Figure 2.15 shows the scope and geographic distribution of the various suitability subclasses within the Panchayat. Particularly around the river watershed in the Panchayat, a majority area of about 400 ha (13.62%) is designated as reasonably suitable (Class S2) for paddy production. Furthermore, 178 ha, or 6.08% of the total, are classified as marginally suitable (Class S3) for paddy cultivation. It is noteworthy that the lack of water facilities prevented additional Panchayat areas from being assessed. This determination of the land's potential for paddy farming offers helpful advice for optimizing agricultural practices and enhancing crop productivity in the specified regions.

2.3.5.2 Maize (*Zea mays* L.)

One important food crop that is grown on a vast expanse of 13.37 lakh hectares, almost all of the state's districts, is maize. A detailed evaluation of soil-site features was carried out in order to match with the specific crop requirements for maize

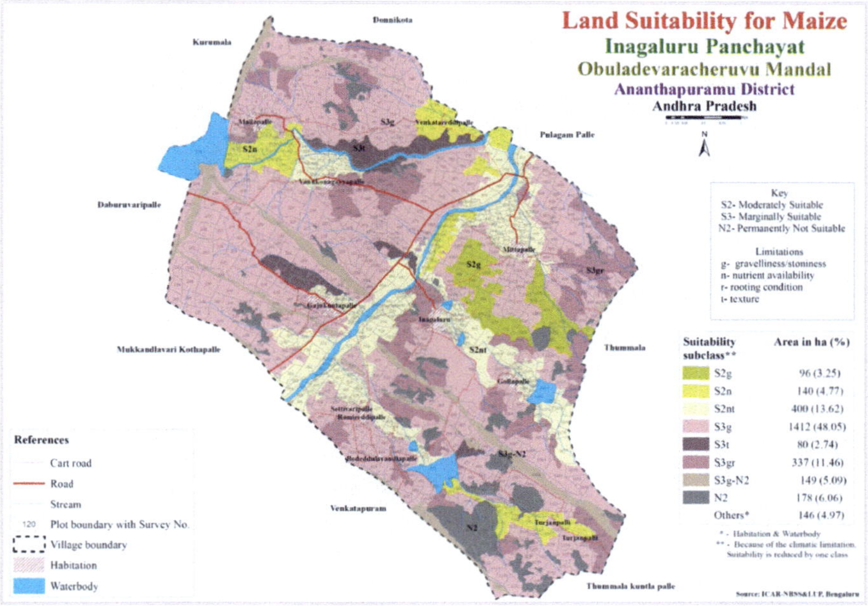

Fig. 2.16 Land suitability map of maize in Inagalur panchayat

cultivation, and the outcome was the creation of a land suitability map. Figure 2.16 shows the range and spatial arrangement of different suitability subclasses in the Panchayat.

About 636 ha, or 21.64% of the total area, fall within the moderately appropriate (Class S2) category for maize farming. This area is mostly found in the Panchayat's north, east, and south and has some minor restrictions on texture, rooting depth, and gravelliness. Conversely, the greatest area, which is roughly 1829 ha (62.25%), is categorized as slightly appropriate (Class S3) for maize cultivation. This category is spread across all parts of the Panchayat, exhibiting moderate limitations related to rooting depth, texture, and gravelliness. Furthermore, an area of about 327 ha (11.15%) falls into the not suitable category (Class N2) for maize cultivation, concentrated in the southern part of the Panchayat. This comprehensive assessment of land suitability aids in pinpointing areas with varying degrees of suitability, allowing for strategic agricultural planning and the implementation of targeted practices to enhance maize cultivation outcomes.

2.3.5.3 Bajra (*Pennisetum glaucum* L.)

The assessment for Bajra cultivation suitability in the Panchayat involved matching crop requirements with soil characteristics, resulting in a land suitability map (Fig. 2.17). The largest contiguous area, approximately 2050 ha (69.69%), falls into

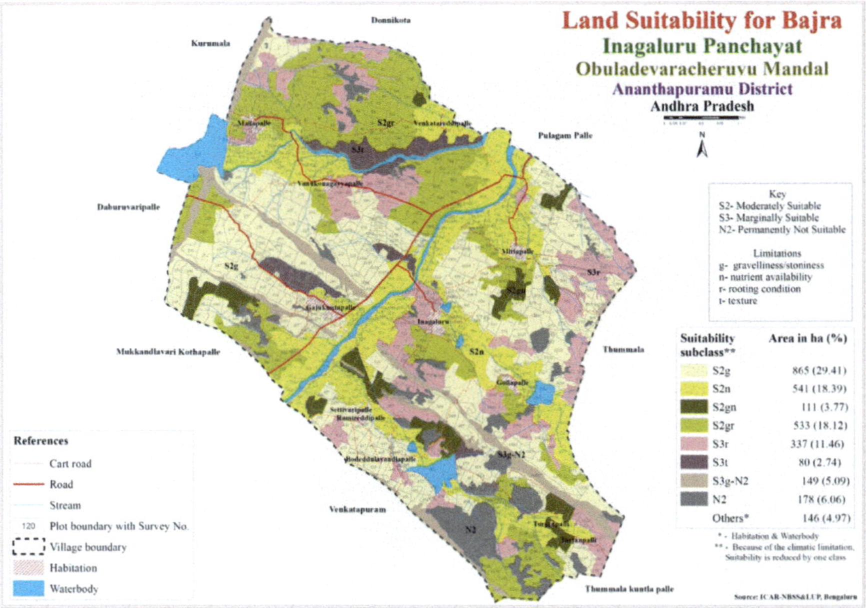

Fig. 2.17 Land suitability map of bajra in Inagalur panchayat

the moderately suitable category (Class S2) for Bajra cultivation. This region, spread across the Panchayat, has minor limitations related to texture, rooting depth, and gravelliness. About 417 ha (14.2%) are classified as marginally suitable (Class S3), mainly in the northern, eastern, and southern parts, with moderate limitations in rooting depth and gravelliness. Additionally, around 327 ha (11.15%) are considered not suitable (Class N2) for Bajra cultivation, primarily in the southern part. This detailed assessment offers insights for strategic planning and targeted agricultural practices to enhance Bajra cultivation in specific regions.

2.3.5.4 Groundnut (*Arachis hypogaea*)

Groundnut, being a major oilseed crop, underwent a meticulous evaluation wherein its crop requirements were carefully considered alongside the soil-site characteristics of the Panchayat. A groundnut land suitability map was produced as a result of this extensive investigation, showing the amount and geographic distribution of different suitability subclasses within the Panchayat (Fig. 2.18).

The majority of the land, or about 1597 ha (54.3%), is classified as somewhat appropriate (Class S2) for groundnut cultivation. This vast area is spread over the Panchayat's northern, eastern, and western sections. It has some minor restrictions regarding texture, rooting depth, gravelliness, and fertilizer availability. Furthermore, 1017 ha, or 34.67% of the total, are classified as somewhat suitable (Class S3) for

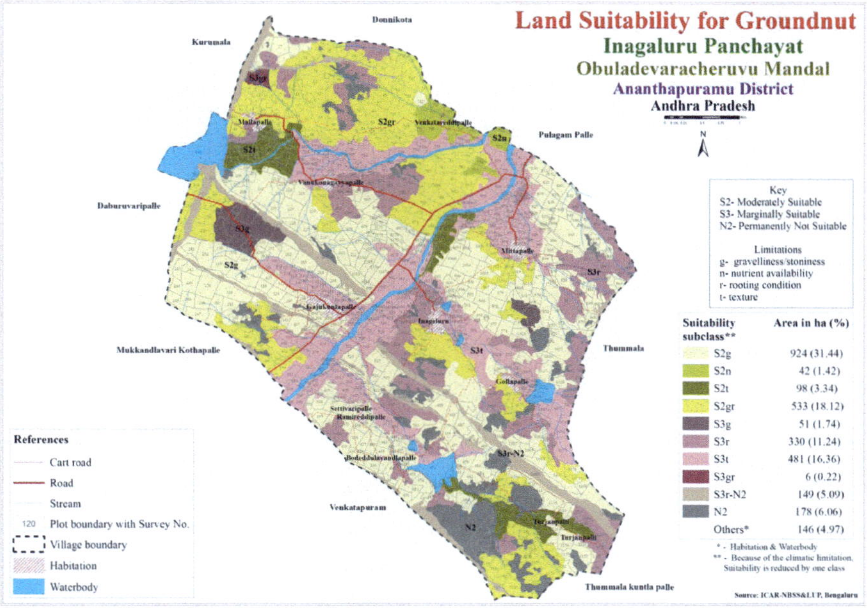

Fig. 2.18 Land suitability map of groundnut in Inagalur panchayat

groundnut production. This category, which has moderate restrictions on texture, rooting depth, and gravelliness, is found in the Panchayat's northern, western, eastern, and southern regions. Conversely, an area of about 178 ha (6.06%) is deemed permanently not suitable (Class N2) for growing groundnut due to the presence of rock outcrops in the Panchayat. These findings align with similar results reported by Srinivasan et al. (2020a). This detailed land suitability assessment contributes valuable insights for informed decision-making, facilitating strategic agricultural practices tailored to enhance groundnut cultivation outcomes in specific regions.

2.3.5.5 Castor (*Ricinus communis*)

The detailed evaluation matching Castor crop requirements with soil characteristics resulted in a land suitability map (Fig. 2.19) outlining various suitability subclasses within the Panchayat. The most extensive area, approximately 2050 ha (69.69%), falls under the moderately suitable category (Class S2) for Castor cultivation. Spread across all parts of the Panchayat, this region has minor limitations related to rooting depth, nutrient availability, and gravelliness. About 431 ha (19.29%) are classified as marginally suitable (Class S3) for Castor cultivation, mainly in the northern, eastern, and southern parts, with limitations in rooting depth, texture, and gravelliness. However, roughly 178 ha (6.06%) are permanently unsuitable (Class N2) for Castor cultivation, primarily in the southern part of the Panchayat. This detailed

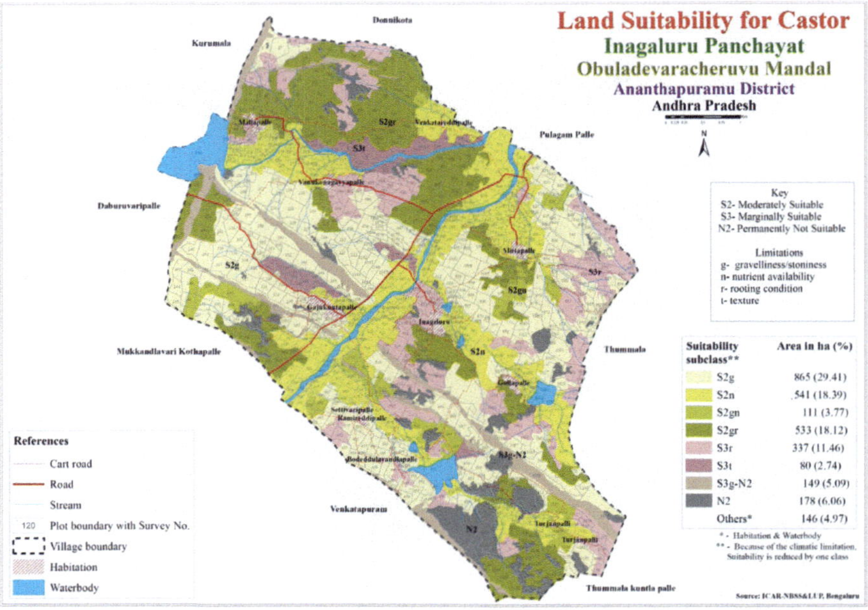

Fig. 2.19 Land suitability map of castor in Inagalur panchayat

assessment aids in strategic decision-making, allowing tailored agricultural practices to enhance Castor cultivation outcomes in specific regions.

2.3.5.6 Red Gram (*Cajanus cajan*)

The assessment aligning red gram crop requirements with soil-site characteristics led to a land suitability map (Fig. 2.20) outlining various suitability subclasses in the Panchayat. A confined area of approximately 23 ha (6%) is highly suitable (Class S1) for red gram cultivation, primarily in the northern part. The largest expanse, covering about 637 ha (21.64%), falls under the moderately suitable category (Class S2). Spread across central, northern, western, southern, northeastern, and eastern areas, this category features minor limitations related to nutrient availability, rooting depth, and gravelliness. Additionally, an extensive area of around 1493 ha (50.79%) is marginally suitable (Class S3) for red gram cultivation, distributed across all parts of the Panchayat. This classification indicates moderate limitations in texture, rooting depth, and gravelliness. The detailed land suitability assessment provides insights for informed decision-making, enabling optimized agricultural practices to enhance red gram cultivation outcomes in specific regions.

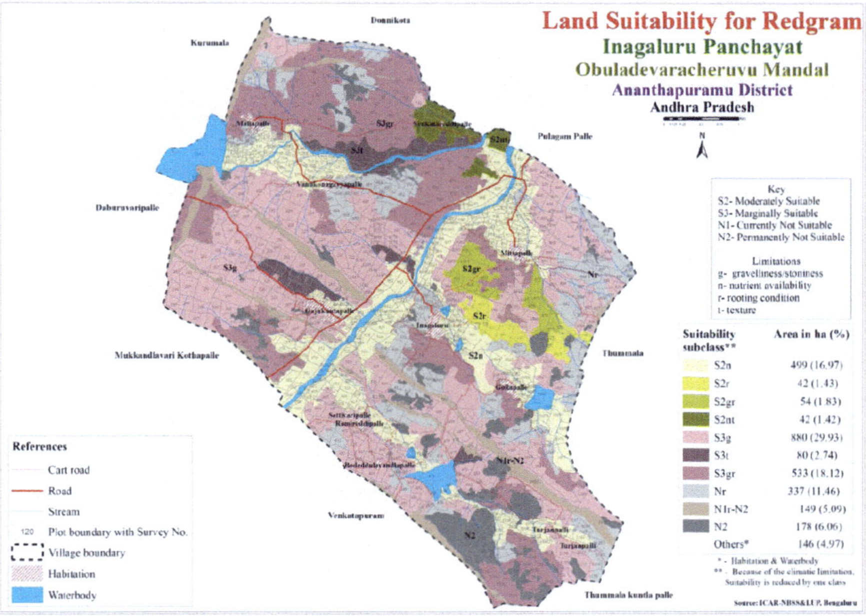

Fig. 2.21 Land suitability map of horse gram in Inagalur panchayat

2.3.5.7 Horse Gram (*Macrotyloma uniflorum*)

The assessment aligning Horse gram crop requirements with soil characteristics generated a land suitability map (Fig. 2.21). The largest area, approximately 2287 ha (77.82%), falls within the moderately suitable category (Class S2) for Horse gram cultivation. Spread across all parts of the Panchayat, this extensive region has minor limitations related to gravelliness, nutrient availability, and rooting depth. Additionally, around 178 ha (6.08%) are marginally suitable (Class S3) for Horse gram, observed in small parts of the northern, central, and southern regions, with limitations in nutrient availability, texture, and gravelliness. Conversely, about 327 ha (11.15%) are classified as not suitable (Class N2) for Horse gram cultivation in Inagalur Panchayat. This detailed land suitability assessment serves as a valuable tool for strategic decision-making, enabling the implementation of tailored agricultural practices to optimize Horse gram cultivation outcomes in specific regions.

2.3.5.8 Tomato (*Solanum lycopersicum*)

The meticulous assessment aligning tomato crop requirements with soil-site characteristics resulted in a land suitability map (Fig. 2.22) delineating various suitability subclasses in the Panchayat. Around 138 ha (4.67%) are moderately suitable (Class S2) for tomatoes, concentrated in the northeastern part, featuring minor limitations in

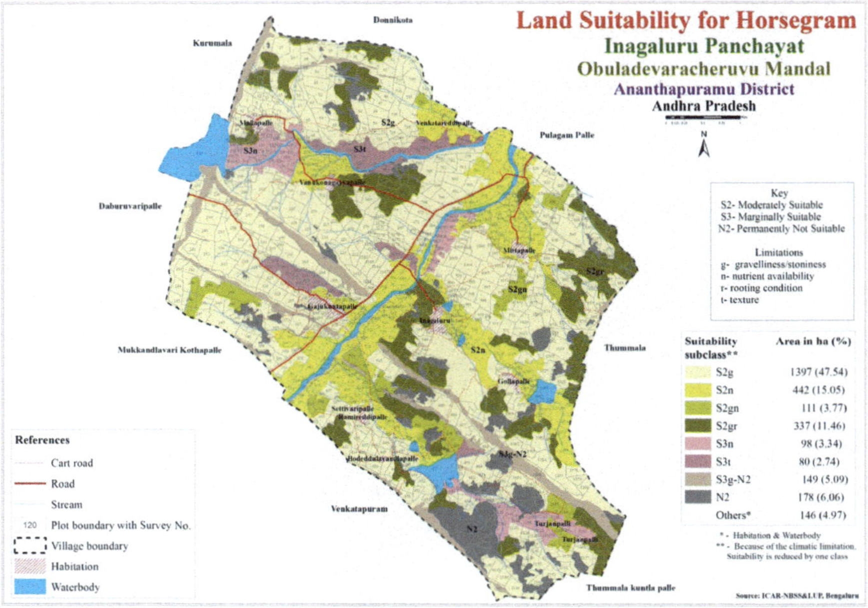

Fig. 2.20 Land suitability map of redgram in Inagalur panchayat

gravelliness and nutrient availability. However, a substantial area of approximately 2230 ha (75.87%) falls within the marginally suitable category (Class S3). This spans across the entire Panchayat, exhibiting moderate limitations in nutrient availability, texture, rooting depth, and gravelliness. Additionally, about 425 ha (14.49%) are unsuitable (Class N2) for tomatoes, primarily in the northern and southern parts. This detailed assessment supports informed decision-making and tailored agricultural practices to enhance tomato cultivation outcomes in specific regions within the Panchayat's northern and southern parts.

2.3.5.9 Mango (*Mangifera indica*)

The State relies significantly on the cultivation of mango, a pivotal fruit crop. Matching the crop's requirements with the soil-site characteristics, an assessment of land suitability was conducted, as depicted in Fig. 2.23. Approximately 531 ha (18.07%) are identified as moderately suitable (Class S2) for mango cultivation, primarily situated in the north-eastern and western regions of the Panchayat. These areas present minor limitations related to nutrient availability and gravelliness.

Marginal suitability (Class S3) encompasses an expanse of 1065 ha (36.24%), concentrated in the southern and western parts of the Panchayat. Challenges such as moderate limitations in rooting depth, nutrient availability, texture, and gravelliness are prevalent in these lands.

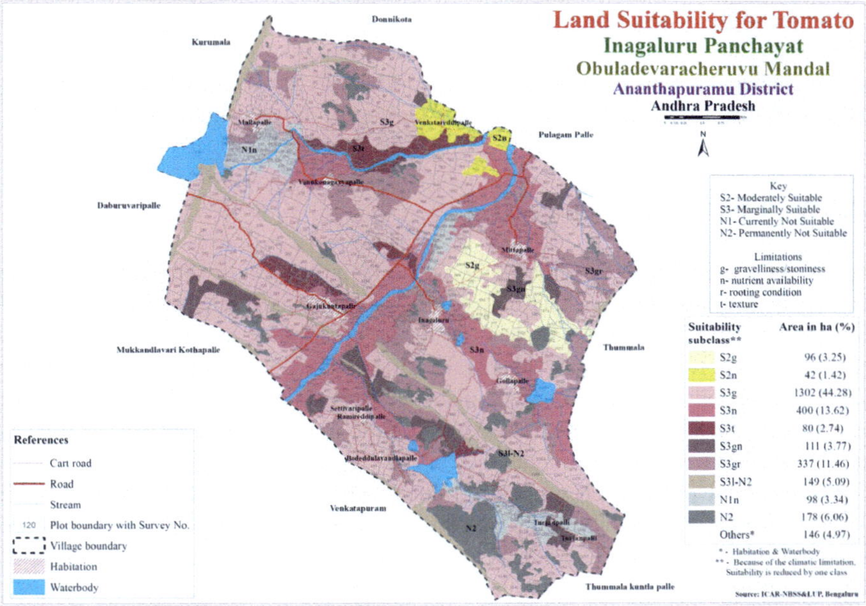

Fig. 2.22 Land suitability map of tomato in Inagalur panchayat

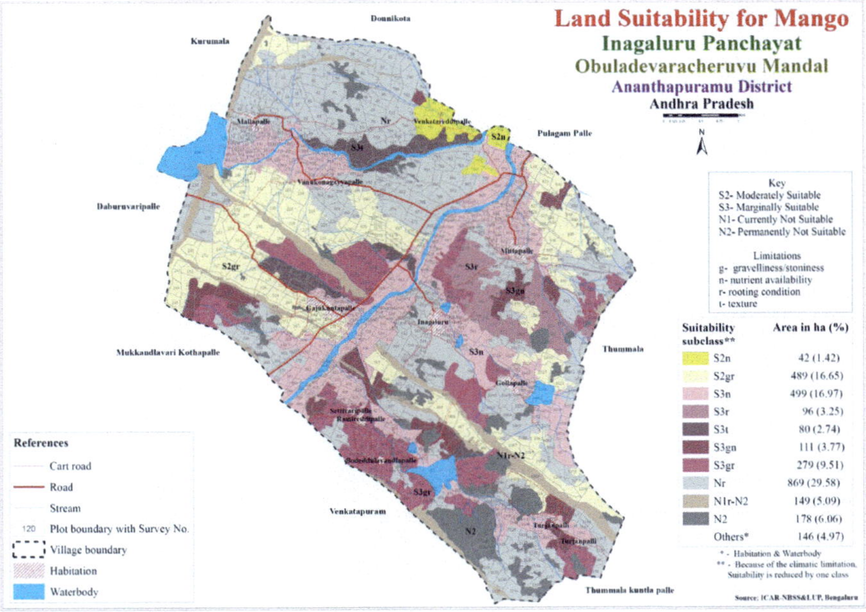

Fig. 2.23 Land suitability map of mango in Inagalur panchayat

Contrastingly, a substantial portion of approximately 1196 ha (40.73%) is deemed unsuitable (Class N1 and N2) for mango cultivation, distributed across all parts of the Panchayat. The unsuitability is attributed to severe limitations in rooting depth. Notably, Srinivasan et al. (2020c) reported similar findings in their research.

2.3.5.10 Sapota (*Manilkara zapota*)

Sapota holds a significant position among the fruit crops cultivated across diverse districts in Andhra Pradesh. The cultivation requirements for sapota were aligned with the soil-site characteristics, leading to the creation of a land suitability map (Fig. 2.24). Approximately 627 ha (21.32%) are identified as moderately suitable (Class S2) for sapota cultivation, with distribution spanning the western and eastern regions of the Panchayat. These areas exhibit minor limitations related to gravelliness, nutrient availability, and rooting depth.

A substantial portion, encompassing about 1403 ha (47.76%), falls under the category of marginal suitability (Class S3) for sapota cultivation. These lands are dispersed throughout the Panchayat, presenting moderate limitations in terms of texture, rooting depth, nutrient availability, and gravelliness.

Conversely, an area comprising 584 ha (25.95%) is deemed unsuitable (Class N1, N2) for sapota cultivation, distributed across all parts of the Panchayat. These lands face severe limitations in rooting depth and nutrient availability.

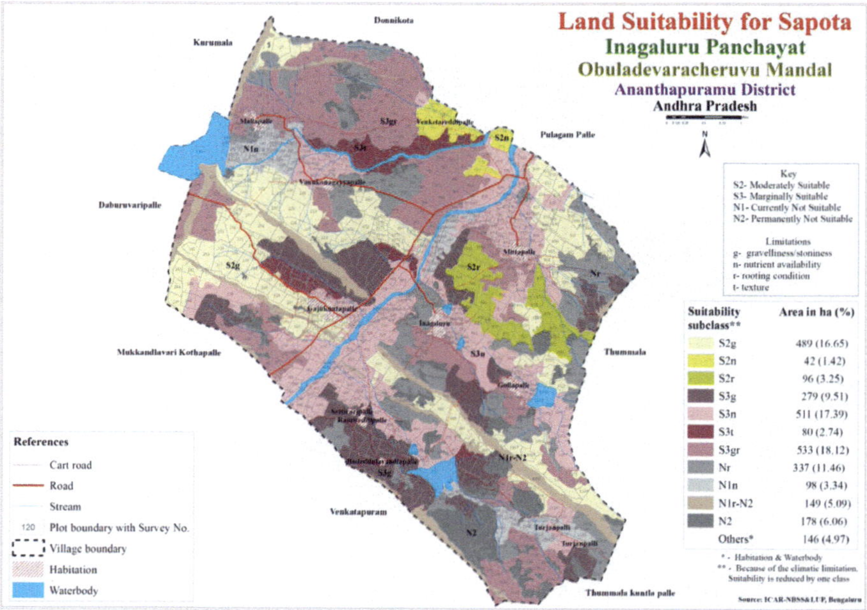

Fig. 2.24 Land suitability map of sapota in Inagalur panchayat

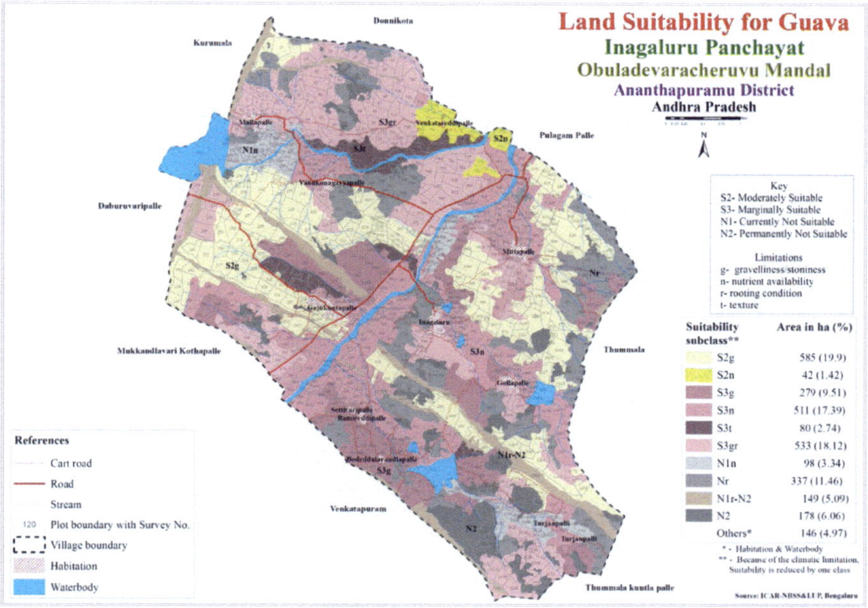

Fig. 2.25 Land suitability map of guava in Inagalur panchayat

2.3.5.11 Guava (*Psidium guajava*)

The thorough matching of guava cultivation requirements with soil characteristics resulted in a land suitability map (Fig. 2.25). Approximately 627 ha (21.32%) are moderately suitable (Class S2) for guava cultivation, found in the central, northern, and western parts. These areas have minor limitations in rooting depth, nutrient availability, and gravelliness. The largest expanse, around 1403 ha (47.76%), is marginally suitable (Class S3), dispersed throughout the entire Panchayat. These lands present moderate limitations in texture, rooting depth, nutrient availability, and gravelliness. Conversely, about 762 ha (25.95%) are unsuitable (Class N1 and N2) for guava cultivation, distributed across all parts of the Panchayat.

2.3.5.12 Jamun (*Syzygium cumini*)

Jamun holds significance as a crucial arid fruit crop cultivated in Andhra Pradesh. After careful examination of the crop needs for farming Jamun in accordance with the soil-site characteristics, a land suitability map was created (Fig. 2.26). Approximately 531 ha (18.07%) are identified as moderately suitable (Class S2) for Jamun cultivation, encompassing a major portion of the Panchayat. These areas exhibit minor limitations related to nutrient availability, rooting depth, and gravelliness. The greatest area, which is around 1499 ha (51.02%), is classified as somewhat

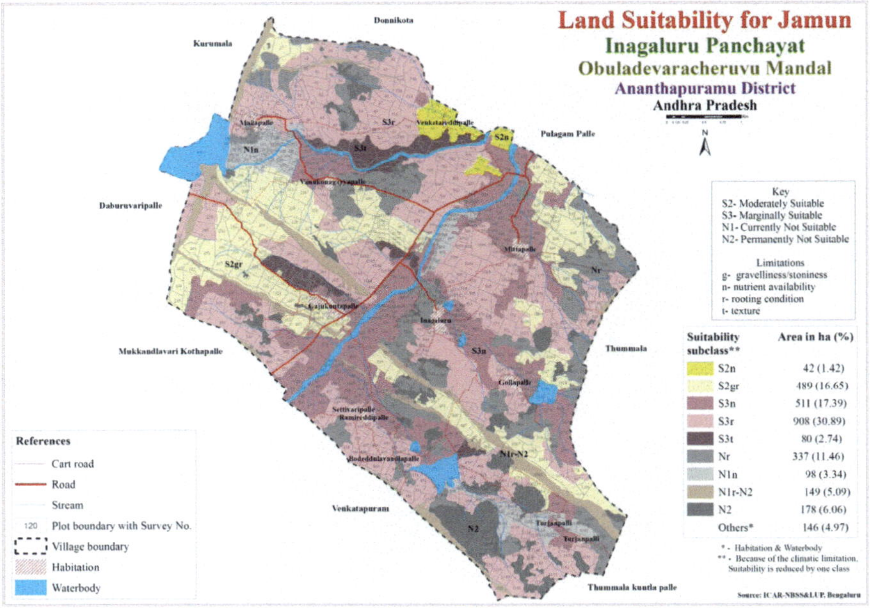

Fig. 2.26 Land suitability map of jamun in Inagalur panchayat

appropriate (Class S3) for the growth of jamun. These areas are scattered across the Panchayat and offer mild constraints for texture, gravelliness, rooting depth, and nutrient availability. On the other hand, a total of 762 ha, or 25.95% of the land, are classified as inappropriate (Class N1 and N2) for the cultivation of jamun and are located throughout the Panchayat. These areas are severely limited in terms of available nutrients and rooting depth.

2.3.5.13 Custard Apple (*Annona reticulata*)

Custard apple stands out as a significant fruit crop cultivated across various districts of the State. The crop requirements for cultivating custard apples underwent a meticulous evaluation in conjunction with soil-site characteristics, resulting in the creation of a land suitability map (Fig. 2.27). A modest area of approximately 96 ha (3.25%) is identified as highly suitable (Class S1) for custard apples, concentrated in the eastern part of the Panchayat. Moderately suitable lands (Class S2) cover a more extensive area of about 1853 ha (63.1%), presenting minor limitations related to gravelliness, rooting depth, and nutrient availability. The Panchayat is covered by these lands. Approximately 515 ha (17.54%) of marginally suitable sites (Class S3) are located in the southern and northern regions of the Panchayat. These lands have moderate restrictions in terms of texture, nutrient availability, and rooting depth. On the other hand, 327 ha (11.15%) of land are not appropriate (Class N1 and N2) for the

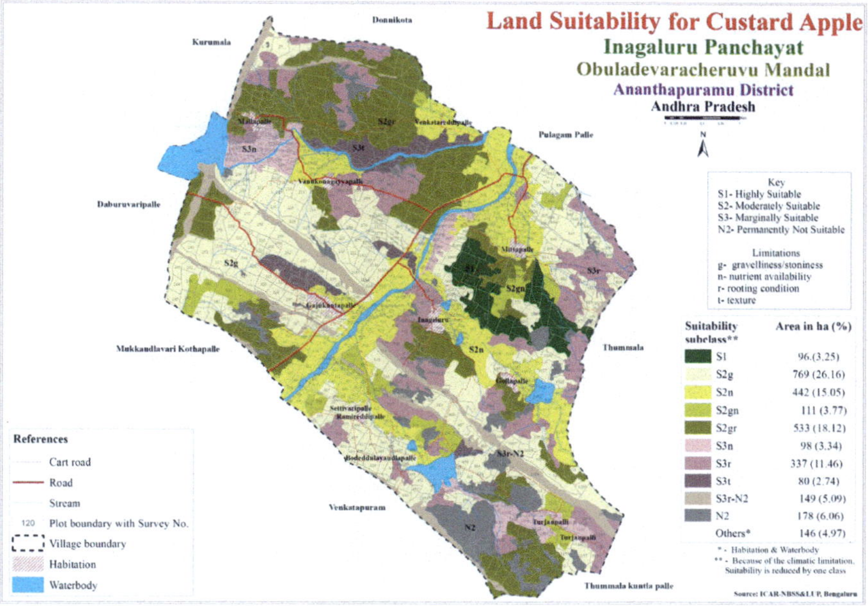

Fig. 2.27 Land suitability map of custard apple in Inagalur panchayat

Panchayat's custard apple farming. These lands are severely limited in their ability to sink roots.

2.3.5.14 Amla (*Phyllanthus emblica*)

Amla holds significant importance as both a fruit and a medicinal crop cultivated in Andhra Pradesh. The geographic distribution and extent of various suitability subclasses in the Panchayat are illustrated in Fig. 2.28. A substantial area, approximately 1520 ha (51.69%), is identified as moderately suitable (Class S2) for amla cultivation. These lands are situated in the north-eastern and southern parts of the Panchayat, exhibiting minor limitations related to rooting depth, nutrient availability, and gravelliness. The southern and western regions of the Panchayat have 848 ha (28.85%) of marginally suitable land (Class S3). The rooting depth and nutrient availability in these regions are moderately limited. On the other hand, 425 ha, or 14.49 percent, of the Panchayat's southern region are categorized as unsuitable (Class N1 and N2) for amla farming. These areas are severely limited in terms of available nutrients and rooting depth.

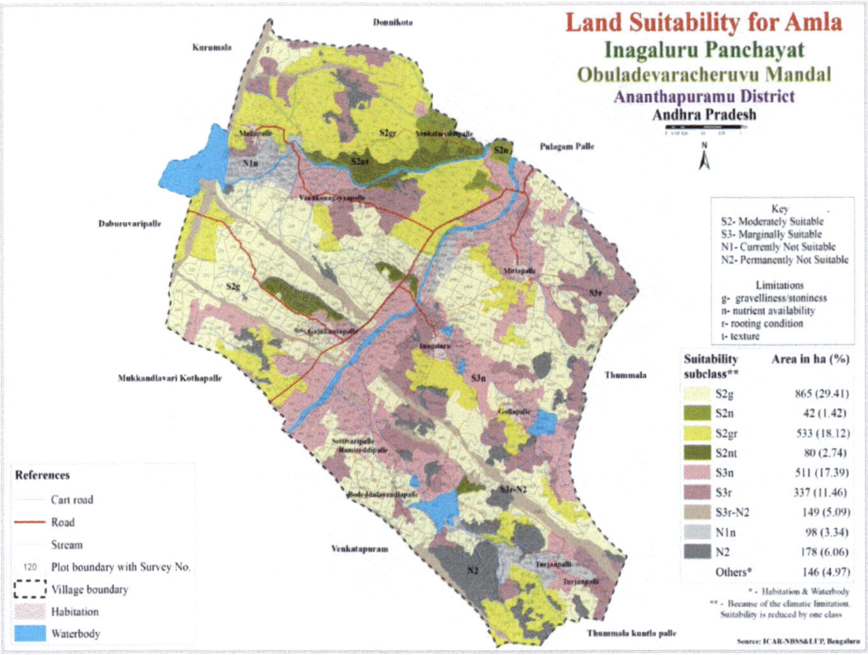

Fig. 2.28 Land suitability map of amla in Inagalur panchayat

2.3.5.15 Tamarind (*Tamarindus indica*)

Tamarind holds significant importance as a spice crop cultivated in various states. The crop requirements for tamarind cultivation underwent a comprehensive assessment aligned with soil-site characteristics, resulting in the creation of a land suitability map (Fig. 2.29). There are about 531 ha (18.07%) in the western, northeastern, and eastern regions of the Panchayat that are classified as somewhat suitable (Class S2) for tamarind cultivation. There are also slight restrictions on these regions in terms of roots depth and gravelliness. The center, western, and southern regions of the Panchayat have 566 ha (19.27%) of marginally suitable lands (Class S3) for tamarind cultivation. The nutrient availability, rooting depth, roughness, and gravelliness of these lands are moderately limited. The greatest area, which covers roughly 1695 ha or 57.7% of the Panchayat, is classified as unsuitable for tamarind cultivation (Class N1 and N2). These areas are severely limited in terms of available nutrients and rooting depth.

2.3.5.16 Ber (*Ziziphus mauritiana*)

The region, known for its drought conditions, prioritizes Ber cultivation. Soil-site traits were aligned with Ber's needs, producing a land suitability map (Fig. 2.30).

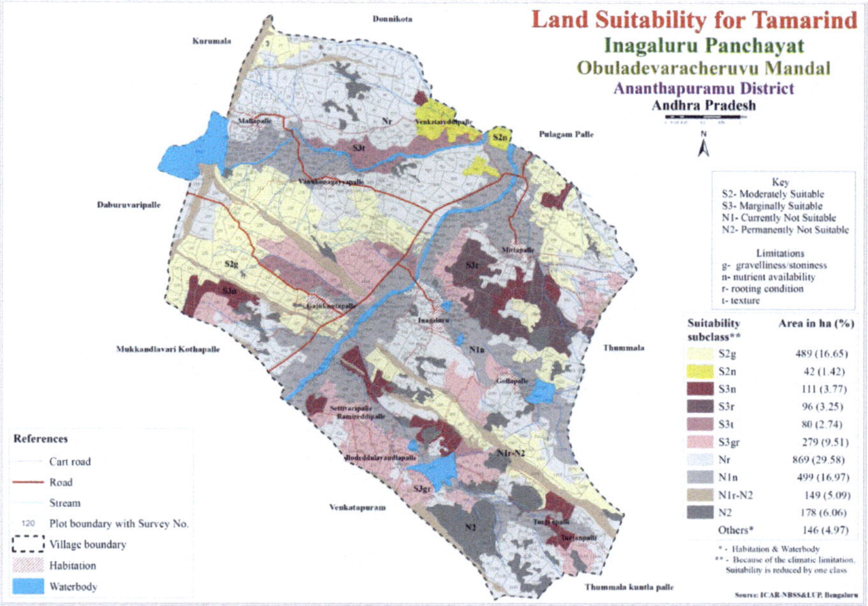

Fig. 2.29 Land suitability map of tamarind in Inagalur panchayat

About 538 ha (18.3%) in the central and eastern Panchayat areas are highly suitable (S1) for Ber. Moderately suitable land (S2) spans 1511 ha (51.39%) in the north and south, with slight limitations in rooting depth, nutrients, and gravel presence. An additional 417 ha (14.2%) in the southern, eastern, and north-eastern parts are marginally suitable (S3) due to moderate rooting depth and texture issues. Areas unsuitable (N1 and N2), covering 327 ha (11.15%), mostly in the south, present severe rooting depth limitations, alongside minor nutrient and gravel concerns. This comprehensive map guides where Ber cultivation thrives, suggests areas with manageable constraints, and highlights regions unsuitable without significant intervention, aiding efficient land use planning.

2.4 Recommendations

Conducting a land resources inventory is essential for identifying soil potential and addressing issues to ensure sustainable agriculture and effective land and crop management. Farm-scale soil information specific to the Inagalur Panchayat is valuable for implementing precision farming practices across various cropping systems, mitigating crop losses in the arid region of Andhra Pradesh. Thematic maps that highlight site variations, soil characteristics, and crop suitability provide farmers with valuable insights. These maps empower farmers to make informed decisions about

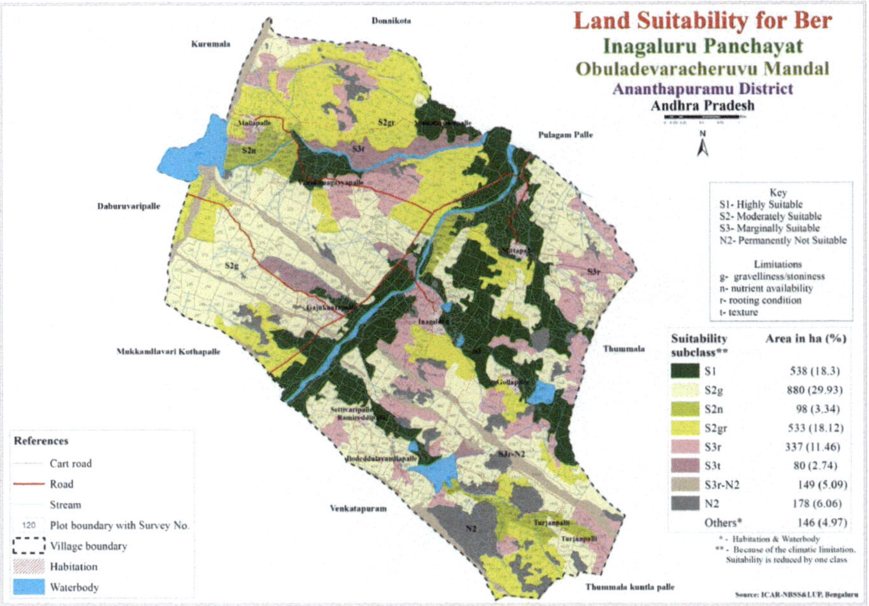

Fig. 2.30 Land suitability map of ber in Inagalur panchayat

crop management practices, ultimately enhancing crop production and promoting sustainability.

2.5 Conclusion

The present investigation focused on adopting an alternate cropping system for the arid region of Andhra Pradesh. Site and soil characteristics directly link to crop productivity, different classes of the slope, soil depth, texture, available water content, pH, EC, and organic carbon help to assess the problem and potential of the soils and degradation status. Suitability evaluation of different field crops, pulses, and plantationshelped to classify highly (S1), moderately (S2), and marginally (S3) suitable and not suitable (N1). The extent and distribution of soil problems and crop suitability can be used to design and implement appropriate site-specific soil and water conservation measures and crop management interventions to cultivate highly suitable crops formitigating climatic change and resilience agriculture.

References

Abdel Rahman MAE, Natarajan A, Hegde R (2016) Assessment of land suitability and capability by integrating remote sensing and GIS for agriculture in Chamarajanagar district, Karnataka, India. Egypt J Remote Sens Space Sci 19(1):125–141. https://doi.org/10.1016/j.ejrs.2016.02.001

Al-Mashreki MH, Akhir JB, Abd Rahim S, Desa K, Lihan T, Haider A (2011) Land suitability evaluation for sorghum crop in the Ibb Governorate, Republic of Yemen using remote sensing and GIS techniques. Aust J Basic Appl Sci 5:359–368

Bandyopadhyay S, Jaiswal RK, Hegde VS, Jayaraman V (2009) Assessment of land suitability potentials for agriculture using a remote sensing and GIS based approach. Int J Remote Sens 30(4):879–895. https://doi.org/10.1080/01431160802395235

Bobade SV, Bhaskar BP, Gaikwad MS, Raja P, Gaikwad SS, Anantwar SG, Patil SV, Singh SR, Maji AK (2010) GIS-based land use suitability assessment in Seoni district, Madhya Pradesh, India. Trop Ecol 51(1):41–54

Chaturvedi RK, Joshi J, Jayaraman M, Bala G, Ravindranath NH (2012) Multi-model climate change projections for India under representative concentration pathways. Current Sci 103:791–802

El Baroudy AA (2016) Mapping and evaluating land suitability using a GIS-based model. CATENA 140:96–104. https://doi.org/10.1016/j.catena.2015.12.010

FAO (1976) A framework for land evaluation. Food and Agriculture Organization of the United Nations, Soils Bulletin No. 32. FAO, Rome

FAO (1983) Guidelines: Land evaluation for rain-fed Agriculture. Soil resources management and conservation services, land and water development division Rome. Food and Agriculture Organization Soil Bulletin 52: 237

FAO (1985) Guidelines: land evaluation for irrigated agriculture. Soil Bulletin No.55. FAO, Rome

FAO (1991) Guidelines for land evaluation for extensive grazing. Food and Agriculture Organization of the United Nations, Soils Bulletin No. 58. FAO, Rome

Godfray HCJ, Beddington JR, Crute IR, Haddad L, Lawrence D, Muir JF, Pretty J, Robinson S, Thomas SM, Toulmin C (2010) Food security: the challenge of feeding billion people. Science 327(5967):812–818

IPCC (2007) Climate change: impacts, adaptation and vulnerability. In: Parry ML, Canziani OF, Palutikof JP, van der Linden PJ, Hanson CE (eds) Contribution of working group I to the fourth assessment report of the intergovernmental panel on climate change. Cambridge University Press, Cambridge, United Kingdom and New York, NY, USA, pp 273–313

Jackson ML (1973) Soil chemical analysis. Prentice Hall of India Pvt. Ltd., New Delhi

Keesstra S, Mol G, Leeuw J de, Okx J, Molenaar C, de Cleen M, Visser S (2018) Soil-related sustainable development goals: four concepts to make land degradation neutrality and restoration work. Land 7: 133. https://doi.org/10.3390/land7040133

Knox J, Hess T, Daccache A, Wheeler T (2012) Climate change impacts on crop productivity in Africa and South Asia. Environ Res Lett:034032

Lei K, Pan H, Lin C (2016) A landscape approach towards ecological restoration and sustainable development of mining areas. Ecol Eng 90:320–325

Mishelia A, Zirra EM (2015) Application of Geographic Information System (G.I.S) in evaluating suitable areas for wheat. Cultivation in Adamawa State Nigeria. Int J Sci Knowl 6:14–22

Mu Y (2006) Developing a suitability index for residential land use: a case study in Dianchi drainage area. University of Waterloo, Canada

NBSS and LUP (1994) Proceeding national meet on soil-site suitability criteria for different crops held at NBSS and LUP (ICAR), Feb. 7–8, 1994, New Delhi

Page AL, Miller RH, Keeney DR (1982) Methods of soil analysis, Part 2, 2nd edn. Soil Science Society of America, Madison, USA

Roudier P, Sultan B, Quirion P, Berg A (2011) The impact of future climate change on West African crop yields: what does the recent literature say? Glob Environ Change 21:1073–1083

Soil Survey Staff (1999) Soil taxonomy: a basic system of soil classification for making and interpreting soil surveys, Natural Resources Conservation Service (2nd ed., 786 p). U.S. Department of Agriculture Handbook

Soil Survey Staff (2003) Soil survey manual. USDA Handbook No. 18. Scientific Publishers, Jodhpur, India

Srinivasan R, Rajendra Hegde, Srinivas S, Niranjana KV, Maddileti N (2020a) Mapping land suitability for groundnut (Arachis hypogaea L.) in arid environment of Andhra Pradesh using geographic information system. Int J Chem Stud 8(2): 201–207. https://doi.org/10.22271/chemi.2020.v8.i2c.8768

Srinivasan R, Vasundhara R, Lalitha M, Kalaiselvi B, Niranjana KV, Hegde R (2020c) Characterization and classification of major mango growing soils in arid part of Anantapur district, Andhra Pradesh, India. Agropedology 30(02):123–132. https://doi.org/10.47114/j.agroped.2020.dec5

Srinivasan R, Karthika KS, Amar Suputhra S, Chandrakala M, Rajendra Hegde (2021) Mapping of soil erosion and probability zones using remote sensing and GIS in arid part of South Deccan plateau, India. J Indian Soc Remote Sens. https://doi.org/10.1007/s12524-021-01396-5

Sys C, Van Ranst E, Debaveye J (1991) Land evaluation, Part I. Principles in land evaluation and crop production calculations. General administration for development cooperation, Brussels, pp 40–80

Sys C, Van Ranst E, Debaveye, J, Beernaert F (1993) Land evaluation part-3. Crop requirement. Agricultural Pub. No. 7, Belgium

Walthall CL, Hatfield J Backlund P, Lengnick, L (2012) Climate change and agriculture in the United States: effects and adaptation. USD

Chapter 3
Approaches and Principles in Agroforestry: A Tool for Land Policy Initiatives to Improve Soil Health and Mitigate Climate Change

Gyanaranjan Sahoo, Pravat Kumar Roul, and Asim Chandra Dash

Keywords Agroforestry systems · Ecosystem services · Integrated landscape management · Land-Use and Land-Cover (LULC) · Multi-stakeholder strategies · Sustainable development goals

3.1 Introduction

Agroforestry is frequently promoted as a sustainable solution to non-sustainable land-use practices that have negative effects on the enterprise, civilization, and the ecosystem. Agroforestry is a modern agricultural strategy that includes growing tree crops, farming in and around forests, cattle ranching, and working with forest land at various sizes. It is more than just "agricultural with trees." One of the best methods for reducing carbon dioxide emanations and addressing the issue of global warming, according to the FAO, is agroforestry (Gedefaw et al. 2014). The deliberate blending of a woody element reduced farming produce per level is known as agroforestry. Even so, despite being quite widely used in tropical nations, agroforestry is only used to a relatively limited extent in temperate regions, as is the case in Europe (Gizachew et al.

G. Sahoo (✉)
Krishi Vigyan Kendra, Odisha University of Agriculture and Technology, Nayagarh, Odisha, India
e-mail: gyanaranjan.sahoo3@gmail.com

P. K. Roul
Odisha University of Agriculture and Technology, Bhubaneswar, Odisha, India

A. C. Dash
College of Agriculture, Odisha University of Agriculture and Technology, Bhubaneswar, Odisha, India

P. P. Adhikary et al. (eds.), *Soil, Water Pollution and Mitigation Strategies*, Environmental Science and Engineering, https://doi.org/10.1007/978-3-031-63296-9_3

2017) or the USA (SD21 2012). This is because farming systems were previously intensified, forest and agricultural land were not integrated, and there were no current policies that were adequate to promote agroforestry practices.

Agroforestry is the integrated farm management practice of compounding trees with agricultural crops or livestock (Fagerholm et al. 2016). Secluded fruit trees, trees for shade in grasslands, plantations, dense forests, or gardens can be used as a crucial part of agroforestry systems to give production and advantages for all land users (Hombegowda et al. 2022; Sahoo et al. 2023). The numerous ecological services that agroforestry offers have been promoted, and these services typically result in more varied income streams for land users (such as nutrition, pasture, firewood, and non-timber items) and an improvement in overall quality of life (Kassie 2018; FAO 2005, 2019). In sub-Saharan Africa, households' yearly gross revenue from trees is 17% higher (Christiaensen and Demery 2018). Moreover, various agroforestry system mechanisms are chronologically and spatially organized to resemble biological and chemical mechanisms with a reduced need for synthetic materials inputs like compost, weed killers, and agrochemicals, providing essential regulatory services like enhancing soil fertility, preventing erosion, regulating and purifying water, conserving biodiversity, and sequestering carbon (Lorenz and Lal 2014; HLPE 2019). They contribute to pest populations eradication, a vastly greater variety of microbial activity, and insects that serve as food for birds and as pollinators, among other things, to the preservation of biodiversity. Moreover, trees on farms improve the landscape's ecological connection, preserving the networks of protected areas. In agricultural soils, agroforestry systems significantly contribute to the carbon pool (Adhikary et al. 2017; Sahoo et al. 2022). Furthermore, agroforestry structures are related to environmental natural systems, such as associations with customs, aesthetic standards, indigenous information, or leisure industry. For example, silvipastoral systems in Honduras, cacao interactions in Peru, indigenous "tembawang" agroforestry systems in Indonesia, or small-scale coffee cultivation in Uganda are typical agro—ecosystems (Getu et al. 2011; Marjokorpi and Ruokolainen 2003). These nations all use agroforestry systems as an approach to counteract the deforestation caused by a growing agricultural border. This is congruent with viewpoints held by people all around the world who consider agroforestry to be an essential component of managing forests, particularly habitat revival (Sharma and Sahoo 2021).

Decision makers must have access to data on the complexities of land use and land cover (hereafter referred to as LULC) in order to integrate community forest practices with growing plants in the agroforestry environment (Hassan et al. 2016). In the broadest sense of the word, ecosystem, "land" states to the topography, weather, biotic and abiotic properties, growth, and water resources. Since prehistory, LULC has changed due to human activity on the combining parts of such assets, both directly and indirectly (Meles et al. 2019). The detrimental shift, which is mostly brought on by the transition of land covered by trees into farming fields, social habitation, and organization, has a significant impact on biodiversity loss and unfavourable climatic change (Duguma et al. 2019). Only 9% more of the world's land surface has been affected by human activity, which has also caused an overall decline in ecosystem services of about 60% (Yahya et al. 2019). The destruction of vegetative areas to advance agriculture could have been the beginning of this shift, which would have resulted in the massive deforestation and deterioration of the Earth's surface that is still happening now, albeit at a faster rate. This is a major issue in Ethiopia, the second-most populous country in Africa, where vegetation reserves are slowly dwindling (Rawat and Kumar 2015).

The LULC dynamics and the expanding the number of people have a variety of effects on the atmospheric concentrations of greenhouse gases (GHG) worldwide (Forestry Commission 2011). The conversion of wetlands, wild plants, and bushes into agricultural land and communities has been the main contributor to rising CO_2 levels related with land usage; this has been made worse by the development of roads and other infrastructure. Along with altered climate conditions and legislative initiatives (Luedeling et al. 2014), the variations in LULC have led to significant challenges in east Africa. After major reforms, protected trees and bushes in Ethiopia were forcibly cut down to make way for habitation and agricultural land (Reubens et al. 2011).

Knowing the state of LULC is essential for choosing the best land-use options and exploring those options in order to satiate rising demand for essential human needs while also protecting the environment (Wulder et al. 2016). Agroforestry is a form of land utilization which frequently mixes farming with crops to be able to combat climate change and provide numerous ecological, financial, and social advantages (Sharma and Sahoo 2021). To discover an elevated productivity of agricultural production in forest growing techniques across different modern agricultural regions, Ethiopia must conduct more study in this area. Nationally, traditional agroforestry management techniques and forest conservation are more dependent on the interaction of the local population and scientific understanding (Meragiaw 2017). However, the deliberate preservation of plant diversity in conventional agroforestry systems may help offset some of the continued deforestation in unprotected forests (Asfaw and Lemenih 2010). The Wonchi District has a wide range of land-use types, biogeochemical contexts and sustainable farming conditions in addition to several indigenous plant species (Meragiaw et al. 2021a, b). Knowing the status of the LULC shift and having a thorough understanding of what caused this change are prerequisites for creating efficient conservationalsafety and ideas for society plantation operations (Venter et al. 2016).

The implications of this multipurpose ground use for society have been revealed by numerous and well-established scientific investigations of agroforestry systems that combine habitat research, economics, and implementation (Sahoo et al. 2021). The research on managing justifiable habitats has also grown significantly (Arts et al. 2017). Nonetheless, the majority of the agroforestry literature naturally focuses on the agricultural sciences and is often site-specific. The scope and geographic focus of the works on sustainable landscape management, in comparison, are wider (Ali et al. 2011; Lemenih 2006). We bridge the gap between the agroforestry and sustainable landscape management groups with the help of this unique characteristic. Through such actions, we draw attention to the vital part that agroforestry plays in the global governance of green infrastructure and provide further context for the notion of ecological sustainability for agroforestry-related land-use policies (Appiah et al. 2016).

3.2 Agroforestry

Agroforestry's core tenet is that, in contrast to monoculture agriculture, combining forests and commodities in specific spatial or temporal patterns outcomes in increased physical and efficient diversity (Jose 2009; Malaviya et al. 2010). As a viable approach to input-intensive "single commodity" farming, its complexity increases the proficiency of gathering and utilizinghydration, nourishment, and illumination, increases food and nutritional security, produces irreplaceable social identities, and mitigates environmental harm (Nair et al. 2017). While promoting agroforestry has primarily focused on improving rural cultural values and living conditions, bettering groundwater resources, reducing soil erosion, and preserving biological diversity in the Global South, it has primarily focused on tumbling poverty, enhancing food security, and stopping deforestation in the Global North (Nair 2007). Agroforestry provides an array of ecological, societal, historical, and potential importance at the urban fringe, according to numerous international and regional analyses (e.g., Rosenstock et al. 2019; Torralba et al. 2016). Agroforestry is a key strategy for the "perennialization" of agriculture, which attempts to produce long-lasting vegetative cover for ecosystem protection (Helen and Gasparatos 2019), in accordance with this line of thinking. This could include converting much larger expanses of land, albeit in carefully chosen landscape regions, or introducing relatively little perennial vegetation in the right places inside crop-dominated agricultural regions (Asbjornsen et al. 2014; Menker and Hassan 2011). Agroforestry has evolved over time to encompass a wide range of world regions, civilizations, and ecosystems. Examples of well-known agroforestry systems include the Parkland system of West Africa (Fifanou et al. 2011), the historical orchard meadows known as Streuobst in Germany, and the Chattisgarh system of multifunctional trees incorporated into rice farming in India (Plieninger et al. 2015b). Some of these regionally specific and traditional agroforestry systems, called "Cinderella systems" (Nair et al. 2017), have lately come to light as probable role models for supportable spatial growth and agricultural utilization. Many of these systems were lost due to agricultural modernization. Despite the fact that some people have extraordinarily high output, this is not always the case (Birhanu et al. 2019; Moges and Tenkir 2014). This sometimes occurs because modern analysis places more focus on a system's output of a single economically significant commodity than on the sum of its outputs from various products and services (BarguésTobella et al. 2014; Brown et al. 2014). Also, these systems haven't received much attention in terms of technology development and study. There are still many traditional agroforestry systems in use around the world, include silvicultural concepts (Plieninger et al. 2015a), even so, due to either crop yields or farmland alienation, they are typically in extinction. From the 1980s, new, enhanced, and more uniform agroforestry systems have been pushed, initially by working together on improvement in the tropics and subtropics, spearheaded by the International Agroforestry Centre (ICRAF), and further subsequently as a diverse land strategy in Europe and North America (Nair 2007). Improved agroforestry systems typically offer a smaller range of ecosystem services than conventional agroforestry systems, though this can vary depending on the system, location, and supervisionpractices. As opposed to that, they are incredibly prolific, more often than distinct agricultural and forestry innovative techniques (Nair 2007). For instance, in Europe, improved agroforestry systems have been shown to have a land-use efficiency that is up to 40% greater than conventional agricultural and forestry grain production according to farming techniques (Graves et al. 2010). These improved systems, which frequently include components of local and indigenous knowledge, have been the subject of extensive research by modern science (Arevalo et al. 2011; Metzger et al. 2006). Because to regional land-use customs and the inclusion of tree products on farms for revenue divergence, they spread throughout huge expanses in the Global South. Due to a confluence of monetary, radical, scientific, and socio-cultural hurdles, they often experience more gradual expansion in the Global North (Garca de Jalón et al. 2018). As a land-use and landscape perception, agroforestry is always changing (Van Noordwijk 2019). First, the term "agroforestry" was used to refer to a variety of practices that involving a land threshold with people and forests, frequently with an emphasis on the interactions between trees, crops, and soil (Bălteanu et al. 2013). The notion of using agroforestry to boost landscape multi-functionality first surfaced in the 1990s, integrating economies and habitats (i.e., the landscape level) front and centre (Appiah et al. 2016; Friis et al. 2010). Agroforestry has recently received more attention at the policy level. Numerous efforts have been made to promote uniform policies for all land uses and to do away with the distinction between agricultural, forestry, and other subregional regulations, particularly those that are environmental. The Sustainable Development Goals are frequently presented as being attained through these initiatives (Van Noordwijk 2019).

Agroforestry is gaining popularity as a method of land management that can aid in both adaptation and mitigation to climate change while solving many of the problems that artisanal growers are now facing (Dragan et al. 2003; FAO 2011a; Fikir et al. 2018). Agroforestry can help farmers cope with extreme and unpredictable weather, so enhancing their livelihoods and the environment (Ray et al. 2021). Agroforestry encourages the ecological services provided by trees, such as the controlling the flow of water and dirt, the driving of nutrients and carbon in soils, and the creation of habitat for biodiversity (Gizachew et al. 2017; Hassan et al. 2016). This has the effect of improving soil fertility, reducing soil erosion, and managing pests and flooding. Artisanal growers' profit from agroforestry because it boosts farm productivity and reduces the demand for external inputs like conventional fertilizers and pesticides, which boosts income. According to Ripple et al. 2019, global warming now is occurring and the need for fast response to keep the increase in global temperature to 1.5 degrees (IPCC 2019). Climate change hazards, such as extreme weather events, storms, and infections, can have a large detrimental effect on farming systems, resulting in land degradation, poor soil fertility, resource depletion, decreased moisture levels, pest infestations, and financial issues. Farmers are already having trouble planning planting and harvesting because of a rise in flash floods and more frequent drier and wetter weather (SIWI 2018), endangering current production systems and the availability of food. To cut carbon emissions and meet the goals set forth in the Paris Agreement, agriculture, forests, and trees are essential. Farmers can adapt to climatic effects by replanting the appropriate tree species in the appropriate location (Pearson et al. 2005; Sabogal et al. 2013).

3.2.1 Multi-stakeholder Landscape Strategies for Widening Agroforestry

Long-term institutional agreements among many stakeholders are necessary for sustainable landscape management in order to debate, promote, and collaboratively resolve issues to changing obstacles and possibilities for managing natural resources (Sayer et al. 2013; Smith and Wollenberg 2012). These alliances could substantially contribute to the advancement of agroforestry at the supra-farm and landscape governance levels. In order to improve sustainability, Andreotti et al. (2020) propose revamping a sophisticated agroforestry system based on coffee in central Nicaragua. The results demonstrate how a multi-stakeholder approach with assistance can facilitate the co-creation of transition paths by offering a forum for exchanging best practices and knowledge as well as debating opportunities and restrictions. Such discussions are crucial for creating a basis for creating a common vision and acting as a team to implement integrated landscape management. They also help stakeholders grasp the present situation and available solutions. In Ecuador's Amazon and Chocó-Andean Bio Corridor, institutional innovations for integrated landscape management have been studied by Buck et al. (2020). The development of agroforestry systems was made possible by these platforms following the initial failure of field-specific advertising initiatives. The writer explains how desired outcomes, cross-sectoral teamwork, social integration, and coordinated regional integration of agroforestry and related land-use systems are five critical elements of ILM—work synergistically to promote the awareness and dynamic preparedness steps required to scale the uptake of agroforestry and to develop an eloquent thresholding approach. According to Zinngrebe et al. (2020), national programmes and multi-stakeholder landscape platforms should adapt their methods in light of a knowledge of the responsibilities that various parties play in the governance of trees on farms. The authors discover that complementary structures for regulation, funding, and information are needed for the implementation of agroforestry-specific policy tools in Honduras, Peru, Indonesia, and Uganda (Rwanga and Ndambuki 2017; Zeledon and Kelly 2009). Through supported learning processes, actor groups must coordinate their various functions. Local governments and farmer associations are examples of "bridging" actors that might be crucial in aiding these processes (Puyravaud 2003; Tefera et al. 2002). To achieve successful results, the authors advise integrating agroforestry support into all pertinent governance processes. At the Ventoux Biosphere

Reserve and Verdon Regional Nature Park in France, Therville et al. (2020) look into the "policy scape" of agroforestry, which is the collaboration among various guidelines that have an impact on a variety of agroforestry systems. 121 policy instruments have been discovered as a consequence of in-depth discussions with professionals and executives of the agencies. Only a small portion of these were agroforestry-specific, although many other sectoral instruments had an indirect effect on agroforestry systems on a landscape scale (FAO 2004, 2006, 2011b). They come to the conclusion that it is crucial to take into account interactions between various forms of agroforestry as well as other agroecological practises, territorial issues, and interconnections of places, people, and networks.

3.2.2 Habitat Loss: A Global Warming Driver

Around 25% of the Earth's ice-free land area has been degraded by anthropogenic activities, and many tropical countries with large agricultural populations are anticipated to be adversely impacted by climate change (IPCC 2019). Biomass in soils, which is easily lost by land degradation, contains nearly three times as much carbon as the atmosphere, according to Jobbágy and Jackson (2000). Soil may potentially operate as a carbon source or sink liable on its properties, the climate where it is located, and how it is used (IPCC 2019). Around 23% of the world's total emissions of greenhouse gases come from agriculture, forestry, and other land uses (GHGs, such as CO_2, N_2O, and CH_4). These land uses include livestock raising, deforestation, soil degradation, and biodiversity loss (IPCC 2019). GHG emissions from traditional agriculture may be large due to the use of plant nutrients, chemicals, energy sources, compost, forest residues, and a rise in the number of livestock.

As the population is growing and there is a greater need for food, artisanal growers find cropping to be more difficult and unpredictable due to the consequences of climate change. Landscape degradation results in missed opportunities, detrimental effects on people, society, economies, ecosystem services, and environmental flows (Gebrehiwot 2013). Additionally, allowing degradation to persist makes land restoration more difficult and expensive (Molin et al. 2018).

3.2.3 Agroforestry Promotes Environmental Resilience as Well as Flexibility

Climate change can raise the risk of pests, flooding, and droughts in agriculture. How successfully farmers can adjust to the hazards of climate change will determine how long they can continue to live on their land (FAO 2012; Mbow et al. 2014). There are numerous ways to use agroforestry for improved landscape resilience and farm-level climate adaptation. For instance, agroforestry can improve warming and cooling of the environment while reducing air pollution, so establishing a dependable microclimate for cattle and crops (Ellison et al. 2017). By conserving water catchments and watersheds, it also improves groundwater and soil infiltration, which increases water security (BarguésTobella et al. 2014). Agroforestry techniques are suitable for landscape restoration because they have the potential to enhance soil qualities and plants' access to water. Furthermore, trees offer a range of ecosystem services, like feed and forage, medicine, lumber, soil fertility, water management, climate buffering, erosion control, and flood control, all of which are essential for reducing local populations' susceptibility to climate change and increasing climate change resilience (Sahoo et al. 2023). Equally significant, agroforestry might enhance agricultural production systems' incomes by expanding revenue from farming endeavors (such cocoa, coffee, and nuts), boosting energy security, and essential treatment to nutritionally products. Moreover, trees on farms could hasten the rehabilitation of the economy after natural disasters (Simelton et al. 2015). Because female local farmers have less access to resources than their male counterparts, climate adaptation is essential for them (Kiptot et al. 2014). Although women frequently lack the same alternatives as men to improve their standard of living, female farmers provide a significant portion of the food in many underdeveloped countries (Agroforestry Network 2018). In addition, women are commonly left in charge of smallholder farms while their husbands hunt for employ abroad, so they must be better equipped to handle the added burden (Leder et al. 2016). In order to reduce gender disparities in resource access while fostering greater control over their benefits, agroforestry may be a beneficial land management strategy.

3.2.4 Agroforestry as a Prospective Preventive Approach

Agroforestry practices can greatly reduce or completely eliminate greenhouse gas emissions by amplifying carbon packing in vegetation from the top to the land in addition to soil organic carbon (Sahoo et al. 2022). Agroforestry has the capacity to greatly enhance the carbon storage capacity of livestock and crop systems. Agroforestry systems that store CO_2 comprise backyard gardening, trees and bushes, river deltas, horticultural crops, boundaries vegetation, forest resources, and coal areas (IPCC 2003; Verchot et al. 2007).

For the purpose of lowering atmospheric GHG concentrations, agroforestry is crucial (IPCC 2003). Among all the land uses considered in the IPCC's Land-Use, Land-Use Change, and Forestry report, agroforestry has the greatest potential to store carbon among non-Annex I countries. Because of the magnitude of the area that is vulnerable to changes in land use, agroforestry has a higher potential than other land uses even if it has the lowest carbon density (630,106 ha). The standards for a qualifying afforestation/reforestation (A/R) activity under the Clean Development Mechanism (CDM) are typically met by improved agroforestry practices since they lowering small farmers' susceptibility and assisting them in climate change adaptation (Tegene 2002; Torres-rojo and Flores-xolocotzi 2001).

Global advancement is stagnant, and national execution of climate targets and other land-related issues is still highly challenging. AFOLU has increased its GHG emissions since 2000 (Birhanu et al. 2019). The amount of CO_2 emissions from AFOLU that have been reduced by restrictions between 2009 and 2019 is just about 8 $GtCO_2$, or 0.5% of all emissions (author calculations available in Additional Information). The existing commitments are more compatible with 2.6–3.1 °C of warming by the end of the century as opposed to the 1.5 and 2 °C promised in the Paris Agreement (Gizachew et al. 2017). The NYDF goals and the Aichi Biodiversity Objectives for 2020, which aimed to reestablish 150 million hectares (Mha) of deforested land by 2020, have not been achieved, and some setbacks have taken place since the goals were established. However, despite some development So, if land-based mitigation is to be scaled up to its full capacity, maximize help, and minimize adaptations, it will take substantially more resources and effort (Sharma and Sahoo 2021).

The nature of the interaction, the configuration approach (including magnitude, technique, substitutability with other initiatives and segments), and the geographic context (including current biogeographical complexities, weather patterns, food chain, and land rights) all have a big impact on the effectiveness and scope of land-based measures (Rosenstock et al. 2019). For the adoption of land-based mitigation techniques to be successful and sustainable, as well as for the right prioritization of those projects, more regional and national data on emission causes, mitigation abilities, co-benefits, and hazards are required (Andreotti et al. 2020). Also, to fully realize the potential for AFOLU mitigation and co-benefits, policies and methods for managing the land and food systems that are location- and time-varying and perspective will be required (Kassie 2018). The implementation of land-based projects and the success of various programmes relies on empoweringcircumstances and limitations, which vary widely by country. Implementation is more or less likely based on the available funding and financial incentives, institutional capacity, governance, geophysical capacity, technological ability, sociocultural context, and environmental and ecological considerations (Sahoo et al. 2022). As a consequence, in the IPCC's Sixth Assessment Report, Parties to the United Nations Framework Convention on Climate Change (UNFCCC) have requested more thorough studies of regional mitigation potential and their practicality (AR6). National and international stakeholders may be better able to concentrate their investment and effort in areas with promise and need with the help of this information.

3.2.5 Agroforestry Implementation and Scaling-Up: Perpetuating Factors and Impediments

Both emerging and industrialized nations use agroforestry extensively, and it is becoming more visible at the both at the federal level and in humanitarian groups (Feoli et al. 2002; Rwanga and Ndambuki 2017). Due to a number of difficulties, such as inadequate economics, ambiguous property ownership, and little information and funding, and a paucity of transdisciplinary integration, agroforestry practices are currently applied on a small scale (Agroforestry Network 2018). The Agroforestry Network (2018) offers the critical empoweringevents listed below for clambering up agroforestry: facilitating the development of flexible valuation forecasting that address protracted investments in farming activities, enhancing farmers' access to credit and other financial resources, developing revenue streams for agroforestry products, and more (Gessesse 2010; Rizvi et al. 2015).

• Enhancing demand-driven, inclusive, and participatory agroforestry-related research; • Improving the availability of superior seedlings and technical assistance for farmers. Despite efforts to increase the use of agroforestry, there are substantial gaps between countries' goals and their capacities to track, record, and substantiate agroforestry activities (Rosenstock et al. 2018). At all levels, strategies, methods, and indicators must be created to assess the benefits of agroforestry's diverse systems for the environment.

3.3 Policy Implications of Agroforestry Measures to Combat Climate Change

Using ecologically friendly farming practices, such agroforestry, to provide numerous social and ecological consequences is crucial, according to the United Nations Framework Convention on Climate Change (UNFCCC), other humanitarian bodies, and academic groups (FAO 2019; IPBES 2019; IPCC 2019). Agroforestry can assist countries achieve their objectives for reforestation, food and energy security, as well as resilience and environmental issues abatement. According to a CGIAR report, as of June 2018, 59 out of 147 poor countries had identified agroforestry as a way to combat climate change and fulfil their NDCs (Nationally Determined Contributions) under the UNFCCC (Rosenstock et al. 2018). Hence, agroforestry has the potential to be a flexible landscape strategy tool with many assistances for both adaptation and mitigation of climate change.

3.3.1 Case Study—Kenya Agricultural Carbon Project

Agroforestry is promoted by the Kenya Agricultural Carbon Project, which has also assisted nearly 30,000 growers in western Kenya in planting over 3 million native species for agroforestry and exploring other methods for managing land sustainably, such as recycling, composting, and distributing organic manure. Agroforestry educates the producers taking part in the initiative in methods to raise the content of organic matter in the soil as part of a partnership funded by the World Bank (Kyalangalilwa et al. 2013; Tumwebaze 2012). Improved yields, tolerance to severe droughts and flooding, reduced erosion, and the storage of carbon are all benefits of more organic matter in the soil, for which farmers are compensated. The programme increased forest resources, nutrition security, and climate change adaptation among both 2010 and 2016, while trapping about 345,000 tonnes of CO_2. The project was a game-changer since it enabled farmers whose crop and livestock yields had been declining over time as well as those who had experienced considerable environmental degradation to increase their yields by over 150% over the course of eight years. Since the project's beginning, rural farmers have saved more money and become more adaptable to a variety of disasters, such as the consequences of climate change. The main monetary reward for the producers to join the initiative was an upsurge in productivity. Other ramifications include a better understanding of climate change and easier access to charcoal, veggies, and animal feed made from trees. By the time the project is finished in 2030, 2 million tonnes of carbon should have been captured (Agroforestry Network 2018).

3.4 Additional Research Goals

The element in nature that directly affects rural livelihoods is climate change. Although the changes cannot be halted, they can be managed through a clever and practical approach (Mandal et al. 2021). Government, experts, and the general public should work together to create a barrier against the dangers posed by climate change. The study was an effort to create a successful system that could support the government of India's great initiatives. The study's future focus could begin with something like:

3.4.1 National Effort

The Government of India can put the findings of the study into practice by implementing the National Watershed Development Project for Rainfed Areas, Integrated Nutrient Management in Rural Areas, and Encouraging Use of Informatics in Agriculture in Rural Areas. Mission for Rural Livelihoods in America: Local governments and Organizations can enhance farming communitieslinked to the national rural livelihoods mission. The organizations take the initiative to increase the self-confidence of farming communitiesby recognizing their value to the community and the country. The fundamental idea behind this project is to group the poor into self-help groups (SHGs) and get them ready for jobs. By offering low-income households' opportunities for lucrative job and self-employment, the technique contributes to the alleviation of poverty by significantly raising their long-term standard of life. Governmental effort to promote rural development The Indian government has taken action to maintain the people' current occupations and pursuits. According to the requirements of farming communities, the financial institutions are allocated a supportive role and provide help.

3.5 Suggestions for Decision- and Strategy

1. Promote the adoption of farming activities in local villages to combat climate change with, for instance, financial, legal, and prospects help to produce multiple economic benefits, such as increased crop production and biodiversity, increased food and water stability, enhanced environmental quality, and increased feminism.
2. Provide rewards for inspiring growers to dedicate time and resources to land usage strategies, such as agroforestry, that have a longer financial return and stable land tenure rights.
3. Promote agroforestry by looking into policy reforms to incorporate it, for example, in budgets, technical aid, and development cooperation initiatives. Making it simpler to determine if support for agroforestry is provided as part of development cooperation is a good place to start.
4. Provide frameworks, methods, and indicators at all scales to track the advancement of silvicultural techniques and their positive effects on the environment.
5. Use agroforestry to integrate economic growth with climate change adaptation and mitigation, additionally to their overlapping strengths, to create effective, affordable, and equitable policies.
6. Relate agroforestry to the climate change agenda and update on developments. Since that states are committed to finishing their National Determined Contributions (NDC) under the United Nations Framework Convention on Climate Change (UNFCCC), supporting agroforestry stands to reason considering both variation and mitigation.

3.6 Conclusion

The majority of current incentives encourage agricultural techniques that disregard vegetation, habitat, and possibly encourage changes in land use and the destruction of main forests. Only a small number of direct policy proposals are supported through programmes and instruments. To promote acceptance and accountability for sustainable farming, agroforestry needs to be included in societal and governmental ideas of "good farming." Nevertheless, there is little information on context-specific best practice solutions. Universities, research facilities, non-governmental organizations (NGOs), nonetheless, forestry and agribusiness agencies, can offer enhancing efficiency and show how agroforestry provides ecosystem services like shade, wind breaks, fruit provision, and (economic) wood.

Second, complementary organizations that provide regulation, funding, and information are necessary for the deployment of agroforestry-specific instruments. Agriculture is supported by private initiatives (such as carbon credits), value chains, and local credit schemes, but the application of agroforestry faces significant obstacles. Short lending terms and challenging access for farmers in distant areas impair long-term tree investments. To fully realize the promise of agroforestry, it is necessary to integrate existing scientific knowledge, relevant financial choices, and supportive regulatory frameworks to socioeconomic themes concerning regional public institutions and agriculture. Lastly, it's important to coordinate the numerous functions that actor groups play in aided learning processes. Extension services have an impact on farmer decisions, but markets and lending institutions also have standards that must be met. The environmental industry, which is in charge of carrying out biodiversity aims and preserving habitat structures, is terribly cut off from regional structures. To organize federal and non-governmental actions for participatory leadership programmes and systemic education with a defined vision and mandate for agroforestry, extant frameworks and partnership agreements should be used. Local governments and farmer associations play a crucial role in fostering such initiatives.

Nothing compares to a tree for increasing agricultural diversity and strengthening landscapes. Agroforestry has the opportunity to encourage synergy between initiatives aimed at slowing climate change and those that assist deprived areas acclimate to its detrimental impact. The research agenda is primarily set in this field. As there is currently a lot of information available, applying these ideas to farm households will allow us to gain important knowledge through practical application. The main factors influencing the area include agroforestry operations, settlement growth, road building, and a shift in tenure policies made worse by an upsurge in the number of people living there. The research on agroforestry systems' activities and LULC modifications may offer guidance for the region's numerous objectives of environmental conservation, poverty alleviation, land-use planning for sustainable development, and adaptation and mitigation of climate change. Based on the LULC data, the authors suggested that every available piece of land be used in the most sensible way possible in order to enhance the local natural resources without further harming the bio-environment. So, to develop modern land-use systems with a more justifiable and comprehensive perspective to rural livelihoods, the potentiality of conventional agroforestry techniques must be addressed.

- For far too long, agroforestry has been seen as a specialized field. Instead, agroforestry ought to develop into a distinct field of study within the realm of sustainability science, leading to the development of "agroforestry sustainability science". The lack of acceptance and upkeep that many agroforestry practices and systems face may be overcome by participating agroforestry's essential plays in the management of justifiable landscapes.
- Route constraints, which frequently evolved over centuries, if not generations or eras, can serve as the driving force behind the governance of natural resources. Awareness of regional farm trends, heritage, and civilizations are also necessary for the implementation of agroforestry for sustainable landscape management.

Since producers are the main players impacting present-day and historical land-use change, this also entails developing a greater understanding of farmers' views regarding agroforestry over time.

- Agroforestry policies shouldn't be restricted to the agriculture and patch dimensions. For agroforestry to significantly subsidize to the Sustainable Development Goals, it must be scaled up to generateprofits at the landscape level. Other actors and their networks in the region, in contrast to the farmers and land managers who are using them, can benefit from using a range of agroforestry techniques in the appropriate areas of a landscape and managing them appropriately.
- Consideration and promotion of agroforestry's many economic, environmental, social, and cultural qualities as well as the connections between them are crucial for maintaining the practise over the long run. This entails actively promoting market and commercial growth that makes agroforestry sustainable, in order to preserve important qualities as these land-use systems and associated landscapes change over time.

References

Adhikary PP, Hombegowda HC, Barman D, Jakhar P, Madhu M (2017) Soil erosion control and carbon sequestration in shifting cultivated degraded highlands of eastern India: performance of two contour hedgerow systems. Agroforest Syst 91:757–771. https://doi.org/10.1007/s10457-016-9958-3

Agroforestry Network (2018) Scaling up agroforestry: potential, challenges and barriers. A review of environmental, social and economic aspects on the farmer, community and landscape level

Ali HO, Descheemaeker K, Steenhuis TS, Pandey S (2011) Land use and land cover changes in the Ethiopian highlands. Cornell-Bahir Dar University, Ethiopia

Andreotti F, Speelman EN, Van den Meersche K, Allinne C (2020) Combining participatory games and back casting to support collective scenario evaluation: an action research approach for sustainable agroforestry landscape management. Sustain Sci 15(5):1383–1399. https://doi.org/10.1007/s11625-020-00829-3

Appiah DO, Bugri JT, Forkuo EK, Yamba S (2016) Agricultural and forest land use potential for REDD+ among smallholder land users in Rural Ghana. Int J For Res:7218305. https://doi.org/10.1155/2016/7218305

Arevalo CBM, Bhatti JS, Chang SX, Sidders D (2011) Land-use change effects on ecosystem carbon balance: from agricultural to hybrid poplar plantation. Agr Ecosyst Environ 141:342–349. https://doi.org/10.1016/j.agee.2011.03.013

Arts B, Buizer M, Horlings L, Ingram V, van Oosten C, Opdam P (2017) Landscape approaches: a state-of-the-art review. Annu Rev Env Resour 42(1):439–463

Asbjornsen H, Hernandez-Santana V, Liebman M, Bayala J, Chen J, Helmers M, Ong CK, Schulte LA (2014) Targeting perennial vegetation in agricultural landscapes for enhancing ecosystem services. Renew Agric Food Syst 29(2):101–125

Asfaw B, Lemenih M (2010) Traditional agroforestry systems as a safe haven for woody plant species: a case study from a topo-climatic gradient in South Central Ethiopia. Forests, Trees and Livelihoods 19:359–377. https://doi.org/10.1080/14728028.2010.9752678C

Bălteanu D, Dragotă C-S, Popovici AM, Dumitraşcu A, Kucsicsa G, Grigorescu I (2013) Land use and crop dynamics related to climate change signals. Proc Roman Acad Ser A 15(3):265–278

BarguésTobella A, Reese H, Almaw A, Bayala J, Malmer A, Laudon H, Ilstedt U (2014) The effect of treeson preferential flow and soil infiltrability in an agroforestry parkland in semiarid Burkina Faso. Water Resour Res 50(4):3342–3354

Birhanu L, Hailu BT, Bekele T, Demissew S (2019) Land use/cover change along elevation and slope gradient in highlands of Ethiopia. Rem Sens Appl: Soc Environ 16. https://doi.org/10.1016/j.rsase.2019.100260

Brown DG, Polsky C, Bolstad P, Brody SD, Hulse D, Kroh R, Loveland TR, Thomson A (2014) Land use and land cover change. In: Melillo JM, Richmond TTC, Yohe GW (eds) Climate change impacts in the United States: the third national climate assessment. U.S. Global Change Research Program, Washington D.C., pp 318–332

Buck L, Scherr S, Mecham J, Trujillo L, Fleming M, Bosco S (2020) Using integrated landscape management to scale agroforestry: examples from Ecuador. Sustain Sci 15(4). https://doi.org/10.1007/s11625-020-00839-1

Christiaensen L, Demery L (2018) Agriculture in Africa: telling myths from facts. Directions in development—agriculture and rural development. World Bank, Washington, DC. https://openknowledge.worldbank.org/handle/10986/28543

Dragan M, Feoli E, Fernetti M, Woldu Z (2003) Application of a spatial decision support system (SDSS) to reduce soil erosion in northern Ethiopia. Environ Model Softw 18:861–868. https://doi.org/10.1016/S1364-8152(03)00104-X

Duguma LA, Atela J, Minang PA, Ayana AN, Gizachew B, Nzyoka JM, Bernard F (2019) Deforestation and forest degradation as an environmental behavior: unpacking realities shaping community actions. Land 8:26. https://doi.org/10.3390/land8020026

Ellison D, Morris CE, Locatelli B, Sheil D, Cohen J, Murdiyarso D, Gutierrez V, Van Noordwijk M, Creed IF, Pokorny J, Gaveau D et al (2017) Trees, forests and water: cool insights for a hot world. Glob Environ Chang 43:51–61

Fagerholm N, Torralba M, Burgess PJ, Plieninger T (2016) A systematic map of ecosystem services assessments around European agroforestry. Ecol Indic 62:47–65. https://doi.org/10.1016/j.ecolind.2015.11.016

FAO (2004) Assessing carbon stocks and modeling win-win scenarios of carbon sequestration through land-use changes. FAO, Rome

FAO (2005) Land cover classification system classification concepts and user manual. Environment and Natural Resources Service (SDRN) Publications, Rome

FAO (2006) Global Forest resource assessment 2005: progress towards sustainable forest management. FAO, Rome

FAO (2011a) Tees outside forests: towards rural and urban integrated resources management. FAO, Rome

FAO (2011b) State of the world's forests. FAO, Rome

FAO (2012) Forests and climate change working paper 10: forest management and climate change: a literature review. FAO/UN; Rome

FAO (2019) The state of the world's biodiversity for food andagriculture. In: Bélanger J, Pilling D (eds) FAO commissionon genetic resources for food and agriculture assessments. Rome, 572 pp. http://www.fao.org/3/CA3129EN/CA3129EN.pdf

Feoli E, Vuerich LG, Woldu Z (2002) Processes of environmental degradation and opportunities for rehabilitation in Adwa, northern Ethiopia. Landscape Ecol 17:315–325. https://doi.org/10.1023/A:1020566801067

Fifanou VG, Ousmane C, Gauthier B, Brice S (2011) Traditional agroforestry systems and biodiversity conservation in Benin (West Africa). Agrofor Syst 82(1):1–13

Fikir D, Tebikew M, Gebremariam Y (2018) Diversity of indigenous woody species in small holder farm lands: comparison across different agroecology and land use types in chilga and dabat district, Northern Ethiopia. Int J Scien Res Manage 6(08):178–189. https://doi.org/10.18535/ijsrm/v6i8.ah01

Forestry Commission (2011) Forests and climate change: UK forestry standard guidelines. Forestry Commission; Edinburgh

Friis I, Demissew S, Breugel VP (2010) Atlas of the potential vegetation of Ethiopia. Copenhagen Biologiske Skrifter 58:3607

Gebrehiwot M (2013) Recent transitions in Ethiopian homegarden agroforestry: driving forces and changing gender relations. Licentiate Thesis Swedish university of agricultural sciences Umeå 2013. Acta Universitatis Agriculturae Sueciae 21:1401–1470

Gedefaw M, Soromessa M, Belliethathan S (2014) Forest carbon stocks in woody plants of Tara Gedam Forest: implication for climate change mitigation. Sci Technol Arts Res J 3(1):101–107. https://doi.org/10.4314/star.v3i1.16

Gessesse B (2010) In: Edwards S (ed) Review of land degradation and land management in Ethiopia up to 2008/09. Ethiopian Environment Review, No. 1. Forum for Environment, Addis Ababa, Ethiopia, pp 187–214

Getu Z, Dale G, Tafa M, James GN, Gonfa T (2011) Carbon stock assessment in different land use for REDD+ in ethiopia: practitioners field guide, yayu forest coffee biosphere reserve. Addis Ababa, Ethiopia.

Gizachew B, Astrup R, Vedeld P, Zahabu EM, Duguma LA (2017) REDD+ in Africa: contexts and challenges. Natural Resources Forum. https://doi.org/10.1111/1477-8947.12119

Graves AR, Burgess PJ, Palma J, Keesman KJ, van der Werf W, Dupraz C, van Keulen H, Herzog F, Mayus M (2010) Implementation and calibration of the parameter-sparse Yield-SAFE model to predict production and land equivalent ratio in mixed tree and crop systems under two contrasting production situations in Europe. Ecol Model 13–14:1744–1756.

Hassan Z, Shabbir R, Ahmad SS, Malik AH, Aziz N, Butt A, Erum S (2016) Dynamics of land use and land cover change (LULCC) using geospatial techniques: a case study of Islamabad Pakistan. Springer plus 5:812. https://doi.org/10.1186/s40064-016-2414-z

Helen JMP, Gasparatos A (2019) Land-use change, carbon stocks, and tree species diversity in green spaces of a secondary city in Myanmar, PyinOo Lwin. PLoS ONE 14(11):e0225331. https://doi.org/10.1371/journal.pone.0225331

HLPE (2019) Agroecological and other innovative approaches for sustainable agriculture and food systems that enhance food security and nutrition. A report by the High Level Panel of Experts on Food Security and Nutrition of the Committee on World Food Security, Rome

Hombegowda HC, Adhikary PP, Jakhar P, Madhu M (2022) Alley cropping agroforestry system for improvement of soil health. In: Shit PK, Adhikary PP, Bhunia GS, Sengupta D (eds) Soil health and environmental sustainability. Environmental Science and Engineering. Springer, Cham. https://doi.org/10.1007/978-3-031-09270-1_23

IPBES (2019). In: Díaz S, Settele J, Brondizio ES, Ngo HT, Guèze M, Agard J, Arneth A, Balvanera P, Brauman KA, Butchart SHM, Zayas CN (eds) The global assessment report on biodiversityand ecosystem services of the Intergovernmental Science-Policy Platform on Biodiversity and Ecosystem Services. IPBES secretariat, Bonn, Germany

IPCC (2003) In: Penman J, Gystarsky M, Hiraishi T, Krug T, Kruger D, Pipatti R, Buendia L, Miwa K, Ngara T, Tanabe K, Wagner F (eds) Good practice guidance for land use, land-use change, and forestry, prepared by the national greenhouse gas inventories programme. IGES; Kamiyamaguchi

IPCC (2019) IPCC special report on climate change, desertification, land degradation, sustainable land management, food security, and greenhouse gas fluxes in terrestrial ecosystems. Intergovernmental Panel on Climate Change

Jobbágy EG, Jackson RB (2000) The vertical distributionof soil organic carbon and its relation to climate andvegetation. Ecol Appl 10(2):423–436

Jose S (2009) Agroforestry for ecosystem services and environmental benefits: an overview. Agrofor Syst 76(1):1–10

Kassie GW (2018) Agroforestry and farm income diversification: synergy or trade-off? The case of Ethiopia. Environ Syst Res 6:8. https://doi.org/10.1186/s40068-017-0085-6

Kiptot E, Franzel S, Degrande A (2014) Gender, agroforestry and food security in Africa. Curr Opin Environ Sustain 6:104–109

Kyalangalilwa B, Boatwright JS, Daru BH, Maurin O, Van der Bank M (2013) Phylogenetic position and revised classification of Acacia s.l. (Fabaceae: Mimosoideae) in Africa, including new combinations in Vachellia and Senegalia. Bot J Linnean Soc 172(4):500–523. https://doi.org/10.1111/boj.12047

Leder S, Das D, Reckers A, Karki E (2016) Participatorygender training for community groups. A manual for critical discussions on gender norms, roles and relations. Report from CGIAR research program on water, land and ecosystems

Lemenih M (2006) Expediting ecological restoration with the help of foster tree plantations in Ethiopia. J Drylands 1(1):72–84

Lorenz K, Lal R (2014) Soil organic carbon sequestration in agroforestry systems. Rev Agron Sustain Dev 34(2):443–454. https://doi.org/10.1007/s13593-014-0212-y

Luedeling E, Kindt R, Huth N, Koenig K (2014) Agroforestry systems in a changing climate—challenges in projecting future performance. Curr Opin Environ Sustain 6:1–7. https://doi.org/10.1016/j.cosust.2013.07.013

Malaviya S, Munsi M, Oinam G, Joshi PK (2010) Landscape approach for quantifying land use land cover change (1972–2006) and habitat diversity in a mining area in Central India (Bokaro, Jharkhand). Environ Monit Assess 170:215–229. https://doi.org/10.1007/s10661-009-1227-8

Mandal U, Sena DR, Dhar A, Panda SN, Adhikary PP, Mishra PK (2021) Assessment of climate change and its impact on hydrological regimes and biomass yield of a tropical river basin. Ecol Ind 126:107646

Marjokorpi A, Ruokolainen K (2003) The role of traditional forest gardens in the conservation of tree species in West Kalimantan Indonesia. Biodivers Conserv 12(4):799–822. https://doi.org/10.1023/A:1022487631270

Mbow C, Smith P, Skole D, Duguma1 L, Bustamante M (2014) Achieving mitigation and adaptation to climate change through sustainable agroforestry practices in Africa. In: Mbow C, Neufeldt H, Minang PA, Luedeling E, Kowero G (Eds) Sustainability challenges. Curr Opin Environ Sustain 6:8–14. https://doi.org/10.1016/j.cosust.2013.09.002

Meles K, Bishaw B, Iiyama M, Emiru B, Negussie N, Davis CM, Bernart B (2019) Climate-smart agriculture: enhancing resilient agricultural systems, landscapes, and livelihoods in Ethiopia and beyond. World Agroforestry ICRAF; Nairobi, Kenya

Menker H, Hassan RM (2011) Drivers of land-use change in the Southern Nations, Nationalities and People's Region of Ethiopia. Afr J Agric Res Econ 9(2):148–164. https://doi.org/10.22004/ag.econ.176515

Meragiaw M (2017) Role of agroforestry and plantation on climate change mitigation and carbon sequestration in Ethiopia. J Tree Sci 36(1):1–15. https://doi.org/10.5958/2455-7129.2017.00001.2

Meragiaw M, Woldu Z, Martinsen V, Singh BR (2021a) Floristic composition and structure of the Kibate Forest along environmental gradients in Wonchi, Southwestern Ethiopia. J for Res 32:2669–2682. https://doi.org/10.1007/s11676-021-01305-z

Meragiaw M, Woldu Z, Martinsen V, Singh BR (2021b) Carbon stocks of above- and belowground tree biomass in Kibate Forest around Wonchi Crater Lake, Central Highland of Ethiopia. PLoS ONE 16(7):e0254231. https://doi.org/10.1371/journal.pone.0254231

Metzger MJ, Rounsevell MDA, Acosta-michlik DS, Leemans LR (2006) The vulnerability of ecosystem services to land-use change. Agr Ecosyst Environ 114:69–85. https://doi.org/10.1016/j.agee.2005.11.025

Moges Y, Tenkir E (2014) Overview of REDD+ process in Ethiopia. Ministry of Environment and Forest; Addis Ababa

Molin PG, Chazdon R, de Barros F, Ferraz S, Brancalion PH (2018) A landscape approach for cost effective large-scale forest restoration. J Appl Ecol 55(6):2767–2778

Nair PK (2007) The coming of age of agroforestry. J Sci Food Agric 87(9):1613–1619. https://doi.org/10.1002/jsfa.2897

Nair PKR, Viswanath S, Lubina PA (2017) Cinderella agroforestry systems. Agrofor Syst 91:901–917

Pearson TRH, Walker S, Brown SL (2005) Source book for LULUCF projects. Winrock International; Arlington

Plieninger T, Hartel T, Martin-Lopez B, Beaufoy G, Bergmeier E, Kirby K, Montero MJ, Moreno G, Oteros-Rozas E, Van Uytvanck J (2015a) Wood-pastures of Europe: geographic coverage, socialecological values, conservation management, and policy implications. Biol Conserv 190:70–79

Plieninger T, Levers C, Mantel M, Costa A, Schaich H, Kuemmerle T (2015b) Patterns and drivers of scattered tree loss in agricultural landscapes: orchard meadows in Germany (1968–2009). PLoS ONE 10(5):e0126178

Puyravaud JP (2003) Standardizing the calculation of the annual rate of deforestation. For Ecol Manage 177:593–596. https://doi.org/10.1016/S0378-1127(02)00335-3

Rawat JS, Kumar M (2015) Monitoring land use/cover change using remote sensing and GIS techniques: a case study of Hawalbagh block, district Almora, Uttarakhand, India. Egypt J Remote Sens Space Sci 18:77–84

Ray M, Prusty M, Sahoo GR, Rout S, Sahoo KC, Mishra P (2021) Enhancing farming profits with agroforestry.agroforestry—prospective, strategies, and future aspects, pp 326–334. ISBN: 9788195339297

Reubens B, Moeremans C, Poesen J, Nyssen J, Tewoldeberhan S, Franzel S, Deckers J, Orwa C, Muys B (2011) Tree species selection for land rehabilitation in Ethiopia: from fragmented knowledge to an integrated multi-criteria decision approach. Agrofor Syst 82:303–330. https://doi.org/10.1007/s10457-011-9381-8

Ripple WR, Wolf C, Newsome TM, Barnard P, Moomaw WR (2019) World scientists' warning of a climate emergency. Bio Science. https://doi.org/10.1093/biosci/biz088

Rizvi AR, Baig S, Barrow E, Kumar C (2015) Synergies between climate mitigation and adaptation in forest landscape restoration. IUCN; Gland, Switzerland

Rosenstock TS, Dawson IK, Aynekulu E, Chomba S, Degrande A, Fornace K, Jamnadass R, Kimaro A, Kindt R, Lamanna C, Malesu M, Mausch K, McMullin S, Murage P, Namoi N, Njenga M, Nyoka I, Paez Valencia AM, Sola P, Shepherd K, Steward P (2019) A planetary health perspective on agroforestry in Sub-Saharan Africa. One Earth 1(3):330–344

Rosenstock T, Wilkes A, Jallo C, Namoi N, Bulusu M, Suber M, Bernard F, Mboi D (2018) Making trees count: Measurement, reporting and verification of agroforestry under the UNFCCC. CCAFS Working Paper no. 240. Wageningen, theNetherlands: CGIAR Research Program on Climate Change, Agriculture and Food Security (CCAFS)

Rwanga SS, Ndambuki JM (2017) Accuracy assessment of land use/cover classification using remote sensing and GIS. Int J Geosci 8:611–622. https://doi.org/10.4236/ijg.2017.84033

Sabogal C, Guariguata MR, Broadhead J, Lescuyer G, Savilaakso S, Essoungou N, Sist P (2013) Multiple-use Forest management in the humid tropics: opportunities and challenges for sustainable forest management. FAO Forestry Paper (173) Rome, FAO/UN, and Bogor, Indonesia, Center for International Forestry Research

Sahoo GR, Mishra PJ, Dash AC, Wani AM, Gupta S. (2021). Agroforestry for adaptation, mitigation to climate change and livelihood security. In: Agroforestry—prospective, strategies, and future aspects, pp 203–207. ISBN: 9788195339297

Sahoo GR, Swamy SL, Wani AM, Mishra A (2022) Agroforestry systems for Carbon sequestration and food security: implications for climate change mitigation. In: Soil health and environmental sustainability application of geospatial technology. Springer Nature, pp 503–528. ISBN 978-3-031-09270-1. https://doi.org/10.1007/978-3-031-09270-1

Sahoo GR, Mishra PJ, Wani AM, Sharma A, Mishra D, Patra D, Mishra I, Behera M (2023) Impact of climate change on livelihood security and biodiversity—issues and mitigation strategies. In: Climate change impacts on natural resources, ecosystems and agricultural systems. Springer Nature, pp 1–27. ISBN 978-3-031-19059-9. https://doi.org/10.1007/978-3-031-19059-9

Sayer J, Sunderland T, Ghazoul J, Pfund J-L, Sheil D, Meijaard E, Venter M, Boedhihartono AK, Day M, Garcia C, van Oosten C, Buck LE (2013) Ten principles for a landscape approach to

reconciling agriculture, conservation, and other competing land uses. Proc Natl Acad Sci USA 110(21):8349–8356
SD21 (2012) (Sustainable development in the 21st century) Sustainable land use for the 21st century. Department of Economic and Social Affairs, USA; United Nations
Sharma A, Sahoo GR (2021) Agroforestry in organic farming. Agricos E-Newsletter. Article No. 37. 2(4): 131–134
Simelton E, Dam BV, Catacutan D (2015) Trees andagroforestry for coping with extreme weather events: experiencesfrom northern and central Viet Nam. Agroforestry Syst 89:1065–1082
SIWI (2018) Water for productive and multifunctional landscapes. Stockholm International Water Institute, Report no. 38.
Smith P, Wollenberg E (2012) Achieving mitigation through synergies with adaptation. In: Wollenberg E, Nihart A, Tapio-Bostrom M-L, Grieg-Gran M (eds) Climate change mitigation and agriculture. ICRAF-CIAT, London-New York, pp 50–57
Tefera B, Ayele G, Atnafe Y, Jabbar MA, Dubale P (2002) Nature and causes of land degradation in the Oromiya Region. Socio-economic and policy research working. ILRI
Tegene B (2002) Land-cover/land-use changes in the Derekolli catchment of the South Welo Zone of Amhara Region, Ethiopia. East Afr Soc Sci Res Rev 18(1):1–20
Therville C, Antona M, de Foresta H (2020) The policyscape of agroforestry within Mediterranean protected landscapes in France. Sustain Sci 15:1435:1448. https://doi.org/10.1007/s11625-020-00821-x
Torralba M, Fagerholm N, Burgess PJ, Moreno G, Plieninger T (2016) Do European agroforestry systems enhance biodiversity and ecosystem services? A meta-analysis. Agric Ecosyst Environ 230:150–161
Torres-rojo JM, Flores-xolocotzi R (2001) Deforestation and land-use change in Mexico. Clim Change Forest Manage Western Hemis 12(1):171–191
Tumwebaze BS (2012) Impact of land use/cover change on soil carbon stocks, livelihoods, and opportunities for mitigation in Mt. Elgon region. Research Application Summary, Third Ruforum Biennial Meeting 24–28 September 2012, Entebbe, Uganda
Venter O, Sanderson EW, Magrach A, Allan JR, Beher J, Jones KR, Possingham HP, Laurance WF, Wood P, Fekete BM, Levy MA, Watson JEM (2016) Sixteen years of change in the global terrestrial human footprint and implications for biodiversity conservation. Nat Commun 7:12558. https://doi.org/10.1038/ncomms12558
Verchot LV, Van Noordwijk M, Kandji S, Tomich T, Ong C, Albrecht A, Mackensen J, Bantilan C, Anupama KV, Palm C (2007) Climate change: linking adaptation andmitigation through agroforestry. Mitig Adapt Strat Glob Change 12(5):901–918
Wulder MA, White JC, Loveland TR, Woodcock CE, Belward AS, Cohen WB, Fosnight EA, Shaw J, Masek JG, Roy DP (2016) The global Landsat archive: status, consolidation, and direction. Remote Sens Environ 185:271–283. https://doi.org/10.1016/j.rse.2015.11.032
Yahya N, Mulat S, Getachew M (2019) Ethiopian Environment and Forest Research Institute; Addis Ababa, Ethiopia: 2019. Technical manual on the use of satellite images in ecological studies
Zeledon EB, Kelly NM (2009) Understanding large-scale deforestation in Southern Jinotega, Nicaragua from 1978 to 1999 through the examination of changes in land use and land cover. J Environ Manage 90:2866–2872. https://doi.org/10.1016/j.jenvman.2008.03.016
Zinngrebe Y, Borasino E, Chiputwa B, Dobie P, Garcia E, Gassner A, Kihumuro P, Komarudin H, Liswanti N, Makui P, Plieninger T, Winter E, Hauck J (2020) Agroforestry governance for operationalizing the landscape approach: connecting conservation and farming actors. Sustain Sci 15:1417–1434 https://doi.org/10.1007/s11625-020-00840-8

Chapter 4
Agroforestry and Diversified Land Management for Sustainable Agriculture and Climate Change

Hemant Kumar, Savita Kumari, Ragini Kumari, and Neha Pareek

Keywords Agroforestry · Climate change · Land management · Sustainable agriculture

4.1 Introduction

Agriculture and related industries are crucial to the economic, social, and industrial development of most countries in Asia and Africa. A number of nations located in the Asia–Pacific area base a significant portion of their economy on agriculture. As a result, these nations are particularly susceptible to variations in agricultural production, prices, and related policy alternatives within the region as well as on worldwide markets. The entire land area of China, which measures 932.64 million hectares, is the largest in all of Asia and the Pacific, followed by Australia, which measures 768.2 million hectares, India, which measures 297.3 million hectares, and Indonesia, which measures 297.4 million hectares (181.1 million ha). There are numerous minor countries that have land areas that are less than 10 million hectares, and a few examples of these countries include Bhutan, Fiji, the Republic of Korea, and Sri Lanka. There has

H. Kumar (✉) · S. Kumari · R. Kumari · N. Pareek
Sam Higginbottom University of Agriculture Technology and Sciences (SHUATS), Prayagraj, Uttar Pradesh, India
e-mail: hemant.kumar@shiats.edu.in

S. Kumari
Mahatma Gandhi Central University (MGCU), Motihari-Bihar, India

R. Kumari
Lalit Narayan Mithila University (LNMU), Darbhanga-Bihar, India

N. Pareek
Bihar Agricultural University (BAU), Sabour, Bhagalpur-Bihar, India

P. P. Adhikary et al. (eds.), *Soil, Water Pollution and Mitigation Strategies*, Environmental Science and Engineering, https://doi.org/10.1007/978-3-031-63296-9_4

not been much of a shift in the share of land that is used for agricultural purposes over the past ten years. Agroforestry is practiced on around 8.2% of India's total land area, which is equivalent to 25.32 million hectares. This compares to the entire acreage of the world, which is approximately 1023 million hectares (Nair et al. 2009a, b). In addition, it has the potential to develop into a significant land use activity in the near future by absorbing 28 million hectares of land from a variety of sources, such as fallows, cultivable fallows, pastures, groves, and the rehabilitation of problem soils. This would make it possible for it to become a significant land use activity in the near future. As a result, it is an important system in India, coming in third place after agriculture (140.86 Mha) and forestry (69.63 Mha) (Dhyani et al. 2013). Because it possesses a substantial amount of arable land, both irrigated (68.38 Mha) and rainfed (86 Mha), India is a frontrunner in this sector of the agricultural industry. Agriculture development has continued to be the primary focus of many developing countries, despite the fact that the majority of these nations are pursuing industrialization. As a consequence, the agricultural sector of any developing country is vitally important for a variety of reasons, including the production of food, employment, earnings in foreign money, the accumulation of capital, and the replacement of labour. The importance of efficient land-use practices and environmentally sound agricultural and food systems will only grow as the world's population and land area continue to expand. Agroforestry systems are becoming more widely acknowledged as an important land-use alternative in a variety of contexts around the world due to their solutions to environmental services and their financial viability and attractiveness (King 1987).

4.2 Land Diversification and Global Population

Apart from Australia, Japan, and New Zealand, the percentage of the population that relies on agriculture is between 20 and 30% in the vast majority of developed nations. In these three countries, however, a decrease in this percentage has been seen. In the vast majority of developing countries, even while the percentage of people working in agriculture is on the decline, it is still much larger than the overall population. The transition from the agricultural to the industrial sectors has been taking place on a vast scale all throughout the world, and not just in the Asia–Pacific region. The movement of people away from agriculture and towards industry in the Region has brought to the forefront two significant challenges: crop intensification and land diversification. Both of these issues relate to the land. The rapid increase in the world's population has put an increased burden on the land that is available. Over the past decade, the global population has increased from 8 billion to its current level of nearly 9 billion. More than 1.4 billion people call China and India home, but India has a slightly larger population. The Asia–Pacific area is home to over 4 billion people, which is approximately half of the world's total population. Several countries have annual population increases that are more than 3.5%. As a direct consequence of this, the typical size of a holding in the Area has decreased, and the proportion of land to people has decreased substantially over the course of history. The amount of land that is suitable for agriculture will continue to decline in many of these countries as a result of urbanization and industrialization, which will render the operation of some farms unprofitable (Webb et al. 1992).

4.2.1 Concept of Land Diversification

In many cases, diversifying a farm land use has been proven as an effective way to escalated output while decreasing the risk of total crop failure. Both the Vertical approach, in which additional crops are added to the existing cropping system, and the Horizontal approach, in which many different types of crops are grown in the landscape, can be used to diversify crop production. This form of crop diversification, on the other hand, entails expanding the system's foundation by growing additional crops within the framework of the current cropping system using innovative concepts and techniques like multiple cropping in conjunction with other effective management practices. Multiple cropping methods have led to a 200–300% increase in cropping intensity, which has increased the potential for food production. (IPBES 2019), Vertical diversification is the other type of crop diversification, in which a variety of activities are carried out in the crop's downstream market, possibly reflecting the degree to which and the stage at which the crop has been industrialized. It's commonly accepted that diversifying crops can increase overall profits by taking into account the monetary benefits of various factors. This is in stark contrast to the idea of multiple cropping, which takes into account the crops grown on a given plot of land during a specific time frame. Also, the degree to which countries diversify will vary. While growing a variety of crops on a single farm can help a community become more self-sufficient in terms of food production and provide other benefits, a national strategy for diversification may take a very different form (MoFED 2010). National crop diversification will call for more resources, as well as the selection and management of a crop that is ideally suited to a particular region and accomplishes multiple goals. Crop diversification has many benefits, the most important ones being a higher net return, more efficient use of resources, more productive use of land, and more job openings. Land diversification should shift away from crop rotations and toward intensive systems like relay cropping and intercropping, or specialise in a few crops, so as to reap the aforementioned benefits.

4.2.2 Land Diversification and Farming Systems

Agricultural intensification and farm consolidation farming practices and agrodiversity have become increasingly standardized worldwide (Batáry et al. 2020; Tscharntke et al. 2005). These diversification techniques include important agricultural practices like land-sharing schemes since they are easy to implement, can have considerable ecological advantages, and do not create excessive financial demands on farmers. Diversified cropping, zero tillage, intercropping, and cover crops are all examples of field-level practices that can be used as part of a land-sharing, nature friendly farming approach (Rosa-Schleich et al. 2019). Practices such as organic farming, which emphasizes soil health and biodiversity, and conservation farming are all examples of varied farming. Conservation agriculture was founded on the principles of minimal soil disturbance, permanent soil cover, and a diverse crop rotation or combination (Vanbergen et al. 2020). As a result of these measures, biodiversity rises, soil fertility improves, and biological pest management is implemented (Palm et al. 2014). Economically, they increase long-term yield consistency while decreasing input costs, such as those related with machinery and labour (Rosa-Schleich et al.2019). More resilient agroecosystems have numerous economic benefits, including enhanced production stability and decreased input costs (Reganold and Wachter 2016).

4.2.3 Land Diversification and Ecosystem Services

The IPCC report makes it clear that managed lands are vital for maintaining carbon sinks and providing renewable resources. If we are going to wean ourselves off of fossil fuels in the foreseeable future, we'll need protected areas just as much as we'll need food and renewable energy to feed the world's expanding population. In theory, all agroforestry systems share three characteristics. Keeping or increasing output is a goal shared by the vast majority (if not all) agroforestry initiatives. There are many ways in which agroforestry can boost output (Miehe 1986). It is possible to increase not only the output of tree products but also the yields of allied crops, the inputs to the cropping system, the efficiency of labour, and so on. Because of its ability to conserve the production capacity of the resources, agroforestry is capable of achieving and indefinitely maintaining conservation and fertility goals. This is mostly accomplished through the favourable impacts that woody perennials have on soils. Adoptability is distinct from the more popular word "adapt," which connotes a change. Adoptability refers to the ability to adopt (Shibu 2009). Agroforestry is already generally acknowledged among farmers despite the fact that the term "agroforestry" has only been around for the past few decades. This is due to the fact that the practises that forms the basis of agroforestry stretch back considerably further in time than the term itself. They are incredibly beneficial in soil and water conservation through the provision of permanent cove; they offer a major pathway for sustainable diversification of agricultural systems and incomes; they can increase the use efficiency of rain water; they are an economically viable and environmentally friendly means of improving soil fertileness; agroforestry systems have the potential to limit carbon emissions and sequester carbon; they play a crucial role in moderating the microclimate; they are highly effective in soil and water conservation through the provision of permanent cove; they have (Kiptot et al. 2014).

4.2.4 *Strategies and Basic Approach in Land Diversification*

It is important to incorporate the knowledge and experience of local farmers and communities that rely on the forest into the basic approach of agricultural diversification. Adaptations to the changing climate that go beyond the crop component are necessary for practicality and widespread adoption. These include livestock, fisheries, and appropriate supporting strategies. Adopting new technologies allows genetic diversification, a central activity in today's farming systems, to be extended to the landscape. The perceived risks associated with climate change, which reduce output on farms as a whole, have influenced this choice. This will be followed by the implementation of cropping systems and farm enterprises that promote resilience to climate change and secure livelihoods. In this process, the production system must adapt to the needs of the market through the creation of value and the provision of appropriate policy support at all levels, from the local to the global.

4.2.5 *Tools and Technology for Increasing Yields*

Although it is possible to increase crop production by expanding agricultural land, this is clearly not possible for the reasons stated above. The other option is to increase the amount of arable land by cultivating marginal lands in different countries, but these areas are nearly depleted and may remain marginal even with significant investment. An increase in cultivation intensity and farm productivity is the only viable options for satisfying the food requirements of a growing population. New technologies are being developed in many countries that could have an effect on agricultural output in the future. As farmland becomes more scarce, most of these tools are geared toward maximising output. Increases in food production in the future, then, to improve food security, will have to be accomplished primarily through intensive cultivation and high yield, and only secondarily through increase in land areas. Increases in crop yield will account for more than 75% of the total, with the remaining increase coming from increased land area, crop diversification, and cropping intensities (FAO, 2005). Future yield improvements may be possible with the help of new technologies. It's more likely that developing nations will see larger yield increases than developed nations. Technology fatigue has caused yields in developed nations to plateau at their maximum possible level, while yields in developing nations are consistently lowered than the potential yield under ideal laboratory conditions (Webb et al. 1992),. There is a clear need for less expensive technologies in light of the yield gaps that exist between expected and actual volumes of most crops grown in different countries. Estimates suggest that in less developed countries, yields are roughly more than half of those obtained in developed countries are observed. As a result, there is a lot of room to improve crop yields (Nair 1987, Nair 2011),.

4.2.6 Transformation of Lives and Landscapes

Agroforestry is the frontrunner in the race to restore healthy functions on the 40% of the world's land that has been degraded to some extent. Trees replenish depleted soils with nutrients, create a habitat for the microorganisms responsible for nutrient management, regulate the water table with their roots, and benefit farmers in numerous other ways, including by providing shade, food, fuel, materials, and income. Sustainable land use systems like agroforestry are being looked at as the way of the future because of the potential they present for making good use of our already abundant resources. Globally, there are vast swaths of uncultivated land that could be productive if farmed using today's methods (FAO, 2012). More and more people are turning to agroforestry as a way to relieve some of the stress on natural landscapes by reaping the benefits of trees in agricultural settings (such as their aesthetic value, their ability to prevent soil erosion, their capacity to collect rainwater, etc.). Involving farmers, livestock, trees, and forests from the household to the landscape scales, agroforestry is more than just a method of growing food sustainably. (Hall et al. 2005)

4.3 Agroforestry and Land Diversification as a Solution to Global Problems

More than two billion people are still going hungry every day around the world and its being a big challenge to increase food production by an estimated 60% to meet the needs of the people who are expected to exist by 2050. These numbers highlight the need for sustainable improvements to our food system management and land diversification strategies (Fernandez and Nair 1986). More sustainable farming practices using natural and modern practices in a hybrid mode are an absolute necessity for any long-term vision and strategies to address food insecurity. Increasing food production while decreasing agriculture's environmental impact may be possible through the use of ecologically functional ecosystems comprised of trees, animals, and crops that are managed in harmony with their surrounding environment. There are some ways in which agroforestry systems can excel over more conventional forms of agriculture and timber harvesting. They have the potential to boost output, food security, economic benefits, and ecological diversity. When it comes to solving the world's problems, agroforestry and land diversification play a huge role. Water quality and availability are improved, agricultural impacts on the climate are lessened, and other positive environmental outcomes are realized as a result (Sintayehu 2013). Agroforestry has the potential to improve farmers' financial situations by generating additional income and allowing them to grow and sell a wider variety of crops, all of which contribute to a healthier diet and longer life (Rajput et al. 1989, Rosenstock et al.2018). Women's empowerment, the validation of indigenous and traditional knowledge, and a contribution to sustainability are just a few of the social benefits and rural livelihoods that can result from a shift toward a more diverse approach to land management.

4.3.1 Agroforestry and Environmental Challenges

Humans have degraded about a quarter of the ice-free land on Earth, and many of the countries most likely to be hit hard by climate change are in the tropics, where a large percentage of the population relies on agriculture for survival (IPCC 2019). Land degradation is a major contributor to atmospheric carbon dioxide levels because organic soil matter contains about three times as much carbon as the atmosphere (Jobbágy and Jackson 2000). Soil, depending on its properties, local climate, and land use, can act as a carbon sink or source over time (IPCC 2019). Data on reflects more than 15% of the world's lands have degraded due to human based activity. Causes of this degradation include intensive farming and forest clearing. As the world's population grows exponentially, so do the pressures on the environment. Attempts to feed a growing population have had a devastating effect on soil quality and fertility. Pressure on land resources is a common result of intensive and unsustainable modern agricultural practices. In addition, the widespread felling of trees is common in systems like mono-cropping because animals, crops, and trees are all seen as rivals rather than complementary. While global deforestation rates have been declining, they are currently at their lowest levels in the tropics and subtropics. Deforestation, tillage, and excessive fertilizer use are all unsustainable agricultural practices that have serious environmental consequences (Mann and Saxena 1980). A third of all greenhouse gas emissions are attributed to agriculture and forest clearing. There are many ways in which agroforestry can improve farmers' and local communities' quality of life and the environment as a whole (IPCC 2019). Trees provide a variety of ecosystem services, including flood control, sediment control, carbon sequestration, and nutrient recycling in soils, and agroforestry helps sustain these services while also providing a place for wildlife to thrive. As a result, soil fertility rises, erosion decreases, and problems like flooding and pests are mitigated. Increased farm productivity and decreased reliance on expensive external inputs like conventional fertilizers and pest management chemicals can be two of agroforestry's biggest advantages for smallholder farmers.

4.3.2 Agroforestry and Economic Stresses

While hunger has been reduced in many regions, malnutrition still affects about 2 billion people, mostly children and women. Majority of the world's poor are living in countries rural areas. However, if we only consider these numbers, we risk ignoring an additional crucial facet of food security and dietary quality. More than half of the world's subsistence farmers, pastoralists, and landless agricultural workers make up the 10% of the rural population that was in extreme poverty worldwide in 2015. Farmers' commercial activities are especially precarious because they frequently center around the production of a single commodity, leaving them open to price swings. To make matters worse, poverty reduction is often hampered by low land and labour productivity. In order to ensure food security, it is crucial to alleviate the pressures that the majority of the population is under economically (Oxfam 2010, Qiao et al. 2019). Root-level tools, especially for rural population and forest-dependent communities, to supply food and goods all year round include agroforestry and diverse land management.

4.3.3 Agroforestry and Social Issues

Most people who call the countryside home are among the country's most economically and socially marginalized, and their land holdings are often severely fragmented. Indigenous peoples are another vulnerable group, making up only 5% of the global population but 15% of all the poor. Because of the slim pickings for a good life in the countryside, many men and their families leave the country for the cities. Due to their greater susceptibility to land degradation and natural disasters, women tend to have less access to resources than men. The situation is especially dire for families in which a woman serves as the primary breadwinner. Traditionally, firewood gathering has been a female-dominated activity. Marginalized land is notoriously difficult to farm, which opens up a wealth of possibilities for land diversification.

4.3.4 Carbon Sequestration Potential in Agroforestry

Atmospheric carbon dioxide is taken up by plants during photosynthesis and subsequently transferred to fixed C pools in the form of vegetation, detritus, and soil. This process is defined in the context of agroforestry as the term "carbon sequestration" (Nair 2010). AGFs have the capacity to sequester carbon at a rate that ranges from 0.29 to 15.21 megagrams per hectare per year aboveground and between 30 and 300 megagrams per hectare per year in the soil down to a depth of 1 m. It has been demonstrated that the application of AGFs can improve soil quality by lowering the rates of erosion and soil disturbance while simultaneously raising the levels of carbon stability and biodiversity. The amount of carbon that can be stored away relies on a variety of things, such as the type of plants that are utilized, the climate conditions, the maturity level of the AGFs system, and the type of soil that is there. The government of India recognizes the potential of agroforestry systems as a strategy to slow down the consequences of climate change by increasing and conserving the amount of carbon that is stored aboveground and in the soil. It is anticipated that agroforestry will produce an average of 25 t of carbon per hectare across 96 million acres (Sathaye and Ravindranath1998). Agroforestry in India has the tremendous capacity to store huge carbon dioxide annually. Although it covers only 8.2% of the total reported geographical area (305.6 million hectares), agroforestry is responsible for 19.3% of the total carbon stock across all land uses. This is despite the fact that agroforestry only accounts for 8.2% of the total reported geographical area (2755.5 million ton C). The enhanced collection of growth resources such as light and water, as well as the enhancement of the fertility of the soil, both have the potential to result in higher levels of productivity in agroforestry (Adhikary et al. 2017). According to the findings of a number of studies carried out all over the United States, agroforestry provides farmers with more returns than either agriculture or forestry does for the same amount of land (Samra et al. 1999). In the not-too-distant future, there will be a considerable expansion in the quantity of land in India that is designated for agroforestry. The potential for carbon sequestration that is offered by various agroforestry systems varies, but this potential can only be realized if certain actions are carried out. There was a wide range of possibilities for carbon sequestration, anywhere from 12 to 228 metric tonnes of carbon dioxide per hectare (Newaj and Dhyani 2008).

4.3.5 Conservation of Soil and Water Through Agroforestry

It is possible to help buffer the impacts of flooding on soil and land by planting trees and ensuring that they are adequately maintained. Agroforestry can also be used to restore and maintain water quality on damaged land since it can collect sediments, absorb nutrients, and immobilize dangerous substances. This can be accomplished by planting trees (Jakhar et al. 2017). Agroforestry systems have the potential of hydraulic lift which transfers the water through the soil from more saturated bottom layers to less saturated upper layers. It has been said that these characteristics are present in natural tree-grass pairings as well as in agroforestry systems (Hawkins and Sutton 2009). The amount of water that was available for the crops was increased thanks to terrace risers planted with *Alnus acuminata* and *Sesbania sesban*. These plants were clipped for their shoots and roots (Siriri et al. 2013). During normal dry seasons, hydrological reorganization may encourage the transpiration of water and the growing of plants. Agroforestry systems are preferable to oceanic and other terrestrial options for mitigating climate change because of the secondary environmental benefits they provide. These benefits include food and land security, increased farm income, restoration and maintenance of above- and below-ground biodiversity, maintenance of watershed and soil conservation. Other options for mitigating climate change include planting trees.

4.4 Climate Change and Food Security

Millions of homes around the world rely on forest products and services. This highlights the significance of evaluating climate change adaptation efforts in the forest sector within the framework of sustainable development and community impact (Garrity et al. 2010). Strong data suggests that agroforestry systems can improve water usage efficiency by lowering wasteful water-related costs. We need alternative land-use strategies that improve people's ability to make a living and make them less susceptible to the negative impacts of climate and environmental change (FAO, WFP 2010). The simultaneous production of food, fodder, and firewood through agroforestry systems and other traditional resource management adaptations may provide opportunities for improved livelihoods and mitigation of climate change's negative effects. When it comes to rehabilitating degraded land and reducing the effects of climate change, bamboo-based agroforestry systems are a promising alternative energy source (Mathukia et al.2016).

4.4.1 Agroforestry in Combating Climate Change

The agroforestry adoption can be strategy for minimizing the consequences of climate change, protecting human life and ensuring economic stability, and building the foundation for sustained economic and social expansion has immense promise. Sustainable forest management provides the foundation for planning at the global and national levels to address this complex and ever-changing obstacle (Charles et al.2014). On the other hand, agroforestry presents an opportunity to contribute to the development of national adaptation plans. It promotes sustainable management and community-based practices that have the potential to protect land and people from negative impacts of rising global temperatures have the potential to create the possibility for greater, more sustainable rural development as well as the possibility of alleviating poverty through the generation of income and the creation of employment opportunities. This possibility is made possible as a result of the existence of this potential. Using these methods, farmers are able to increase their revenue, improve food security, and lessen the risks associated with their farms. In addition to satisfying business demands for timber, they are good for the environment. Biomass from trees and plants provides 70–80% of the energy in rural areas (Carlsson et al. 2017). Agroforestry's permanent tree cover and diverse ecological niches serve as a buffer against climate variability, and the sector's resilience—or ability to bounce back after a disruption like extreme weather or a market collapse—is high thanks to the wide range of management strategies it employs. Soil carbon reserves are increased and soil quality is improved by a forest's permanent canopy. Non-harvested components serve a significant protective role, management flexibility is consistent with variations in labour availability, and commodity diversification allows for adaptation to market needs. The current climate change scenario validates agroforestry's promise. Agroforestry's sustainable practices are invaluable tools in the fight against climate change (Network, A., 2018). Agroforestry has the potential to address both Greenhouse Gas-Mitigation Strategy through Carbon Sequestration and Sustainable Adjustment to Changing Conditions in relation to climate change. This is because integrated systems are more efficient at resource capture and utilization than single-species systems. With the maintenance and growth of their root and woody biomass, they continue to supply food, fiber, energy, and vegetative cover for the soil throughout the entire year. Agroforestry is one strategy that can be used to improve the synergy between adapting to the consequences of climate change and mitigating those effects (Kumar and Nair 2011, Leder et al. 2016). A mental shift is required in order to design agroforestry networks that are resistant to the effects of climate change. There is a need for a new paradigm in agroforestry, one in which trees and crops are chosen and managed together. These unique associations will be able to solve challenges of food security and climate change concurrently (Bewket and Alemu, 2011). A crop system and a tree system grown side by side have been the primary components of agroforestry up until this point.

4.4.2 Agroforestry Adaptation for Ecosystem Resilience

As a result of climate change, many agricultural challenges, such as droughts, floods, and pests, may become even more difficult to manage. There is a risk that farmers will not be able to adapt to the consequences of climate change and maintain their level of self-sufficiency in their farms (Shankarnarayan et al. 1987, Thorlakson and Neufeldt 2012). Both the ability of an entire landscape to withstand the climate change impacts and the capability of individual farms to do so can be improved by the application of agroforestry in a broad variety of different methods. For example, agroforestry has been shown to reduce air pollution while simultaneously boosting atmospheric warming and cooling, resulting in a more consistent local environment that is conducive to the growth of both crops and cattle (Ellison et al. 2017). It does this by enhancing infiltration into soils and groundwater, which in turn protects water catchments and watersheds (BarguésTobella et al. 2014). Because of their potential to improve soil properties and increase the amount of water that is available to plants, agroforestry practices are ideally suited for the landscape restoration. Ecosystem services of tree are water regulation, climate buffering, soil fertility, erosion and flood control, food, fodder, medicine, and wood, all of which contribute to human and environmental resilience in the face of climate change and lessen people's susceptibility to its effects. Just a few of these services include: water regulation, climate buffering, soil fertility, erosion and flood control, food, fodder, and wood (Mbow et al. 2014). Agroforestry has the ability to improve food security and extend people's access to nutritious meals, as well as increase the incomes of smallholder farmers and the cash crop systems they use. In the aftermath of natural disasters, the presence of trees on farms might help speed up the process of economic recovery for the farmers (Simelton et al. 2015). The Carbon sequestration is significantly enhanced with the implementation of agroforestry practices into agricultural and animal production systems. A wide variety of agroforestry practices, such as backyard gardens, boundary planting, fruit orchards, riverine planting, hedgerows, woodlots, and fuel lots, all have the potential to capture carbon dioxide (CO_2) and store it for future use (Nair 1985a, Nair 1985b).

4.4.3 Global Climate Goals and Agroforestry

Already, the global average temperature is 1.1 °C higher than it was before industrialization. This increase in temperature is causing glaciers to melt, which in turn is raising sea levels. Climate change has many negative impacts on flooding and drought, which have resulted in the displacement of millions of people, the push of millions more people into poverty and hunger, the cutting off of their access to fundamental services such as healthcare and education, the slowing of economic progress, and even the initiation of conflict (Huxley 1996). By the year 2030, it is anticipated that drought will put the lives of around 700 million people all over the world in danger of being uprooted. Taking prompt action to combat climate change and its destructive repercussions is therefore a necessity to save lives and livelihoods, and it is key to achieving the 2030 Agenda for Sustainable Development and its 17 goals—the blueprint for a better future—in order to make them a reality. This is because taking prompt action to combat climate change and its destructive repercussions is a necessity to save lives and livelihoods. In the year 2020, the concentrations of greenhouse gases across the globe reached new highs, and real-time figures point at rises that are continuing. The temperature of the globe will continue to rise if this accumulation of gases is allowed to continue. Between 2015 through 2021, the years that were each and every one of the seven warmest years ever recorded. If global warming is to be limited to 1.5 °C over pre-industrial levels, the Paris Agreement stipulates that the peak in greenhouse gas emissions must occur by the year 2025. They need to then drop by 43% by the year 2030 and completely disappear by the year 2050. In order to mitigate global warming‘s effects and adapt to its changes, countries are drafting climate action plans, known as nationally defined contributions. However, national commitments as they stand now are not enough to achieve the 1.5 °C goal (Kumar and Nair2011). Agroforestry can help the globe satisfy international obligations on the climate of developing additional carbon sink by 2030 and net zero by mid-century. As a result, we can better combat desertification and achieve other sustainable development goals (SDGs) (Molin et al.2018).

4.4.4 Strengthening of Agroforestry Infrastructure for Mitigation

With the right mix of trees/shrubs and crops/livestock, an agroforestry system can give economic and environmental benefits to the landowner over the short and long term. However, a comprehensive agroforestry infrastructure for research, education, extension, and policy is necessary for widespread adoption of agroforestry in the United States (Schoeneberger et al. 2012). The landowner is the end user, thus the infrastructure must be built with them in mind and optimized for quick dissemination of systems research results. The interdisciplinary and multi-dimensional, inter-sectoral nature of effective technical assistance and technology transfer concerning land resource managementand empowered institutional mechanisms for planning and managing agroforestry development activities; Now is the time to act to keep global warming below 1.5 °C (Ripple et al. 2019, IPCC 2019). Soil erosion, crop failure, loss of biodiversity, decreased soil moisture, pest damages, and economic losses can all be triggered by the climate change hazards (such as severe droughts, flooding, illnesses). Current production systems and food security are threatened by the occurrence of extreme events and the higher occurrence of drier and wetter weather (SIWI 2018). Reducing carbon emissions and meeting the goals of the Paris Agreement are dependent on trees, forests, and agriculture. Farmers can better cope with climate change if they replant the appropriate tree species in the appropriate locations (Torquebiau 1992).

4.4.5 Changing Trends for Agroforestry

Agriculture must adapt to new requirements. The scope of the problem has widened to include sustainable agriculture. The public expects agriculture to maintain its current level of productivity and profitability, maintain its current level of food security, and maintain its current level of environmental sustainability. Future farm assistance programmes are anticipated to place greater emphasis on the supporting elements of the agroecosystem than on the crop itself. Therefore, it is necessary to take an ecosystem-based approach to programme design and rollout across agricultural, hydrologic, and ecological scales (IPCC Climate Change 2014). Teams tackling the notion of sustainable development in agro ecosystems should include foresters, wildlife biologists, agronomists, etc., as agroforestry is an essential component of such systems. Management of ecosystems is a procedure rather than a final deliverable. It is absolutely necessary to utilize multi-agency and multi-disciplinary teams in order to construct a planning strategy that is based on ecosystems and to provide assistance on private properties. Natural resource management moves towards an approach at the watershed scale, it is becoming increasingly obvious that trees are an essential component of the agroecosystem infrastructure. Agroforestry depends on trees to protect the soil and water quality over the long term, strengthen the resilience of the system, improve the habitat for wildlife, contribute to the protection of biodiversity, and provide for people's physical and economic well-being. Because it is the primary source of biodiversity, the agroforestry component of the system is very important to the process of keeping the ecosystem in a state of equilibrium. It is essential for developing an ecosystem-based approach to place emphasis on the economic, environmental, and social interactions between agricultural systems and natural resources (Vandenbeldt 1992), Verchot et al.2007). This, in turn, necessitatesmulti-disciplinary pilot projects in priority land that are both diversified and strike a balance between technologies and expertise in order to optimise integrated production and conservation systems.

4.4.6 Emerging Challenges for Agroforestry

A rising volume of scientific research highlights the advantages of agroforestry, yet this approach still encounters several barriers to implementation. FAO's efforts are based on a commitment to resolving these issues. The breakeven point for some agroforestry systems may be several years away, despite the fact that trees become lucrative over time by delivering positive net present values. Agroforestry value chains are not supported as heavily as those for agricultural and livestock commodities, and the market for tree products is less efficient and less developed (Dhyani et al. 2009). Agricultural policies may include tax breaks and other forms of financial assistance that are targeted toward industrial agricultural output, as well as agricultural incentives that support particular agricultural types like monoculture. Adoption of agroforestry is hindered by policies that provide financial incentives to farmers for growing certain crops while rarely doing so for planting trees. Because of their familiarity with just conventional farming practices and their lack of understanding of sustainable alternatives, policymakers are not as invested in agroforestry growth as they could be. This has a knock-on effect on investments in areas like education, communication, market intelligence, and germplasm propagation, all of which are vital to agroforestry's eventual success. Land tenure issues, such as insecurity or ambiguity, are frequent in poor countries and might discourage the adoption of agroforestry practices by making it difficult for individuals to clearly define their ownership interests in their land. The lack of secure property rights can prevent long-term investments like agroforestry from being made in many parts of the world. For other situations, however, collecting, cutting, and selling tree products is restricted due to forest restrictions that forbid tree growing on farms. Agricultural policies often discourage small-scale and specialized farming methods like agroforestry in favour of mass production of food, fiber, and fuel. It's possible that the tax system treats woods unfairly compared to farmland. Sometimes the bureaucratic procedures associated with a programme or policy that is meant to promote agroforestry end up being too cumbersome or not elaborated in a way that is relevant to agroforestry and its potential users. Agroforestry, an intervention that has far-reaching consequences across many fields and industries, such as agriculture, forestry, livestock, rural development, the environment, energy, health, water, and trade, is frequently hampered by policy conflicts and omissions that leave gaps or create adverse incentives. Some of these fields and industries include agriculture, forestry, livestock, rural development, the environment, and trade. It is also likely that agroforestry will not be able to be implemented because of variations in the organizational culture and the goals of the different departments.

4.5 Conclusion

By providing a platform for a diverse group of stakeholders to work together on a shared agenda for conservation and land-use practices, agroforestry presents an opportunity to tackle some of the most intractable challenges facing our country and the globe today. While a number of critical conditions for its growth are converging, the benefits of agroforestry will remain unrealized until targeted policies, programmes, funding, and interdisciplinary teams are put in place.

References

Adhikary PP, Hombegowda HC, Barman D, Jakhar P, Madhu M (2017) Soil erosion control and carbon sequestration in shifting cultivated degraded highlands of eastern India: performance of two contour hedgerow systems. Agroforest Syst 91:757–771. https://doi.org/10.1007/s10457-016-9958-3

Agroforestry Network (2018) Scaling up agroforestry: potential, challenges and barriers. A review of environmental, social and economic aspects on the farmer, community and landscape level

BarguésTobella A, Reese H, Almaw A, Bayala J, Malmer A, Laudon H, Ilstedt U (2014) The effect of trees on preferential flow and soil infiltrability in an agroforestry parkland in semiarid Burkina Faso. Water Resour Res 50(4):3342–3354

Batáry P, Báldi A, Ekroos J, Gallé R, Grass I, Tscharntke T (2020) Biologia futura: landscape perspectives on farmland biodiversity conservation. Biol Futur 71(1):9–18

Bewket W, Alemu D (2011) Farmers' perceptions of climate change and its agricultural impacts in the Abay and Baro-Akobo River Basins, Ethiopia. Ethiopian J Dev Res 32(2)

Carlsson G, Martensson LM, Prade T, Svensson SE, Jensen ES (2017) Perennial species mixtures for multifunctional production of biomass on marginal land. GCB Bioenergy 9:191–201

Charles RL, Nzunda EF, Munishi PKT (2014) Agroforestry as a resilient strategy in mitigating climate change in mwanga district, kilimanjaro, Tanzania. Glob J Biol Agri Heal Sci 3(2):11–17

Devi SV, Dhyani SK, Gupta VK, Kumar RV, Ahlawat SP, Kumar S, Datta A (2015) TBOs Program at ICAR-CAFRI, Jhansi. Lead Pap 115

Dhyani SK, Handa AK, Uma (2013) Area under agroforestry in India: an assessment for present status and future perspective. Ind J Agrofor 15(1): 1–11

Dhyani SK, Newaj R, Sharma AR (2009) Agroforestry: Its relation with agronomy, challenges and opportunities. Ind J Agron 54(3):70–87

Ellison D, Morris CE, Locatelli B, Sheil D, Cohen J, Murdiyarso D, Gutierrez V, Van Noordwijk M, Creed IF, Pokorny J, Gaveau D et al (2017) Trees, forests and water: cool insights for a hot world. Glob Environ Chang 43:51–61

FAO Commission on Genetic Resources for Food and Agriculture Assessments, Rome, 572 pp. http://www.fao.org/3/CA3129EN/CA3129EN.pdf
FAO (2019). The state of the world's biodiversity for food and agriculture (pp. xli+-529).
FAO (2005) State of the world's forests. Realizing the economic benefits of agroforestry: experiences, lessons and challenges. Part II Selected current issues in the forest sector, pp 85–95. ftp://ftp.fao.org/docrep/fao/007/y5574e/y5574e09.pdf
Fao/Stat (2012) Crop production statistics
FAO/WFP [Food and Agriculture Organisation and World Food Programme] (2010) State of food insecurity in the world. FAO/WFP, Rome
Fernandez ECM, Nair PKR (1986) An evaluation of the structure and function of tropical home gardens. Agric Syst 21(4):279–310
Garrity D, Akinnifesi FK, Ajayi OC, Weldesemayat SG, Mowo JW et al (2010) Evergreen agriculture: a robust approach to sustainable food security in Africa. Food Sec 2(3):197–214
Hall NM, Kaya B, Dick J (2005) Effect of improved fallow on crop productivity, soil fertility and climate-forcing gas emissions in semi-arid conditions. Biol Fertil Soils 42: 224–230
Hawkins E, Sutton R (2009) The potential to narrow uncertainty in regional climate predictions. Bull Amer Meteor Soc 90(8):1095–1108
Huxley P (1996) Tropical agroforestry. Blackwell Science, pp 371
IPCC (2019). Land: an IPCC Special Report on Climate Change. Desertification, Land Degradation, Sustainable Land Management, Food Security, and Greenhouse Gas Fluxes in Terrestrial Ecosystems, 41.
IPBES (2019) The global assessment report on biodiversity and ecosystem services of the intergovernmental science policy platform on biodiversity and ecosystem services
IPCC (2000) Land use, land-use change, and forestry. A special report of the IPCC. Cambridge University Press Cambridge, UK
IPCC Climate Change (2014) Synthesis report. Contribution of working groups I, II and III to the fifth assessment report of the intergovernmental panel on climate change. IPCC, Geneva, Switzerland, 2014, p 151
Jakhar P, Dass A, Adhikary PP, Sudhishri S, Naik BS, Hombegowda HC, Madhu M, Lenka NK, Chaudhary PR, Panda RK (2017) Multitier agroforestry system for integrated resource conservation on uplands of Eastern Ghats region in India. Agrofor Syst 91(4):697–712
Jobbágy EG, Jackson RB (2000) The vertical distribution of soil organic carbon and its relation to climate and vegetation. Ecol Appl 10(2):423–436
King KFS (1987) The history of agroforestry. In: Steppler HA, Nair PKR (eds) Agroforestry: a decade of development, ICRAF, Nairobi, Kenya, pp 1–11
Kiptot E, Franzel S, Degrande A (2014) Gender, agroforestry and food security in Africa. Curr Opin Environ Sustain 6:104–109
Kumar BM, Nair PKR (2011) Carbon sequestration potential of agroforestry systems: opportunities and challenges. In: Kumar BM, Nair PKR (eds) Advances in agroforestry
Leder S, Das D, Reckers A, Karki E (2016) Participatory gender training for community groups. A manual for critical discussions on gender norms, roles and relations. Report from CGIAR research program on water, land and ecosystems
Mann HS, Saxena SK (1980) "Kherji" (Prosopis cineraria) in the Indian Desert: it's role in agroforestry. In: Mann HS, Saxena SK (eds) CAZRI Monograph 11. Central Arid Zone Research Institute, Jodhpur, India
Mathukia RK, Sagarka BK, Panara DM (2016) Fodder production through agroforestry: a boon for profitable dairy farming. Innovare J Agri Sci 4(2):13–19
Mbow C, Smith P, Skole D, Duguma L, Bustamante M (2014) Achieving mitigation and adaptation to climate change through sustainable agroforestry practices in Africa. Curr Opin Environ Sustain 6:8–14
Miehe S (1986) Acacia albida and other multipurpose trees on the Fur farmlands in the Jebel Marra highlands, Western Dafur, Sudan. Agrofor Syst 4(2):89–119

MoFED (2010) The federal democratic Republic of Ethiopia: growth and transformation plan (2010–2014/15), Draft Document: Addis Ababa
Molin PG, Chazdon R, de Barros F, Ferraz S, Brancalion PH (2018) A landscape approach for costeffective large-scale forest restoration. J Appl Ecol 55(6):2767–2778
Nair KP (2010) The agronomy and economy of important tree crops of the developing world
Nair PKR (1985a) Classification of agroforestry systems. Agrofor Syst 3(2):97–128
Nair PKR (1985b) Agroforestry in the context of land clearing and development in the tropics
Nair PKR (1987) Agroforestry and firewood production. In: Hall DO, Overend RP (eds) Biomass. John Wiley, Chichester, UK, pp 367–386
Nair PKR (2011) Carbon sequestration studies in agroforestry systems: a reality-check. Agrofor Syst 86(2):243–253
Nair PKR, Kumar BM, Nair VD (2009a) Agroforestry as a strategy for carbon sequestration. J Plant Nutr Soil Science 172(1):10–23
Nair PK, Kumar BM, Nair VD (2009b) Agroforestry as a strategy for carbon sequestration. J Plant Nutr Soil Sci 172:10–23
Newaj R, Dhyani SK (2008) Agroforestry systems for carbon sequestration: present status and scope. Indian J Agrofor 10(1)
NMA (National Meteorological Agency) (2007) National meteorology agency: final report on evaluation criteria for identifying high priority adaptation activities prepared by B and M development consultants for NMA. Addis Ababa, Ethiopia
Oxfam (2010) Poverty, vulnerability and climate variability in Ethiopia. International Research report. Addis Ababa, Ethiopia
Palm C, Blanco-Canqui H, DeClerck F, Gatere L, Grace P (2014) Conservation agriculture and ecosystem services: an overview. Agric Ecosyst Environ 187:87–105
Qiao X, Sai L, Chen X, Xue L, Lei J (2019) Impact of fruit-tree shade intensity on the growth, yield, and quality of intercropped wheat. PLoS ONE 14(4): 203–238. https://doi.org/10.1371/journal.pone.0203238
Rajput MS, Biswas PP, Joshi PP, Srivastava KC (1989) Mango(Mangiferaindica) based cereal–pulse intercropping system. Indian J AgrilSci. 59(3):149–153
Reganold JP, Wachter JM (2016) Organic agriculture in the twenty-first century. Nat Plants 2(2):1–8
Ripple WR, Wolf C, Newsome TM, Barnard P, Moomaw WR (2019) World scientists' warning of a climate emergency. BioScience. https://doi.org/10.1093/biosci/biz088
Rosa-Schleich J, Loos J, Mußhoff O, Tscharntke T (2019) Ecological-economic trade-offs of diversified farming systems–a review. Ecol Econ 160:251–263
Rosenstock T, Wilkes A, Jallo C, Namoi N, Bulusu M, Suber M, Bernard F, Mboi D (2018) Making trees count: Measurement, reporting and verification of agroforestry under the UNFCCC. CCAFS Working Paper No. 240
Samra SK, Stanley JC, Zelenock GB, Dorje P (1999) An assessment of contributions made by extracranial tissues during cerebral oximetry. J Neurosurg Anesthesiol 11(1):1–5
Sathaye JA, Ravindranath NH (1998) Climate change mitigation in the energy and forestry sectors of developing countries. Annu Rev Energy Environ 23(1):387–437
Schoeneberger M, Bendrub G, Gooijer H, Soolanayakanahal R, Sauer T et al (2012) Branching out: agroforestry as a climate change mitigation and adaptation for agriculture. J Soil Water Conserv 67(5):128–136
Shankarnarayan KA, Harsh LN, Kathju S (1987) Agroforestry systems in the arid zones of India. Agrofor Syst 5(1):69–88
Shibu J (2009) Agroforestry for ecosystem services and environmental benefits: an overview. Agrofor Syst 76:1–10
Simelton E, Dam BV, Catacutan D (2015) Trees and agroforestry for coping with extreme weather events: experiences from northern and central Viet Nam. Agrofor Syst 89:1065–1082
Sintayehu W (2013) National greenhouse gas emissions baseline scenarios learning from experiences in developing countries.The Danish Energy Agency (DEA), the Organization for Economic Co-operation and Development (OECD) and the UNEP Risø Centre (URC)

Siriri D, Wilson J, Coe R, Tenywa MM, Bekunda MA, Ong CK, Black CR (2013) Trees improve water storage and reduce soil evaporation in agroforestry systems on bench terraces in SW Uganda. Agrofor syst 87:45–58

SIWI (2018) Water for productive and multifunctional landscapes. Stockholm International Water Institute, Report no. 38

Thorlakson T, Neufeldt H (2012) Reducing subsistence farmers' vulnerability to climate change:evaluating the potential contributions of agroforestry in western Kenya. Agric Food Secur. 1:15

Torquebiau E (1992) Are tropical agroforestry homegardens sustainable. Agr Ecosyst Environ 41(2):189–207

Tscharntke T, Klein AM, Kruess A, Steffan-Dewenter I, Thies C (2005) Landscape perspectives on agricultural intensification and biodiversity–ecosystem service management. Ecol Lett 8(8):857–874

Vanbergen AJ, Aizen MA, Cordeau S, Garibaldi LA, Garratt MP, Kovács-Hostyánszki A, Young JC (2020) Transformation of agricultural landscapes in the Anthropocene: nature's contributions to people, agriculture and food security. In: Advances in ecological research, vol 63. Academic Press, pp 193–253

Vandenbeldt RJ (1992) Faidherbiaalbida in the West African Semi-Arid Tropics.ICRISAT, Hyderabad, India and ICRAF, Nairobi, Kenya

Verchot LV, Van Noordwijk M, Kandji S, Tomich T, Ong C, Albrecht A, Mackensen J, Bantilan C, Anupama KV, Palm C (2007) Climate change: linking adaptation and mitigation through agroforestry. Mitig Adapt Strat Glob Change 12(5):901–918

Webb P, von Braun J, Yohannes Y (1992) Famine in Ethiopia: policy implications of coping failure at national and household levels. International Food Policy Research Institute. Research Report No. 92. IFPRI, Washington, D.C.

Woldeamlak B (2009) Rainwater harvesting as a livelihood strategy in the drought-prone areas of the Amhara regions of Ethiopia. Organization for Social Science Research in Southern and Eastern and Southern Africa, Addis Ababa

Chapter 5
Turmeric Yield, Soil Properties, Nutrient Uptake and Nutrient Use Efficiency as Affected Under Irrigation Scheduling and Nutrient Sources at Aeric Haplaquept

Sandeep Kumar Tripathi, Babloo Sharma, Reena Kumari, Pratibha Kumari, Adyant Kumar, and R. Ray

Keywords Irrigation regimes · Nutrient supplements · PFP · Soil attributes · Turmeric yield

---

S. K. Tripathi
Soil and Land Use Survey of India, Rajendra Nagar, Hyderabad 500030, India

S. K. Tripathi · R. Ray
Department of Soil and Water Conservation, Faculty of Agriculture, Bidhan Chandra Krishi Viswavidyalaya, Mohanpur, Nadia, West Bengal 741252, India

B. Sharma
Krishi Vigyan Kendra Jaisalmer-1, Swami Keshwanand Rajasthan Agricultural University, Jaisalmer, Rajasthan 345001, India

R. Kumari (✉)
Department of Agricultural Engineering, N. M. College of Agriculture, Navsari Agricultural University, Navsari, Gujarat 396450, India
e-mail: reenakumari.bhu@gmail.com

P. Kumari
Krishi Vigyan Kendra Aurangabad, Bihar Agricultural University, Sirish Agricultural Farm, Aurangabad, Bihar 824112, India

A. Kumar
Department of Agronomy, Dr. Kalam Agricultural College, Bihar Agricultural University, Kishanganj, Bihar 855107, India

P. P. Adhikary et al. (eds.), *Soil, Water Pollution and Mitigation Strategies*, Environmental Science and Engineering, https://doi.org/10.1007/978-3-031-63296-9_5

5.1 Introduction

In tropical and subtropical regions turmeric is mainly grown. Nair (2013) reported that in India about 80% of world turmeric production comes. Turmeric is largely produced, consumed, and exported by India. Per year 11 lakh tonnes of turmeric are produced by India, which contributes 78% of the world turmeric production, whereas the contribution of China is 8%, Myanmar 4%, Nigeria and Bangladesh are 6% (Singh et al. 2014). In most important spice crops in India, it comes on third place (Pandey et al. 2011). In the country, Andhra Pradesh, Tamil Nadu, Odisha, Karnataka and West Bengal are major turmeric-producing states because these states are producing 90% of the total country production.

For crop production, irrigation water is a key input under water-scarce conditions and is also pricey. Therefore, proficient use of this input is necessary which can be achieved through cautious water management practices. To make the most efficient use of the irrigation system, scheduling of water application is very crucial, as excessive irrigation reduces yield, while insufficient irrigation causes water stress and reduces production. For the maximum returns to the farmer and the intensity of the operation requires that the water supply be kept at the optimum level (Yuan et al. 2003). Many studies on the irrigation scheduling of many crops have been completed based on the US Weather Service Class A pan evaporation (Tripathi et al. 2015, 2018). Yields of many crops would be maximized by keeping the total quantity of irrigation water constant and increasing the frequency of irrigation (El-Hendawy et al. 2008).

For turmeric, a large range of soil types as Inceptisols, Entisols, Vertisols, Alfisols, and Ultisols, varying in fertility, have been found appropriate. For turmeric cultivation, good fertile soils with high organic matter content as well as the availability of major and secondary nutrients are the most suitable (Sahu and Mitra 1992). Under different soil and weather conditions, the use of manure and fertilizer is necessary to produce a flourishing crop. In turmeric, nutrients are used exogenously which generally involves either exclusive application of chemical fertilizers or organic manures or an integrated approach. On turmeric effect of organic manures was studied (Sanwal et al. 2007), organic manures and biofertilizers in combination with different rates of chemical fertilizers (Kamble et al. 2009), and in different combinations of chemical fertilizers alone (Akamine et al. 2007) have been reported. Hence, the objective of this study is (a) to find out the effect of water regimes and nutrient sources on the production of turmeric; (b) to evaluate the effect of water regimes and nutrient sources on nutrient content, its uptake, and PFP of turmeric crop.

5.2 Materials and Methods

5.2.1 Study Area

The paper is based on the data from the experiments on turmeric crop carried out at the research farm of All India Coordinated Research Project (AICRP) on Water Management, Eastern India during two consecutive years of 2011 and 2012. The site lies at 89° 31′ E longitude and 23° 58′ N latitude with altitude 9.75 m above mean sea level. The Research Farm is located in the tropical hot sub-humid climatic zone. Normal annual rainfall is 1650 mm and 85% of it received during rainy season (June to September). January is the coldest month with a normal mean temperature value ranges from 15.5 to 21.3 °C, whereas, maximum normal mean temperature (27.6–31.7 °C) is recorded during May. The normal mean relative humidity remains high (82–95%) during June to October and is at its lowest level (70%) in February. The soil of the experimental site is coming under Aeric Haplaquept and soil texture is sandy loam (Tripathi et al. 2019; NBSSLUP 2001). Basic chemical properties of the surface soil were: pH 6.7, electrical conductivity 0.39 dSm^{-1} and organic carbon 5.8 g kg^{-1}, respectively. The initial levels of N, P and K in soil of experimental site were recorded 118.0, 17.8 and 124.3 kg ha^{-1}, respectively.

The trial was laid out four levels of irrigation schedules as main factor ($I_1 = 0.6$ IW:CPE, $I_2 = 0.9$ IW:CPE, $I_3 = 1.2$ IW:CPE and $I_4 =$ rainfed) and three sub-factor of integrated nutrient management [$N_1 = 100\%$ RDF as inorganic source, $N_2 = 75\%$ RDF + 25% organic source (Farm Yard Manure (FYM)) and $N_3 = 50\%$ inorganic + 50% organic source (25% FYM + 25% vermicompost (VC))] which was three time replicated in a split-plot design. Plot size was 16 m^2 and total number of treatments was twelve. The uniform healthy turmeric rhizome cultivar *cv.* Suguna was planted at a spacing of 30 cm × 20 cm on second week of April month. The rhizome was harvested on last week of November in each year. The RDF for the crop was 150 N: 75 P_2O_5:150 K_2O kg ha^{-1}. Irrigation scheduling was based on IW:CPE ratio approach with 50 mm depth of water applied in each irrigation. Each irrigation water was 50 mm depth as measured with a Parshall flume and cumulative pan evaporation was monitored from a standard USWB class A pan evaporimeter installed in the research station.

5.2.2 Soil Properties

Soil samples were collected after harvesting of crop each year. Soil pH, EC, OC and CEC were determined by using glass electrodes pH meter (Jackson 1973), digital E.C. meter (Piper 1966), 1N potassium dichromate by Walkley and Black method (Jackson 1973) and neutral normal ammonium acetate solution (Chapman 1965), respectively. The major nutrients in soil were determined using the following standard procedure i.e. available Nitrogen by Alkaline potassium permanganate method (Subbiah and

Asija 1956), available phosphorus by Olsen method using 0.5 M sodium bicarbonate extractant at pH 8.5 (Olsen et al. 1954) and available potassium by neutral normal ammonium acetate method (Hesse 1971).

5.2.3 Plant Nutrient Content and Its Uptake

The sample of turmeric rhizome was collected at 90, 150 DAS and harvest. The N, P and K content in the rhizome were estimated by modified Kjeldahl's method (Jackson 1973), vanadomolybdo-phosphoric acid yellow color method using spectrophotometer at 660 nm wavelength (Jackson 1973) and flame photometry method (Jackson 1973), respectively. The uptake of nutrient was calculated using the following formula.

$$Nutirent\ uptake\left(\mathrm{kg\,ha^{-1}}\right) = \frac{Nutrient\ concentration(\%) \times rhizome\ yield\left(\mathrm{kg\,ha^{-1}}\right)}{100} \tag{5.1}$$

5.2.4 PFP

The nutrient use efficiency was calculated in terms of PFP. For that reason, the applied N, P and K through different source (organic and inorganic) in turmeric crop were calculated after precise estimations following standard analytical procedures (Rana et al. 2014). Then, the turmeric rhizome yield was divided by total applied nutrient to estimate their respective PFP by using following expression (Choudary and Suri 2018; Kumari et al. 2021):

$$PFP\text{–}N = \frac{Turmeric\ rhizome\ yield\left(\mathrm{kg\,ha^{-1}}\right)}{Total\ applied\ N\left(\mathrm{kg\,ha^{-1}}\right)} \tag{5.2}$$

$$PFP\text{–}P = \frac{Turmeric\ rhizome\ yield\left(\mathrm{kg\,ha^{-1}}\right)}{Total\ applied\ P\left(\mathrm{kg\,ha^{-1}}\right)} \tag{5.3}$$

$$PFP\text{–}K = \frac{Turmeric\ rhizome\ yield\left(\mathrm{kg\,ha^{-1}}\right)}{Total\ applied\ K\left(\mathrm{kg\,ha^{-1}}\right)} \tag{5.4}$$

5.2.5 Statistical Analysis

The collected data of the study were statistically analyzed by SPSS 16.0 version (SPSS Inc., Chicago, IL). Statistical analysis and interpretation were done by calculation value of standard error of mean [SEm(±)] and critical difference (CD) at 5% level of significance.

5.3 Results and Discussion

5.3.1 Rhizome Yield

There was a significant difference in rhizome yield due to application of different sources of nutrient supplements and moisture regimes. The yields of fresh and cured rhizome were reported highest in I_2 followed by I_3. The yield of fresh rhizomes under I_2 increased by 13.92 and 0.78% during first year and 15.67 and 2.98% during second year over I_1 and I_3, respectively (Fig. 5.1). During both years, the yield of cured rhizomes showed a similar trend to fresh rhizomes. There was the least productivity was obtained under I_4. Under rainfed and irrigation systems, there was variation in yield mainly occur due to changes in wetting front, distributions of soil moisture and amount of water use by turmeric. In this study, treatment I_4 showed poor performance which might be attributed to the relatively low water content in the soil compared with other experimental fields. The water deficiency is the major factor to affected the yield of the crop and crop yield reduces significantly when the drought stress occurred during critical stages of crops (Sarhadi and Sharif 2014; Hailu et al. 2018). The maximum yield of fresh rhizome was recorded in N_3 while the minimum was in N_1. During first and second year, yield of cured rhizome was recorded at 7.12 and 8.25% respectively higher under N_3 as compared to N_2 (Fig. 5.1). These results showed that in determining the yield of turmeric rhizome, INM is a key factor. Srinivasan et al. (2016) reported that in the integrated nutrient management treatment, rhizome yield was 26.4% greater compared to the organic nutrient management whereas it was 12.4% greater in the conventional nutrient management. Saleque et al. (2004) observed that the RDF along with FYM gave superior yield as compared to the RDF only, which is in accordance with our findings. Tripathi et al. (2021) reported that the turmeric rhizome yield increased by using organic and inorganic fertilizer compared to 100% RDF. The data showed that during both years, yields of fresh and cured rhizome of I_2N_3 were superior compared to the other treatments. Actually, in the first and second year, in the treatment of I_2N_3 yield of fresh turmeric was 43.44 and 44.16% higher while the yield of cured rhizome was 47.65 and 44.00% higher as compared to the I_4N_1 (Fig. 5.1). Application of adequate amount of irrigation water along with nutrient supplements ensured a better absorption of nutrients from the soil and subsequently it's promoting phonological growth of plant which resulted in superior production of turmeric. Rathod et al. (2010) also result corroborated that the turmeric rhizome yield was greatly influenced by different moisture and nutrient regimes in soil.

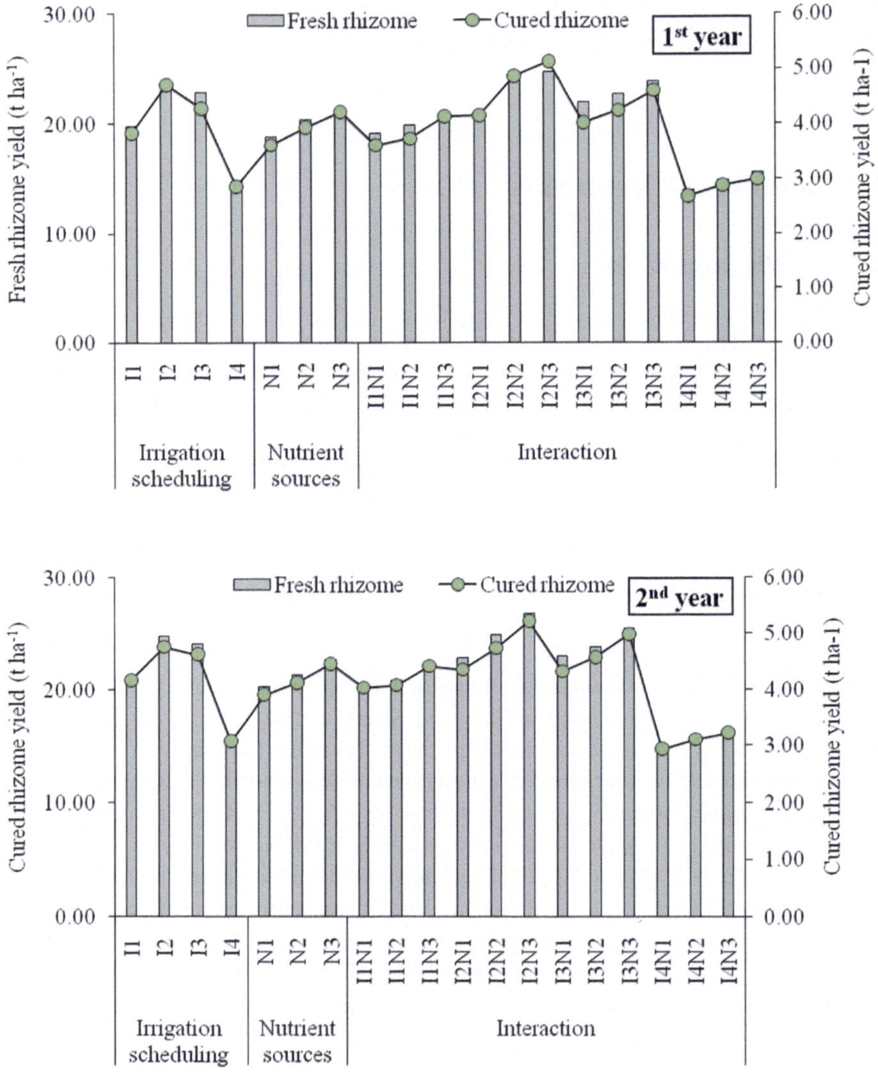

Fig. 5.1 Productivity of turmeric under different irrigation schedules and nutrition source during 2011 and 2012

5.3.2 Soil pH, EC, OC and CEC

The change in soil pH, EC, OC and CEC were non-significantly affected by different irrigation schedules during first year whereas during second year, soil EC, OC and CEC were also significantly affected but soil pH was non-significantly under different irrigation schedules in the experimental plots of turmeric crop (Table 5.1). Highest soil pH, EC, OC and CEC was recorded with I_2 followed I_3. The crop grown under I_4 recorded the lowest of value soil pH, EC, OC and CEC. Soil pH is important aspect in term of nutrients absorption by the plant from the soil because it's influences solubility and availability of nutrients in soil (Bendra et al. 2012). The values for this parameter were higher for the irrigation levels than for the conventional rainfed situation in the same year. This was caused by the application of additional amounts of soluble salts in the soil matrix with irrigation water. The application of 100% RDF either alone or, integrating application of fertilizers and manures was significantly influenced to EC, OC and CEC of soil except pH. Maximum pH, EC, OC and CEC of soil was observed with the addition of nutrition at N_3 in both the years. Lowest pH, EC, OC and CEC of soil was recorded with N_1 (Table 5.1). Citak and Sonmez (2011) reported that the increased of soil pH with the application of nutrition through organic sources and decrease with inorganic sources of nutrition. Nevertheless, integration of inorganic and organic sources of nutrients influenced microbial activity which is responsible for decomposition of organic compound and chemical transformations of nutrients in the soil matrix. Organic matter in soil is also play a vital role for enhancing the plant nutrients supply, CEC, soil aggregation, water holding capacity, soil biological activity (Dudal and Deckers 1993) and created a favorable environment for plant growth but also influenced various processes which are governing the establishment of soil-based environmental services (Vanlauwe 2004). An explanation, soil OC is responsible for the variable charge of the soil particles. Thus, with reduced soil organic carbon, CEC decreased as well. Interaction effect among irrigation and nutrient showed no significant difference with respect to soil pH and CEC except EC and OC of the second-year experiment.

5.3.3 Status Available Primary Nutrients in Soil

In general, the availability of primary nutrients in soil was significantly increasing up to 150 DAS with different irrigation schedules-based treatments; thereafter it was declining significantly at harvest of turmeric. Irrigation scheduling of I_2 and I_3 indicated a higher amount of available N, P and K content in soil at end of experiment. Significantly higher availability of N, P and K were recorded under irrigation schedules of I_2, which was at par with irrigation schedules of I_3. However, the least amount of available N, P and K content in soil were observed under I_4 (Table 5.2). The application of nutrient management practices were significantly influenced the available N, P and K content in soil. On an average, the available N, P and K content

Table 5.1 Effect of different levels of irrigation scheduling and nutrients sources on pH, EC, OC and CEC of soil

Treatments	pH		EC (dS m^{-1})		OC (g kg^{-1})		CEC (cmol (p^+) kg^{-1})	
	2011	2012	2011	2012	2011	2012	2011	2012
Irrigation scheduling								
I_1	6.49	6.54	0.36	0.39	5.54	5.57	19.57	20.05
I_2	6.55	6.58	0.44	0.42	5.81	5.86	21.01	22.27
I_3	6.52	6.57	0.42	0.40	5.56	5.72	19.85	20.67
I_4	6.46	6.53	0.35	0.34	5.20	5.29	18.69	19.91
SEm (±)	0.04	0.12	0.01	0.01	0.01	0.01	0.23	0.24
CD at 5%	NS	NS	NS	0.04	NS	0.02	NS	0.81
Nutrient sources								
N_1	6.47	6.53	0.34	0.37	5.51	5.45	18.28	20.04
N_2	6.50	6.56	0.40	0.39	5.49	5.57	20.42	20.71
N_3	6.54	6.57	0.43	0.41	5.58	5.81	20.63	21.42
SEm (±)	0.03	0.03	0.01	0.01	0.01	0.01	0.38	0.16
CD at 5%	NS	NS	0.01	0.01	NS	0.02	1.14	0.47
Interaction								
I_1N_1	6.46	6.51	0.36	0.37	5.43	5.43	17.51	18.68
I_1N_2	6.47	6.53	0.40	0.39	5.53	5.53	20.72	20.35
I_1N_3	6.55	6.57	0.40	0.42	5.67	5.73	20.47	21.10
I_2N_1	6.53	6.55	0.40	0.39	6.07	5.63	19.22	21.79
I_2N_2	6.51	6.62	0.44	0.41	5.67	5.70	21.79	22.13
I_2N_3	6.56	6.61	0.47	0.46	5.70	6.23	22.02	22.88
I_3N_1	6.47	6.53	0.39	0.38	5.40	5.63	19.90	20.79
I_3N_2	6.49	6.60	0.40	0.42	5.60	5.73	19.46	20.19
I_3N_3	6.60	6.57	0.41	0.40	5.67	5.80	20.19	21.03
I_4N_1	6.45	6.50	0.34	0.32	5.13	5.10	16.49	18.88
I_4N_2	6.47	6.51	0.35	0.34	5.17	5.30	19.73	20.18
I_4N_3	6.47	6.55	0.37	0.35	5.30	5.47	19.85	20.66
SEm (±) I × N	0.06	0.05	0.01	0.01	0.02	0.01	0.76	0.32
SEm (±) N × I	0.06	0.12	0.01	0.01	0.02	0.01	0.66	0.35
CD (5%) I × N	NS	NS	NS	0.02	NS	0.04	NS	NS
CD (5%) N × I	NS	NS	NS	0.04	NS	0.04	NS	NS

in soil were increasing significantly with the age of crop up to 150 DAS and thereafter it was reducing significantly at harvest stage irrespective of treatments. N_3 significantly increased the availability of N, P and K content in soil, which was superior to the other treatments (Table 5.2). The above results showed that integrated application of nutrients *i.e.,* 50% through fertilizer and rest through organic manure were more effective for improving available N, P and K content in soil. It might be due to increasing soil organic matter, porosity and water movement into soil (Obi and Ebo 1995). The integration of organic manures in the nutrient management practices has been found to increase the microbial population and microbial activity in the soil matrix leading to enhance the labile pool of carbon and N in soil (Shanthi et al. 2012). Krishnamoorthy et al. (2012) and Karthikeyan et al. (2012) have recorded similar trend in turmeric. The combination of irrigation and nutrient application showed significant difference with respect to available N, P and K content in soil at different growth stage during the both years except at 90 DAS during second year of experiment. The results indicated that interaction of application of water and INM treatment I_3N_3 are necessary for higher storage of available N, P and K content in soil and also buildup soil fertility status (Table 5.2).

5.3.4 Nutrient Content and It's Uptake in Turmeric

The data clearly showed that there was a considerable increment in concentrations of nutrients in rhizome with the use of various irrigation levels (Table 5.3). Concentrations of N, P, and K content were highest with the treatment I_2. Though, under I_4 nutrient concentration was found lowest. From the results, it is clear that under I_4, the crop grown in the prevailing moisture regimes was not sufficiently favorable to support the higher growth of rhizomes and the resulting buildup of N, P and K content in rhizome. In contrast, with irrigation scheduling at I_2, the quantity of applied water and use were fulfilled the crop water requirements at the optimal level, afterward better development of rhizome portions of the turmeric and accordingly the higher concentrations of N and K were influenced (Tripathi et al. 2018). Under the irrigation regimes, the nutrient concentration was high due to the higher amount of available moisture with soluble forms of nutrients in soil matrix and high available to plants (Hamoud et al. 2019). During both the years, the N, P, and K content of rhizome were statistically increased in N_3 treatment as compared to other nutrient sources-based treatments. Results revealed that for higher accumulation of N in crops appeared under incorporation of 50% of RDF + 50% organic manure (FYM + VC). The nutrient content in turmeric by the integrated application of Inorganic and organic source of nutrient supplement was enhanced possibly due to a reduction in the loss of nutrients and higher nutrient availability (He et al. 2014; Zhang et al. 2012). During the first year of the experiment at 90 DAS, the combined effect of irrigation and nutrient supplements was significantly affected N and P content in rhizome. However, K concentration in rhizome was found non-significant. Maximum N, P and K concentrations were observed in the rhizome of the crop in I_2N_3, followed

Table 5.2 Effect of different levels of irrigation scheduling and nutrients sources on soil available N, P and K content

Treatments	Available N content (kg ha^{-1})						Available P content (kg ha^{-1})						Available K content (kg ha^{-1})					
	90 DAS		150 DAS		At harvest		90 DAS		150 DAS		At harvest		90 DAS		150 DAS		At harvest	
	2011	2012	2011	2012	2011	2012	2011	2012	2011	2012	2011	2012	2011	2012	2011	2012	2011	2012
Irrigation																		
I_1	135.77	140.21	150.95	156.28	112.71	114.21	27.05	29.07	39.63	43.44	16.50	18.11	152.27	157.77	178.16	181.17	120.44	124.46
I_2	145.53	148.62	159.36	168.52	118.34	123.07	30.11	32.36	47.76	49.64	19.88	22.78	158.54	165.61	185.22	189.40	127.75	132.85
I_3	141.91	144.49	155.23	164.10	115.77	118.64	29.17	31.15	42.87	45.73	18.33	20.67	155.69	162.42	182.26	185.25	124.20	129.66
I_4	126.8	129.7	140.44	145.14	102.78	105.47	23.46	25.55	33.96	35.90	13.67	15.42	142.16	148.33	NS	170.51	NS	116.55
SEm (±)	0.36	0.54	0.70	2.09	0.65	0.89	0.58	0.56	1.45	2.09	0.77	0.67	0.41	0.15	0.61	0.39	0.46	0.29
CD at 5%	1.26	1.87	2.41	7.23	2.25	3.09	2.00	1.93	5.00	7.23	2.67	2.33	1.40	0.52	2.13	1.33	1.59	1.01
Nutrient sources																		
N_1	132.52	135.34	146.09	151.61	110.80	109.59	24.54	26.73	34.90	37.35	14.67	16.92	147.26	152.32	173.62	175.97	115.12	119.25
N_2	137.19	140.47	151.21	158.25	110.47	115.13	27.24	29.53	41.36	43.88	17.08	19.00	152.17	159.64	178.55	182.02	121.10	126.24
N_3	142.80	146.44	157.19	165.67	115.93	121.32	30.58	32.33	46.91	49.81	19.54	21.82	157.06	163.64	183.21	186.77	126.55	132.16
SEm (±)	0.22	0.20	0.19	0.81	0.21	0.26	0.23	0.35	0.47	0.59	0.44	0.19	NS	0.19	0.26	NS	0.31	0.31
CD at 5%	0.66	0.61	0.57	2.44	0.62	0.77	0.68	1.04	1.40	1.77	1.31	0.56	0.48	0.56	0.78	0.81	0.93	0.92
Interaction																		
I_1N_1	130.29	133.70	144.44	148.70	113.32	107.90	24.00	25.88	33.85	36.78	12.67	15.00	147.82	151.31	174.07	177.33	115.09	118.47
I_1N_2	135.86	140.69	151.43	156.68	109.32	114.72	27.27	29.03	39.09	42.64	17.07	17.67	151.60	159.45	179.18	181.08	120.85	124.53
I_1N_3	141.16	146.24	156.99	163.45	115.47	120.00	29.88	32.29	45.95	50.91	19.77	21.67	157.40	162.57	181.24	185.10	125.39	130.39
I_2N_1	138.95	141.76	152.50	158.95	112.86	115.94	25.93	28.68	36.96	39.94	17.33	21.33	151.64	156.17	179.13	179.84	119.54	123.12
I_2N_2	144.43	147.39	158.14	167.14	117.26	121.45	29.67	32.28	48.00	48.98	19.59	21.33	158.96	168.35	185.31	190.69	127.89	134.29
I_2N_3	153.21	156.70	167.44	179.46	124.91	131.82	34.74	36.12	58.31	60.00	22.71	25.67	165.01	172.30	191.22	197.67	135.81	141.14

(continued)

Table 5.2 (continued)

Treatments	Available N content (kg ha^{-1})						Available P content (kg ha^{-1})						Available K content (kg ha^{-1})					
	90 DAS		150 DAS		At harvest		90 DAS		150 DAS		At harvest		90 DAS		150 DAS		At harvest	
	2011	2012	2011	2012	2011	2012	2011	2012	2011	2012	2011	2012	2011	2012	2011	2012	2011	2012
I_3N_1	136.59	139.01	149.75	156.96	113.01	113.19	26.21	28.66	38.33	40.30	16.67	18.00	150.91	158.05	177.59	181.06	118.93	124.00
I_3N_2	142.11	144.62	155.36	164.42	114.60	118.83	28.64	31.05	44.27	47.58	18.00	21.00	155.35	161.91	182.15	185.12	123.95	129.53
I_3N_3	147.03	149.84	160.58	170.90	119.68	123.91	32.65	33.73	46.00	49.31	20.33	23.00	160.80	167.29	187.05	189.57	129.74	135.46
I_4N_1	124.25	126.91	137.66	141.82	103.99	101.34	22.00	23.71	30.44	32.38	12.00	13.33	138.67	143.76	163.67	165.63	106.93	111.41
I_4N_2	126.35	129.18	139.92	144.74	100.70	105.49	23.37	25.77	34.08	36.31	13.67	16.00	142.76	148.84	167.57	171.18	111.70	116.61
I_4N_3	129.79	133.00	143.74	148.84	103.64	109.56	25.03	27.17	37.36	39.02	15.33	16.93	145.04	152.39	173.34	174.73	115.26	121.63
SEm (±) I × N	0.44	0.41	0.38	1.63	0.41	0.51	0.45	0.69	0.93	1.18	0.87	0.37	0.32	0.37	0.52	0.54	0.62	0.62
SEm (±) N × I	0.51	0.63	0.76	2.48	0.73	0.99	0.69	0.80	1.63	2.30	1.05	0.74	0.48	0.34	0.75	0.59	0.68	0.58
CD (5%) I × N	1.31	1.21	1.15	4.88	1.24	1.53	1.35	NS	2.80	3.53	1.80	1.12	0.96	1.11	1.56	1.62	1.85	1.85
CD (5%) N × I	1.65	2.11	2.58	8.24	2.47	3.33	2.28	NS	5.49	7.78	2.14	2.50	1.60	1.05	2.47	1.87	2.19	1.81

by I_3N_3 (Table 5.3). These results, thereby, signified that I_2N_3 is helpful for higher amount of N accumulation in different parts of the turmeric crop. In soil, fertilizer placement improves fertilizer acquisition by plants (Nkebiwe et al. 2016) and the risk of nutrient loss diminishes throughout the environment.

Except for K uptake, uptake of N and P showed a rising trend up to the harvest stage. In all the crop growth stages notably maximum N, P, and K uptake were registered with I_2 which was intimately followed by I_3 (Table 5.4). In contrast under the conventional rainfed situations significantly the lowest N, P, and K uptake by rhizome portion were obtained. The efficient consumption of nutrients under irrigation scheduling could be enhanced by better moisture regimes, which improved root growth, thus, resulting in higher nutrient uptake (Kumar and Dey 2011). In N_3 treatment, addition of nutrition statistically enhanced N, P, and K uptake of rhizome as compared to N_1 and N_2 (Srinivasan et al. 2016). Sheshbahreh et al. (2019) reported positive and significant correlation between absorption of N and P uptake and was not correlated with K (Dobermaann and Fairhurst 2000a). Among different interaction levels the I_2N_3 provided greater accumulation of N, P, and K by the rhizome portion of turmeric, followed closely by I_3N_3. Under I_4N_1 treatment significantly lower uptake of N, P and K were observed (Table 5.4). Under the rainfed condition or higher interval of two conjunctive irrigations, the uptake of nutrients was decreased with the drying of surface soil (Ramireddy et al. 1982).

5.3.5 *PFP*

During both years of study irrigation scheduling significantly affected the PFP of applied nutrients. Under I_2, during first and second year, the maximum value of PFP-N was recorded 153.27 and 165.44 kg kg^{-1}. During first year, it was recorded statistically at par with I_3 while, it was significantly superior to I_3 during second year (Fig. 5.2). The PFP-P and PFP-K also have similar trends to PFP-N. Under different water regimes, the PFP was improved due to the higher production (Rahaman and Sinha, 2013). PFP increased with increasing water application which favors nutrient uptake and a higher rate of photosynthesis. Under nutrient management, during both years, the maximum PFP-N, P, and K were accounted for in N_3 followed by N_2. While, during both years the least PFP of nutrients was obtained under N_1 (Fig. 5.2). Through organics, partial nutrients supplementation might be achieved higher PFP (Dobermann and Fairhurst 2000b). Under balanced fertilizer, the PFP of applied nutrient was significantly higher as compared to the controlled fertilizer under maize (Zhang et al. 2016). The interaction effect of Irrigation and nutrient level statistically affected PFP-N, P and K. The PFP-N, P and K were recorded lowest with I_4N_1 followed by I_4N_2 and I_4N_3 (Fig. 5.2). Results indicated that under rainfed conditions, the PFP of applied nutrients was lower and it was increased by using the addition of organic manure with fertilizer. During both years of the study the maximum PFP was found with I_2N_3 treatment, and it was significantly superior to other interaction treatments of irrigation and nutrient. The reduction in PFP during both years can be

Table 5.3 Nutrient content by turmeric rhizome as influenced by water regime and nutrient management

Treatment	N content (%)						N content (%)						K content (%)					
	At 90 DAS		At 150 DAS		At harvest		At 90 DAS		At 150 DAS		At harvest		At 90 DAS		At 150 DAS		At harvest	
	2011	2012	2011	2012	2011	2012	2011	2012	2011	2012	2011	2012	2011	2012	2011	2012	2011	2012
Irrigation schedules																		
I_1	1.00	1.06	1.22	1.23	1.39	1.43	0.42	0.46	0.48	0.50	0.54	0.59	1.61	1.65	1.72	1.79	1.43	1.47
I_2	1.17	1.22	1.36	1.39	1.50	1.55	0.56	0.59	0.57	0.57	0.67	0.74	1.69	1.75	1.82	1.85	1.55	1.60
I_3	1.10	1.18	1.29	1.32	1.47	1.51	0.55	0.57	0.55	0.56	0.62	0.65	1.65	1.71	1.78	1.82	1.49	1.53
I_4	0.92	0.97	1.12	1.15	1.27	1.31	0.31	0.36	0.48	0.50	0.44	0.50	1.50	1.53	1.60	1.69	1.33	1.36
SEm (±)	0.020	0.010	0.014	0.023	0.006	0.018	0.011	0.014	0.003	0.010	0.014	0.015	0.014	0.014	0.014	0.019	0.012	0.014
CD (5%)	0.068	0.034	0.049	0.080	0.020	0.061	0.039	0.048	0.009	0.033	0.050	0.053	0.049	0.047	0.050	0.064	0.042	0.050
Nutrient source																		
N_1	0.98	1.03	1.19	1.22	1.35	1.39	0.39	0.43	0.49	0.51	0.50	0.57	1.54	1.60	1.68	1.75	1.40	1.43
N_2	1.03	1.09	1.23	1.26	1.39	1.43	0.43	0.46	0.50	0.52	0.55	0.61	1.60	1.65	1.72	1.77	1.44	1.48
N_3	1.09	1.19	1.31	1.34	1.48	1.53	0.55	0.59	0.56	0.59	0.64	0.68	1.69	1.74	1.79	1.84	1.51	1.56
SEm (±)	0.005	0.004	0.007	0.009	0.006	0.011	0.007	0.006	0.004	0.004	0.006	0.007	0.007	0.010	0.010	0.012	0.007	0.009
CD (5%)	0.015	0.013	0.021	0.027	0.019	0.032	0.021	0.017	0.012	0.011	0.019	0.022	0.020	0.030	0.031	0.037	0.022	0.026
Interaction																		
I_1N_1	0.96	0.98	1.17	1.19	1.36	1.37	0.39	0.43	0.45	0.50	0.48	0.53	1.55	1.62	1.74	1.81	1.39	1.43
I_1N_2	1.01	1.09	1.23	1.23	1.40	1.44	0.42	0.44	0.47	0.48	0.56	0.61	1.60	1.66	1.71	1.77	1.43	1.48
I_1N_3	1.03	1.10	1.25	1.26	1.41	1.47	0.45	0.50	0.52	0.53	0.58	0.63	1.68	1.67	1.73	1.79	1.45	1.49
I_2N_1	1.04	1.10	1.26	1.28	1.40	1.45	0.45	0.46	0.52	0.54	0.56	0.65	1.61	1.66	1.73	1.78	1.48	1.51
I_2N_2	1.09	1.17	1.30	1.33	1.46	1.49	0.50	0.51	0.53	0.54	0.62	0.68	1.66	1.71	1.80	1.83	1.51	1.54
I_2N_3	1.11	1.38	1.52	1.55	1.65	1.71	0.62	0.64	0.65	0.67	0.82	0.88	1.80	1.88	1.92	1.94	1.67	1.74

(continued)

Table 5.3 (continued)

Treatment	N content (%)						N content (%)						K content (%)					
	At 90 DAS		At 150 DAS		At harvest		At 90 DAS		At 150 DAS		At harvest		At 90 DAS		At 150 DAS		At harvest	
	2011	2012	2011	2012	2011	2012	2011	2012	2011	2012	2011	2012	2011	2012	2011	2012	2011	2012
I_3N_1	1.06	1.11	1.25	1.28	1.40	1.44	0.46	0.48	0.48	0.50	0.57	0.62	1.57	1.63	1.70	1.78	1.43	1.48
I_3N_2	1.11	1.14	1.28	1.32	1.45	1.48	0.50	0.51	0.49	0.51	0.60	0.65	1.64	1.70	1.77	1.82	1.49	1.53
I_3N_3	1.14	1.28	1.33	1.37	1.56	1.59	0.69	0.72	0.53	0.56	0.71	0.74	1.73	1.80	1.87	1.88	1.54	1.60
I_4N_1	0.88	0.93	1.08	1.11	1.23	1.28	0.27	0.33	0.47	0.47	0.41	0.47	1.44	1.48	1.56	1.64	1.29	1.32
I_4N_2	0.92	0.96	1.11	1.15	1.26	1.31	0.31	0.37	0.48	0.50	0.43	0.49	1.48	1.52	1.60	1.68	1.33	1.36
I_4N_3	0.96	1.01	1.15	1.19	1.31	1.35	0.35	0.38	0.51	0.52	0.48	0.53	1.56	1.59	1.65	1.75	1.38	1.40
SEm (±) I × N	0.010	0.009	0.014	0.018	0.013	0.022	0.014	0.012	0.008	0.007	0.013	0.015	0.013	0.020	0.021	0.025	0.015	0.017
SEm (±) N × I	0.021	0.012	0.018	0.027	0.012	0.025	0.016	0.017	0.007	0.011	0.018	0.019	0.018	0.021	0.022	0.028	0.017	0.020
CD (5%) I × N	0.030	0.026	0.043	0.054	0.038	0.065	0.042	0.035	0.025	0.022	0.038	0.044	NS	0.060	0.063	0.075	0.044	0.052
CD (5%) N × I	0.072	0.040	0.060	0.091	0.037	0.081	0.051	0.056	0.022	0.038	0.059	0.064	NS	0.068	0.071	0.088	0.055	0.065

Table 5.4 Nutrient uptake by turmeric rhizome as influenced by water regime and nutrient management

Treatment	N uptake (kg ha^{-1})						P uptake (kg ha^{-1})						K uptake (kg ha^{-1})					
	At 90 DAS		At 150 DAS		At harvest		At 90 DAS		At 150 DAS		At harvest		At 90 DAS		At 150 DAS		At harvest	
	2011	2012	2011	2012	2011	2012	2011	2012	2011	2012	2011	2012	2011	2012	2011	2012	2011	2012
Irrigation schedules																		
I_1	20.20	25.51	51.91	55.73	70.53	77.80	5.40	8.73	20.47	22.82	27.52	32.27	32.44	39.88	73.53	81.00	72.63	79.86
I_2	26.78	38.14	60.92	70.53	83.32	93.17	7.12	11.63	24.45	27.96	37.31	44.30	41.99	54.78	86.12	95.63	81.47	93.56
I_3	26.55	33.87	56.92	63.48	78.63	88.36	5.10	11.19	22.25	24.90	33.14	38.14	38.50	49.09	79.56	89.97	78.91	87.21
I_4	12.43	13.92	44.75	46.70	58.20	67.79	4.70	6.46	19.39	20.21	20.22	25.77	20.27	22.10	64.20	70.21	61.54	68.56
SEm (±)	0.67	0.74	1.61	1.81	0.83	1.11	0.16	0.39	0.47	0.65	0.75	0.89	2.11	0.89	1.78	2.18	0.92	1.24
CD (5%)	2.29	2.56	5.58	6.25	2.87	3.84	0.19	1.35	1.62	2.24	2.59	3.08	7.30	3.07	6.16	7.54	3.20	4.30
Nutrient source																		
N_1	15.09	16.81	48.94	51.48	65.17	72.74	4.42	7.62	19.94	21.34	24.58	29.79	23.64	26.12	69.15	73.96	67.79	71.14
N_2	21.39	27.60	53.13	59.23	71.18	79.71	6.45	8.89	21.43	23.92	28.47	33.92	33.20	41.65	74.23	83.17	73.70	82.31
N_3	27.51	38.59	58.80	66.62	81.67	92.90	8.74	12.00	23.54	26.66	35.60	41.64	42.87	56.14	83.40	94.29	80.20	90.61
SEm (±)	0.34	0.46	0.65	0.78	0.77	0.69	0.18	0.21	0.31	0.27	0.52	0.42	1.26	0.67	0.82	1.09	0.96	0.64
CD (5%)	1.13	1.38	1.95	2.34	2.31	2.06	0.53	0.64	0.93	0.81	1.55	1.25	3.76	2.02	2.47	3.28	2.89	1.91
Interaction																		
I_1N_1	12.97	13.18	47.37	48.47	65.44	70.44	6.05	7.87	18.17	20.44	23.16	27.13	20.82	21.87	70.18	73.66	67.41	73.12
I_1N_2	20.89	27.60	52.51	57.27	70.48	75.33	6.16	8.37	20.03	22.31	28.16	31.95	33.16	41.94	72.99	82.02	72.09	77.35
I_1N_3	27.21	37.06	55.84	61.44	75.68	87.64	6.35	9.94	23.20	25.71	31.24	37.73	44.18	56.38	77.44	87.32	72.40	85.11
I_2N_1	18.80	22.36	53.12	58.13	70.63	79.49	6.63	7.88	22.83	24.71	28.37	35.56	29.17	33.81	73.22	80.54	71.77	79.56
I_2N_2	26.91	36.76	58.29	67.39	80.77	89.16	8.17	10.85	24.54	28.08	34.61	40.39	41.20	53.48	80.94	92.28	78.83	92.02
I_2N_3	35.37	58.43	71.36	86.08	98.57	110.86	10.48	15.15	25.97	31.09	48.96	56.94	56.90	79.48	99.74	112.31	90.26	107.88

(continued)

Table 5.4 (continued)

Treatment	N uptake (kg ha^{-1})						P uptake (kg ha^{-1})						K uptake (kg ha^{-1})					
	At 90 DAS		At 150 DAS		At harvest		At 90 DAS		At 150 DAS		At harvest		At 90 DAS		At 150 DAS		At harvest	
	2011	2012	2011	2012	2011	2012	2011	2012	2011	2012	2011	2012	2011	2012	2011	2012	2011	2012
I_3N_1	18.29	20.86	52.52	55.25	73.13	78.00	5.67	8.97	20.36	21.48	29.61	33.31	27.17	30.62	74.64	79.74	71.78	76.80
I_3N_2	27.01	34.95	57.17	65.18	76.47	86.64	7.82	10.18	22.03	24.95	31.79	37.84	40.12	51.96	78.84	89.63	77.55	88.49
I_3N_3	35.23	47.25	61.06	70.02	86.30	100.45	9.66	14.43	24.37	28.26	38.03	43.25	49.15	66.17	86.09	95.19	83.48	91.68
I_4N_1	10.93	11.84	42.75	44.08	51.49	63.03	3.80	5.76	18.41	18.74	17.20	23.16	17.95	18.86	61.41	64.86	54.32	63.14
I_4N_2	12.12	13.37	44.55	47.07	56.98	67.70	3.97	6.16	19.14	20.32	19.31	25.51	19.60	21.22	64.15	68.75	60.34	65.40
I_4N_3	14.34	16.72	46.94	48.94	66.12	72.64	4.38	7.46	20.63	21.58	24.16	28.63	23.43	26.43	69.96	75.07	67.03	72.05
SEm (±) I × N	0.78	0.92	1.30	1.56	1.54	1.38	0.26	0.43	0.62	0.54	1.04	0.83	1.31	1.35	1.65	2.19	1.93	1.28
SEm (±) N × I	0.93	1.05	1.93	2.21	1.51	1.58	0.34	0.52	0.69	0.78	1.13	1.12	1.28	1.42	2.23	2.82	1.83	1.62
CD (5%) I × N	2.15	2.75	3.91	4.67	4.63	4.13	0.82	1.28	1.58	1.61	3.10	2.50	3.53	4.05	4.94	6.56	5.78	3.82
CD (5%) N × I	3.07	3.39	6.41	7.30	4.73	5.10	0.96	1.70	1.96	2.60	3.61	3.69	4.12	4.50	7.35	9.23	5.69	5.30

due to some factors like the loss of nutrients by leaching, low soil moisture, fixation of nutrients and their immobilization, and high suction pressure (Vasileva and Pachev 2015; Sheshbahreh et al. 2019).

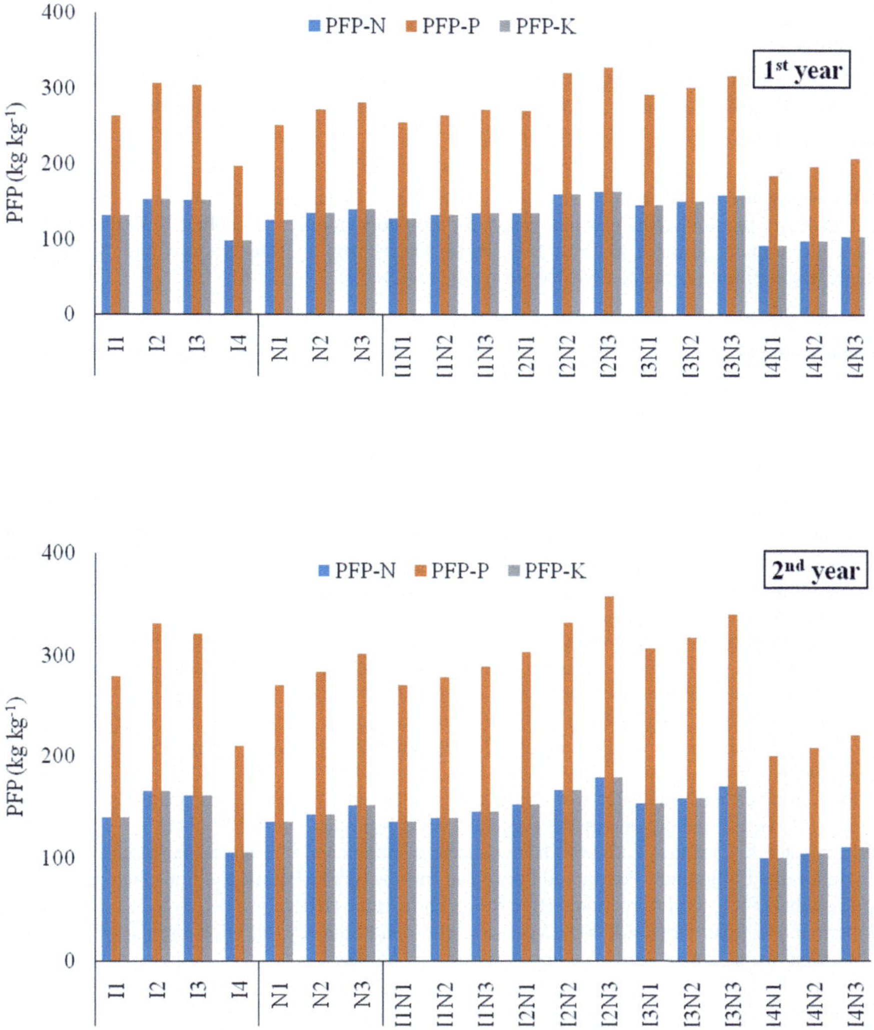

Fig. 5.2 Effect of different levels of irrigation and nutrients on PFP of applied nutrients during 2011 and 2012

5.4 Conclusion

This study showed that the yield, accumulation of nutrients in the rhizome, and PFP were enhanced with the application of irrigation (0.9 IW: CPE) and INM (50% chemical fertilizer + 50% organic fertilizer). This result develops a better understanding of irrigation management with the INM of turmeric crops under the Aeric haplaquept Zone, of eastern India. The result will be helpful for future planning of efficient use of crucial inputs i.e. irrigation water and nutrient management for spices crops in Aeric haplaquept zone of eastern India.

Acknowledgements We would like to thank to faculty of Department of Soil and Water Conservation, Bidhan Chandra Krishi Viswavidyalaya for their cooperation and facilitated for this research. We are extremely grateful to Prof. S.K. Patra, Department of Soil Science and Agricultural Chemistry, for their technical advisory during the research.

Declaration of Competing Interests All the authors agree that there are no interests to declare.

References

Akamine H, Hossain MA, Ishimine Y, Yogi K, Hokama K, Iraha Y, Aniya Y (2007) Effects of application of N, P and K alone or in combination on growth: yield and curcumin content of turmeric (*Curcuma longa* L.). Plant Prod Sci 10:151–154

Bendra B, Senaa F, Xavier L, Marnik V, Mohamed S, Lofti A (2012) Effect of irrigation on soil physico-chemistry; A case study of the Triffa plain (Morocco). Irrig Drain 4:507–519

Chapman HD (1965) Cation-exchange capacity. In: Black CA (ed) Methods of soil analysis—chemical and microbiological properties. Agron J 9:891–901

Choudhary AK, Suri VK (2018) System of rice intensification in promising rice hybrids in north-western Himalayas: crop and water productivity, quality and economic profitability. J Plant Nutr 41:1020–1034

Citak S, Sonmez S (2011) Effects of chemical fertilizer and different organic manures application on soil pH, EC and organic matter content. J Food Agric Environ 9:739–741

Dobermaann A, Fairhurst T (2000a) Rice nutrient disorders and nutrient management, IRRI, Los Banos, the Philippines, Potash and Phosphate Institute (PPI), Potash and Phosphate Institute of Canada (PPIC) and International Rice Research Institute (IRRI), p 191

Dobermann A, Fiarhusrt T (2000b) Assessing nitrogen efficiency. In: Rice nutrient disorder and nutrient management. IRRI Handbook series, pp 155–160

Dudal R, Deckers J (1993) Soil organic matter in relation to soil productivity. In: Mulongoy K, Merckx R (eds) Soil organic matter dynamics and sustainability of tropical agriculture. John Wiley and Son, West Sussex, United Kingdom

El-Hendawy SE, Hokam EM, Schmidhalter U (2008) Drip irrigation frequency: the effects and their interaction with nitrogen fertilization on sandy soil water distribution, maize yield and water use efficiency under Egyptian conditions. J Agron Crop Sci 194:180–192. https://doi.org/10.1111/j.1439-037X.2008.00304.x

Hailu EK, Urga YD, Sori NA, Borona FR, Tufa KN (2018) Sesame yield response to deficit irrigation and water application techniques in irrigated agriculture, Ethiopia. Int J Agron:1–6. Article ID 5084056. https://doi.org/10.1155/2018/5084056

Hamoud YA, Hamoud, Wang Z, Guo X, Shaghaleh H, Sheteiwy M, Chen S, Qiu R, Elbashier MMA (2019) Effect of irrigation regimes and soil texture on the potassium utilization efficiency of rice. Agronomy 9:1–17. https://doi.org/10.3390/agronomy9020100

He FF, Liang YS, Yi ZY, Rong XM, Wu AP, Liu Q (2014) Effect of combined application of manure and chemical fertilizer on the nitrification in acid vegetable soil. Plant Nutr Fert Sci 20:534–540

Hesse PR (1971) A text book of soil chemical analysis. John Murray (Publishers) Ltd., p 50

Jackson (1973) Soil chemical analysis, 2nd ed. Prentice hall of India Pvt. Ltd., New Delhi, pp 111–182

Kamble BM, Rathod SD, Phalke DH (2009) Effect of organic and inorganic fertilizers on yield of soybean: wheat and turmeric and available nutrient status of soil. Adv Plant Sci 22:511–551

Karthikeyan PK, Muthuraman R, Patricia I, Menachem A (2012) Application of potassium and magnesium on turmeric (*Curcuma longa*) to increase productivity in Inceptisols. Arch Agron Soil Sci 58(1):47–50

Krishnamoorthy C, Soorianathasundaram K, Rajamani K, Selvaraj PK (2012) Impact of fertigation on leaf nutrient status and yield attributes in turmeric (*Curcuma longa* L.) cv.BSR 2. Madras Agric J 99(10–12),:739–742

Kumar S, Dey P (2011) Effects of different mulches and irrigation methods on root growth, nutrient uptake, water-use efficiency and yield of strawberry. Sci Hortic 127:318–324

Kumari P, Sharma PK, Kumari R, Singh UP (2021) Effect of tillage and crop establishment practices on performance of rice (*Oryza sativa*) under rice-wheat cropping system. Ann Plant Soil Res 23(2):170–174. https://doi.org/10.47815/apsr.2021.10051

Nair KPP (2013) Turmeric. In: The agronomy and economy of turmeric and ginger, pp 1–5. https://doi.org/10.1016/b978-0-12-394801-4.00001-6

NBSSLUP (2001) Soil series of West Bengal, National Bureau of Soil Survey and Land Use Planning (NBSSLUP), Nagpur, India, pp 260

Nkebiwe PM, Weinmann M, Müller T (2016) Improving fertilizer-depot exploitation and maize growth by inoculation with plant growth-promoting bacteria: from lab to field. Chem Biol Technol Agric 3:1–16

Obi ME, Ebo PO (1995) The effect of organic and inorganic amendments on soil physical properties and maize production in a severely degraded sandy soil in southern Nigeria. Biores Technol 51:117–123

Olsen SR, Cole CV, Watanabe FS (1954) Estimation of available phosphorus in soil by extraction with sodium bi-carbonate. U.S.D.A. Circular Nr 939, US Gov. Print. Office, Washington, D.C.

Pandey DK, Mishra DP, Mustafa M (2011) Response of various level of nitrogen and plant spacing on yield and yield attributing characters of turmeric (*Curcuma longa* L.) at farmers field of Azamgarh, U.P. Progress Hortic 43:320–322

Piper CS (1966) Soil and plant analysis. Academic Press, New York

Rahaman S, Sinha AC (2013) Effect of water regimes and organic sources of nutrients for higher productivity and nitrogen use efficiency of summer rice (*Oryza sativa*). Afr J Agric Res 8:6189–6195

Ramireddy S, Chalam PS, Sankarareddi GH, Raju AP (1982) Effect of irrigation frequency and nitrogen on groundnut yield and nutrient uptake. Plant Soil 65:257–263

Rana KS, Choudhary AK, Sepat S, Bana RS, Dass A (2014) Methodological and analytical agronomy. Post Graduate School, IARI, New Delhi, p 276

Rathod SD, Kamble BM, Pawar VP (2010) Effect of irrigation and levels of fertilizer on yield of turmeric in vertisols irrigated through micro sprinkler. Adv Plant Sci 23:201–203

Sahu SK, Mitra GN (1992) Influence of physico-chemical properties of soil on yield of ginger and turmeric. Fertilizer News 37:59–63

Saleque MA, Abedin MJ, Bhuiyan NI, Zaman SK, Panaullah GM (2004) Long-term effects of inorganic and organic fertilizer sources on yield and nutrient accumulation of lowland rice. Field Crop Res 86:53–65

Sanwal SK, Laxminarayana K, Yadav RK, Rai N, Yadav DS, Mousumi B (2007) Effect of organic manures on soil fertility growth, physiology, yield and quality of turmeric. Indian Journal of Horticulture 64:444–449

Sarhadi J, Sharif M (2014) The effect of deficit irrigation on seasam growth, yield and yield components in drought conditions on base of sustainable agriculture. Int J Farm Allied Sci 3:1061–1064

Shanthi G, Somasundaram E, Muthukrishnan P, Thiyageshwari S (2012) Influence of organic sources of nutrients on soil fertility and quality parameters of turmeric. Madras Agric J 99(7/9):553–555

Sheshbahreh MJ, Dehnavi MM, Salehi A, Bahreininejad B (2019) Effect of irrigation regimes and nitrogen sources on biomass production, water and nitrogen use efficiency and nutrients uptake in coneflower (*Echinacea purpurea* L.). Agric Water Manag 213:358–367

Singh S, Sahoo S, Dash S, Nayak S (2014) Association of growth and yield parameters with bioactive phytoconstituents in selection of promising turmeric genotypes. Ind Crops Prod 62:373–379

Srinivasan V, Thankamani CK, Dinesh R, Kandiannan K, Zachariah TJ, Leela NK, Hamaz S, Shajina O, Ansha O (2016) Nutrient management system in turmeric: effects on soil quality, rhizome yield and quality. Ind Crops Prod 85:241–250

Subbiah BV, Asija GL (1956) A rapid procedure for estimation of available nitrogen in soils. Curr Sci 28(8):259–260

Tripathi SK, Sharma B, Meena SK, Ray R, Patra SK (2015) Influence of irrigation schedules and nutrient sources on yield, quality, water use and water productivity of turmeric under alluvial soils of West Bengal. J Pure Appl Microbiol 9:3109–3114

Tripathi SK, Sharma B, Ray R, Raha P, Denis AF (2018) Performance of turmeric and soil moisture depletion pattern under different water regimes and nutrient sources at New Alluvial Zone of Indo-Gangetic Plains, India. Commun Soil Sci Plant Analysis 49:995–1008. https://doi.org/10.1080/00103624.2018.1448405

Tripathi SK, Sharma B, Ray R, Devi M, Mishra P (2019) Studies on growth and yield of turmeric under different irrigation and nutrient management strategies at West Bengal, India. Bull Environ Pharmacol Life 6:14–20

Tripathi SK, Sharma B, Kumari P, Deb P, Ray R, Denis AF (2021) Evaluation of productivity, quality and economics of turmeric under different moisture regime and integrated nutrient management at Eastern Indo-Gangetic Plains, India. Agric Res. https://doi.org/10.1007/s40003-020-00524-w

Vanlauwe B (2004) Integrated soil fertility management research at TSBF: the framework, the principles, and their application. In: Bationo A (ed) Managing nutrient cycles to sustain soil fertility in Sub-Saharan Africa. Academy Science Publishers, Nairobi, Kenya

Vasileva V, Pachev I (2015) Nitrogen use efficiency and life cycle of root nodules in Alfalfa after different mineral fertilization and soil cultivation. GJESM 1:333–339

Yuan BZ, Nishiyama S, Kang Y (2003) Effects of different irrigation regimes on the growth and yield of drip-irrigated potato. Agric Water Manag 63:153–167

Zhang F, Cui Z, Chen X, Ju X, Shen J, Chen Q, Liu X, Zhang W, Mi G, Fan M, Jiang R (2012) Integrated nutrient management for food security and environmental quality in China. Adv Agron 116:1–40

Zhang D, Pan G, Wu G, Kibue GW, Li L, Zhang X, Zheng J, Zheng J, Cheng J, Joseph S, Liu X (2016) Biochar helps enhance maize productivity and reduce greenhouse gas emissions under balanced fertilization in a rainfed low fertility Inceptisol. Chemosphere 142:106–113. https://doi.org/10.1016/j.chemosphere.2015.04.088

Chapter 6
Rhizospheric Soil: Role on Stress Mitigation, Pollution Control and Crop Enhancement

Pritish Mitra, Asish Mandal, and Sabyasachi Chatterjee

Keywords Plant growth promoting rhizobacteria · Plant–microbe interactions · Rhizosphere · Stress · Sustainable agriculture

6.1 Introduction

Rapid growth in industrial sector is increasing day by day to fulfill the demands of large production of goods for increasing world population. With rapid urbanization and industrialization, anthropogenic activity led climate changes are occurring which increases the chances of production and release of various noxious compounds (Kanwal et al. 2019). And these toxic chemicals can affect the environment as well as daily life of human beings (Singh et al. 2020). These chemical compounds also become threats for organic farming as they input stress of soil composition and disturb the fertile nature of soil. Uncontrolled use of synthesized fertilizers, pesticides, herbicides in agricultural land as well as irrigation practice with untreated polluted or waste water causes soil degradation of the agricultural land (Purakayastha et al. 2008). Crop production has been facing serious problems due to the effect of these stresses including biotic and abiotic. Sustainable strategies should be developed to enhance crop yield per unit area without further hampering the natural resources (Saeed et al. 2021).

In this context use of soil microorganisms more specifically rhizobacteria, resident of rhizosphere has gained momentum. Soil comprised of complex composition of organic and inorganic compounds with various microflora. Soil is a reservoir for microbial community including bacteria, fungi etc. (Fig. 6.1), dispense benefits to

P. Mitra · A. Mandal · S. Chatterjee (✉)
PG Department of Botany, Ramananda College, Bishnupur, Bankura, West Bengal, India
e-mail: schatterjeebiotech@gmail.com

P. P. Adhikary et al. (eds.), *Soil, Water Pollution and Mitigation Strategies*,
Environmental Science and Engineering, https://doi.org/10.1007/978-3-031-63296-9_19

plants, ecosystem with both direct and indirect mechanisms (Ahkami et al. 2017) (Fig. 6.2). Rhizodeposits like sugars, organic acids, amino acids secreted by plants are valuable source of nutrition for soil habitants' microbes (Nautiyal et al. 2008). And these rhizodeposition of nutrients attracts both beneficial and pathogenic strains (Nannipieri et al. 2020). That is why, a "recognition mechanism" has been evolved by plant community to discriminate and select microorganisms and impair around rhizosphere (Chandra et al. 2022). Root exudates of specific molecules, felicitated the shape of microbial community structure and form "rhizosphere microbiome" (Yin et al. 2020). Secondary metabolites like strigolactones, flavonoids and terpenoids are some of plant root exudates used for chemical communication and attract specific microbes, play an important role in plant–microbe interaction (Dessaux et al. 2016; Pascale et al. 2020). Plant roots exudates create vicinity around the plant root region, known as rhizosphere and resident microbes of these region called as Plant Growth Promoting Rhizobacteria (PGPRs).

A mutualistic symbiotic relationship between plant and rhizobacteria directly or indirectly assist the growth of plants and increased per area production. PGPRs significant in promoting plant growth through mobilizing and or solubilising nutrient like fixation of atmospheric nitrogen, solubilisation of soil deposited phosphorous, secretion of EPS, ACC deaminase (1-aminocyclopropane-1-carboxylate) production of siderophore, phytohormones such as auxin, cytokinin. Auxin or IAA functions in cell division and elongation, leads to plant growth. Auxin also helps in slackening of cell wall of root to felicitate increased amount of secretion of root metabolites for better colonization of PGPR microbes (Gupta et al. 2015). PGPRs directly enhance plant tolerance against both biotic and abiotic stress including salinity, drought, nutrient deficiency, metal stress. ACC deaminase, an enzyme produced by PGPR bacteria

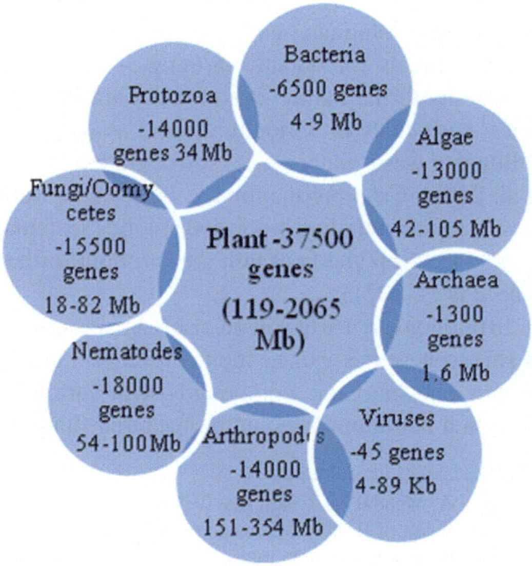

Fig. 6.1 Microorganisms present in Rhizosphere zone (Mendes et al. 2013)

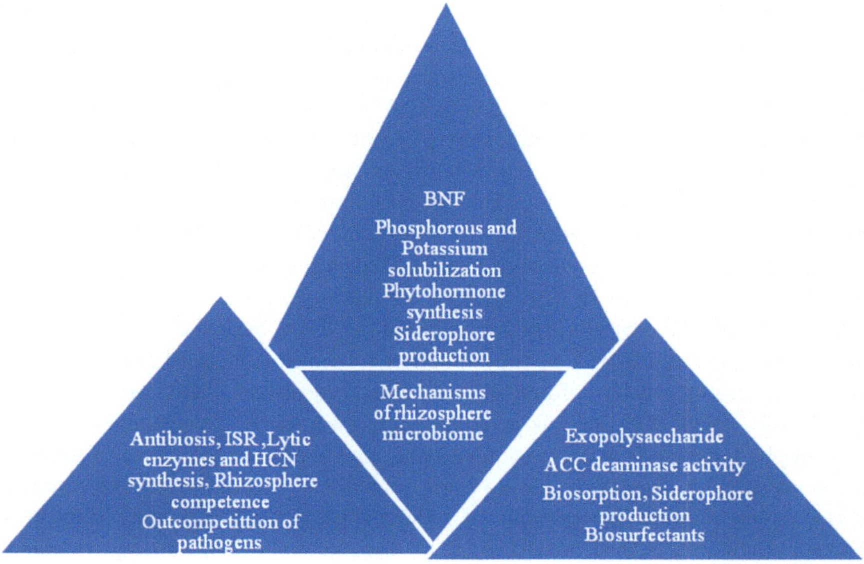

Fig. 6.2 Schematic view of mechanism employed by Rhizobacteria and impact on plant growth by biocontrol, bioremediation, solubilization of unavailable nutrients (Saeed et al. 2021)

aids plants in allaying the stress by ceasing or reducing the synthesis and accumulation of ethylene through converting the synthesis of precursor ACC deaminase into α-butanone acid and ammonia (Danish et al. 2020). PGPR strains like *Pseudomonas* sp., *Azospirillum* sp., *Azotobacter* sp., *Enterobacter* sp., *Bacillus* sp., *Alcaligenes* sp., *Arthrobacter* sp. *Serratia* sp. *Burkholderia* sp. etc. have been isolated from various crop field including rice, wheat, maize, sorghum, medicinal plant, horticultural and vegetable crops and subjected for crop enhancement (Chandra et al. 2022; Kumar et al. 2014; Swain et al. 2007).

6.2 Rhizosphere—A Plant Root Zone

German physiologist Hiltner coined the term "Rhizosphere" and describe about plant-root interface. Rhizosphere is known as close proximity of plant roots with a unique population of microorganisms. Metabolites secreted by plant roots help to increase the population of microorganism. This zone has been divided into three regions on the basis of their relative distance, are named as endorhizosphere, rhizoplane and ectorhizosphere. Endorhizosphere is the nearest zone expanded up to the cortex and endodermic with apoplastic space, is a free space for beneficial microbes and cations. Rhizoplane the second region consist epidermal region of root and mucilage around the root. Ectorhizosphere resides after rhizoplane and ends inside the bulk soil (Chandra et al. 2022). Rhizodeposits are the photosynthetic product of plants, are flushed by plant roots and deposited in the soil. 10–40% of total photosynthetically fixed carbon used by microbes as energy source. Both organic (low molecular weight) and inorganic (such as HCO_3) form of carbon helps to improve the ecological process of the rhizosphere region. Different factors such as plant types, environments and availability of nutrients, physicochemical properties of soil greatly influenced the secretion process of rhizodeposits (Farrar et al. 2003). Root exudates serve as chemical attractant for beneficial soil microorganisms and "bait" for pathogenic bacteria (Hayat et al. 2017). Depending on molecular weight, root exudates are classified as HOC (higher molecular weight organic compounds) and LOC (low molecular weight organic compounds. HOC molecules are not easily utilized by soil microbiota i.e., they function for improvement of soil quality by enhancing aeration, water holding capacity and aggregation of soil particles (Badri and Vivanco 2009). Mucilage is gelatinous and viscous HOC substances excreted through plant roots aids nutrient acquisition to plants, protection of root from desiccation and dry. Organic acids, amino acids, phenolics, proteins, sugar and different secondary metabolites are example of LOC molecule, helps felicitate solubilization and occupancy of both macro (N, P, K) and micro (Fe, Zn) nutrients in plants. Through chemotaxis mechanism LOC function to attract to the soil microbiota to establish a symbiotic colonization between beneficial microbes and plant root. These molecules also buffered the allelopathy process which has detrimental effect against pests, nematodes, harmful for crop.

6.3 Characteristics of Rhizosphere

Apart from biological properties of root exudates of plants, microbe plays an important role in changing the physical and chemical nature of soil (Fig. 6.3). Root exudates like organic acids, secondary metabolites, polysaccharides secreted in rhizosphere region work as a chemoattractant for microorganisms. Soil properties like moisture content, pH greatly changed along with colonization of microbes and released root exudates. Mucilage is a mixture of hexose or pentose sugar and uronic acid help to form rhizosheath and which improves the water holding capacity and movement of nutrient in rhizosphere (Chandra et al. 2022). Mucilage also maintains the hydraulic conductivity of soil and promotes soil stability through resisting soil erosion. Availability of nutrient and their absorption in the rhizosphere region also get promoted by production of mucilage. Root exudate in rhizospheric region affects the pH by acidification or alkalinisation and in turn immobilized nutrients to make them available to the plants. Organic acids released by wheat and maize roots in osmotic stress condition attracts signal stress removing microorganisms and help to uptake nutrient from soil. Changes in chemical condition in the rhizosphere region also dependent of dissolution of insoluble minerals which usually desorbed from clay matter (Pii et al. 2015). Nitrogen availability in soil is generally low but microorganisms make it available by fixing atmospheric nitrogen to ammonia, nitrite and nitrate through

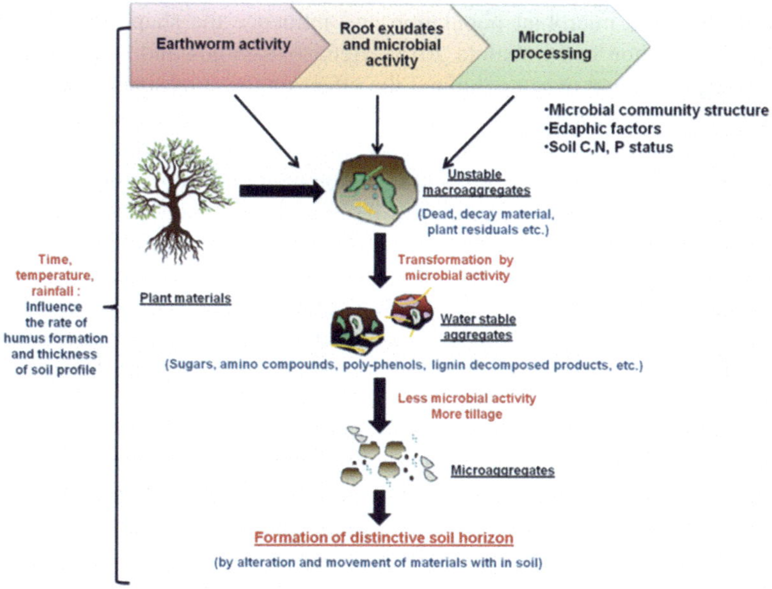

Fig. 6.3 Role of microbes in soil formation (Kaviya et al. 2019)

dinitrogenase activity. Ammonium ions in the rhizospheric region induce the root to excrete proteins which combined with NH_4^+ available in the soil environment and reduce the rhizospheric pH. Conversely the available NO_3^- of rhizospheric region helps to release of bicarbonate ion from root exudates and increases the pH. Some organic acids, amino acids like malic acid, lactic acid, citric acid effectively changes the pH of soil and help to chelates mineral ions to make them available for plants (Rai et al. 2020; Pantigoso et al. 2020).

6.4 Rhizosphere Microbiome and Its Dynamics

Rhizospheric microbiome's characteristics are found to be strictly specific to plant types because different plant secretes different root exudates and attract beneficial microbes. Different genotype of plants leads to variation in microbiome according to liberation of root exudates. Several researchers reported about how root exudates influences (Table 6.1) the microbial community around the roots in rhizospheric zone, leads to rhizosphere microbiome. It's also been founded that microbial diversity is much greater in rhizospheric zone than bulk soil without roots. Both HOC and LOC exudates from roots which acts as signaling molecules, antimicrobial agents are accountable for rhizospheric microbiome dynamics (Rohrbacher and St-Arnaud 2016). Root exudates secreted by different plants acts as chemo-attractant and attracts PGPRs. Coumarins are secondary metabolites released by root in the rhizosphere region enhance the microbial population and influence the shape of rhizosphere microbiome (Igiehon and Babalola 2018). Benazoxazinoids and canavanine from crops like legumes and maize exhibited the effect on the rhizospheric microbiome (Hayat et al. 2017). *Brachypodium* on barley root system architecture define structure of root including types, root lengths, root hairs, which had influence on shaping of rhizospheric microbial communities. Different pattern of root exudates secretion also varies with distribution of microbial biomass in the plant root zone (Pascale et al. 2020). Chemotaxis helps to attract microorganism by the secretion of signaling molecule as root exudates and form colony near the root surface area, are impactful for root elongation. Dynamicity of microbial adhesion near root surface and their movement along the elongated root also restricted for each species. Bacterial dynamicity generally is less in matured roots but higher in elongation zone. Dissemination and chemotactic movement of rhizosphere residence bacteria governs diversity and shift in the microbiome community of rhizosphere (Mendes et al. 2013). Root exudation processes and temporal changes affect curve of dynamics of rhizosphere microbiome according to the different stages of plants (Dessaux et al. 2016). Idiosyncratic rhizospheric microorganisms at different developmental stages of different crops like rice, oats, barley etc. also discussed (Berendsen et al. 2012). In seedling stage oats root secretes sucrose to confer symbiotic relationship with rhizospheric microorganisms while it secretes aromatic compounds, amino acids during vegetative phase to improve its defense system whereas in developmental stage, *Arabidopsis* releases amino acids and enhances root colonization. Nutrient acquisition process, defense system (ISR–induce systemic resistance) also correlated with changes in rhizospheric microbial community in response to the root exudation process (Lu et al. 2020). PGPRs residing inside the root or near root surface area can easily flatter transcriptomes and metabolic variation of roots and shoots of host plants (Chandra et al. 2022). Various secondary metabolites as root exudates help to reform or reshape the architect of rhizospheric microbiome. Chemo-attractant like malic acid also characterized as signaling molecule for disease in infected plants (Doornbos et al. 2012). Root exudates like rosmarinic acid, a derivative of caffeic acid released by *Ocimum basilicum* functions as antimicrobial agent against multiple soil borne pathogens. Rhizospheric microbiome of strawberry provides upgraded nutrient uptake with resistance against soil-borne pathogenic strains (Lazcano et al. 2021). Plants also have the specificity to harbor specific types of microbes. Rhizosphere region of wheat proliferate the colonization of *Pseudomonas*, have the capability to produce antibiotic like 2,4-diacetylphloroglucinol in response to fungal disease caused by *Gaeumannomyces graminis* var. *tritici* (Paulin et al. 2017). *Bacillus subtilis* association got flourish when *Arabidopsis* infected with *Pseudomonas syringae* (Chandra et al. 2022). Bacterial quorum sensing process involved in both positive and negative interaction between microbes and rhizosphere region, modulated by secondary metabolites like naphthoquinones, rosemarinic acid.

Table 6.1 Different type of root exudates with their function

Sl. No.	Root exudates	Functions	Compounds
1	Alanine, Arginine, Asparagine, Aspartic acid, alpha alanine, β-alanine, cystathionine, cystine/cysteine, glycine, homoserine, lysine, leucine/isoleucine, methionine, phenylalanine, proline, serine, threonine, tryptophan, tyrosine	Efficient in metal binding through complexing with metal ions	Amino compounds
2	Biotin, choline, inositol, niacin, N-methyl nicotinic acid, pantothenate, pyridoxine	These components enhance the mobilization of plant micronutrients in soils	Amino acids/ compounds
3	Arabinose, fructose, galactose, glucose, maltose, oligosaccharides, raffinose, rhamnose, ribose, sucrose, xylose	Provide energy source and supports growth of soil microflora	Carbohydrates/ Sugars
4	Amylase, Invertase, Phosphatase, Polygalacturonase, protease	Improved and enhanced nutrient availability	Enzymes
5	Campesterol, cholesterol, linolenic acid, linolenic, oleic, palmitic, stearic acid, sitosterol, stigmasterol	Acts as signaling molecules	Fatty acids and sterols
6	Flavonols, flavonine, flavones, flavanones, isoflavones, anthocyanidins	Improved communicative strategies through chemical components	Flavonoids
7	Different organic acids like butyric, citric, fumaric, glycolic, malic, malonic, oxalic, succinic, tartaric, valeric	Helps in solubilising mineral nutrients such as phosphorus	Organic acids
8	Caffeic acid, Cinnamic acid, Ferulic acid, Glycosides, Saponins, Salicylic acid, Syringic acid, Vanillic acid	Functions as a signaling molecules for plants	Phenolic compounds
9	Auxins, Gibberellins	Supports the growth of plants and microorganisms through phytohormone secretion	Plant hormones

6.5 Plant and Microbe Interaction in Rhizosphere Region

Colonization of microbes around or inside the plant roots in the rhizosphere zone occurred due to the interaction between plants and microorganisms which is one of the most important ecological events consequently induce plants growth (Philippot et al, 2013) (Fig. 6.4). Interaction can be neutral, competitive, antagonistic, parasitic or mutualistic (Vishwakarma et al. 2020) (Fig. 6.5, Table 6.2). Several microorganisms of plant's rhizospheric zone are beneficial for plants growth and immunity, known as plant growth promoting rhizobacteria (PGPR). This kind of beneficial interaction among plants and microbes can be both non-symbiotic and symbiotic

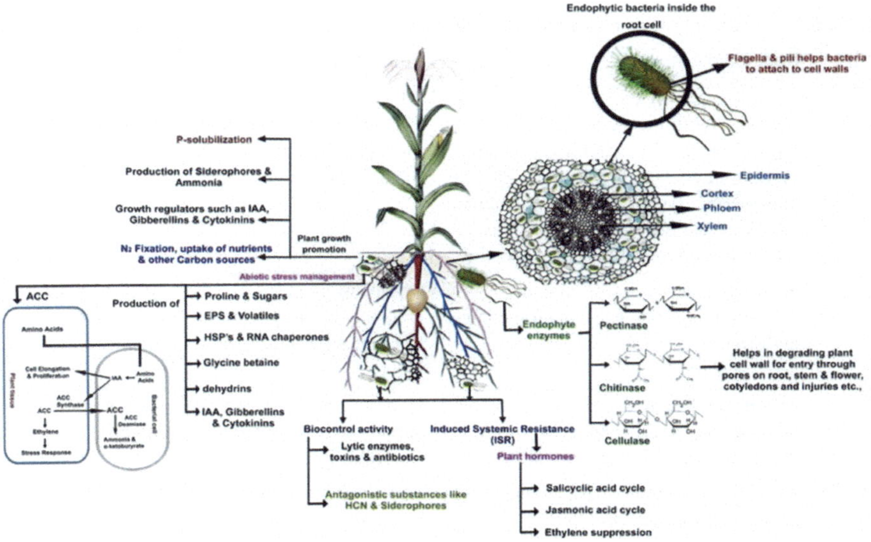

Fig. 6.4 Possible plant–microbe interaction in Rhizospheric region (Vurukonda et al. 2018)

(Bhattacharyya and Jha 2012). Symbiotic interaction involves diazotrophic Rhizobium bacteria, helps to induce the root nodules formation by entering into the root zone and help to fix atmospheric nitrogen to ammonia through nitrogenase enzyme complex and make it available for plants (Terpolilli et al. 2012). *Azospirillum*, a non-symbiotic diazotroph in the rhizospheric zone also fix nitrogen but without entering into the plant root tissues (Chandra et al. 2022). Non-symbiotic bacteria *Azospirillum* promotes growth through nitrogen fixation and also secretes phytohormones like auxin, cytokinin, gibberellin to promote plants growth (Bashan et al. 2004). PGPR aids plants growth through both direct and indirect means. In direct effect they help for acquition of macro and micronutrients from soil via phosphate solubilization, nitrogen fixation, siderophore formation and release of phytohormones (Bhattacharyya and Jha 2012). Indirectly they help to get relief from stress, pathogen attack by enhancing ISR, secretion of stress-related phytohormones like ACC deaminase, cadaverine, and jasmonic acid (Tahir et al. 2019; Singh and Jha 2016, Priyanka Chandra et al. 2022).

6.6 PGPR and Its Characteristics

Plant growth promoting Rhizobacteria, resident of rhizosphere promote growth in plants including various crops by symbiotic association with or on plants. Rhizobacteria not only improve nutrient uptake of plants from environment but also prevents various plant diseases (Zhuang et al. 2007). PGPR bacteria are a new promising

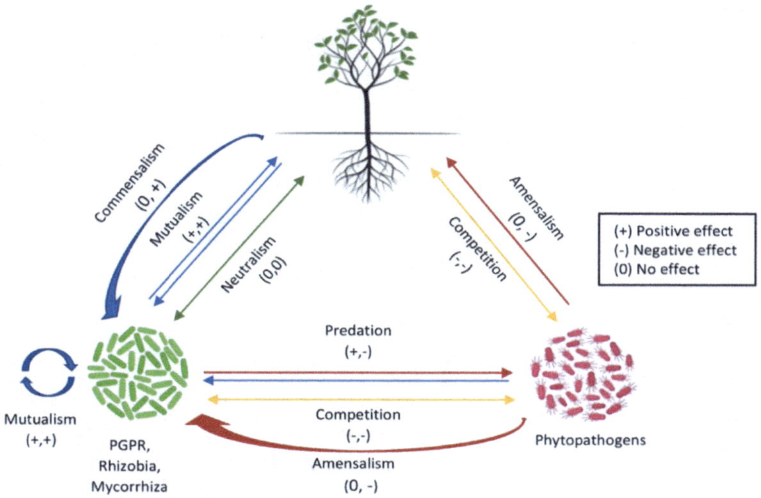

Fig. 6.5 Rhizospheric microbes interacting with plants (Jamil et al. 2022)

Table 6.2 Plant–microbe interaction and their impact

Sl. No.	Type of association	Interaction	Impact on association
1	Antagonistic	Negative	Also known as commensalism where one is producing few substances which are inhibitory to others
2	Competition	Negative	For survival, nutrient resources both plants and microbes put adverse affect on each other
3	Mutualistic	Positive	Both plants and microbes gets benefitted from each other and shows obligatory relationship for metabolism
4	Neutral	–	Both microbes and plant lives in a association but neither microbes nor plant gets affected by each other
5	Parasitic	Negative	One gets benefits from this association as it derive nutrition from living host organism and causes harm to other

and sustainable approach to remediate contaminants like heavy metals from the environment. Characteristics of ideal PGPR candidate are listed in brief in Fig. 6.6.

6.6.1 *Biological Nitrogen Fixation*

Nitrogen is one of the vital macro elements for plant growth and for their metabolic activities. Although nitrogen (N_2) accounts for 78% of all atmospheric gases but amount of nitrogen present in soil is very low and plants are unable to utilize diatomic nitrogenous gases. In biological nitrogen fixation certain convert nitrogen to ammonia

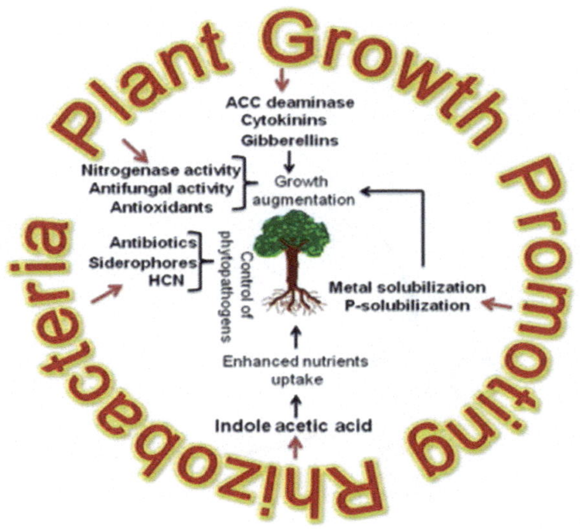

Fig. 6.6 Characteristics of PGPR bacteria (Ahemad and Kibret 2010)

using nitrogenase protein complex of bacteria and archaea involving *nif* genes (Saeed et al. 2021). Various nitrogen fixers like Azotobactor and Rhizobacteria, found naturally are able to fixed nitrogen in soil and prepare for plant use by converting diatomic nitrogen into nitrate and ammonia. These nitrogen fixers can be used as a substitute for commercial fertilizers (Wilson et al. 2006). This group of bacteria classified as non-symbiotic (free living and endophytic organisms) and symbiotic (mutualistic association between microbes and non-leguminous and leguminous plants).

6.6.2 Phosphate Solubilisation

Phosphorous is the 2nd most important macromolecule after nitrogen, plays pivotal role in plant nutrition, photosynthesis, energy transfer by breaking sugar molecule, cell division of meristematic tissue and other metabolic processes. Phosphorus content in soil is excess but not accessible for plant as phosphate ions are highly reactive and becomes immobilized through interacting with cations like Fe^{3+}, Al^{+} in acidic soils and Ca^{+}, Mg^{+} in alkali soils (Ali et al. 2017). Only 0.1% of soil phosphorous is ready to utilize for plant use (Adnan et al. 2019). Rhizobacteria decomposed various organic compounds to available phosphorous for plant growth. *Paenibacillus* and *Bacillus* bacteria are phosphate solubilizing bacteria enhance the phosphorous elements in soils and provide for plant use. In modern agriculture system, rhizobacteria like *Rhizobium* sp., *Pseudomonas* sp., *Bacillus* sp. are found to be beneficial for their phosphate solubilizing activity.

6.6.3 Potassium Solubilization

Potassium is also a required macronutrient for plant growth. On Earth's crust it is the seventh most abundantly found element. Potassium plays a vital role in activation of various enzymes (more than 80 enzymes) involved in different metabolic processes of plants such as nitrate reduction, photosynthesis activity, starch synthesis etc. (Wang et al. 2017). Adequate supply of K in plants increases the activity of plants against diseases, biotic stresses. 98–99% of total soil potassium content is fixed in mineral form known as mineral K and only 1–2% is available to plants (Sparks et al. 1985). Potassium in soil can be present in solution K, mineral K, exchangeable K and non-exchangeable form (Sparks et al. 1985). Microbial community can enhance the nutrient acquisition via mineralization, decomposition. Potassium-solubilizing bacteria (KSB) are able be to release the fixed K from mineral K through production of organic acids (Parmar, Sindhu 2013). Organic acids help to release K from mineral complex by forming complex between K minerals and chelating agent like Ca^{2+}, Fe^{2+}, Al^{3+}, Si^{4+} (Meena et al. 2014). IAA production, characteristics of PGPR increases the root exudate formation, play a vital role in providing potassium to plants (Saeed et al. 2021).

6.6.4 Siderophore Production

Siderophores are iron-chelating small organic molecules formed by microorganisms, which improve the iron uptake of plant in iron-deficient conditions, can be divided into four groups such as hydroxamates, catecholates, carboxylates, and salicylates (Saha et al. 2016; Grobelak and Hiller 2017). Iron is an important micronutrient for plants growth. The amount of Fe available for living organisms is very scanty as predominant form of iron Fe^{3+} is poorly soluble. Both microbes and plants require a fixed amount of Fe for their growth for which they can each other in the root regions (Hider and Kong 2010). PGPR bacteria help to generate low molecular weight organic molecules, known as siderophores which provides iron for plants in iron-limiting environment and increases plants growth (Rasouli-Sadaghiani et al. 2014). Siderophore production by PGPRs in the rhizospheric zone helps to reduce the pathogenic microbes' abundance by retaining free iron and showing better affinity towards rhizospheric Fe^{3+}(Saha et al. 2016; Hakim et al. 2021). *Pseudomonas* strain GRP3, a siderophore-producing bacteria applied in mung bean cultivated in iron deficient soil which results in increased chlorophyll level as well as reduction of the chlorotic symptoms (Braud et al. 2006).

6.6.5 Phytohormone Production

Phytohormones or plant growth regulators are positive organic constituents help to promote plants growth and developments. PGPR help to synthesize and produce phytohormones in plants (Damam et al. 2016). Important phytohormones like auxin, cytokinin, gibberellins, abscisic acid, and ethylene help in different function of plants like roots growth, blooming of root cells by overproduction of adjacent roots which in turn increases nutrient absorption form soil, water holding capability, enduring stress (Sureshbabu et al. 2016). Molecular and biochemical mechanisms of plant growth promoting Rhizobacteria greatly induced by phytohormones which in turn improves plant growth by controlling plant pathogens of Rhizospheric zone. Secondary metabolite like rhizobitoxine improves and enhances plant growth in stress conditions by restricting ethylene production. Induced systematic resistance activated by PGPRs and facilitated via phytohormone signaling, stimulate the manifestation of defense regulatory proteins, pathogenesis-related genes to protect pathogen attacks in plants (Pieterse et al. 2014).

6.6.6 Production of Antibiotics in Soil

Plant growth promoting Rhizobacteria plays a key role in plants growth by combating effect of phytopathogens through the producing of antibiotics in soil. Rhizobacteria associated with development of one or more antibiotics and functions as biocontrol candidate against plant pathogens. Recently the concept of antibiosis or biocontrol-based synthesis of molecules functions as bacteriostatic or bacteriocidal against phytopathogens has become well defined. Antimicrobial substances produced by soil bacteria restrict pathogens by inhibiting cell wall synthesis of pathogens, influencing the structure of cell membrane as well as inhibition of development of small subunit of the ribosome (Maksimov et al. 2011). Phenazines, pyrrolnitrin, phlorogluvinols, pyoluteorin, cyclic lipopeptides in diffusible form, HCN are classified antibiotics functions as biocontrol agents to suppress plant diseases (de Bruijn et al. 2007; Raaijmakers JM et al. 2010). PGPR like *Pseudomonas, Streptomyces, Bacillus,* and *Stenotrophomonas* spp. produces various antimicrobial agents (antibacterial, antifungal, anthelminthic, antiviral) like amphisin, hydrogen cyanide, phenazine, tropolone, oomycin A, tensin, kanosamine, xanthobaccin, oligomycin A, zwittermicin, are effective in killing both gram positive and gram-negative bacteria and pathogenic fungi (Silo-Suh et al. 1994; He et al. 1994).

6.7 Impact on PGPR on Plant Growth, Development and Health

6.7.1 Enhanced Nutrient Acquisition

Accurate amount of nutrient is required for each plant to complete every biological parameter. Rhizospheric microbiome have significant influencing role on nutrient uptake and maintaining status of plants just like conversion of atmospheric nitrogen to utilizable form of nitrogen gas by nitrogen-fixing rhizobia. Ionic form of potassium are mobilised by mycorrhizal fungi and facilitate the potassium uptake (Miransari 2011). Symbiotic associations like mycorrhizal fungi help to translocate nutrients, minerals from soil to the plants (Johnson and Graham, 2013). Its documented and recognized that PGPRs help to generate stable soil aggregates and structuring soil physiology with suppressing soilborne plant pathogens (Whipps 2001; R. Mendes et al. 2013). *Bradyrhizobium*, *Rhizobium* and other various nitrogen-fixing rhizospheric bacteria have identified function towards plants. In Western Amazon the Cowpea plant root region showed higher genetic diversity of symbiont Rhizobial species. Guimaraes et al. (2012) experimented about the nodule formation capability of *Bradyrhizobium, Rhizobium, and Achromobacter* in cowpea plants and their efficiency in nitrogen fixation. Rhizospheric microbiome also played important function in nutrient acquisition of trace elements such as iron. Some bacteria make availability of iron to the plants by producing siderophores. Fluorescent pseudomonads are able to promote iron uptake for both monocotyledonous plants (Gramineae) and dicotyledonous plant species via siderophores formation (Mendes et al. 2013). Rhizoferrin is a siderophore produced by *Rhizopus arrhizus* able to chelates iron and functions as carrier of iron transport to the plants (Yehuda et al. 2000). Bacteria isolated from rhizosphere and ectomycorrhizosphere contribute to plants growth by releasing nutritive cations from soil minerals (both macro and micro minerals) in nutrient poor soils (Fig. 6.7).

6.7.2 Support Under Biotic Stress Condition

PGPRs provide the frontline protection to plants against attack of soilborne pathogens (Cook et al. 1995). Rhizospheric microbiome members' shows antagonistic activity against soilborne pathogens in both before and during the primary, secondary infection spread on/in root tissues. Rhizospheric microorganisms provide various mechanisms like competition for trace elements, nutrient uptake and microsites, antibiosis, parasitism, induced systemic resistance to ward off the plant pathogens (Mendes et al. 2013). Growth or activities of competing microorganisms are inhibited by Rhizobacteria as they are prolific producers of antibiotic (Brakhage and Schroeckh, 2011). Most of the bacterial or fungal strains act as biocontrol agent by producing antibiotic having different degrees of antimicrobial activity (Mendes et al. 2013). Agrocin 84, a

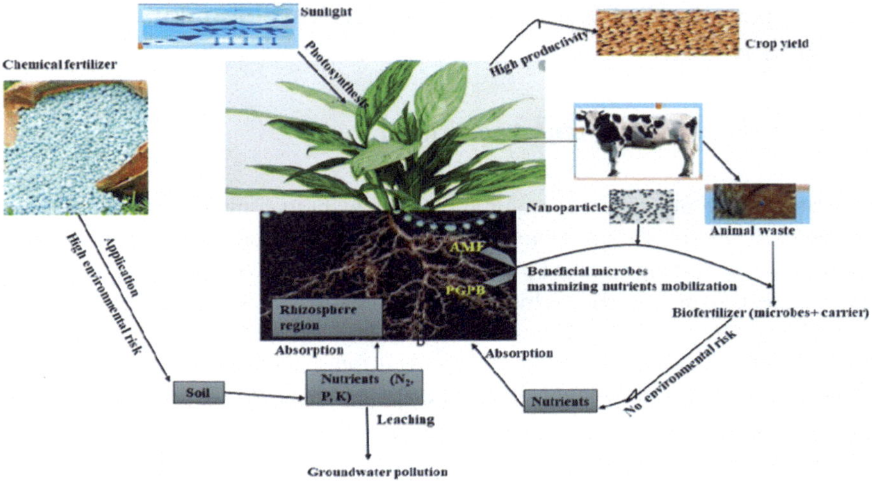

Fig. 6.7 Impact of PGPRs on nutrient acquisition from soil (Fasusi et al. 2021)

bacteriocin produced by *Agrobacterium radiobacter* showed narrow spectrum activities whereas nonribosomal peptide and polyketide are broad-spectrum antibiotic activities (Raaijmakers et al. 2010). Antibiotic compounds have other natural functions like defense against predator protozoa, biofilm production, nutrient uptake etc. (Raaijmakers and Mazzola 2012). Volatile organic compounds (VOCs) are one kind of bacterial metabolites (<300 Da), are gaining its attention for performing growth promoting function and mediating intricate communicative dialogues between plants and rhizosphere microorganisms in soil ecosystem (Effmert et al. 2012; Mendes et al. 2013). Highly vaporised VOCs can diffuse within soil through water and gas mediated pores. Volatile product of various bacterial species like *Bacillus subtilis*, *Pseudomonas trivialis*, *Pseudomonas fluorescens*, *Stenotrophomonas maltophilia*, *Serratia plymuthica* capable to inhibit the mycelial growth of fungal plant pathogens (Zou et al. 2007; Mendes et al. 2013). Bacterial volatile products amended with root exudates without amino acids have mild antifungal activity and strong antibacterial effects but with amino acid it had strong antifungal activity (Mendes et al. 2013). Bacterial volatile compounds are important for promoting growth of ectomycorrhizal fungi and plays role in tripartite interaction between fungi, bacteria and nematodes. For example, *Paenibacillus lentimorbus*, *Paenibacillus polymyxa*, have antifungal activities and during the interference with disease caused by *Fusarium oxysporum*, *Meloidogyne incognita* in tomato plants they also interact with concurrent nematodes infestation (Son et al. 2009). Chermin et al. (2011) reported about the interference of bacterial volatiles against quorum sensing process by suppressing N-acyl-homoserine lactone synthase genes of phylogenetically different bacterial by restricting. VOCs can modulate plant immune system by inducing systemic resistance (ISR) in plants (Cho et al. 2008; Bailly and Weisskopf 2012). Systematic resistance is beneficial for plant growths which are regulated by phytohormones like jasmonic acid, ethylene and sometimes via salicylic acid (Barriuso et al. 2008; Mendes et al. 2013) whereas *Bacillus cereus* AR156 strain activate induce systemic resistance (ISR) by activating both signaling pathways (Niu et al. 2011). Quorum sensing molecules released by rhizobacteria provoked a wild range of plant responses including activation of various defense-related genes such as MPK3, MPK6, WRKY22, WRKY29 (Hartmann and Schikora, 2012). Metabolic and transcriptional changes in plants are made by Rhizobacterial activity via JA/ET or SA pathways. For example, Weston et al. in 2012 showed transcriptional, metabolic profiles of two *P. fluorescens* strains and their reducing effects on carbon content of host plants but in challenged condition with *Pseudomonas* syringae, those two strains provide fitness benefits to the host. Various studies founded that Rhizobacteria help to remove biotic stress by providing immune responses and inducing physiology/ metabolism processes to the host plants (Table 6.3).

6.7.3 Mitigation of Abiotic Stress and Plant Growth

Just like biotic stress rhizospheric microbes also help to mitigate the abiotic stress (Table 6.4) like drought stress, salt stress, and metal stress and provide plants growth by improving physiochemical parameters, immune responses etc.

6.7.3.1 Drought Stress

Photosynthesis, cell proliferation is two most important activities for plants, also affected by drought. Effect of water scarcity on carbon metabolism is likely to cause changes in sugars content used in various cellular signaling processes. Water scarcity as a result of drought condition misdirects electron transfer during light reaction of photosynthetic process which arise ROS formation (Chandra et al. 2021). Nitrogen-fixing bacteria like *Azotobacter*, *Azospirillum* inoculated plants showed a hike in antioxidant enzymatic activities to settle drought stress in stressed plants (Zakikhani et al. 2012) (Table 6.4). PGPR also alleviate the drought by forming antioxidants, cytokinins or by breaking down the ethylene precursor by ACC deaminase (Leontidou et al. 2020a, b). During drought condition PGPR also functions for accumulation of abscisic acid (ABA) in leaves improvement in cytokinin formation to closure of stomata. Several microorganisms synthesize and release Exoploysaccharide, produce proline, improve water circulation in plants, stimulate resistance genes to survive the drought stress (Tiwari et al. 2016). Biofilm production of PGPR bacteria under drought prevents desiccation to improvise the survival rate of plants. Inoculation of microbes like *Pseudomonas*, *Rhizobium* in *Zea maize*, and *Vigna radiata* under drought condition maintained relative water content as well as conferred salt tolerance (Ahmad et al. 2013; Chandra et al. 2022).

Table 6.3 Impact of plant growth promoting Rhizobacterial strains against biotic stress

Sl. No.	PGPRs	Biotic stress	Mechanism to control disease	Crop (s)
1	*Azotobacter* sp., *Azospirillum* sp.	*Rhizoctonia solani*	Antibiosis or antagonistic association between two organisms	Wheat
2	*Bacillus amyloliquefaciens*	*Fusarium graminearum*	Antifungal activity through Iturins, a lipopeptide antibiotic	Wheat
3	*Bacillus* sp.	*Plasmopara viticola*	Antagonistic activity against fungi	Grapevine
4	*Bacillus* spp.	*Pyricularia oryzae*	Antioxidant activity	Rice
5	*Bacillus subtilis*	*Fusarium* sp.	Siderophore activity	Pepper
6	*Bacillus subtilis*	*Bipolaris sorokiniana*	Antioxidant defense enzymes	Wheat
7	*Bacillus siamensis, Bacillus tequilensis*	*Pestalotiopsis versicolor*	Secretion of lipopeptides like iturin, surfactin, mycosubtilin, protease and chitinase activity	Bayberry
8	*Bacillus subtilis, Bacillus simplex*	*Fusarium oxysporum*	Antifungal activity	Pea
9	*Bacillus subtilis, Bacillus amyloliquefaciens*	*Penicillium crustosum, Penicillium digitatum Sacc., Penicillium italicum Wehmer*	Secretion of volatile compounds	Citrus fruit
10	*Pseudomonas* sp.	*Fusarium oxysporum f. Sp. Cepa*	Systematic resistance	Onion
11	*Pseudomonas aeruginosa* and *P. Fluorescence*	*Meloidogyne javanica*	Systematic resistance through salicylic acid secretion	Tomato

(continued)

Table 6.3 (continued)

Sl. No.	PGPRs	Biotic stress	Mechanism to control disease	Crop (s)
12	*P. aeruginosa* and *B. subtilis*	*Pythium*	Secreting antifungal substance such as phenazine, pyochelin, pyoverdins, surfactins, 2,4-DAPG and AHLs	Arugula
13	*Bacillus, Brevibacillus, Pseudomonas*	*Ralstonia solanacearum*	Smooth and improved secretion of defense-related enzymes	Tomato
14	*B. subtilis, B. Amyloliquefaciens, Pseudomonas aeruginosa* and *P. Fluorescence*	*Clavibacter michiganensis subsp. michiganensis*	Production of hydrogen cyanide (HCN), indole acetic acid (IAA) and siderophores	Tomato
15	*Stenotrophomonas maltophilia*	*Fusarium graminearum*	Plant growth enhancing potential and oxidative stress tolerance activity	Wheat
16	*Streptomyces* spp.	*Sclerotium rolfsii*	Secretion and induction of defense regulatory genes	Chickpea

6.7.3.2 Salt Stress

Microorganisms found to help in plant growth under higher salt concentration by adapting salinity stress (Chandra et al. 2020; Rai et al. 2020) (Table 6.4). Halophilic microorganisms maintain enzyme activity for various metabolic pathways, protein structure at higher salt concentrations. The PGPRs has a vital role in removal of salt toxicity in the rhizospheric zone by assisting osmotic balance, ion homeostasis, and also through maintaining the turgor pressure of the plant cells (Dakora and Phillips, 2002). Under the salt toxicity PGPR improve the growth of plants by ameliorating the activity of ACC deaminase, modulating antioxidant enzymes, biosynthesis of biofilms, production of exoploysaccharides (EPS), release of osmoprotectants and phytohormones as well as improving the nutrient availability (Singh and Jha 2016). ACC deaminase enzyme of PGPR bacteria in rhizosphere escalate the synthesis of IAA, zeatin, gibberellins and lowers down the synthesis of ethylene to increase the root cell division and their elongation under salinity stress. EPS in rhizospheric zone help to form biofilms around the plant roots and bind with excess Ca^{2+}, Na^{+}, and

Table 6.4 PGPRs with growth promoting function and potentiality towards abiotic stress mitigation

Sl. No.	Bacterial strain(s)	Mechanism against abiotic stresses	Abiotic stresses
1	*Azospirillum brasilense*	Production of auxin hormone and solubilization of phosphate	Salt
2	*Azospirillum brasilense*	Improved photosynthetic pigments and improved antioxidant enzymes	Water
3	*Azospirillum brasilense*	Production of phytohormones and accumulation of abscisic acid	Drought
4	*Azospirillum lipoferum*	Phytohormones production	Salt
5	*Azospirillum brasilense*, *Azospirillum lipoferum*	Fixation of nitrogen	Salt
6	*Azospirillum* sp.	Production of osmoprotectants and phytohormones	Salt
7	*Azospirillum chroococcum*	Increased K^+/Na^+ ratio as well as polyphenol compound	Salt
8	*Azospirillum chroococcum*	Accumulation of ABA, and production, secretion of growth promoting phytohormones	Drought
9	*Azospirillum* sp.	Nitrogen fixation, Phosphate solubilization, Indole acetic acid production	Salt stress
10	*Bacillus amyloliquefaciens*	Formation of rhizosheath by secreted EPS (exoploysaccharides) to inhibit the influx of sodium	Salinity
11	*Bacillus cereus*	Protease activity	Heavy metal
12	*Bacillus cereus*	Induction of cAPX, rbcL, and rbcS genes	Drought
13	*Bacillus insolitus*	Increased antioxidant enzyme activity in plants to reduce ROS activity	Salinity
14	*Bacillus megaterium*	Induction and incensement of IAA hormone production, solubilization of phosphorus, antioxidant enzymes	Drought
15	*Bacillus pumilus*	Increased ROS- scavenging enzymatic activity and level of proline accumulation	Heavy-metal toxicity
16	*Bacillus* spp.	Production of IAA and ACC deaminase	Drought
17	*Bacillus* sp.	Biosorption and bioaccumulation technique	Chromium
18	*Bacillus safensis*	Increase of anti-oxidant responses	Drought
19	*Bacillus subtilis*	Elevation of Na^+/K^+ ratio	Salt stress
20	*Bacillus thuringiensis*	Biosorption technique	Heavy metals
21	*Bacillus thuringiensis*	Biofilm production	Drought
22	*Enterobacter*	Biosorption	Heavy metal
23	*Enterobacter hormaechei*	Activity of EPS and ACC deaminase	Drought

(continued)

Table 6.4 (continued)

Sl. No.	Bacterial strain(s)	Mechanism against abiotic stresses	Abiotic stresses
24	*Enterobacter* spp.	Production of IAA phytohormone and increase of ACC deaminase enzyme activity	Drought
25	*Pseudomonas aeruginosa*	Introduction of oxidative stress tolerance activity	Zinc toxicity
26	*Pseudomonas aeruginosa*	EPS, Biofilm production as well as bioremoval of metal toxicity	Cadmium toxicity
27	*Pseudomonas aeruginosa*	Increased antioxidant enzyme activity such as peroxidase, catalase, and superoxide dismutase. Up regulating activity of DHN, CAT1 AND DREB2A as well as production of phytohormones like IAA	Drought
28	*Pseudomonas chlororaphis*	Increased IAA production	Salinity
29	*Pseudomonas aurantiaca, Pseudomonas extremorientalis*	Phytohormones like Auxin production	Salinity
30	*Pseudomonas fluorescens*	Phosphate and auxin solubilization	Salinity
31	*Pseudomonas putida*	Adsorption activity	Heavy metal
32	*Pseudomonas putida*	ACC deaminase activity	Salt
33	*Pseudomonas fluorescens, Pseudomonas putida*	Enzymatic activity like ACC deaminase	Drought
34	*Pseudomonas pseudoalcaligenes*	Improved antioxidant activity	Salt
35	*Pseudomonas* sp	Biosorption activity to reduce the toxic state of heavy metal	Heavy metal
36	*Pseudomonas syringae*	Inhibition of sodium by formation of rhizosheath with EPS	Salinity
37	*Pseudomonas syringae, Pseudomonas fluorescens*	Enhanced ACC deaminase enzymatic activity	Salinity
38	*Serratia fonticola*	Secretion of phytohormones and activity of biofilm, ACC deaminase	Normal
39	*Serratia marcescens*	Introduction of systematic resistance, ACC deaminase activity. Modulation of antioxidant enzymatic activity with PGP properties	Salinity
40	*Serratia* sp.	Induction of cAPX, rbcL, and rbcS genes	Drought

K^+ and prevent their uptake in plants. PGPRs present in rhizospheric soil neutralizes the pernicious effect of ROS (Reactive oxygen species) in plants under salt stress by modulating the activity of various antioxidant enzyme such as peroxides, catalase, superoxide dismutase, nitrate reductase glutathione reductase. PGPRs aids accumulation of osmolytes like polyamines, proline, betaine, glycine, and other amino acids to control stress through salt efflux via modifying intracellular environment and cell wall structure to drain out toxic ions (Tahir et al. 2019).

6.7.3.3 Heavy Metal Tolerance

Urbanization, industrialization increases the anthropogenic activities, out-turned the hike of heavy metal concentrations in environment including soil. Rise of HM adversely impactful for plant growth and their survival. Heavy metals cannot be rooted out from environment permanently but toxicity can be reduced or decreased by switching their oxidation state. PGPRs also physiologically evolved to restrict or resist the heavy metal stress (Khanna et al. 2019). There are several machineries used by PGPR candidate to detoxify the effect of heavy metal like capsule formation by polysaccharide layer to entraps the HM, bioaccumulation of heavy metals in the cell wall, use of microbial methylation process to neutralize the HM or precipitation of heavy metal through secreting various neutralizing compounds, oxidation–reduction of heavy metals (Tak et al. 2013; Chandra et al. 2022) (Table 6.4). PGPRs bacteria like *Pseudomonas*, *Bacillus*, and *Streptomyces* are also known as heavy metal tolerant microbes, able enhance the production and growth of plants in heavy metal concentrated areas (Tiwari and Lata 2018).

6.8 Redirecting the Rhizosphere Microbiome in Crop Field by Stimulating Microorganism and Genetic Engineering

Use of rhizospheric beneficial microorganism specifically Rhizobacteria is comparable with human probiotics as it helps to reconstruct rhizosphere microbiome and protect host plant against pathogen attack (Mendes et al. 2013). Numerous strategies proposed for constructing the microbial community of rhizosphere and redirect their activity such as introduction of beneficial microorganisms onto seeds, planting materials and in soil as well as stimulation of rhizosphere resident microorganisms for soil and crop management. Introduction of beneficial fungal and bacterial strains into soil or planting material or onto seeds boost the growth performance of plant (Bhattacharyya and Jha, 2012). Plant probiotics could also eradicate plant pathogens from soil by introducing siderophore-producing bacterium. Short or one time exposure of control agent on agriculture system may not enough to remove the pathogens even if they are in low well counts. But proliferation of beneficial microorganisms in rhizosphere can attain higher cell densities to restrict soil borne pathogens. When isolated PGPRs are introduced into a new environmental condition microorganisms are unable to establish or survive in rhizosphere region, necessary for controlling soil borne plant pathogens. Although there are concerns with recombinant microorganisms one approach can be develop to design so-called plant probiotics through genetical engineering of microbial strains (Picard et al. 2008; Berlec, 2012). Genetic engineering can be a promising tool but time consuming. Various studies revealed application of genetic engineering to improve organic anion efflux, modification of rhizosphere pH or to facilitate citrate efflux (Sasaki et al. 2004; Li et al. 2005; Wang et al. 2007). Instead of one-microorganism policy, treatment with assemblages of more than one complementary or synergistic microbe provides consistent and more effect result in antagonism. Consortia of PGPRs are able to eradicate soil borne pathogens in the rhizosphere region by achieving niche saturation in less time (Mendes et al. 2013). Rhizobacteria always distinguish among the pathogenic and beneficial microbes and increases the biosynthesis of antimicrobial substances to control plant pathogens. Mendes et al. in 2011 used PhyloChip based metagenomic study and detects 33,000 archaeal and bacterial species in root soil region of sugar beet plants and the responsible *Rhizoctonia* for sugar beet damping-off diseases (Mendes et al. 2013). Kinkel et al. in 2011 studied about the effect of coevolutionary framework to manage natural disease in less nutritive soil. They also proposed that *'effective antagonistic populations or phenotypes are likely to vary for different plant pathogens suggesting that distinct evolutionary and coevolutionary trajectories may be significant to disease suppression in different cropping systems.*

6.9 Conclusion

Although there are so many available data about importance of rhizospheric microbiome in growth promotion of plants but in detail knowledge about role of numerous rhizospheric organisms and their approaches are still limited. Advance techniques like next-generation sequencing helps to study the ecological, physiological and molecular insights of microbes in rhizosphere. Characterization of root exudates, signaling molecules and other key players of rhizosphere microbiome will help to elucidate how plants stimulate or recruit beneficial microorganisms. These root exudates are nothing but chemical markers which plays important role in improving soil community and plants health by tolerating the abiotic and biotic stress, improving nutrient availability and soil buffering capacity. Selective accumulation of stress tolerant microbes in rhizosphere also helps to improve defense system of plants through soil-plants-microbes' interactions. Interactions between plants-microbes are gaining their momentum as a sustainable, less hazardous system for crop development in agriculture. Detailed study about interactions among plants roots and microbes are required to understand their exact mechanisms before using PGPRs for agricultural practices.

Acknowledgements The authors gratefully acknowledge the PG Department of Botany, Ramananda College, Bishnupur, Bankura for providing the necessary infrastructure to carry out this review work.

Authors Contribution Each author has made substantial contributions to complete this work. Sabyasachi Chatterjee and Asish Mandal help in conception building and Pritish Mitra involved in data collection and designing this whole review work Manuscript preparation was done by Pritish Mitra, Sabyasachi Chatterjee, and Asish Mandal.

Funding No funding was received for conducting this study.

Competing Interests The authors have no relevant financial or non-financial interests to disclose.

References

Adnan M, Fahad S, Khan IA, Saeed M, Ihsan MZ, Saud S, Riaz M, Wang D, Wu C (2019) Integration of poultry manure and phosphate solubilizing bacteria improved availability of Ca bound P in calcareous soils. 3 Biotech. 9, 1–10

Ahemad M, Kibret M (2010) Mechanisms and applications of plant growth promoting rhizobacteria: current perspective. J King Saud Univ Sci 26(1):1–20, ISSN 1018–3647. https://doi.org/10.1016/j.jksus.2013.05.001

Ahkami AH, White RA, Handakumbura PP, Jansson C (2017) Rhizosphere engineering: enhancing sustainable plant ecosystem productivity. Rhizosphere 3(2):233–243

Ahmad M, Zahir ZA, Khalid M, Nazli F, Arshad M (2013) Efficacy of Rhizobium and *Pseudomonas* strains to improve physiology, ionic balance and quality of mung bean under salt-affected conditions on farmer's fields. Plant Physiol Biochem 63:170–176

Ali MA, Naveed M, Mustafa A, Abbas A (2017) The good, the bad, and the ugly of rhizosphere microbiome. In Probiotics and Plant Health, Springer. Singapore, pp 253–290

Badri DV, Vivanco JM (2009) Regulation and function of root exudates. Plant Cell Environ 32:666–681

Bailly A, Weisskopf L (2012) The modulating effect of bacterial volatiles on plant growth. Plant Signal Behav 7:1–7

Barriuso J, Ramos Solano B, Gutierrez Manero FJ (2008) Protection against pathogen and salt stress by four plant growth-promoting rhizobacteria isolated from *Pinus* sp on *Arabidopsis thaliana*. Phytopathology 98:666–672

Bashan Y, Holguin G, de-Bashan LE (2004) Azospirillum-plant relationships: physiological, molecular, agricultural, and environmental advances (1997–2003). Can J Microbiol 50:521–577

Berendsen RL, Pieterse CMJ, Bakker PAHM (2012) The rhizosphere microbiome and plant health. Trends Plant Sci 17:478–486

Berlec A (2012) Novel techniques and findings in the study of plant microbiota: search for plant probiotics. Plant Sci 193:96–102

Bhattacharyya PN, Jha DK (2012) Plant growth-promoting rhizobacteria (PGPR): emergence in agriculture. World J Microbiol Biotechnol 28:1327–1350

Brakhage AA, Schroeckh V (2011) Fungal secondary metabolites—strategies to activate silent gene clusters. Fungal Genet Biol 48:15–22

Braud A, Jezequel K, Leger MA, Lebeau T (2006) Siderophore production by using free and immobilized cells of two pseudomonads cultivated in a medium enriched with Fe and/ or toxic metals (Cr, Hg, Pb). Biotechnol Bioeng 94:1080–1088

Chandra P, Dhuli P, Verma P, Singh A, Choudhary M, Prajapat K, Rai AK, Yadav RK (2020) Culturable microbial diversity in the rhizosphere of different biotypes under variable salinity. Trop Ecol 61:291–300. https://doi.org/10.1007/s42965-020-00089-3

Chandra P, Wunnava A, Verma P, Chandra A, Sharma R K (2021) Strategies to mitigate the adverse effect of drought stress on crop plants—influences of soil bacteria: a review Pedosphere 31(3):496–509

Chandra P, Rai AK, Sundha P, Basak N, Kaur H (2022) Rhizospheric soil–plant-microbial interactions for abiotic stress mitigation and enhancing crop performance. In: Shit PK, Adhikary PP, Bhunia GS, Sengupta D (eds.) Soil health and environmental sustainability. Environmental Science and Engineering. Springer, Cham. https://doi.org/10.1007/978-3-031-09270-1_26

Chernin L, Toklikishvili N, Ovadis M, Kim S, Ben-Ari J, Khmel I, Vainstein A (2011) Quorum-sensing quenching by rhizobacterial volatiles. Environ Microbiol Rep 3:698–704

Cho SM, Kang BR, Han SH et al (2008) 2R,3R-butanediol, a bacterial volatile produced by *Pseudomonas chlororaphis* O6, is involved in induction of systemic tolerance to drought in *Arabidopsis thaliana*. Mol Plant Microbe Interact 21:1067–1075

Cook RJ, Thomashow LS, Weller DM, Fujimoto D, Mazzola M, Bangera G, Kim DS (1995) Molecular mechanisms of defense by rhizobacteria against root disease. P Natl Acad Sci USA 92:4197

Dakora FD, Phillips DA (2002) Root exudates as mediators of mineral acquisition in low-nutrient environments. Plant Soil 245:35–47

Damam M, Kaloori K, Gaddam B, Kausar R (2016) Plant growth promoting substances (phytohormones) produced by rhizobacterial strains isolated from the rhizosphere of medicinal plants. Int J Pharm Sci Rev 37:130–136

Danish S, Zafar-ul-Hye M, Mohsin F, Hussain M (2020) ACC-deaminase producing plant growth promoting *rhizobacteria* and biochar mitigate adverse effects of drought stress on maize growth. PLoS ONE 15(4):e0230615

de Bruijn I, de Kock MJD, Yang M, de Waard P, van Beek TA, Raaijmakers JM (2007) Genome-based discovery, structure prediction and functional analysis of cyclic lipopeptide antibiotics in *Pseudomonas* species. Mol Microbiol 63:417–428. https://doi.org/10.1111/j.1365-2958.2006.05525.x

Dessaux Y, clémentC G, Faure D (2016) Engineering the rhizosphere. Trend Plant Sci 21(3):266–278

Doornbos RF, van Loon LC, Bakker PAHM (2012) Impact of root exudates and plant defense signaling on bacterial communities in the rhizosphere a review. Agron Sustain Dev 32:227–243

Effmert U, Kalderas J, Warnke R, Piechulla B (2012) Volatile mediated interactions between bacteria and fungi in the soil. J Chem Ecol 38:665–703

Farrar J, Hawes M, Jones D, Lindow S (2003) How roots control the flux of carbon to the rhizosphere. Ecology 84:827–833

Fasusi OA, Cruz C, Babalola OO (2021) Agricultural sustainability: microbial biofertilizers in rhizosphere management. Agriculture 11:163. https://doi.org/10.3390/agriculture11020163

Grobelak A, Hiller J (2017) Bacterial siderophores promote plant growth: Screening of catechol and hydroxamate siderophores. Int J Phytorem 19:825–833. https://doi.org/10.1080/15226514.2017.1290581

Guimaraes AA, Jaramillo PMD, Nobrega RSA, Florentino LA, Silva KB, de Souza Moreira FM (2012) Genetic and symbiotic diversity of nitrogen-fixing bacteria isolated from agricultural soils in the western Amazon by using cowpea as the trap plant. Appl Environ Microbiol 78:6726–6733

Gupta G, Parihar SS, Ahirwar NK, Snehi SK, Singh V (2015) Plant Growth Promoting *Rhizobacteria* (PGPR): current and future prospects for development of sustainable agriculture. J Microb Biochem Technol 7:096–102

Hakim S, Naqqash T, Nawaz MS, Laraib I, Siddique MJ, Zia R, Mirza MS, Imran A (2021) Rhizosphere engineering with plant growth-promoting microorganisms for agriculture and ecological sustainability. Front Sustain Food Syst 5:617157. https://doi.org/10.3389/fsufs.2021.617157

Hartmann A, Schikora A (2012) Quorum sensing of bacteria and trans-kingdom interactions of N-acyl homoserine lactones with eukaryotes. J Chem Ecol 38:704–713

Hayat S, Faraz A, Faizan M (2017) Root exudates: composition and impact on plant–microbe interaction. In: Ahmas I, Husai FM (eds) Biofilms in plant and soil health, pp 179–193

He H, Silo-Suh LA, Handelsman J, Clardy J, Zwittermicin A (1994) An antifungal and plant protection agent from *Bacillus cereus*. Tetrahedron Lett 35:2499–2502. https://doi.org/10.1016/S0040-4039(00)77154-1

Hider RC, Kong X (2010) Chemistry and biology of siderophores. Nat Prod Rep 27:637–657. https://doi.org/10.1039/b906679a

Igiehon NO, Babalola OO (2018) Rhizosphere microbiome modulators: contributions of nitrogen fixing bacteria towards sustainable agriculture. Int J Environ Res Public Health 15(4):574

Jamil F, Mukhtar H, Fouillaud M, Dufossé L (2022) Rhizosphere signaling: insights into plant-rhizomicrobiome interactions for sustainable agronomy. Microorganisms 10(5):899

Johnson NC, Graham JH (2013) The continuum concept remains a useful framework for studying mycorrhizal functioning. Plant Soil 363:411–419

Kanwal MS, Mukherjee S, Joshi R, Rai S (2019) Impact assessment of changing environmental and socio-economical factors on crop yields of central Himalaya with emphasis to climate change. Environ Ecol 37(1B):324–332

Kaviya N, Upadhayay VK, Singh J, Khan A, Panwar, M, Singh AV (2019) Role of microorganisms in soil genesis and functions. In: Varma A, Choudhary D (eds) Mycorrhizosphere and Pedogenesis. Springer, Singapore. https://doi.org/10.1007/978-981-13-6480-8_2

Khanna K, Jamwal VL, Gandhi SG et al (2019) Metal resistant PGPR lowered Cd uptake and expression of metal transporter genes with improved growth and photosynthetic pigments in *Lycopersicon esculentum* under metal toxicity. Sci Rep 9:5855

Kinkel LL, Bakker MG, Schlatter DC (2011) A coevolutionary framework for managing disease-suppressive soils. Annu Rev Phytopathol 49:47–67

Kumar A, Maurya BR, Raghuwanshi R (2014) Isolation and characterization of PGPR and their effect on growth, yield and nutrient content in wheat (Triticum aestivum L.). Biocatal Agri Biotechnol 3:121–128

Lazcano C, Boyd E, Holmes G et al (2021) The rhizosphere microbiome plays a role in the resistance to soil-borne pathogens and nutrient uptake of strawberry cultivars under field conditions. Sci Rep 11:3188

Leontidou K, Genitsaris S, Papadopoulou A et al (2020a) Plant growth promoting *rhizobacteria* isolated from halophytes and drought-tolerant plants: genomic characterisation and exploration of phyto-beneficial traits. Sci Rep 10:14857

Leontidou K, Ifigeneia M, Papadopoulou A, Genitsaris S, Karamanoli K, Kamou N, Vokou D, Bosmali I, Madesis P, Matsi T (2020) Plant growth promoting rhizobacteria isolated from halophytes and drought-tolerant plants: genomic characterisation and exploration of phyto-beneficial traits. Scientific Reports. 14857. https://doi.org/10.1038/s41598-020-71652-0

Li JY, Xu RK, Xiao SC, Ji GL (2005) Effect of low-molecular-weight organic anions on exchangeable aluminum capacity of variable charge soils. J Colloid Interface Sci 284:393–399

Lu P, Yang T, Li L, Zhao B, Liu J (2020) Response of oat morphologies, root exudates, and rhizosphere fungal communities to amendments in saline-alkaline environment. PLoS ONE 15(12):e0243301

Maksimov IV, Abizgildina RR, Pusenkova LI (2011) Plant growth promoting rhizobacteria as alternative to chemical crop protectors from pathogens (Review). Appl Biochem Microbiol 47:333–345. https://doi.org/10.1134/S0003683811040090

Meena VS, Maurya BR, Verma JP (2014) Does a rhizospheric microorganism enhance K+ availability in agricultural soils? Microbiol Res 169:337–347. https://doi.org/10.1016/j.micres.2013.09.003

Mendes R, Garbeva P, Raaijmakers JM (2013) The rhizosphere microbiome: significance of plant beneficial, plant pathogenic and human pathogenic microorganisms. FEMS Microbiol Rev 37(5):634–663. https://doi.org/10.1111/1574-6976.12028. Epub 2013 Jul 22 PMID: 23790204

Mendes R, Kruijt M, de Bruijn I, Dekkers E, van der Voort M, Schneider JH, Piceno YM, DeSantis TZ, Andersen GL, Bakker PA, Raaijmakers JM (2011) Deciphering the rhizosphere microbiome for disease-suppressive bacteria. Science 332(6033):1097–1100. https://doi.org/10.1126/science.1203980. Epub 2011 May 5 PMID: 21551032

Miransari M (2011) Arbuscular mycorrhizal fungi and nitrogen uptake. Arch Microbiol 193:77–81
Nannipieri P, Judith A-J, Ceccherini MT, Pietramellara G, Renella G, Schloter M (2020) Beyond microbial diversity for predicting soil functions: a mini review. Pedosphere 30:5–17
Nautiyal CS, Srivastava S, Chauhan PS (2008) Rhizosphere colonization: molecular determinants from plant-microbe coexistence perspective. In: Nautiyal CS, Dion P (eds) Molecular mechanisms of plant and microbe coexistence. Soil Biology, vol 15. Springer, Berlin, Heidelberg
Niu DD, Liu HX, Jiang CH, Wang YP, Wang QY, Jin HL, Guo JH (2011) The plant growth-promoting *rhizobacterium Bacillus cereus* AR156 induces systemic resistance in *Arabidopsis* thaliana by simultaneously activating salicylate-and *jasmonate*/ethylene-dependent signaling pathways. Mol Plant Microbe Interact 24:533–542
Pantigoso HA, Yuan J, He Y, Guo Q, Vollmer C, Vivanco JM (2020) Role of root exudates on assimilation of phosphorus in young and old *Arabidopsis thaliana* plants. PLoS ONE 15(6):e0234216
Parmar P, Sindhu SS (2013) Potassium solubilization by rhizosphere bacteria: influence of nutritional and environmental conditions. J Microbiol Res 3:25–31
Pascale A, Proietti S, Pantelides IS, Stringlis IA (2020) Modulation of the root microbiome by plant molecules: the basis for targeted disease suppression and plant growth promotion. Front Plant Sci 10:1741
Paulin MM, Novinscak A, Lanteigne C, Gadkar VJ, Filion M (2017) Interaction between 2,4-Diacetylphloroglucinol- and Hydrogen Cyanide-producing *Pseudomonas brassicacearum* LBUM300 and *Clavibacter michiganensis* subsp. michiganensis in the Tomato Rhizosphere. Appl Environ Microbiol 83(13): e00073–17
Philippot L, Raaijmakers JM, Lemanceau P, Van Der Putten WH (2013) Going back to the roots: the microbial ecology of the rhizosphere. Nat Rev Microbiol 11:789–799
Picard C, Baruffa E, Bosco M (2008) Enrichment and diversity of plant-probiotic microorganisms in the rhizosphere of hybrid maize during four growth cycles. Soil Biol Biochem 40:106–115
Pieterse CM, Zamioudis C, Berendsen RL, Weller DM, Van Wees SC, Bakker PA (2014) Induced systemic resistance by beneficial microbes. Annu Rev Phytopathol 52:347–375. https://doi.org/10.1146/annurev-phyto-082712-102340
Pii Y, Penn A, Terzano R, Crecchio C, Mimmo T, Cesco S (2015) Plant-microorganism-soil interactions influence the Fe availability in the rhizosphere of cucumber plants. Plant Physiol Biochem 87:45–52
Purakayastha TJ, Viswanath T, Bhadraray S, Chhonkar PK, Adhikari PP, Suribabu K (2008) Phytoextraction of zinc, copper, nickel and lead from a contaminated soil by different species of Brassica. Int J Phytorem 10(1):61–72
Raaijmakers J, Mazzola M (2012) Diversity and natural functions of antibiotics produced by beneficial and pathogenic soil bacteria. Annu Rev Phytopathol 50:403–424
Raaijmakers JM, de Bruijn I, Nybroe O, Ongena M (2010) Natural functions of lipopeptides from *Bacillus* and *Pseudomonas*: more than surfactants and antibiotics. FEMS Microbiol Rev 34:1037–1062. https://doi.org/10.1111/j.1574-6976.2010.00221.x
Rai AK, Johri SN, Kaur H, Basak N, Sundha P (2020) Salinity and Sodicity influence mutualistic association of beneficial microorganism in arid soils. J Soil Salin Water Qual 12(1):15–21
Rasouli-Sadaghiani M, Malakouti MJ, Khavazi K, Miransari M (2014) Siderophore efficacy of fluorescent pseudomonades affecting labeled iron (^{59}Fe) Uptake by Wheat (*Triticum aestivum* L.) Genotypes Differing in Fe Efficiency; pp. 121–132, Use of Microbes for the Alleviation of Soil Stresses. Springer, New York, NY, USA
Rohrbacher F, St-Arnaud M (2016) Root exudation: the ecological driver of hydrocarbon rhizoremediation. Agronomy 6:19
Saeed Q, Xiukang W, Haider FU, Kucerik J, Mumtaz MZ, Holatko J, Naseem M, Kintl A, Ejaz M, Naveed M, Brtnicky M, Mustafa A (2021) Rhizosphere bacteria in plant growth promotion, biocontrol, and bioremediation of contaminated sites: a comprehensive review of effects and mechanisms Int J Mol Sci 22(19):10529. https://doi.org/10.3390/ijms221910529. PMID: 34638870; PMCID: PMC8509026

Saha M, Sarkar S, Sarkar B, Sharma BK, Bhattacharjee S, Tribedi P (2016) Microbial siderophores and their potential applications: a review. Environ Sci Pollut Res Int 23(5):3984–3999. https://doi.org/10.1007/s11356-015-4294-0. Epub 2015 Mar 12 PMID: 25758420

Sasaki T, Yamamoto Y, Ezaki B, Katsuhara M, Ahn SJ, Ryan PR, Delhaize E, Matsumoto H (2004) A wheat gene encoding an aluminum-activated malate transporter. Plant J 37(5):645–653. https://doi.org/10.1111/j.1365-313x.2003.01991.x. PMID: 14871306

Silo-Suh LA, Lethbridge BJ, Raffel SJ, He H, Clardy J, Handelsman J (1994) Biological activities of two fungistatic antibiotics produced by *Bacillus cereus* UW85. Appl Environ Microbiol 60(6):2023–2030. https://doi.org/10.1128/aem.60.6.2023-2030.1994.PMID:8031096;PMCID: PMC201597

Singh AP, Singh SK, Rai S, Kumar M (2020) Soil carbon dynamics in relation to soil surface management and cropping system. In: Carbon management in tropical and sub-tropical terrestrial systems. Springer, pp 159–172

Singh RP, Jha PN (2016) The multifarious PGPR *Serratia marcescens* CDP-13 augments induced systemic resistance and enhanced salinity tolerance of wheat (*Triticum aestivum* L.). PLoS One 11: e0155026.

Son SH, Khan Z, Kim SG, Kim YH (2009) Plant growth-promoting rhizobacteria, Paenibacillus polymyxa and Paenibacillus lentimorbus suppress disease complex caused by root-knot nematode and fusarium wilt fungus. J Appl Microbiol 107:524–532

Sparks DL, Huang PM (1985) Potassium in agriculture. Soil science society of America; Madison, WI, USA: Physical chemistry of soil potassium; pp 201–276

Sureshbabu K, Amaresan N, Kumar K (2016) Amazing multiple function properties of plant growth promoting rhizobacteria in the rhizosphere soil. Int. J. Curr. Microbiol. Appl. Sci. 5:661–683. https://doi.org/10.20546/ijcmas.2016.502.074

Swain M, Samir KN, Ray R (2007) Indole-3-acetic acid production and effect on sprouting of Yam (*Dioscorea rotundata* L.) minisetts by *Bacillus subtilis* isolated from culturable cowdung microflora. Polish J Microbiol 56:103–110

Tahir M, Ahmad I, Shahid M, Shah GM, Farooq ABU, Akram M et al (2019) Regulation of antioxidant production, ion uptake and productivity in potato (*Solanum tuberosum* L.) plant inoculated with growth promoting salt tolerant Bacillus strains. Ecotox Environ Safe 178:33–42

Tak HI, Ahmad F, Babalola OO (2013) Advances in the application of plant growth-promoting rhizobacteria in phytoremediation of heavy metals. Rev Environ Contam Toxicol 223:33–52

Terpolilli JJ, Hood GA, Poole PS (2012) What determines the efficiency of N(2)-fixing Rhizobium legume symbioses? Adv Microb Physiol 60:325–389

Tiwari S, Lata C (2018) Heavy metal stress, signaling, and tolerance due to plant-associated microbes: an overview. Front Plant Sci 9:452

Tiwari S, Lata C, Chauhan PS, Nautiyal CS (2016) Pseudomonas putida attunes morphophysiological, biochemical and molecular responses in *Cicer arietinum* L. during drought stress and recovery. Plant Physiol Biochem 99:108–117

Vishwakarma K, Kumar N, Shandilya C, Mohapatra S, Bhayana S, Varma A (2020) Revisiting plant microbe interactions and microbial consortia application for enhancing sustainable agriculture: a review. Front Microbiol 11:560406

Vurukonda SSKP, Giovanardi D, Stefani E (2018) Plant Growth Promoting and Biocontrol Activity of *Streptomyces* spp. as Endophytes. Int J Mol Sci 19:952. https://doi.org/10.3390/ijms19040952

Wang J, Raman H, Zhou M et al (2007) High-resolution mapping of the Alp locus and identification of a candidate gene HvMATE controlling aluminium tolerance in barley (*Hordeum vulgare* L.). Theor Appl Genet 115:265–276

Wang Y, Wu WH (2017) Regulation of potassium transport and signaling in plants. Curr Opin Plant Biol 39:123–128

Weston DJ, Pelletier DA, Morrell-Falvey JL et al (2012) *Pseudomonas fluorescens* induces strain-dependent and strain-independent host plant responses in defense networks, primary metabolism, photosynthesis, and fitness. Mol Plant Microbe Interact 25:765–778

Whipps JM (2001) Ecological and biotechnological considerations in enhancing disease biocontrol. Enhancing biocontrol agents and handling risks, vol 339 (Vurro M, Gressel J, Butt T, Harman GE, Pilgeram A, St. Leger RJ, Nuss DL eds.), pp 43–51. IOP Press, Amsterdam

Wilson MK, Abergel RJ, Raymond KN, Arceneaux JEL, Byers BR (2006) Siderophores of Bacillus anthracis, Bacillus cereus, and *Bacillus thuringiensis*. Biochem Biophys Res Commun 34(8):320–325

Yehuda Z, Shenker M, Hadar Y, Chen YN (2000) Remedy of chlorosis induced by iron deficiency in plants with the fungal siderophore rhizoferrin. J Plant Nutr 23:1991–2006

Yin L, Liu B, Wang H, Zhang Y, Wang S, Jiang F, Ren Y, Liu H, Liu C, Wan F, Wang H, Qian W, Fan W (2020) The Rhizosphere Microbiome of *Mikania micrantha* Provides Insight into adaptation and invasion. Front Microbiol 11:1462. https://doi.org/10.3389/fmicb.2020.01462. PMID:32733410;PMCID:PMC7359623

Zakikhani H, Ardakani MR, Rejali F, Gholamhoseini M, Joghan AK, Dolatabadian A (2012) Influence of diazotrophic bacteria on antioxidant enzymes and some biochemical characteristics of soybean subjected to water stress. J Integr Agric 11(11):1828–1835

Zhuang X, Chen J, Shim H, Bai Z (2007) New advances in plant growth-promoting rhizobacteria for bioremediation. Environ Int 33(3):406–413. https://doi.org/10.1016/j.envint.2006.12.005

Zou CS, Mo MH, Gu YQ, Zhou JP, Zhang KQ (2007) Possible contributions of volatile-producing bacteria to soil fungistasis. Soil Biol Biochem 39:2371–2379

Chapter 7
The Role of Microbes in Mitigating Salinity Stress: Insights and Implications for Sustainable Agriculture

Krishnendu Sen, Anik Mandal, Subhadip Dhabal, Sk Saruk Islam, Subrata Dutta, and Sujoy Midya

Keywords Ion Channel · Osmoprotectants · PGPRs · Phytohormones · Salinity · Sustainable agriculture · VOCs

7.1 Introduction

The shrinking agricultural area left less room to develop plant agriculture as urbanisation grew. In such cases, boosting soil fertility is necessary to ensure that there is enough food for everyone under the present-day paradigm of worldwide food security (Godfray et al. 2010). Now, in mother Earth the estimated human population is more than 7 billion and in the near future, with the growing rate of 1.05% per year, it would likely increase more than 10 billion. In this adverse situation, the soil texture and water accessibility play a crucial role in sustainable agro-ecosystems and crop productivity (Singh et al. 2022). Any disproportion of salinity in both water and soil causes a loss in plant productivity as well as, eventually, their abandonment when the

K. Sen (✉) · S. Dutta
Department of Plant Pathology, Bidhan Chandra Krishi Viswavidyalaya, Mohanpur, West Bengal 741252, India
e-mail: krishnendusen10@gmail.com

K. Sen
Department of Microbiology, Vidyasagar University, Midnapore, West Bengal 721102, India

A. Mandal · S. Dhabal
Department of Microbiology, Bankura Sammilani College, Bankura, West Bengal 722102, India

S. S. Islam · S. Midya
Department of Zoology, Raja Narendra Lal Khan Women's College (Autonomous), Midnapore, West Bengal 721102, India

P. P. Adhikary et al. (eds.), *Soil, Water Pollution and Mitigation Strategies*, Environmental Science and Engineering, https://doi.org/10.1007/978-3-031-63296-9_20

land shape shifts from productive to marginal (Bhat et al. 2020). In the habitat, plants are perpetually uncovered to numerous biotic and abiotic factors, which may affect their growth increment when the factors are in right amount or, in abundant quantity it may introduce drought. Although plant endurance to salt stress can differ with few species, even variations, along with both abiotic and biotic components (Rai et al. 2023). Stress is represented as any extrinsic abiotic (water, salinity, heat) or any biotic (herbivore) restraint that control the amount of photosynthesis & reduces a plant's ability to change energy to bio-mass. By 2050, it is estimated that 50% of farming land will be damaged by salinity (Bharti et al. 2016). Salinization of farming land take place primarily as a result of salt buildup in the soil, especially chloride (Cl^-) and sodium (Na^+) ions. High Na^+ aggregation restrict water conductance (Table 7.1), soil porousness, and aeration. In addition, soil salinization has downward effects on the microbial variety inside & surrounding plant roots. Numerousstructural, physiological, and molecular alterations occur in plants. under the salinity stress, which hinder its proliferation, Numerous structural, Additionally, a large salt dilution has an impact on the rate of stomatal conductance, photosynthesis, and enzymatic activity. Reactive oxygen species (ROS), which damage cell membranes, proteins, lipids and nucleic acids (RNA& DNA), as well as perhaps triggering programmed cell death, are generate more readily under salinity stress (Kumar et al. 2020a, b).

To withstand with the only modest success, technologically driven genetically modified organisms (GMO) have been introduced into the field, unconventional ways, such as the introduction of salt-resistant microbes are more analyzed for acceptance in augmenting and, as such, improving crop growth in salty soils (Etesami 2018). Plant growth-promoting rhizobacteria (PGPR) are a significant group of these microbes

Table 7.1 Salinity stress across some states of India according to ICAR-Central Soil Salinity Research Institute (CSSRI), 2022

Sr. No	States & Union territories	Saline soil (ha)	Costal saline soil (ha)	Total Saline soil (ha)
1	Andhra Pradesh	0	77,598	77,598
2	A & N Islands	0	77,000	77,000
3	Bihar	47,301	0	47,301
4	Gujarat	1,218,256	462,315	1,680,571
5	Haryana	49,157	0	49,157
6	Karnataka	1307	586	1893
7	Kerala	0	20,000	20,000
8	Maharastra	177,093	6996	184,089
9	Orissa	0	147,138	147,138
10	Rajasthan	196,571	0	196,571
11	Tamilnadu	0	13,231	13,231
12	Uttar Pradesh	21,989	0	21,989
13	West Bengal	0	441,272	441,272

that have been successfully used to help plants develop systemic tolerance to abiotic stressors (Etesami and Beattie 2018). On the other hand, it was found that PGPR from high saline soils (halotolerant PGPR) expressed plant growth-promoting (PGP) traits are not as much ofpretentious by environmental variables like climate, soil characteristics, etc., and are therefore highly effective at increasing salt resistance in plants than PGPR from non-saline environment (Khan et al. 2016). It was discovered that plants control the development of the microbiota to promote adaptation to shifting environmental conditions as part of the interaction between bacteria and plants at the rhizospheric plane (Berendsen et al. 2012). Study claims that the use of PGPR against saline stress optimises the maturation and output of French bean (*Phaseolus vulgaris*) (Kumar et al. 2020a, b). Rhizobacteria that produce siderophores may be a promising substitute for chemical fertilisers since they can cut down the effects of salinization while increasing the amount of iron that is accessible in salty soils (Ferreira et al. 2019). A shift in membrane transportion characteristics, such as the control of Na^+/H^+ antiporters and other ion channels regulators, is achieved in effect to increasing salinity stress (Kumar et al. 2017). Additionally, through regulating the functioning of stress-related genes, transcription factors (TFs) show a critical role in promoting tolerance to salt stress. AP2/ERFBP, bZIP, NAC, WRKY, and MYB have been recognised as possible contributors for enhancing crop resistance to saline stress among the several TF families (Barnawalet al. 2017; Tang et al. 2019). Additionally, PGPR manages plant pathogens by producing antibiotics and competing with other organisms for nutrients and ecological niches (Kumar and Verma 2018). The purpose of the current study is to emphasize the significance of plant-bacterial interactions and provide in-depth information about the mechanical vision that direct at the plant level in reducing salt stress and improving crop production as part of climate-smart farming practises intended to feed the world's expanding population.

7.2 Microbe–Plant Interactions: Mechanisms and Implications for Salinity Stress Mitigation

Unlike many other species, plants have developed unique defence mechanisms to protect them from stressful situations, foster growth, and prevent stressful situations from occurring in the first place (Raza et al. 2019). By creating phytohormones or by causing signalling in the host, PGPR either directly or indirectly influences plants. Most frequently, a direct involvement for phytohormones such as gibberellins, IAA, ABA, cytokinin, and ethylene as well as biological nitrogen fixation (BNF) and phosphate solvabiliy is assumed (Gupta et al. 2021a, b). However, indirect processes that inhibit phytopathogens include the creation of hydrogen antibiotics, cyanide, volatile organic compounds (VOCs), ammonia, and siderophores. These soil microorganisms alter the ion balance and enhance crop water relations in addition to modulating the management of abiotic stress through several routes (Sehrawat et al. 2022) (Fig. 7.1).

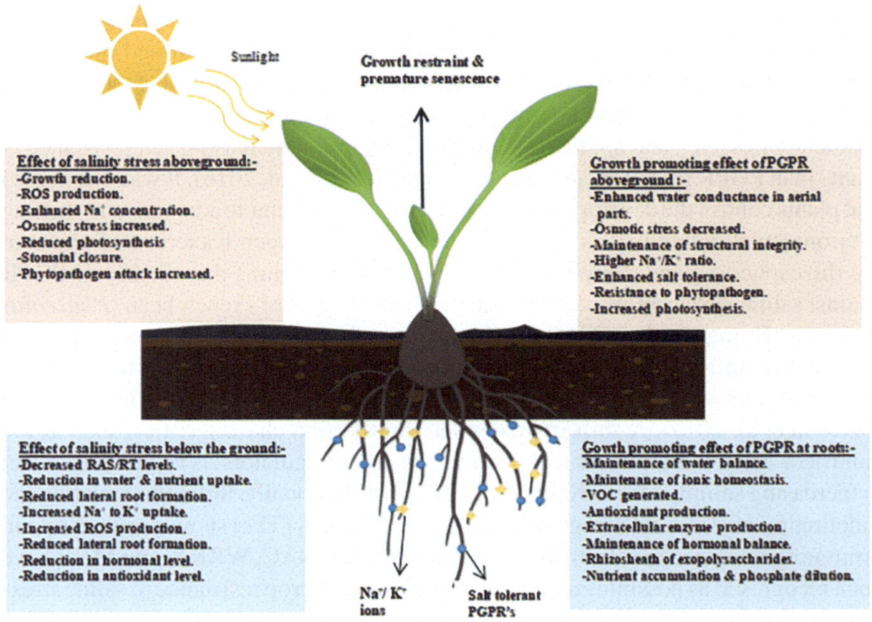

Fig. 7.1 Adverse effects and PGPR mediated plant growth enhancement in salinity stress

Fast water fluxes along the osmotic gradient leave the cell when bacteria are exposed to high osmotic conditions, lowering turgor and drying the organism to balance the water outflow. K^+ is kept at a constant osmotic pressure by an imbalance of ions caused by high levels of ions in the cytoplasm, stimulates osmotic processes, and upregulates the genes for amino acid transport, metabolic, and adaptive pathways. The organic solutes production (such protein, choline, sucrose, glutamic acid, betaine, and serine) is another important process (Shabala and Pottosin 2014), a rise in intracellular osmotic pressure that stabilises cellular macromolecules in salt stress. According to reports, Suitable solutes operate as osmoprotectants and have the ability to stabilize proteins, which helps the actual folding of polypeptides under difficult conditions both in vitro and in vivo (Kumawat et al. 2022).

Exo-polysaccharide production, which accelerates water retention by changing the nature of the membrane, including the number of periplasmic glucans, protein molecules, fatty acids, interpeptide, and shorter peptidoglycans bridges, The second mechanism of halo-tolerant bacteria used in salt stress regulates the usage of carbon sources in microbial cells and shields microbes from altering water and osmotic strain (Shultana et al. 2022). Low electrolyte concentrations allow the microbial cell's flexible surface appendages to act as a preventive barrier, reducing osmotic strain and the detrimental effects of ionic strength alters (Kumawat et al. 2022).

The third way by which PGPR tolerates excessive salt levels is the reduction of salt absorption by gene regulation changing root structure via broad ion affinity transporter rhizo-sheath, as well as Exo-polysaccharides trap cations. Microbes regulate the balance of ions by boosting the K^+/Na^+ ratio in the shoot and decreasing Na^+ and Cl^- in the leaves, respectively, by enhancing the K^+ transporter and rejecting Na^+. Salt stress changes the expression of genes like NHX2, KT1, HKT1, and SOS1, and these molecular changes lead to salt resistance (Pinedo et al. 2015; Kumawat et al. 2022).

Several helpful microbes from various genera, including *Azospirillum, Bacillus, Pantoea, Burkholderia, Rhizobium, Pseudomonas, Paenibacillus, Azotobacter, Enterobacter, Enterococcus, Variovorax, Stenotrophomonas* and, *Serratia,* have been reported to have positive effects on plant proliferation and stimulate stress resistance in various crops, including chickpea, wheat, groundnut, soyabean under unfavourable environmental condition (Gupta et al. 2015). When isolated from the halophytic plant *Arthrocnemum indicum*, salt-tolerant native species of *Agrobacterium, Bacillus, Pseudomonas, Ochrobactrum*, and *Klebsiella*, demonstrated tolerance to 4%–8% salt concentration and increased groundnut productiveness in saline soil (Sharma et al. 2016). Numerous salt-tolerant bacterial types have been discovered in wheat and rice (Zhang et al. 2018). More than 50% of the isolates, according to a phylogenetic study, belonged to the orders *Bacillales*(72%), *Actinomycetales*(22%), *Oceanospirillales*(4%), &*Rhizobiales*(1%) (Zhang et al. 2018). Numerous *Bacillus* strains possessed significant levels of salt (4% NaCl) tolerance and plant growth-promoting (PGP) characteristics (Sharma et al. 2015).

7.3 Siderophores: Iron-Chelating Compounds for Improved Iron Uptake in Plants

Metal nutrient availability for plants in particular iron (Fe), is drastically reduced (Abbas et al. 2015) when crops are grown on salt containing soils with high pH (above 6) (M.Q Nawaz et al. 2020), and Na^+ concentration. For several metabolic pathways, respiration, the generation of chlorophyll, DNA synthesis, and photosynthesis, iron is an important micro-nutrient for plants. It is also an essential component of many enzymes, so it is required for an extensive number of biological functions (Kumar et al. 2020a, b). The majority of living microorganisms, including bacteria, need iron to flourish, just like plants do. It accomplishes this by serving as a catalyst for a wide range of processes, as well as the synthesis of DNA and RNA, oxygen consumption, electron transfer and the creation of biofilms (Aguado-Santacruz et al. 2012). Consequently, a deficiency in iron may act as a growth restriction. Fe is primarily found in soil in the unavailable form of ferric oxidation (Fe^{3+}). In general, both plants and rhizospheric bacteria are unable to access it because it is likely to generate unsoluble hydroxides with a highly depressed solubility constant (10^{-36} M) (Rajkumar et al. 2010). Comparatively much soluble & reachable to plants is the ferrous (Fe^{2+}) state, However, in nature, it easily oxidizes to Fe^{3+}, which then precipitates (Ghosh et al. 2015). Most microorganisms have evolved a strong-affinity iron (Fe^{3+}) assimilation mechanism in the form of organic compounds with a small molecular mass (iron chelators) called siderophores in order to thrive in iron-poor environments. Siderophores function as iron dissolving agents when iron is in short supply by forming a complex with Fe^{3+} on microbial membranes then reduced it to Fe^{2+}, making it accessible to both microbes and plants. After being freed from the cells, siderophores are expelled and regenerated for iron transport (Kermeur et al. 2023).

According to Sadeghi et al. 2012, *Streptomyces sp.* produces siderophores, which promote wheat development in salinized environments. It has been discovered that tomato plants grown at 2% NaCl contained PGPR that could dissolve undissolved phosphorus and produce siderophore. Even in saline environments, *Bacillus halodenitrificans PU62* and *Halobacillus sp. SL3* produce siderophores (Ramadoss et al. 2013). In direct contrast to the control management, (Parrayet al. 2016) showed that *Pseudomonassp.* GRP-3, which produces siderophores, increased the iron intake of *Vigna radiata* by decreasing chlorosis and enhancing chlorophyll content. Rhizobia and rhizobacteria that produce siderophores enhance nitrogen uptake and have a favourable impact on plant height (Shen et al. 2013). Certain *Ensifermeliloti* siderophore-generating strains have been reported to reduce *Macrophominaphaseolina*, the activating agent of groundnut charcoal rot (Rajkumar et al. 2010).

7.4 Ion Channel Regulation: Maintaining Ionic Balance in Plants

Ion toxicity is brought on by the buildup of many ions such as Na^+, Cl^-, Ca2$^+$, Mg2$^+$, $SO_3{}^{2-}$, or $CO_3{}^{2-}$ in salts. Long-term exposure to high levels of salt causes the rate of exclusion to lag behind the rate of ion flux (Zulfiqar et al. 2020). Plants initially compartmentalize excess salts in their vacuoles to prevent them from accumulating in the cytosolic and intracellular spaces., This would restrict photosynthesis andrespiration (Ilangumaran et al. 2017; Zulfiqar et al. 2020). Na^+ substitutes K^+ in biological reactions, causing protein production and structural changes (Shrivastava et al. 2015).Ionic equilibrium disruption is found in plants that are bad Na^+ excluders (rice, beans, etc.,) and highly sensitive to Cl^- ions (soybean, citrus, etc.) grown in high salt soils (Chakraborty et al. 2022). As a result, excluding NaCl from crops becomes hard because only a tiny fraction travels through the phloem, limiting its migration to aerial portions and producing toxicity. Raising Na^+ levels disrupt the K^+/Na^+ ratio, which impedes cytosolic activity and affects respiration (Bhat et al. 2020). As the concentration of Na^+ in the cytosol increases, the Na^+/H^+ antiporter (also known as salt overlay sensitive channel 1, or SOS1) effluxes Na^+(Qiu et al. 2002). The plants improve K^+ uptake over Na^+ ions by actuating high-affinity K^+ transporters, which increases salinity resistance (Chakraborty et al. 2022). Additionally, a Ca^{2+} signal that is produced by membrane-bound Ca2$^+$ channel activation can be utilized to determine whether salt stress is present in plants in response to a discharge event. The calcineurin B-like protein (CBL4; also known as SOS3), which forms a complex with the CBL-interacting protein kinase, detects the Ca^{2+} signal. By retaining K^+ transporters, CIPK24 (also known as SOS2) makes it possible for SOS1 to be phosphorylated for stimulation. This action is crucial for maintaining the Na^+/K^+ ratio (Zhu 2002). Microbes help plants maintain ion homeostasis by reducing the formation of ions by raising Na^+ restriction at the roots and increasing the working capacity of K^+ transporters, which indirectly reduces their boost in aerial portions. In addition to encouraging the development of a biofilm at root surface that blocks the entry of Na^+ into the root system, PGPR strains produce EPS that traps cations inside the matrix and prevents the plants from absorbing them (Sehrawat et al. 2022). When *Arabidopsis thaliana* roots are exposed to *B. subtilis* GB03, *HKT1* is down-regulated, which decreases the plant's ability to take up Na^+ (Qin et al. 2016). Raising the production of HKT1 in stems makes it easier for Na^+ to be recycled from the stem to the roots, which helps plants maintain an elevated K^+/Na^+ ratio by preventing Na^+ uptake at the root surface (Ali et al. 2019). *Arabidopsis thaliana* becomes more susceptible to salinity stress as aoutcome of the Ca^{2+}dependent protein kinase CPK12 mutation caused by RNA interference (Zhang et al. 2018).

7.5 Osmoprotectants: Protecting Plants from Osmotic Stress

The first effect of salinity is osmotic stress, which interrupt the water equilibrium and leads to stomatal closure (Mukhopadhyay et al. 2021). All living organisms, including higher plants and animals, as well as archaebacteria, contain osmoprotectants, also known as appropriate solutes. When concentrations of salt are adverse, osmoprotectants aid to raise cytoplasmic osmotic pressure and can also serve to stabilize proteins and membranes. Osmoprotectants are essential in assisting cells in adapting to a variety of hazardous environmental circumstances. some osmoprotectant compounds are proline, total soluble sugar (TSS), trehalose, and betaine, that decrease osmotic strain (Kumar et al. 2021; Gupta et al. 2021a, b). betaine, Glycine, as well as trehalose serve as osmoprotectants by stabilizing the quaternary structure of proteins and the extremely structured states of membranes. Mannitol functions as a scavenger of free radicals (Table 7.2). Proline is a carbon and nitrogen storage sink as well as a neutralizer of free radicals. Additionally, it buffers cellular redox potential under stress and stabilizes the subcellular structures. These organic osmolytes referredd as osmoprotectants(Chen and Murata 2002). This defense against osmotic stress, which occurs more frequently in salty soils, is also used by ST-PGPR. Many plants and animals deal with osmotic stress by synthesizing and stockpiling osmoprotectants (or suitable solutes). These are small, electrically neutral molecules that are safe in molar amounts and protect membranes and proteins from the distorting action of high level of salts and other noxious solutes (Gupta et al. 2021a, b). When subjected to salt stress, plants respond to ionic Na^+ and osmotic strain signalshigh concentration of Na^+ can be detected by a transmembrane protein on the plasma membrane's surface, Na^+ sensitive enzymes ormembrane proteins inside the cell (Zhu 2003). Chemically, there are three categories: polyols and sugars (such as mannitol and trehalose), amino acids like proline, and betaines and related substances. Exclusion from protein surfaces is likely a component of the protective mechanism in circumstances where osmoprotectants simultaneously have cryoprotective and heat-protective characteristics. Therefore, the levels of naturally occurring osmoprotectants in cytoplasmic compartments can equal or above 200 mM. Such concentrations are vital for sustaining cell turgor and the driving gradient for water uptake during stressful conditions because they are osmotically significant (Surowka et al. 2021). Exclusion from protein surfaces is likely a component of the protective mechanism in circumstances where osmoprotectants simultaneously have cryoprotective and heat-protective characteristics.Ectoine,proline, betaine, trehalose and glycine, are accumulated under salinity stress by *Azospirillum* spp. And*Halomonas* sp. SBS 10, providing maize tolerant to salinity stress (Kushwaha et al. 2019). Salinity tolerance is provided by the significant osmoprotectant sugar trehalose, and in halo-tolerating PGPR, multiple trehalose-producing genes were discovered (Shim et al. 2019).

7.6 Gene Regulation: Modulating Plant Stress Responses

Plants use a number of methods to endure under salinity stress, including physiological, morphological and biochemical changes. These numerous modifications demand alterations in numerous stress-related genes engaged in regulating and signalling networks (Numan et al. 2018). It is crucial to comprehend the mechanisms underlying salinity stress signaling in plants, and within plants and microorganisms. Plant signal transduction typically begins with the detection of signals by surface receptors, followed by the production of secondary messengers like the ROS (Reactive Oxygen Species) and inositol phosphates (Deinlein et al. 2014). Through the conserved macro-autophagy system, Active indole-3-acetic acids (IAA) are destroyed by a disparity of various ROS species or a buildup of ROS brought on by high glucose levels (Huang et al. 2012), prevents root development, and affects root meristem activity. Changes in outside factors alter the metabolic level, which leads to the development of ROS in plant cells. ROS generation is unavoidable in higher plants due to their various cellular metabolic processes. Various abiotic stress conditions, such as metal/metalloid toxicity, dehydration, and salt, significantly boosted ROS generation, disrupting the equilibrium of ROS with antioxidant enzymes (Hasanuzzaman et al. 2014). Under normal settings, the reactive oxygen species are required for an organism's ability to control its gene activity. because they govern multiple activities within the cell. Proteins including protein phosphatase, calcium-dependent protein kinases (CDPK) andmicrotubule associated protein kinase (MAPK) (Fig. 7.2), which are involved in the regulation of gene activity via Ca^+ concentration modulation, are the targets of secondary messengers. In addition to other biological processes, other genes also regulate plant hormones (Miller et al. 2010). Transcription factors (TFs) have a special function in salt stress

Table 7.2 Production of Osmoprotectent by PGPRstrains in salinity stress

Sr. No.	PGPR/PGPF strain	Crop rizosphere	Osmoprotectents	Effect on plants	References
1	*Arthrobacter pascens*	Maize	Proline	The buildup of osmolytes, such as proline and sugar	Ullah and Bano 2015
2	*B. aquimaris*	T. aestivum L	Proline and TSS	Enhanced leaf nutrition, weight and biomass	Upadhyay and Singh 2015
3	*Rhizobium*sp. &*Pseudomonas* sp.	Zea mays L	Betaine, proline	a rise in the biomass, growth, and nutrition uptake	Bano and Fatima 2009
4	*B. aquimaris*	T. aestivum L	TSS and PRP	Enhanced leaf nutrition, weight and biomass	Upadhyay and Singh 2015
5	*Pseudomonas*sp.	S. lycopersicum L	Trehalose and ACCD	increased total dry weight, chlorophyll content and shoot and root length	Orozco-Mosqueda et al. 2019
6	*G. etunicatum*	Glycine max	Proline	Proline levels increased in roots but fell in shoots	Sharifi et al. 2007
7	*B. megaterium*	Glycine max	Betaine, proline	It modulates salt tolerance and maintains osmotic balance	Khan et al. 2019
8	*Zymomonasmobilis*	Rosaceae family	Sorbitol	Regulate inositol biosynthesis for osmotic modulation	Bieleski 1982, Sheveleva et al. 1998
9	*E. coli*	Tobacco, potato	Mannitol	increase ROS scavenging capacity	Shen et al. 1997
10	*Bacillus subtilis*	Rice, Tobacco	Myoinositol	increase photosynthesis of plant	Yamaoka et al. 2011
11	*P.pudita*KT2440	Rice	Trehalose	Regulate ABA pathway	Vilchez et al. 2016

(continued)

Table 7.2 (continued)

Sr. No.	PGPR/PGPF strain	Crop rizosphere	Osmoprotectents	Effect on plants	References
12	Salicornia bigelovii	Arabidopsis	L-Arginine, L-Ornithine, Agmatine, Putrescine, Spermidine, Spermine	Regulate plant stomatal closure through voltage gated inward k^+channels of gurd cell	EL-Tarabily et al. 2020
13	Arthobacter spp.	Rice, Tomato	Glycine betaine	Regulate ROS scavenging, ferric reductase, trancription factors and reducing lipid peroxidation	Einset et al. 2007

tolerance in plants because of their particular actions on several stress-responsive genes. Many TF families, including bZIP, AP2/ERF, MYB, WRKY, and NAC, have an impact on the expression and operation of numerous genes (Kumar et al. 2020a, b). The transition of the genes linked to salt stress is impacted by the post-transcriptional alteration of Transcription Factors. Multiple studies have examined how TFs affect the functioning of multiple genes and the ability to withstand salt stress. *Tamarix-hispida*'sbZIP gene expression (Wang et al. 2010) and in *C.korshinskii*, the CkdREB gene (Wang et al. 2011) provided defence against salinization stressUnder conditions of salt stress, the induction of plant signaling is significantly influenced by PGPB and endophytes. For instance, PGPR like *Dietzianatronolimnaea* (STR1) and *Arthrobacter protophormia*e (SA3) increased the ability of wheat plants to withstand saline stress by changing the gene expression of a controlling element of the ethylene signalingpathway, CTR1 (Constitutive Triple Response1), and DREB2 TF (Barnawalet al. 2017). Under saline stress, *Enterobacter spp.* increased the expression of salt stress-responsive genes such as RD29A, RAB18, DREB2b, and RD29B, in *Arabidopsis* (Kim et al. 2014).

Furthermore, epigenetic processes like methylation of DNA and post-translational modifications to histones have an impact on how effectively stress-induced gene expression under salt stress (Golldack et al. 2011). The translation of proteins and the functioning of genes related to salinity are, however, influenced by the kind of plant. For instance, the regulation of TF bZIP24 in the NaCl sensitive variety of Arabidopsis boosted transcription while inhibiting it in the salt-resistant species (Golldacket al. 2011). Future research will contribute to a deeper knowledge of how plants respond to salinity stress through the use of emerging techniques including genomes, transcriptomics, and proteomics (Kumar et al. 2020a, b).

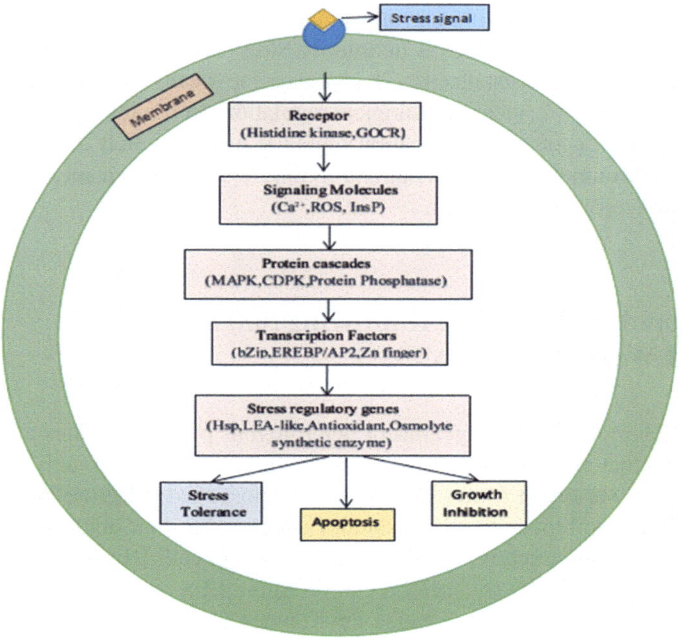

Fig. 7.2 Salinity stress signalling and regulatory intermediates

7.7 Microbial Volatile Organic Compounds (VOCs)

Numerous Volatile Organic Compounds (VOC) that PGPR secretes are expert biocontrol agents for specific nematodes and bacteria. Low molecular weight substances known as volatile organic substances (VOCs; lipophilic in nature) serve as indicators for plant development and cellular response in similar or neighboring plants (Niinemets 2010). Benzene, cyclohexane, tetradecane, and 2-(benzyloxy)-1-ethanamine are a few examples of VOCs (Table 7.3). HCN is one of the VOCs that can biocontrol some phytopathogens when provided by rhizospheric bacteria (Kanchiswamyet al. 2015). Some pathogenic growths are able to be suppressed by *Pseudomonas* sp. HCN (Hamid et al. 2021). Along with boosting systemic resistance to disease and managing plant infections, the PGPR-mediatedsynthesis of VOCs causes a various physical modification in plants that drive growth (raising shoot biomass) (Tahir et al. 2017). VOCs promote the formation of osmo-protectants like glycine betaine, whose development safeguards PS-II and maintains the functioning of enzymes and membrane toughness of cells under condition of osmotic stress (Jagendorf and Takabe 2001). *B. subtilis* VOCs alleviate salt stress by boosting tissue-specific production of the HKT1/K^+ transporter, which promotes the absorption of nutrients in root exterior while reducing the input of Na^+ to the roots (Zhang et al. 2008). The vegetative storage (which controls Na^+ homeostasis), γ-glutamyl hydrolase, and RUBISCO large-chain (This is connected to a shift in the amount of chlorophyll that leads to photosynthesis) are all induced by the VOCs released by *P. simiae*. Proteins called PGPR in Salinity Stress are thought to be important in developing a systemic resistance in *Glycine max* (soybean) (Vaishnav et al. 2015). In addition to its significance for biological regulation, VOCs may impact pollinator attraction by sending out communication signals (Liu and Brettell 2019).In addition, VOCs are environmentally beneficial and might be utilized as a substitute method for sustainable crop enhancement.

7.8 Exopolysaccharides: Improving Soil Structure and Water Retention

Exopolysaccharides are homogenous or heterologous polysaccharides produced by rhizobacteria that allow them to survive in harsh conditions. Although polysaccharides come in a variety of forms, D-glucose, mannose, and galactose have plenty of monomers that combine with other subunits, such as alduronic acids, amino sugars, etc. to form capsular protective biofilm on the exterior of cells (Rossi and De Philippis 2015). Production of EPS by bacteria can be Slime EPS or Capsule EPS (Fukami et al. 2017).However, in such adverse environmental circumstances, microorganisms, especially the PGPR, exude large molecular weight chemicals into the environment known as exopolysaccharides (EPSs), which are predominantly made up of complex organic macromolecules; Polysaccharides, along with lower amounts of protein and uronic acid (Ojuederieet al. 2017), make up about 40%-95% of the bacterial weight. Adsorption of exopolysaccharide on soil via cation bridges and Van der Waals forces which stabilize the soil structure and aggregation, that is formed under unfavorable circumstances (Shultana et al. 2022). By fusing soil fragments to aggregates, EPS generates an enclosed matrix that raises the amount of root tissue per root-adhering soil (RT/RAS), providing resistance to environmental changes. The protecting EPS capsule has sturdy water retention, which helps the plant to absorb nutrients in addition to maintaining a higher water potentiality around the plant root, protecting the plant from drying out and ensuring the survival and develoupment of plant under drought and salinity pressure (Balsanelliet al. 2014). *Bacillus* sp. and *Enterobacter* sp. MN17 *Chenopodium quinoa* MN54 improves the water relationship of plants after growth athigher-than-average salt levels (400 mM NaCl) (Yang et al. 2016). *Bacillus subtilis* was discovered to be able to combat salt stress by downregulating the expression of the HKT1/K^+ transporter when it was inoculated into *Helianthus annus* (Zhang et al. 2008). Therefore, it is essential to select bacteria strains that produce huge amounts of exopolysaccharide when developing bio-inoculants for drought-tolerant plants.

Table 7.3 PGPR generated VOC compounds effects in plant stress tolerance

Sr. No.	PGPR/PGPF Strain	Crop Rhizosphere	VOC Compounds	Signaling in plants	Effects on plant	References
1	*B. subtilis*	Rice (*Oryza sativa*)	GRB03(2–3 butanediol)	Pathogenesis-Related (PR) genes are such as CaPR2, CaPAL, and CaSAR8.2, in pepper root	Reduction in salinitystress through an Improvement in the tissue-specific produtionofthe HKT1/K + transporter that improved nutrient uptake at the root exterior whilemini- mizing the supply of Na + to the root	Yi et al. 2016
2	*P. chlororaphis O6*	*Arabidopsis thaliana*	3R-butanediol, 2R	Aba-1 and OST-1 pathways	Minimize water loss by triggering stomata closures	Cho et al. 2008
3	*Alcaligens faecalis*	Maize (*Zea mays L.*)	Butanoic acid	Auxin and gibberellin pathways	Attribute salt tolerance to plants	Bhattacharyya et al. 2015
4	*Bacillus ambifaria*	*Arabidopsis thaliana*	Dimethyl disulfide	Auxin mediated pathway	Enhanced lateral root formation and biomass	Bitas et al. 2013
5	*Paenibacilluspolymyxa E681*	Cucumber (*C. sativas L.*)	Jasmonic acid, salicylic acid	PR1, ChiB, VSP2	Suppress early seedling damping off	Iee et al. 2012
6	*Rahnellaaquatilis JZ-GX1*	*Acacia Sp*	2R, 3R-butanediol	RpSOS1, RpNHX1	Improve salt tolerance & pathogenic attack	Kong et al. 2020
7	*Trichoderma viride(Fungi)*	Tomato (*Lycopersicon esculentum Mill.*)	HCN	ACC synthase (ACS) and Cyanide-insensitive lternative oxidase (AOX) genes	Enhance defense against plant pathogen	Prasad et al. 2017
8	*B. vallismortis*	*Nicotiana tabacum*	Acetoin	NPR1 activated induced systemic resistance (ISR) and protected plants from DC3000 pathogenesis, and it requires SA/ ET	Resistance against *Pectobacteriumcarotovorum*subsp.	Ann et al. 2013

7.9 Phytohormones: Enhancing Plant Growth and Stress Tolerance

Salt stress cause plants to respond in a range of ways, include the creation of certain proteins and the production of metabolites as well as the modification of genetic expression. Several PGPR's have been shown to boost plant growth through producing hormones such as cytokinin, gibberellins (GA), and auxinas well as dropping ethylene by the use of ACC deaminase (Kumar et al. 2020a, b). Phytohormones such abscisic acid (ABA), Indole-3-acetic acid (IAA),salicylic acid (SA) and jasmonic acids (JA) also play an important role in plant salinity response (Santner et al. 2009).When plants are stressed by salinity and drought, these phytohormones activate various signalling pathways, increasing the synthesis of HSP, secondary metabolites, andantioxidant enzymes. To boost plants' tolerance to drought & salinity, numerous phytohormone-related tactics must be explored and developed (Ullah et al. 2017) (Table 7.4).It was discovered that exogenous indole-3-acetic acid (IAA) administration was successful in promoting the escalation of roots and leaves, consequently reducing salinity-induced decline in plant yield (Sehrawat et al. 2022).Production of IAA, a characteristic of PGPR, has been found to improve the efficiency of plants cultivated in salinity stress. Rhizobacteria use the tryptophan in exudates from root systems is convert in to IAA via a variety of mechanisms so that plant roots may quickly absorb it (Ilangumaran and Smith. 2017). IAA can also support plant defence; for example, exogenous IAA administration reduced tomato infection by *F.oxysporum f. sp. radicis-lycopersici* (Sharaf et al. 2004).

The function of the stress hormone abscisic acid in plant development and leaf abscission is well known. It is created when there is a water deficit, and as part of the plants' survival strategy, it triggers a set of stress-resistance genes, which results in an adaptive response (Pliego et al. 2011). Due to its formation in the roots, which happens in reaction to low water potential, roots develop and lateral roots form, which helps to increase water uptake at the root exterior (Vaishnav et al. 2016). It moves from the roots to the leaves simultaneously, controlling closing of stomatal pores actions to limit water loss by lowering water loss at the leaf surface (Kaushal and Wani, 2016). Plant-PGPR relations depend heavily on PGPRs that can produce ABA (Sehrawat et al. 2022).*A. thaliana* has ability to withstand salt stress is improved by the introduction of *Phyllobacteriumbrassicacearum STM196* because the increased ABA levels lower transpiration at the leaf surface (Bresson et al. 2013). A few PGPR species (*Novosphingobium sp. and Rhodococcus sp.*) found in rhizospheric zones with the ability of metabolizing ABA in vitro suggest another stress-reducing strategy for plants (Belimov et al. 2014). ABA aids in the build-up of Ca^{2+}, K^{+}, and compatible solutes like sugars and proline in the root vacuoles, which neutralize the negative effects of excessive salinity (Numan et al. 2018).

Another crucial group of phytohormones that promotes plant development as well as resistance to diverse stresses is cytokinin (CK) (O'Brien and Benkova 2013). In the both the root & shoot apical meristems of plants, CK maintains the cells totipotency (Howell et al. 2003). Three receptor sitesAHK2,CRE1/AHK4/WOL, and

Table 7.4 Production of secondary metabolites & Enzymes by PGPR/PGPF strains in salinity stress

Sr. No.	PGPR/PGPF Strain	Crop rhizosphere	Secondary Metabolites& Enzymes	Effects on plant	References
1	*Pseudomonas putida*Rs-198	Cotton (*Gossypium hirsutum*)	Indole-3-acetic acid (IAA) &Abscisic acid (ABA)	Promoting germination rate & enhancing plant biomass	Yao et al. 2010
2	*Bacillus sp.*RhStr_71	Pea (*Pisum sativum*)	IAA, ammonia, siderophore production and ACC deaminase activity	Increased plant biomass, enhanced root structural design, higher photosynthetic capability, and modulated synthesis of anti-oxidant enzymes	Gupta et al. 2021a, b
3	*Bacillus sp.*CHR JH 203	Pea (*Pisum sativum*)	ACC deaminase activity	Increased plant biomass, enhanced root structural design, higherphotosynthetic capability, and modulated synthesis of anti-oxidant enzymes	Gupta et al. 2021a, b
4	*Bacillus megaterium*BOFC15	Thale cress (*Arabidopsis thaliana*)	ABA production	Higher photosynthetic rate, higher photosynthetic biomass, and enhanced root architecture	Zhou et al. 2016
5	*Bacillus pumilus*	Rice (*Oryza sativa*)	IAA & ACC deaminase, phosphate	Amplified growth and Phosphate solubilization, biomass production, IAA production, enhanced ACC deaminase activity	Zhang et al. 2018
6	*Azospirillum sp.*	Wheat (*Triticum aestivum*)	Cytokinin, IAA, Gibberellic acid (GA)	Development of the pigment system, improved solute absorption, improved development of seedlings, and increased crop yield	Nia et al. 2012
7	*Bacillus subtilis Rhizo*SF 48	Tomato (*Solanum lycopersicum*)	ACC deaminase	Increases plant growth while preventing oxidative damage & drouth stress	Gowtham et al. 2020

(continued)

Table 7.4 (continued)

Sr. No.	PGPR/PGPF Strain	Crop rhizosphere	Secondary Metabolites& Enzymes	Effects on plant	References
8	*Bacillus polymyxa*	Tomato (*Lycopersicon esculentum*)	Proline	Plant physiological as well as biochemical indices were improved by proline inflation	Shintu and Jayaram 2015
9	*P. putida*H-2–3	Soybean (*Glycine max*)	ABA, salicylic acid, and gibberellic acid	Water loss reduction & promoting plant growth	Kang et al. 2014
10	*Cladosporium sphaerospermum*	Cucumber (Cucumissativus*L.*)	Gibberellic acid (GA)	Plant growth promotion	Hamayun et al. 2010
11	*Glomus intraradices*	*Medicago truncatula L*	Jasmonic acid (JA)	Enhance plant antimicrobial properties	Stumpe et al. 2005
12	*Paecilomycesformosus*	Deficient mutant rice waito-C *Oryza sativa L.cv*	GA'sandIAA	Increased the escalation	Khan et al. 2012
13	*Penicillium*MH7	Crown Daisy (*C. coronarium*)	All physiologically active GA's	Enhanced the Crown daisy's growth traits significantly	Hamayun et al. 2010

AHK3are in charge of CK signaling in plants (Kumar et al. 2020a, b). Greater amounts of cytokinin are associated with greater plant growth. When lettuce plants & *Platycladusorientalis* were inoculated with the *B. subtilis* strain, there was an improvement in the root-to-shoot signalling of CK, which aided plant escalation under stress (Arkhipova et al. 2007).

Gibberellins (GA) are another vital category of phytohormones that govern meristematic activities at the leaves & roots, as well as the elongation and division of cells as part of plant physiological and developmental processes (Wang et al. 2015). PGPRs were shown to increase their levels in plants, which suggests that plants have a forbearance mechanism for proliferation under salt stress. This factor is a major contributor to the suppression of plant development under stress conditions (Kang et al. 2014). The effects of stress on *A. thaliana* are reduced when *ZNP1* and *P. aeruginosa PM389*are inoculated with *B. tequilensis J12 & B. endophyticus J13* (Ghosh et al. 2019).

Aside from other stress regulatory hormones, the making of another stress hormone, ethylene, has been shown to promote tolerance or hasten aging. Ethylene, a critical stress hormone, is generated in stressed plants and modulates various metabolic processes (e.g., organ abscission and senescence, seed emergence) as well as increases the responses that are required for both abiotic and biotic stress tolerance (Etesami and Beattie 2017). Low amounts of ethylene have been shown to protect plants against abiotic stress, whereas large levels of ethylene assembly may generate ethylene stress, which inhibits plant growth. Ethylene, a gaseous hormone, dramatically improves plant responsiveness to stressful situations. Ethylene inhibits root growth and changes a plant's ability to fix nitrogen, performing as a negative controller of plant growth (Mahajan and Tuteja. 2005). Ethylene is synthesized in plants in a biphasic paradigm with two production peaks, where the first minimal peak is in charge of the up-regulation of genes that code for plant guarding proteins (protective proteins) and the second greater peak, known as "stress ethylene", has a negative impact on plant development and results in chlorosis, senescence, and leaf abscission (Glick et al. 2007). Because ethylene-mediated suppression of the auxin response factor limits plant growth, PGPR production of 1-aminocyclopropane-1- carboxylase (ACC) deaminase inhibits its production in plants (Glick et al. 2007). In addition to changing the appearance of genes encoding ACC oxidase and ACC synthase, which are mixed up in the manufacture of ethylene, ACC deaminase secreted by PGPR metabolizes ACC (forebearof ethylene in plants) into ammonia and α-ketoglutarate (Etesami and Beattie. 2017). The promoter regions for adhering regulatory proteins such as the Lrp box for wrap the Lrp protein, the AcdB box for wraping the regulatory protein AcdB, the FNR box for wraping the nitrate reductase protein as well as fumarate, and the CRP box for wraping the cAMP receptor protein make up the regulatory elements for the expression of the ACC deaminase gene. LRP generates an active octamer in the presence of ACC that adheres to a complex of ACC and another protein, AcdB (Cheng et al. 2008). Research found that the production rate of maize cultivated under salt stress conditions was greatly boosted by *Enterobacter sp.* and *P. fluorescens* strains that produce ACC deaminase (Nadeem et al. 2009). Chickpeas are inoculated with *Pantoeadispera*PSB3, which increases the production of IAA and ACC deaminases. This results in an upgrading in seed weight, pod size, seed quantity, and overall plant biomass (Panwar et al. 2016).

7.10 Extracellular Enzyme Production

Numerous plant diseases have been discovered to be inhibited by the development of hydrolytic enzymes including cellulases, chitinases, glucanases, proteases, etc., which hydrolyze polymeric materials like cellulose, chitin, hemicellulose, mannoproteins (cell wall), etc., (Pal and Gardener 2006). It was discovered that the protease assemble by several PGPR compounds was successful in lessening the infections caused by *Fusarium sp* & *M. phaseolina* (Gohel et al. 2004). The synthesis of β-1,3-glucanases and chitinases for use as a carbon source by PGPR such as *B. subtilis BSK17* is crucial because, it forms a large enzyme set capable of digesting the laminarin and chitin components of fungal cell wall (Kumar et al. 2012). The biocontrol capability of chitinase generated by *P.illinoissensis* spp. protects pepper (*Capsicum annuum*) against damping off & blight infections caused by *P. capsica*and *R. solani* (Jung et al. 2003).

7.11 Implications for Sustainable Agriculture: Promoting Crop Productivity and Resilience

Organic farming mostly relies on the soil's usual microflora, which contains a variety of beneficial bacteria and fungi, with arbuscular mycorrhiza fungi (AMF) are also recognized as plant growth-promoting rhizobacteria (PGPR). During nitrogen fixation, biofertilizers maintain a soil ecosystem rich in various micro- and macronutrients, the release of chemicals that control plant development, the generation of antibiotics, the solubilization or mineralization of potassium and phosphate, and the disintegrationof organic matter into the soil (Sinha et al. 2010). In general, 60–90% of the applied fertilizer is wasted, with the left over 10–40% being blot up via plants. In this point, microbial inculcates are critical in incorporated nutrient managing systems for maintaining agricultural manufacture and healthy surroundings (Adesemoye and Kloepper 2009). Rhizosphere, which is the small area of soil around by plant roots, can hold up to 10^{11}/gm of microbial cells of roots surface (Egamberdievaet al. 2008) and over 30,000 in prokaryotic species, which boost plant production in general (Mendes et al. 2013). Plant development stimulating N2-fixing cyanobacteria, rhizobacteria, mycorrhiza, plant disease oppressive valuable bacteria, bio-degrading microbes, stress tolerance endophytesare all examples of agriculturally useful microbial populations (Singh et al. 2011). *Azotobacter*, *Azospirillum*, *cyanobacteria, Rhizobium*, potassium and phosphorus solvability microbes, and mycorrhizae are a few of the PGPRs observed to proliferate in soil under no tillage or minimal tillage conditions (Aziz et al. 2012). *Azotobacter* is an important component of the nitrogen cycle in natural world because it performs a number of metabolic tasks (Sahoo et al. 2014). Apart from nitrogen fixation, *Azotobacter* can produced two vitamins such as thiamine ($C_{12}H_{17}N_4OS^+$) and riboflavin (Revillaset al. 2000), as well as plant hormones such as IAA, CK and GA, (Abd El-Fattahet al. 2013). *Azospirillum*, a free-living, aerobic, gram-variable, motile bacteria that thrives in flooded circumstances (Sahoo et al. 2014) and supports numerous aspects of plant development and improvement (Bhattacharyya and Jha 2012). *Azospirillum* has been proven to improve plant growth and agricultural production in both field trials and greenhouse (Saikia et al. 2012). Several *Azospirillum*species, together with *A. amazonense, A. brasilense*and*A. halopraeferens, A. lipoferum*, have been observed to boost crop productivity (Sahoo et al. 2014). It also extension the rate of lateral roots and improves root hair production to provide greater root surface zone for nutrient assimilation (Mehdipour-Moghaddam et al. 2012). The co-inoculation of *Azospirilliumbrasilense* and *R.meliloti*plus 2,4D developed grain production and N, P, K content in *T.aestivum* (Askary et al. 2009). Various crops have benefited from phosphate-solubilizing PGPRs are *Pseudomonas, Arthrobacter, Bacillus, Rhodococcus, Erwinia, Microbacterium, Flavobacterium, Serratia, Enterobacter*, and*Burkholderia*, (Ditta et al. 2018).Researcher also achievedthat when inoculated with phosphate solubilizing bacteria, soil aggregate stability increased by 37% and phosphorous release from rock phosphate increased by 2.35-fold and also increase the nodule number per plant by 23%,17% in shoot length and A 15% increase in chickpea grain production was also noted. *Bacillus, Acidothiobacillus, Pseudomonas*, and *Burkholderia*are examples of potassium solubilizing PGPRs (Xiao et al. 2017).A formulation based on *P.pinophilum*led to a 35% increase in the fruit yield of pomegranates, according to (Maity et al. 2019). Many different genera of bacteria, including *Aeromonas, Azadirachta, Pseudomonas, Azotobacter, Streptomyces* sp., *Serratia*, and *Burkholderia*, are known to secrete bacterial siderophores that sequester iron and promote plant growth (Sabet and Mortazaeinezhad 2018). On the 32 day of incubation in the groundnut rhizosphere, various species of Bacillus that promoted up to 82% iron release were found by (Sarwar et al. 2020).

7.12 Future Directions: Opportunities and Challenges for Microbe—Based Strategies in Agriculture

Abiotic stress in plants, such as salt and drought, results in environmental problems, but also ones that are social and economic. Plant growth and yields are declining as a result of the changing environmental circumstances. Plant cells create ROS as abiotic stressors increase, which reduces ecological fitness. In the ensuing decades, the situation will worsen, endangering plant life. Plants enlist antioxidants to maintain a steadiness between the creation and quenching of ROS. The plant's reaction to these occurrences is so quick that it is unable to resist challenging circumstances. Phosphate solubilization, and nutrient mobilisation, creation of phytohormones, volatile organic compounds (VOCs), vitamins, and modulation of antioxidant machinery are just a few of the mechanisms by which PGPB promote plant growth. Other mechanisms include osmotic regulation, retaining ion homeostasis, metabolic reprogramming, and modulation of biochemical & molecular pathways. Therefore, PGPB are a good substitute for conventional fertilisers due to their affordability, sustainability, and capacity to boost plant tolerance to a variety of stressors, such as salinity and drought.

Future research initiatives targeted at advancing sustainable agriculture should look into PGPBs that may increase plant resilience to a range of abiotic stresses, mainly those that correspond to field conditions. Scientists must do field research and educate farmers about the advantages that helpful microorganisms can have on plant physical condition and soil value if beneficial bacteria are to be used widely and effectively. In addition, recently developed nanoencapsulation technology is applicable to field testing. The method can be used to spread PGPR more widely, protect them from sudden environmental shocks, and assist control microbial liberate in the rhizosphere and field. If tripartite "plant-fungal-bacterial" symbioses can provide synergistic effect on plants, more investigation is required. Future agriculture in saline and drought-prone areas seems to benefit from these strategies. With the information provided above, it is clear that we are headed in the correct direction to accomplishing our objective of sustainable food manufacture in a climate that is always changing. Governments as well as federal agencies must encourage the use of biofertilizers made from PGPB as eco-friendly substitutes for agricultural development. Entrepreneurs should invest in biofertilizer businesses more frequently and allocate funding for start-ups. Public education is also required to convey to farmers and customers the advantages of PGPB-based biofertilizers for suitable agriculture.

Acknowledgements None.

Author Contribution All authors are equally contributed to prepared the manuscript. All authors read and approved the final manuscript.

Funding No funding was received to assist with the preparation of this manuscript.

Ethical Approval Not applicable.

Conflict of Interest The authors declare that they have no conflict of interest.

References

Abbas G, Saqib M, Akhtar J, Haq MAU (2015) Interactive effects of salinity and iron deficiency on different rice genotypes. J Plant Nutr Soil Sci 178(2):306–311

Abd El-Fattah DA, Eweda WE, Zayed MS, Hassanein MK (2013) Effect of carrier materials, sterilization method, and storage temperature on survival and biological activities of *Azotobacter chroococcum* inoculant. Annals of Agricultural Sciences 58(2):111–118

Adesemoye AO, Kloepper JW (2009) Plant–microbes interactions in enhanced fertilizer-use efficiency. Appl Microbiol Biotechnol 85:1–12

Aguado-Santacruz GA, Moreno-Gómez B, Jiménez-Francisco B, García-Moya E, Preciado-Ortiz RE (2012) Impact of the microbial siderophores and phytosiderophores on the iron assimilation by plants: a synthesis. Revistafitotecniamexicana 35(1):9–21

Ali A, Maggio A, Bressan RA, Yun DJ (2019) Role and functional differences of HKT1-type transporters in plants under salt stress. Int J Mol Sci 20(5):1059

Ann MN, Cho YE, Ryu HJ, Kim HT, Park K (2013) Growth promotion of tobacco plant by 3-hydroxy-2-butanone from *Bacillus vallismortis* EXTN-1. Korean J Pestic Sci 17(4):388–393

Arkhipova TN, Prinsen E, Veselov SU, Martinenko EV, Melentiev AI, Kudoyarova GR (2007) Cytokinin producing bacteria enhance plant growth in drying soil. Plant Soil 292:305–315

Askary, M., Mostajeran, A., Amooaghaei, R., &Mostajeran, M (2009) Influence of the co-inoculation *Azospirillumbrasilense* and *Rhizobium meliloti* plus 2, 4-D on grain yield and N, P, K content of *Triticum aestivum* (cv. *Baccros and Mahdavi*). Am Eurasian J Agric Environ Sci 5(3):296–307

Azziz G, Bajsa N, Haghjou T, Taule C, Valverde A, Igual JM, Arias A (2012) Abundance, diversity and prospecting of culturable phosphate solubilizing bacteria on soils under crop–pasture rotations in a no-tillage regime in Uruguay. Appl Soil Ecol 61:320–326

Balsanelli E, de Baura VA, Pedrosa F, de Souza EM, Monteiro RA (2014) Exopolysaccharide biosynthesis enables mature biofilm formation on abiotic surfaces by *Herbaspirillumseropedicae*. PLoS ONE 9:e110392

Bano A, Fatima M (2009) Salt tolerance in Zea mays (L). following inoculation with *Rhizobium* and *Pseudomonas*. Biol Fertil Soils 45:405–413

Barnawal D, Bharti N, Pandey SS, Pandey A, Chanotiya CS, Kalra A (2017) Plant growth promoting rhizobacteria enhance wheat salt and drought stress tolerance by altering endogenous phytohormone levels and TaCTR1/TaDREB2 expression. Physiol Plant 161:502–514

Belimov AA, Dodd IC, Safronova VI, Dumova VA, Shaposhnikov AI, Ladatko AG, Davies WJ (2014) Abscisic acid metabolizing rhizobacteria decrease ABA concentrations in planta and alter plant growth. Plant Physiol Biochem 74:84–91

Berendsen RL, Pieterse CM, Bakker PA (2012) The rhizosphere microbiome and plant health. Trends Plant Sci 17(8):478–486

Bharti N, Pandey SS, Barnawal D, Patel VK, Kalra A (2016) Plant growth promoting rhizobacteria *Dietzianatronolimnaea* modulates the expression of stress responsive genes providing protection of wheat from salinity stress. Sci Rep 6(1):1–16

Bhat MA, Kumar V, Bhat MA, Wani IA, Dar FL, Farooq I, Jan AT (2020) Mechanistic insights of the interaction of plant growth-promoting rhizobacteria (PGPR) with plant roots toward enhancing plant productivity by alleviating salinity stress. Front Microbiol 11:1952

Bhattacharyya D, Yu SM, Lee YH (2015) Volatile compounds from *Alcaligenes faecalis* JBCS1294 confer salt tolerance in Arabidopsis thaliana through the auxin and gibberellin pathways and differential modulation of gene expression in root and shoot tissues. Plant Growth Regul 75:297–306

Bhattacharyya PN, Jha DK (2012) Plant growth-promoting rhizobacteria (PGPR): emergence in agriculture. World J Microbiol Biotechnol 28:1327–1350

Bieleski, RL (1982) Sugar alcohols In: Loewus FA, Tanner W (eds) Plant Carbohydrates I. Intracellular carbohydrates. Encyclopaedia of Plant Physiology New Series vol 13A

Bitas V, Kim HS, Bennett JW, Kang S (2013) Sniffing on microbes: diverse roles of microbial volatile organic compounds in plant health. Mol Plant Microbe Interact 26(8):835–843

Bresson J, Varoquaux F, Bontpart T, Touraine B, Vile D (2013) The PGPR strain *Phyllobacterium-brassicacearum* STM 196 induces a reproductive delay and physiological changes that result in improved drought tolerance in a rabidopsis. New Phytol 200(2):558–569

Chakraborty, K, Mondal, S, Bhaduri, D, Mohanty, A, Paul A (2022) Interplay between sodium and chloride decides the plant's fate under salt and drought stress conditions. In Plant nutrition and food security in the era of climate change (pp 271–314). Academic Press

Chen TH, Murata N (2002) Enhancement of tolerance of abiotic stress by metabolic engineering of betaines and other compatible solutes. Curr Opin Plant Biol 5(3):250–257

Cheng Z, Duncker BP, McConkey BJ, Glick BR (2008) Transcriptional regulation of ACC deaminase gene expression in *Pseudomonas putida* UW4. Can J Microbiol 54(2):128–136

Cho SM, Kang BR, Han SH, Anderson AJ, Park JY, Lee YH, Kim YC (2008) 2R, 3R-butanediol, a bacterial volatile produced by Pseudomonas chlororaphis O6, is involved in induction of systemic tolerance to drought in Arabidopsis thaliana. Mol Plant Microbe Interact 21(8):1067–1075

Deinlein U, Stephan AB, Horie T, Luo W, Xu G, Schroeder JI (2014) Plant salt-tolerance mechanisms. Trends Plant Sci 19(6):371–379

Ditta A, Imtiaz M, Mehmood S, Rizwan MS, Mubeen F, Aziz O, Tu S (2018) Rock phosphate-enriched organic fertilizer with phosphate-solubilizing microorganisms improves nodulation, growth, and yield of legumes. Commun Soil Sci Plant Anal 49(21):2715–2725

Egamberdieva D, Kamilova F, Validov S, Gafurova L, Kucharova Z, Lugtenberg B (2008) High incidence of plant growth-stimulating bacteria associated with the rhizosphere of wheat grown on salinated soil in Uzbekistan. Environ Microbiol 10(1):1–9

Einset J, Nielsen E, Connolly EL, Bones A, Sparstad T, Winge P, Zhu JK (2007) Membrane-trafficking RabA4c involved in the effect of glycine betaine on recovery from chilling stress in *Arabidopsis*. Physiol Plant 130(4):511–518

El-Tarabily KA, ElBaghdady KZ, AlKhajeh AS, Ayyash MM, Aljneibi RS, El-Keblawy A, AbuQamar SF (2020) Polyamine-producing actinobacteria enhance biomass production and seed yield in *Salicornia bigelovii*. Biol Fertil Soils 56:499–519

Etesami H (2018) Can interaction between silicon and plant growth promoting rhizobacteria benefit in alleviating abiotic and biotic stresses in crop plants? Agr Ecosyst Environ 253:98–112

Etesami, H, Beattie GA (2017) Plant-microbe interactions in adaptation of agricultural crops to abiotic stress conditions. Probiotics and plant health, 163–200

Etesami H, Beattie GA (2018) Mining halophytes for plant growth-promoting halotolerant bacteria to enhance the salinity tolerance of non-halophytic crops. Front Microbiol 9:148

Ferreira MJ, Silva H, Cunha A (2019) Siderophore-producing rhizobacteria as a promising tool for empowering plants to cope with iron limitation in saline soils: A review. Pedosphere 29(4):409–420

Fukami J, de la Osa C, Ollero FJ, Megías M, Hungria M (2017) Co-inoculation of maize with Azospirillumbrasilense and Rhizobium tropici as a strategy to mitigate salinity stress. Funct Plant Biol 45(3):328–339

Ghosh D, Gupta A, Mohapatra S (2019) A comparative analysis of exopolysaccharide and phytohormone secretions by four drought-tolerant rhizobacterial strains and their impact on osmotic-stress mitigation in *Arabidopsis thaliana*. World J Microbiol Biotechnol 35:1–15

Ghosh SK, Pal SUJOY, Chakraborty NILOY (2015) The qualitative and quantitative assay of siderophore production by some microorganisms and effect of different media on its production. Int J Chem Sci 13(4):1621–1629

Glick BR, Cheng Z, Czarny J, Duan J (2007) Promotion of plant growth by ACC deaminase-producing soil bacteria. New Perspect Approaches Plant Growth-Promot Rhizobacteria Res, pp 329–339

Godfray, HCJ, Beddington JR, Crute IR, Haddad L, Lawrence D, Muir JF, Toulmin C (2010) Food security: the challenge of feeding 9 billion people. Science 327(5967);812–818

Gohel V, Chaudhary T, Vyas P, Chhatpar HS (2004) Isolation and identification of marine chitinolytic bacteria and their potential in antifungal biocontrol. Indian J Exp Biol 42:715–720

Golldack D, Lüking I, Yang O (2011) Plant tolerance to drought and salinity: stress regulating transcription factors and their functional significance in the cellular transcriptional network. Plant Cell Rep 30:1383–1391

Gowtham HG, Singh B, Murali M, Shilpa N, Prasad M, Aiyaz M, Niranjana SR (2020) Induction of drought tolerance in tomato upon the application of ACC deaminase producing plant growth promoting rhizobacterium Bacillus subtilis Rhizo SF 48. Microbiol Res 234:126422

Gupta A, Bano A, Rai S, Dubey P, Khan F, Pathak N, Sharma S (2021a) Plant Growth Promoting Rhizobacteria (PGPR): A sustainable agriculture to rescue the vegetation from the effect of biotic stress: A Review. Lett. Appl. Nanobiosci 10:2459–2465

Gupta A, Bano A, Rai S, Kumar M, Ali J, Sharma S, Pathak N (2021) ACC deaminase producing plant growth promoting rhizobacteria enhance salinity stress tolerance in Pisum sativum. 3 Biotech 11(12):514

Gupta G, Parihar SS, Ahirwar NK, Snehi SK, Singh V (2015) Plant growth promoting rhizobacteria (PGPR): current and future prospects for development of sustainable agriculture. J MicrobBiochem Technol 7(2):096–102

Hamayun M, Khan SA, Khan AL, Rehman G, Kim YH, Iqbal I, Lee IJ (2010) Gibberellin production and plant growth promotion from pure cultures of Cladosporium sp. MH-6 isolated from cucumber (Cucumis sativus L.). Mycologia 102(5):989–995

Hamid B, Zaman M, Farooq S, Fatima S, Sayyed RZ, Baba ZA, Suriani NL (2021) Bacterial plant biostimulants: a sustainable way towards improving growth, productivity, and health of crops. Sustainability 13(2856):1–24

Hasanuzzaman M, Alam MM, Nahar K, Ahamed KU, Fujita M (2014) Exogenous salicylic acid alleviates salt stress-induced oxidative damage in Brassica napus by enhancing the antioxidant defense and glyoxalase systems. Aust J Crop Sci 8(4):631–639

Howell SH, Lall S, Che P (2003) Cytokinins and shoot development. Trends Plant Sci 8(9):453–459

Huang CJ, Tsay JF, Chang SY, Yang HP, Wu WS, Chen CY (2012) Dimethyl disulfide is an induced systemic resistance elicitor produced by Bacillus cereus C1L. Pest Manag Sci 68(9):1306–1310

Ilangumaran G, Smith DL (2017) Plant growth promoting rhizobacteria in amelioration of salinity stress: a systems biology perspective. Front Plant Sci 8:1768

Jagendorf AT, Takabe T (2001) Inducers of glycinebetaine synthesis in barley. Plant Physiol 127(4):1827–1835

Jung WJ, An KN, Jin YL, Park RD, Lim KT, Kim KY, Kim TH (2003) Biological control of damping-off caused by *Rhizoctonia solani* using chitinase-producing Paenibacillusillinoisensis KJA-424. Soil Biol Biochem 35(9):1261–1264

Kanchiswamy CN, Malnoy M, Maffei ME (2015) Chemical diversity of microbial volatiles and their potential for plant growth and productivity. Front Plant Sci 6:151

Kang SM, Radhakrishnan R, Khan AL, Kim MJ, Park JM, Kim BR, Lee IJ (2014) Gibberellin secreting rhizobacterium, Pseudomonas putida H-2-3 modulates the hormonal and stress physiology of soybean to improve the plant growth under saline and drought conditions. Plant Physiol Biochem 84:115–124

Kaushal M, Wani SP (2016) Plant-growth-promoting rhizobacteria: drought stress alleviators to ameliorate crop production in drylands. Annals of Microbiology 66:35–42

Kermeur N, Pédrot M, Cabello-Hurtado F (2023) Iron availability and homeostasis in plants: a review of responses, adaptive mechanisms, and signaling. Plant Abiotic Stress Signaling, pp 49–81

Khan AL, Hamayun M, Kang SM, Kim YH, Jung HY, Lee JH, Lee IJ (2012) Endophytic fungal association via gibberellins and indole acetic acid can improve plant growth under abiotic stress: an example of Paecilomycesformosus LHL10. BMC Microbiol 12:1–14

Khan A, Zhao XQ, Javed MT, Khan KS, Bano A, Shen RF, Masood S (2016) Bacillus pumilus enhances tolerance in rice (Oryza sativa L.) to combined stresses of NaCl and high boron due to limited uptake of Na+. Environ Exp Bot 124:120–129

Khan MA, Asaf S, Khan AL, Adhikari A, Jan R, Ali S, Lee IJ (2019) Halotolerant rhizobacterial strains mitigate the adverse effects of NaCl stress in soybean seedlings. BioMed Res Int

Kim K, Jang YJ, Lee SM, Oh BT, Chae JC, Lee KJ (2014) Alleviation of salt stress by Enterobacter sp. EJ01 in tomato and Arabidopsis is accompanied by up-regulation of conserved salinity responsive factors in plants. Mol Cells 37(2):109

Kong WL, Rui L, Ni H, Wu XQ (2020) Antifungal effects of volatile organic compounds produced by Rahnellaaquatilis JZ-GX1 against Colletotrichum gloeosporioides in Liriodendron chinense× tulipifera. Front Microbiol 11:1114

Kumar A, Verma JP (2018) Does plant—microbe interaction confer stress tolerance in plants: a review? Microbiol Res 207:41–52

Kumar A, Singh S, Gaurav AK, Srivastava S, Verma JP (2020a) Plant growth-promoting bacteria: biological tools for the mitigation of salinity stress in plants. Front Microbiol 11:1216

Kumar M, Giri VP, Pandey S, Gupta A, Patel MK, Bajpai AB, Siddique KH (2021) Plant-Growth-Promoting Rhizobacteria emerging as an effective bioinoculant to improve the growth, production, and stress tolerance of vegetable crops. Int J Mol Sci 22(22):12245

Kumar P, Dubey RC, Maheshwari DK (2012) Bacillus strains isolated from rhizosphere showed plant growth promoting and antagonistic activity against phytopathogens. Microbiol Res 167(8):493–499

Kumar V, Khare T, Sharma M, Wani SH (2017) ROS-induced signaling and gene expression in crops under salinity stress. React Oxyg Species Antioxid Syst Plants: Role Regul Abiotic Stress, pp 159–184

Kumar V, Singhal RK, Kumar N, Bose B (2020) Micro-nutrient seed priming: a pragmatic approach towards abiotic stress management. New Front Stress Manag Durable Agric, pp 231–255

Kumawat KC, Nagpal S, Sharma P (2022) Potential of plant growth-promoting rhizobacteria-plant interactions in mitigating salt stress for sustainable agriculture: a review. Pedosphere 32(2):223–245

Kushwaha B, Jadhav I, Verma HN, Geethadevi A, Parashar D, Jadhav K (2019) Betaine accumulation suppresses the de-novo synthesis of ectoine at a low osmotic concentration in Halomonassp SBS 10, a bacterium with broad salinity tolerance. Mol Biol Rep 46(5):4779–4786

Lee B, Farag MA, Park HB, Kloepper JW, Lee SH, Ryu CM (2012) Induced resistance by a long-chain bacterial volatile: elicitation of plant systemic defense by a C13 volatile produced by Paenibacilluspolymyxa. PLoS ONE 7(11):e48744

Liu H, Brettell LE (2019) Plant defense by VOC-induced microbial priming. Trends Plant Sci 24(3):187–189

Mahajan S, Tuteja N (2005) Cold, salinity and drought stresses: an overview. Arch Biochem Biophys 444(2):139–158

Maity A, Sharma J, Pal RK (2019) Novel potassium solubilizing bio-formulation improves nutrient availability, fruit yield and quality of pomegranate (Punica granatum L.) in semi-arid ecosystem. Sci Hortic 255:14–20

Mehdipour-Moghaddam MJ, Emtiazi G, Salehi Z (2012) Enhanced auxin production by Azospirillum pure cultures from plant root exudates. J Agr Sci Tech 14:985–994

Mendes R, Garbeva P, Raaijmakers JM (2013) The rhizosphere microbiome: significance of plant beneficial, plant pathogenic, and human pathogenic microorganisms. FEMS Microbiol Rev 37(5):634–663

Miller GAD, Suzuki N, Ciftci-Yilmaz SULTAN, Mittler RON (2010) Reactive oxygen species homeostasis and signalling during drought and salinity stresses. Plant Cell Environ 33(4):453–467

Mukhopadhyay R, Sarkar B, Jat HS, Sharma PC, Bolan NS (2021) Soil salinity under climate change: challenges for sustainable agriculture and food security. J Environ Manage 280:111736

Nadeem SM, Zahir ZA, Naveed M, Arshad M (2009) Rhizobacteria containing ACC-deaminase confer salt tolerance in maize grown on salt-affected fields. Can J Microbiol 55(11):1302–1309

Nawaz MQ, Ahmed K, Qadir G, Rizwan M, Nawaz MF, Sarfraz M (2020) Growth and yield of turnip (Brassica rapa L.) in response to different sowing methods and nitrogen levels in salt-affected soils. Pakistan J Agric Res 33(1):126–134

Nia SH, Zarea MJ, Rejali F, Varma A (2012) Yield and yield components of wheat as affected by salinity and inoculation with Azospirillum strains from saline or non-saline soil. J Saudi Soc Agric Sci 11(2):113–121

Niinemets Ü (2010) Mild versus severe stress and BVOCs: thresholds, priming and consequences. Trends Plant Sci 15(3):145–153

Numan M, Bashir S, Khan Y, Mumtaz R, Shinwari ZK, Khan AL, Ahmed AH (2018) Plant growth promoting bacteria as an alternative strategy for salt tolerance in plants: a review. Microbiol Res 209:21–32

O'Brien JA, Benková E (2013) Cytokinin cross-talking during biotic and abiotic stress responses. Front Plant Sci 4:451

Ojuederie OB, Babalola OO (2017) Microbial and plant-assisted bioremediation of heavy metal polluted environments: a review. Int J Environ Res Public Health 14(12):1504

Orozco-Mosqueda MDC, Duan J, DiBernardo M, Zetter E, Campos-García J, Glick BR, Santoyo G (2019) The production of ACC deaminase and trehalose by the plant growth promoting bacterium Pseudomonas sp. UW4 synergistically protect tomato plants against salt stress. Front Microbiol 10:1392.

Pal KK, McSpadden Gardener B (2006) Biological Control of Plant Pathogens. Plant Heal Instr

Panwar M, Tewari R, Gulati A, Nayyar H (2016) Indigenous salt-tolerant *rhizobacterium Pantoead-ispersa* (PSB3) reduces sodium uptake and mitigates the effects of salt stress on growth and yield of chickpea. Acta Physiol Plant 38:1–12

Parray JA, Jan S, Kamili AN, Qadri RA, Egamberdieva D, Ahmad P (2016) Current perspectives on plant growth-promoting rhizobacteria. J Plant Growth Regul 35:877–902

Pinedo I, Ledger T, Greve M, Poupin MJ (2015) BurkholderiaphytofirmansPsJN induces long-term metabolic and transcriptional changes involved in Arabidopsis thaliana salt tolerance. Front Plant Sci 6:466

Pliego C, Ramos C, de Vicente A, Cazorla FM (2011) Screening for candidate bacterial biocontrol agents against soilborne fungal plant pathogens. Plant Soil 340:505–520

Prasad RM, Sagar BV, Devi GU, Triveni S, Rao SK, Chari DK (2017) Isolation and screening of bacterial and fungal isolates for plant growth promoting properties from tomato (Lycopersicon esculentum Mill.). Int J Curr Microbiol App Sci 6(8):753–761

Qin Y, Druzhinina IS, Pan X, Yuan Z (2016) Microbially mediated plant salt tolerance and microbiome-based solutions for saline agriculture. Biotechnol Adv 34(7):1245–1259

Qiu QS, Guo Y, Dietrich MA, Schumaker KS, Zhu JK (2002) Regulation of SOS1, a plasma membrane Na+/H+ exchanger in Arabidopsis thaliana, by SOS2 and SOS3. Proc Natl Acad Sci 99(12):8436–8441

Rai GK, Kumar P, Choudhary SM, Singh H, Adab K, Kosser R, Rouphael Y (2023) Antioxidant potential of glutathione and crosstalk with Phytohormones in enhancing abiotic stress tolerance in crop plants. Plants 12(5):1133

Rajkumar M, Ae N, Prasad MNV, Freitas H (2010) Potential of siderophore-producing bacteria for improving heavy metal phytoextraction. Trends Biotechnol 28(3):142–149

Ramadoss D, Lakkineni VK, Bose P, Ali S, Annapurna K (2013) Mitigation of salt stress in wheat seedlings by halotolerant bacteria isolated from saline habitats. Springer plus 2(1):1–7

Raza A, Razzaq A, Mehmood SS, Zou X, Zhang X, Lv Y, Xu J (2019) Impact of climate change on crops adaptation and strategies to tackle its outcome: a review. Plants 8(2):34

Revillas JJ, Rodelas B, Pozo C, Martínez-Toledo MV, González-López J (2000) Production of B-group vitamins by two Azotobacter strains with phenolic compounds as sole carbon source under diazotrophic and adiazotrophic conditions. J Appl Microbiol 89(3):486–493

Rossi F, De Philippis R (2015) Role of cyanobacterial exopolysaccharides in phototrophic biofilms and in complex microbial mats. Life 5(2):1218–1238

Sabet H, Mortazaeinezhad F (2018) Yield, growth and Fe uptake of cumin (Cuminum cyminum L.) affected by Fe-nano, Fe-chelated and Fe-siderophore fertilization in the calcareous soils. J Trace Elem Med Biol 50:154–160

Sadeghi A, Karimi E, Dahaji PA, Javid MG, Dalvand Y, Askari H (2012) Plant growth promoting activity of an auxin and siderophore producing isolate of Streptomyces under saline soil conditions. World J Microbiol Biotechnol 28:1503–1509

Sahoo RK, Ansari MW, Dangar TK, Mohanty S, Tuteja N (2014) Phenotypic and molecular characterisation of efficient nitrogen-fixing Azotobacter strains from rice fields for crop improvement. Protoplasma 251:511–523

Saikia SP, Bora D, Goswami A, Mudoi KD, Gogoi A (2012) A review on the role of Azospirillum in the yield improvement of non leguminous crops. Afr J Microbiol Res 6(6):1085–1102
Santner A, Estelle M (2009) Recent advances and emerging trends in plant hormone signalling. Nature 459(7250):1071–1078
Sarwar S, Khaliq A, Yousra M, Sultan T, Ahmad N, Khan MZ (2020) Screening of siderophore-producing PGPRs isolated from groundnut (Arachis hypogaea L.) rhizosphere and their influence on iron release in soil. Commun Soil Sci Plant Anal 51(12):1680–1692
Sehrawat A, Sindhu SS, Glick BR (2022) Hydrogen cyanide production by soil bacteria: biological control of pests and promotion of plant growth in sustainable agriculture. Pedosphere 32(1):15–38
Shabala S, Pottosin I (2014) Regulation of potassium transport in plants under hostile conditions: implications for abiotic and biotic stress tolerance. Physiol Plant 151(3):257–279
Sharaf EF, Farrag AA (2004) Induced resistance in tomato plants by IAA against Fusarium *oxysporumlycopersici*. Pol J Microbiol 53(2):111
Sharifi M, Ghorbanli M, Ebrahimzadeh H (2007) Improved growth of salinity-stressed soybean after inoculation with salt pre-treated mycorrhizal fungi. J Plant Physiol 164(9):1144–1151
Sharma A, Singh P, Kumar S, Kashyap PL, Srivastava AK, Chakdar H, Sharma AK (2015) Deciphering diversity of salt-tolerant bacilli from saline soils of eastern Indo-gangetic plains of India. Geomicrobiol J 32(2):170–180
Sharma A, Vaishnav A, Jamali H, Srivastava AK, Saxena AK, Srivastava AK (2016) Halophilic bacteria: potential bioinoculants for sustainable agriculture and environment management under salt stress. Plant-Microbe Interact: Approach Sustain Agric, pp 297–325
Shen B, Jensen RG, Bohnert HJ (1997) Mannitol protects against oxidation by hydroxyl radicals. Plant Physiol 115(2):527–532
Shen X, Hu H, Peng H, Wang W, Zhang X (2013) Comparative genomic analysis of four representative plant growth-promoting rhizobacteria in Pseudomonas. BMC Genomics 14:1–20
Sheveleva EV, Marquez S, Chmara W, Zegeer A, Jensen RG, Bohnert HJ (1998) Sorbitol-6-phosphate dehydrogenase expression in transgenic tobacco: high amounts of sorbitol lead to necrotic lesions. Plant Physiol 117(3):831–839
Shim JS, Seo JS, Seo JS, Kim Y, Koo Y, Do Choi Y, Jung C (2019) Heterologous expression of bacterial trehalose biosynthetic genes enhances trehalose accumulation in potato plants without adverse growth effects. Plant Biotechnol Rep 13:409–418
Shintu PV, Jayaram KM (2015) Phosphate solubilising bacteria (Bacillus polymyxa)-An effective approach to mitigate drought in tomato (Lycopersicon esculentum Mill.). Trop Plant Res 2(1):2349–9265
Shrivastava P, Kumar R (2015) Soil salinity: A serious environmental issue and plant growth promoting bacteria as one of the tools for its alleviation. Saudi J Biol Sci 22(2):123–131
Shultana R, Zuan ATK, Naher UA, Islam AM, Rana MM, Rashid MH, Hasan AK (2022) The PGPR mechanisms of salt stress adaptation and plant growth promotion. Agronomy 12(10):2266
Singh JS, Pandey VC, Singh DP (2011) Efficient soil microorganisms: a new dimension for sustainable agriculture and environmental development. Agr Ecosyst Environ 140(3–4):339–353
Singh S, Rajput VD, Upadhyay SK, Minkina T (2022) Arsenic contamination in rice agro-ecosystems: mitigation strategies for safer crop production. J Plant Growth Regul, pp 1–12
Sinha RK, Valani D, Chauhan K, Agarwal S (2010) Embarking on a second green revolution for sustainable agriculture by vermiculture biotechnology using earthworms: reviving the dreams of Sir Charles Darwin. J Agric Biotechnol Sustain Dev 2(7):113
Stumpe M, Carsjens JG, Stenzel I, Göbel C, Lang I, Pawlowski K, Feussner I (2005) Lipid metabolism in arbuscular mycorrhizal roots of Medicago truncatula. Phytochemistry 66(7):781–791
Surowka E, Latowski D, Dziurka M, Rys M, Maksymowicz A, Żur I, Miszalski Z (2021) ROS-scavengers, osmoprotectants and violaxanthin de-epoxidation in salt-stressed Arabidopsis thaliana with different tocopherol composition. Int J Mol Sci 22(21):11370

Tahir HAS, Gu Q, Wu H, Niu Y, Huo R, Gao X (2017) Bacillus volatiles adversely affect the physiology and ultra-structure of *Ralstonia solanacearum* and induce systemic resistance in tobacco against bacterial wilt. Sci Rep 7(1):40481

Tang Y, Bao X, Zhi Y, Wu Q, Guo Y, Yin X, Liu K (2019) Overexpression of a MYB family gene, OsMYB6, increases drought and salinity stress tolerance in transgenic rice. Front Plant Sci 10:168

Ullah A, Mushtaq H, Fahad S, Hakima SA, Chaudhary HJ (2017) Plant growth promoting potential of bacterial endophytes in novel association with Olea ferruginea and Withaniacoagulans. Microbiology 86:119–127

Ullah S, Bano A (2015) Isolation of plant-growth-promoting rhizobacteria from rhizospheric soil of halophytes and their impact on maize (Zea mays L.) under induced soil salinity. Can J Microbiol 61(4):307–313

Upadhyay SK, Singh DP (2015) Effect of salt-tolerant plant growth-promoting rhizobacteria on wheat plants and soil health in a saline environment. Plant Biol 17(1):288–293

Vaishnav A, Kumari S, Jain S, Varma A, Choudhary DK (2015) Putative bacterial volatile-mediated growth in soybean (Glycine max L. Merrill) and expression of induced proteins under salt stress. J Appl Microbiol 119(2):539–551

Vaishnav A, Varma A, Tuteja N, Choudhary DK (2016) PGPR-mediated amelioration of crops under salt stress. Plant-microbe interaction: an approach to sustainable agriculture, pp 205–226

Vílchez JI, García-Fontana C, Román-Naranjo D, González-López J, Manzanera M (2016) Plant drought tolerance enhancement by trehalose production of desiccation-tolerant microorganisms. Front Microbiol 7:1577

Wang GL, Que F, Xu ZS, Wang F, Xiong AS (2015) Exogenous gibberellin altered morphology, anatomic and transcriptional regulatory networks of hormones in carrot root and shoot. BMC Plant Biol 15:1–12

Wang X, Chen X, Liu Y, Gao H, Wang Z, Sun G (2011) CkDREB gene in Caragana korshinskii is involved in the regulation of stress response to multiple abiotic stresses as an AP2/EREBP transcription factor. Mol Biol Rep 38:2801–2811

Wang Y, Gao C, Liang Y, Wang C, Yang C, Liu G (2010) A novel bZIP gene from Tamarixhispida mediates physiological responses to salt stress in tobacco plants. J Plant Physiol 167(3):222–230

Xiao Y, Wang X, Chen W, Huang Q (2017) Isolation and identification of three potassium-solubilizing bacteria from rape rhizospheric soil and their effects on ryegrass. Geomicrobiol J 34(10):873–880

Yamaoka M, Osawa S, Morinaga T, Takenaka S, Yoshida KI (2011) A cell factory of Bacillus subtilis engineered for the simple bioconversion of myo-inositol to scyllo-inositol, a potential therapeutic agent for Alzheimer's disease. Microb Cell Fact 10(1):1–6

Yang A, Akhtar SS, Iqbal S, Amjad M, Naveed M, Zahir ZA, Jacobsen SE (2016) Enhancing salt tolerance in quinoa by halotolerant bacterial inoculation. Funct Plant Biol43(7):632–642

Yao L, Wu Z, Zheng Y, Kaleem I, Li C (2010) Growth promotion and protection against salt stress by Pseudomonas putida Rs-198 on cotton. Eur J Soil Biol 46(1):49–54

Yi HS, Ahn YR, Song GC, Ghim SY, Lee S, Lee G, Ryu CM (2016) Impact of a bacterial volatile 2, 3-butanediol on Bacillus subtilis rhizosphere robustness. Front Microbiol 7:993

Zhang H, Kim MS, Sun Y, Dowd SE, Shi H, Paré PW (2008) Soil bacteria confer plant salt tolerance by tissue-specific regulation of the sodium transporter HKT1. Mol Plant Microbe Interact 21(6):737–744

Zhang H, Zhang Y, Deng C, Deng S, Li N, Zhao C, Chen S (2018) The Arabidopsis Ca^{2+} dependent protein kinase CPK12 is involved in plant response to salt stress. Int J Mol Sci 19(12):4062

Zhou C, Ma Z, Zhu L, Xiao X, Xie Y, Zhu J, Wang J (2016) Rhizobacterial strain Bacillus megaterium BOFC15 induces cellular polyamine changes that improve plant growth and drought resistance. Int J Mol Sci 17(6):976

Zhu JK (2002) Salt and drought stress signal transduction in plants. Annu Rev Plant Biol 53(1):247–273
Zhu JK (2003) Regulation of ion homeostasis under salt stress. Curr Opin Plant Biol 6(5):441–445
Zulfiqar F, Akram NA, Ashraf M (2020) Osmoprotection in plants under abiotic stresses: new insights into a classical phenomenon. Planta 251:1–17

Chapter 8
Plant Growth Promoting Rhizobacteria: Future Direction for Mitigating Risk of Crop Production

Susanta Dutta, Bappa Paramanik, Puja Singh, Sumana Balo, Shyamashree Roy, Malay Kanti Modak, Gopal Dutta, and Animesh Mandal

Keywords Rhizobacteria · Inorganic fertilizers · Crop production · Microorganisms · Plant growth

---

S. Dutta (✉)
Agronomy, Horticulture College, Birsa Agricultural University, Khuntpani, Ranchi, Jharkhand, India
e-mail: susantadutta19@gmail.com

B. Paramanik
Soil Science, Dakshin Dinajpur KVK, Uttar Banga Krishi Viswavidyalaya, Cooch Behar, West Bengal, India

P. Singh
Soil Science and Agricultural Chemistry, Horticulture College, Birsa Agricultural University, Khuntpani, Ranchi, Jharkhand, India

S. Balo
Soil Science and Agricultural Chemistry, School of Agriculture, Gandhi Institute of Education and Technology University (GIETU), Gunupur, Rayagada, Odisha, India

S. Roy
Agronomy, Regional Research Station (OAZ), Uttar Banga Krishi Viswavidyalaya, Majhian, Dakshin Dinajpur, West Bengal, India

M. K. Modak
Department of Seed Science and Technology, Institute of Agricultural Science, University of Calcutta, Kolkatta, West Bengal, India

G. Dutta
Department of Agricultural Meteorology, Bidhan Chandra Krishi Viswavidyalaya, Haringhata, West Bengal, India

A. Mandal
Dakshin Dinajpur KVK, Uttar Banga Krishi Viswavidyalaya, Cooch Behar, West Bengal, India

P. P. Adhikary et al. (eds.), *Soil, Water Pollution and Mitigation Strategies*,
Environmental Science and Engineering, https://doi.org/10.1007/978-3-031-63296-9_6

8.1 Introduction

By utilizing synthetic pesticides and chemical fertilizers, agriculture generates themost chemical pollutants of all industries. This leads to significant ecological devastation andhealth issues. Nitrous oxide (N_2O), a greenhouse gas that contributes to global warming, isproduced as a result of the use of nitrogen fertilizer. Moreover, the soil's quality has declined, and some nitrogen is no longer being fixed by the soil. Ammonium nitrate causes a decline insoilmicrobe symbiosis because farmers apply nitrogen fertilizers excessively. Excessammonium is converted by nitrogen-fixing bacteria into nitrate, which is then devoured bydenitrifying bacteria to release N_2O, and some nitrate also reaches groundwater (Galloway et al. 2008). The generation of natural N_2O is boosted via nitrification and denitrificationprocesses. Nitrogen oxides are transformed into gaseous products in the atmosphere by microorganisms through denitrification. Furthermore, soil bacteria transform ammonium (NH_4) into nitrate (NO_3) (Butterbach-Bahl et al. 2013). For the development of sustainable agriculture, crops must be more nutritious and resistant of biotic and abiotic stress. Soil microorganisms including bacteria, algae, fungi, and other organisms with the ability to ingest nutrients and metabolites while utilizing more water can be used to achieve the required crop attributes (Armada et al. 2014). The most potent soil microorganisms now known to increase plant productivity and stimulate plant growth rate without harming the environment are plant growth-promoting rhizobacteria (PGPR) (Calvo et al. 2014). Many plants growth-promoting bacterial strains, some of which are commercial PGPRs, have been documented for many years. These strains include Bacillus, Klebsiella, Pseudomonas, Azotobacter, Enterobacter, Serratia, Variovorax, and Azospirillum (Glick 2012). Very few PGPR are used in the agriculture sector due to varying qualities around the world (Bashan et al. 2014). The interactions between soil microbes and crops, as well as the environment, all play a key role in the significant usage of PGPR (Martinez-Viveros et al. 2010). Since each PGPR in the soil has a unique method of action, their application is restricted and varies from soil to soil (Choudhary et al. 2011; Dey et al. 2004). Due to the worrisome population growth anticipated to reach 8 billion by 2025 and 9 billion by 2050, the world's gross agricultural production is insufficient to feed the population. As a result, at that time, PGPR and chemical fertilizers will not be comparable (Kour et al. 2020). It was shown that a PGPR's broad range of a mode of action, competency with the rhizosphere and other soil bacteria, eco-friendliness, and tolerance of abiotic stress made them suited for the applications (Nakkeeran et al. 2005). In the conventional agricultural system, overuse of synthetic chemical fertilisers is expensive, depletes nonrenewable resources, negatively impacts soil fertility, pH, and exchangeable gases, harms the environment, and reduces crop output (Joshi et al. 2006; Youssef and Eissa 2014). Farmers use chemical fertilisers as a remedy to the aforementioned issues. Since a few years ago, India's agriculture policy has placed a strong emphasis on the cultivation of healthy crops in a sustainable manner. Hiltner (1904) coined the term "rhizosphere," which refers to the physiologically significant microorganisms that are present around a plant's rhizosphere and that are therefore also known as growth-promoting rhizobacteria. PGPRs improve the soil's fertility for nutrient availability, boosting crop productivity and turning over (Ahemad and Khan 2009). According to Sagar et al. (2012), the interaction of PGPRs in the soil environment is a unique mechanism for enhancing plant production and applies to a variety of businesses, including the pharmaceutical and agro-based ones. PGPRs decrease the use of synthetic fertilizers used in conventional crop production methods, increasing crop yield and improving the soil's fertility and environmental friendliness for crops to grow sustainably (Maheshwari et al. 2012).

8.2 Mechanism of Plant Growth Promotion

In general, PGPR contributes to the improvement of the plant by manufacturing advantageous substances for the plants (Dobbelaere et al. 2003; Khalid et al. 2004), making specific nutrients from the soil available (Backer et al. 2018; Lucas Garcia et al. 2004), and shielding the plants from diseases (Guo et al. 2004; Jetiyanon and Kloepper 2002; Raj et al. 2003; Saravanakumar et al. 2008). Beneficial microorganisms can directly interact with their host plants to promote plant growth, and they can also indirectly influence plant pathogens through their antagonistic behavior. Many studies and increased use of PGPR demonstrate that it has a variety of advantages over traditional chemical pesticides, including the ability to boost plant health, maintain environmental quality, be effective at small doses, multiply by themselves, and decay more quickly. Hence, it may also be utilised in integrated pest management systems (Berg 2009; Chadha et al. 2014; Prasad et al. 2014). However the many agrochemicals used in crop production have a number of drawbacks, including being dangerous by nature, lowering environmental quality, and possibly posing a risk to human health. Yet, PGPR has a significant impact on plant productivity through a combination of direct and indirect mechanisms.

8.2.1 Direct Mechanism

8.2.1.1 Production of Plant Growth Regulators

Auxin, gibberellic acid, cytokinin, ethylene, and ABA are among the phytogrowth regulators that various PGPRs can produce, and these substances enhance plant growth and development through various mechanisms (Khalid et al. 2004). Many plants' root surface area and number of root tips increase as a result of the auxin, or IAA and gibberellin, synthesis (Bhattacharyya and Jha 2012), allowing them to absorb more nutrients from the soil. By the development of lateral and adventitious roots, cytokinin, another plant growth regulator, promotes cell proliferation, cell enlargement, and an increase in the root surface area of crop plants (Werner et al. 2003). Most bacterial species can produce ethylene, which has a variety of effects on plant growth, development, and senescence (Reid 1987) and is also known as a ripening hormone. In addition, ethylene increases germination, breaks seed dormancy, and encourages the development of adventious roots and root hair (Esashi 1991). Production of ABA, which controls stomatal opening and closing activity to reduce leaf transpiration, is credited with helping plants survive in water-stressed situations. in 1995 (Frankenberger and Arshad). In addition to them, Table 8.1 lists the crop response and the bacterial strain's production of plant growth regulators (PGRs).

8.2.1.2 Production of ACC Deaminase

Even though ethylene has positive impacts on plant growth and development, its high concentration causes defoliation, which lowers crop productivity. Hence, PGPR prevented 1 amino cyclopro-pane-1-carboxylic acid (ACC) deaminase from producing ethylene by diverting ACC from the roots of the Arabidopsis thaliana plant (Desbrosses et al. 2009). ACC deaminase activity is present in a number of PGPRs, including Achromobacter, Azospirillum, Bacillus, Enterobacter, Pseudomonas, and Rhizobium (Ghosh et al. 2003; Govindasamy et al. 2008; Duan et al. 2009). Use of ACC deaminase, for instance, increases the plant's resilience to environmental stress in tomato seeds infected with Enterobacter cloacae and Pseudomonas putida (Grichko and Glick 2001). Pseudomonas fluorescens TDK1, an improved bacterial strain, has ACC deaminase enzyme, which improves salinity resistance and total yield in groundnut compared to Pseudomonas strains without ACC deaminase activity (Govin-dasamy et al. 2008).

8.2.1.3 PGPR that Induce Systemic Resistance

One of the best weapons for controlling plant diseases and pests is resistance. An interaction with an external element might cause plants to activate an induced resistance defensive system. This element may have chemical or biological roots. Plants have been found to have two different types of non-specific defence mechanisms: pathogen-induced systemic acquired resistance (SAR) and advantageous microbe-induced systemic resistance (ISR) (Schuhegger et al. 2006). In both types, systemic protection is provided to the plant's nonexposed areas (Kuc 1982). In ISR, plants are prepared by advantageous bacteria that offer defence against a variety of pest and disease attacks (Alstrom 1991; Van Peer et al. 1991; Wei et al. 1991; Walters et al. 2013). ISR doesn't include the buildup of pathogenesis-related (PR) proteins like SAR does. Hoffland et al. initially illustrated this in radish (1995). Components like ethylene, salicylic acid, and jasmonic acid control ISR (De Meyer et al. 1999; Thomma et al. 2001; Audenaert et al. 2002; Verhagen et al. 2004).

Table 8.1 Production of PGRs by different plant growth promoting rhizobacteria (PGPR) and crop response

PGPRs	PGRs	Crops	Response	References
Rhizobium leguminosarum	Indole-3-acetic acid	Rice	Significant growth promoting effects	Biswas et al. (2000)
Azotobacter sp.	Indole-3-acetic acid	Maize	Significant growth enhanced	Zahir et al. (2000)
Pseudomonas fluorescens	Siderophores, Indole-3-acetic acid	Groundnut	Involvement of ACC deaminase and siderophore production promoted nodulation and yield	Dey et al. (2004)
Pseudomonas tolaasii ACC23, *P. fluorescens* ACC9, *Alcaligenes* sp. ZN4	Siderophores, Indole-3-acetic acid	Brassica	Protect canola plant against the inhibitory effects of cadmium	Dell'Amico et al. (2008)
Bacillus sp. *Paenibacillus* sp.	Indole-3-acetic acid	Rice	Significant increase in the root and shoot parts of plants	Beneduzi et al. (2008)
Streptomyces acidiscabies E13	Hydroxamatesiderophores	Cowpea	Promoted cowpea growth under nickel stress	Dimkpa et al. (2008)
Pseudomonas denitrificans Pseudomonas rathonis	Auxin	Wheat, maize	Increase plant growth of wheat and maize in pot experiments	Egamberdiyeva (2005)

8.2.1.4 Protection by Metabolites

Antibiotics are a significant family of secondary metabolites produced by microorganisms that, even at low concentrations, are toxic to the growth of other pathogenic germs. The discovery of antibiotics is seen as a significant turning point in the history of medicine. Actinobacteria make up a significant portion of the makers of antibiotics among microorganisms, which are important sources for them. Actinobacteria are the source of around 45% of the antibiotics that are now in use, according to Liu et al. (2012). Over 33,500 bioactive metabolites have been documented thus far, of which 13,700 are found in actinobacteria (Berdy 2012). The metabolites are a diverse class of chemically diverse substances that include anthracyclines, aminoglycosides, streptothricins, cyclopolylactones, macrolides, polyether antibiotics, and quinoxaline peptides. Glycopeptides and orthosomycins are mostly produced by Non-Streptomyces Actinobacteria (Berdy 2012). Microbes that produce antibiotics have an advantage over those that don't because it increases their chances of surviving. Several researchers have taken advantage of and investigated this property for the biological control of plant pathogens and disease suppression (Agbessi et al. 2003; Khamna et al. 2009; Meschke et al. 2012; Chamberlain and Crawford 1999; Crawford et al. 1993). In comparison to nonmutants of the strain, a study by Beausejour et al. (2001) found that mutants incapable of producing a particular antibiotic did not control sickness. For the control of plant diseases, commercial formulations with the antibiotic or microbe as the active component are available. Commercial products include Mycostop® by Streptomyces griseoviridis K61 (Figueiredo et al. 2010), Actinovate® and Actino-Iron® by Streptomyces lydicus WYEC 108 (Crawford et al. 2005), ArzentTM by four distinct strains of S. hygroscopicus, and others (Hamby and Crawford 2000). Other antibiotics of Actinomycete origin include cycloheximide from Streptomyces griseus, kasugamycin from Streptomyces kasugaensis, blasticidin S from Streptomyces griseochromogenes, and Rhizovit from Streptomyces rimosus. Actinobacteria produce several secondary metabolites that are harmful to a variety of insects, including tetranectin, avermectins, faerifungin, macrotetrolides, and flavonoids. Streptomyces avermitilis, a new species isolated from soil, produces avermectins as a byproduct. It was initially noted to be a potent antihelminthic substance (Burg et al. 1979), but later research revealed that it was also an efficient insecticide, acaricide, and nematicide (Putter et al. 1981). Two species of Saccharopolyspora spinosa include a wide family of novel chemicals known as spinosyn. Spinosyn A and spinosyn D are two metabolites that are created during the fermentation of S. spinosa. They have a new chemical structure, and they work by influencing postsynaptic cells' nicotinic acetylcholine receptors. They have relatively low specificity against many helpful insect predators and nontarget species, and they are highly selective towards target insects like lepidoptera and diptera (Thompson et al. 2000; Salgado and Sparks 2005). The kind of species, their stage of development, exposure period, and administration technique all affect spinosad's effectiveness. Spinosad has the notable benefit of being less harmful to mammals, birds, and aquatic species than other insecticides, making it safer to employ (Thompson and Sparks 2002).

8.2.1.5 Competition

Many microorganisms that support soil's structure and stability live there. There is a vast pool of prospective rivals, and numerous causes may be to blame for the dominance of particular populations (Hibbing et al. 2010). By conducting studies with nutrients that are limited, Monod was able to show how important availability is to microbial competitiveness (Monod 1949, 1950).Exudates from plant roots are important in identifying the particular microbe community that exists nearby. Other components including sugars, amino acids, vitamins, and organic acids serve as crucial nutrition for microbes, and the availability of phenolic and flavonoid chemicals regulates plant symbiosis with beneficial rhizobacteria (Palaniyandi et al. 2013; Dakora and Phillips 2002). One of the strategies used by microorganisms is the synthesis of siderophores (Sontag et al. 2006; Macagnan et al. 2008), lytic enzymes (Potgieter and Alexander 1996), volatile chemicals (Wan et al. 2008), and antibiotics (Agbessi et al. 2003). One strategy for outliving harmful bacteria and suppressing disease is competing for available nutrients (Siddikee et al. 2010; Palaniyandi et al. 2013a).

8.2.1.6 Volatile Substances

Several biocontrol agents are known to create volatile chemicals as pathogen inhibitory strategies in addition to producing antibiotics. Acids like hydrocyanic acid (HCN), other acids, alcohols, ketones, aldehydes, and sulphides are examples of common volatile substances (Bouizgarne 2013). There are reports on the creation of HCN by helpful bacteria to reduce the harmful impact of pathogenic fungus (Ahmad et al. 2008). According to De'fago et al. (1990), the mechanism by which HCN synthesis produces resistance in plants. The production of volatile chemicals by actinobacteria for the prevention of plant diseases has been shown by Moore-Landecker and Stotźky (1973), Wang et al. (2013), and Boukaew et al. (2013).

8.2.1.7 Pathogen Control

In order to attain global food security, it's crucial to concentrate on reducing the threatof yield-limiting variables, which affect both the quantity and quality of crops (Atkinson and Urwin 2012). The degree of devastation brought on by disease-causing pathogens stands outamong these parameters. The famines in Ireland (1845) and Bengal (1943) are two instancesof plant disease-induced yield reductions that had a devastating effect on human lives and thenation's economy.It has been observed that a number of diseases hinder grain legumeproduction. Sterility mosaic, Fusarium wilt, and Phytophthora blight of pigeon pea, as well asBotrytis grey mould, Ascochyta blight, and Fusarium wilt of chickpea, are significantdiseases of grain legumes. A long-term approach to managing illness is biological control ofplant pathogens. Tu (1988) described the parasitization of Colletotrichum lindemuthianum bythe soil Actinobacteria *S. griseus*. On the surface of the hyphae, the pathogen producedswellings resembling appersoria. It was also noted that internal bulb development caused thedegradation of pathogen hyphae. Streptomyces cyaneofuscatus ZY-153, Streptomyceskanamyceticus B-49, Streptomyces rochei X-4, Streptomyces flavotricini Z-13, and Streptomyces phaeopurpureus ExPro138 were all found to have hyperparasitism as a mechanism (Xue et al. 2013; Palaniyandi et al. 2013b). This particular bacterial group's capacity to produce antibiotics is a trait that can be used to combat disease. Rothrock and Gottlieb observed that the antibiotic geldanamycin from S. hygroscopicuspv. geldonus inhibited the growth of Rhizoctonia solani (1984). *Streptomyces* spp. were reported for direct growth improvement and antibiotic synthesis to suppress root rot caused by Phytophthora sojae in soybean. Pathogen population densities were successfully decreased in controlled settings using naturally contaminated soil (Xiao et al. 2002). Certain Actinobacteria have been discovered to produce enzymes that break down cell walls as a biocontrol mechanism. S. cavourensis SY224 reported to generate chitinase and glucanase against Colletotrichum gloeosporioides (Lee et al. 2012b). According to Shekhar et al. (2006) and some Streptomyces species, the synthesis of glucanase by S. violaceusniger XL-2 inhibits the growth of phytopathogenic fungi.

The proteases of *Streptomyces* sp. A6 inhibited the spore germination of Fusarium udum (Singh and Chhatpar 2011). Certain non-streptomycete Actinobacteria were also reported to exert biological control against *Fusarium* spp. *F. oxysporum* f. sp. albedinis' growth was suppressed by Nocardiopsisdassonvillei (Sabaou et al. 1983), while Micromonosporaglobosa used mycoparasitism to manage *F. udum* populations (Upadhyay and Rai 1987). Field trials with the *Streptomyces* sp. P4 isolate successfully decreased the occurrence of Pisum sativum powdery mildew (Sangamee et al. 2009). The antagonistic isolates' spore suspensions used in foliar spray prevented the development of appersoria (Sangmanee et al. 2009). Streptomyces albovinaceus, S. griseus, and 2 Indirect Plant Growth Promotion in Grain Legumes: Role of Actinobacteria 23 S. virginiae were found to produce siderophores as a method of pathogen management (Macagnan et al. 2008). Rhizoctonia solani, Pyricularia grisea, Bipolarisoryzae, and Fusarium fujikuroi populations were successfully managed by the siderophore-producing Streptomyces philanthi RM-1–138 (Boukaew et al. 2013). Another significant mechanism investigated in Actinobacteria is antibiotic synthesis for pathogen control and illness suppression. One of the most important steps in the treatment of plant diseases is the use of metabolites with broad range activity. The phytopathogenic *Fusarium oxysporum, Rhizoctonia solani, Cladosporium cladosporioides, Fusarium chlamydosporum, Colletotrichum gloeosporioides, Alternaria mali*, and *Pestalotia* spp., which are reported to produce azalomycin (Jinhua et al. 2010), were significantly less inhibitive (Jinhua et al. 2010). According to Wang et al. (2013), in vitro growth of commercially significant fungi *Fusarium moniliforme, Aspergillus flavus, Aspergillus ochraceus, Aspergillus niger*, and *Penicillium citrinum* was inhibited by Streptomyces alboflavus TD-1.Actinomycin X2 and fungichromin were reported to be produced by the endophytic bacterium Streptomyces galbus R-5 (Hasegawa et al. 2006). To completely combat the infection, two or more of these processes typically work in conjunction with one another. There are numerous reports on the effective use of Actinobacteria for disease control in various plant species (Neeno-Eckwall et al. 2001; Conn et al. 2008; El-Tarabily et al. 2009; Palaniyandi et al. 2013b).

8.2.1.8 Insect Pest Control

In commercially significant food crops, insect pests and other pathogenic organisms are known to result in yield losses of 20–40% (Zhou 2001). Around 70,000 distinct bug species are known to harm agricultural crops. Lepidoptera, an order of insects, is home to many of the major pests of cultivated crops (Pimental 2009). Several pests, including Helicoverpaarmigera and Spodoptera litura, seriously damage crops like peanuts, pigeon pea, chickpea, soybean, cowpea, pepper, tobacco, caster, and okra (Armes et al. 1996). In addition to the losses caused by Maruca (another significant pod borer) and bruchids, an important storage pest, these insects constitute a serious danger to chickpea yield with yearly losses of up to 40% (Rao et al. 2010). Insecticides are currently used to control these pests' potentially destructive behaviour. Reports on the establishment of resistance make them harder to handle and make it necessary to find cleaner, more environmentally friendly alternatives, which is crucial. Many biological management strategies, such as the utilisation of bacteria, fungi, and viruses, are being researched (Deshpande 1999). Researchers from Gopalakrishnan et al. (2011), Arasu et al. (2013), Vijayabharathi et al. (2014), and Sathya et al. (2015) investigated the insecticidal ability of Actinobacteria against certain significant pests (2015). Extracellular metabolites from three different strains of Streptomyces—*S. griseoplanus, S. bacillaris*, and *S. albolongus*—have been demonstrated to be efficient broad-spectrum entomopathogens with activity against Helicoverpaarmigera, S. litura, and Chilo partellus. Under both in vitro and in vivo conditions, the evaluated biocontrol strains produced insecticidal compounds that clearly suppressed the proliferation of these pests (Vijayabharathi et al. 2014).

8.2.1.9 Abiotic Stress Management

Abiotic challenges include nutrient shortage, salt, metal toxicity, heat, and waterlogging are other factors that reduce productivity (Shao et al. 2008). These abiotic stressors pose a threat to the overall amount of arable land, which, when combined with the expanding population, compromises the sustainability of agriculture (Shahbaz and Ashraf 2013). Particularly in the semiarid zone, heat stress and drought prevent chickpea crop from reaching its full potential (Rao et al. 2010). Similar to how drought, water logging, and salinity dramatically reduce pigeon pea output. Abiotic pressures have a greater financial influence on the production of pigeon peas than biotic stresses (Ryan 1995).The selection of resistant types through breeding and genetic engineering largely overcomes these 24 S. Jacob and H.K. Sudini variables (Rai et al. 2011). Changes in the plant's molecular, cellular, and physiological makeup result from such stress situations (Rejeb et al. 2014). The synthesis of phytohormones, which stimulate plant development, is one method known for helping plants cope with stress in their natural environment. Particularly, plants under water or nutrient stress situations experience physiological changes in their root system that entail the release of hormones that promote development (Potters et al. 2007). Low amounts of phytohormones called auxins encourage cell elongation (Martin and Elliott 1984).

Using microorganisms' resistance to harsh environmental conditions and synthesis of specific enzymes, one may help plants develop abiotic stress tolerance through microbial assistance (Shrivastava and Kumar 2015). This strategy offers a less expensive and long-lasting resolution. The precise mechanisms underlying the many research on the use of bacteria in abiotic stress tolerance are still unknown (Roman et al. 1995; O'Donnell et al. 1996; Penninckx et al. 1998). The stress hormone ethylene is produced by plants in response to all types of stress. Under stressful circumstances, the synthesis of ethylene is considerably boosted, which has a negative impact on plants by accelerating cell senescence (Abeles et al. 1992; Woltering and Van Doorn 1988; Nayani et al. 1998; Ali et al. 2012). By producing 1-aminocyclopropane-1-carboxylate (ACC) deaminase, certain bacteria can control the biosynthesis of ethylene. This enzyme converts ACC, an ethylene precursor, into -ketobutyrate and ammonia. Researchers have recently taken advantage of this characteristic of helpful bacteria to promote plant growth and development under abiotic stress situations (Arshad et al. 2008; Bal et al. 2013; Chookietwattana and Maneewan 2012). ACC deaminase positive strains function by lowering ethylene levels, which encourages root growth (Glick et al. 1998; Burd et al. 2000; Belimov et al. 2007, 2009; Long et al. 2008). In a study using endophytic Actinobacteria that produce ACC deaminase, this was demonstrated experimentally. Mung bean plants treated with PGPA showed improved root and shoot growth as well as an increase in chlorophyll content. In comparison to uninoculated and ACC deaminase-deficient mutants, these physiological alterations allowed the bean plants to survive high salty (100 mM NaCl) and waterlogging conditions better (Jaemsaeng et al. 2013). There have been reports of certain Actinobacteria producing the auxin indoleacetic acid (IAA) (Lin and Xu 2013; Manulis et al. 1994; Kaur et al. 2013). Abiotic stresses have been shown to enhance plant development when IAA is produced by helpful microorganisms. Plant exudates containing tryptophan serve as a substrate for microbes, which then produce IAA, which is absorbed by plants. Gopalakrishnan et al. (2014) shown that rice plants saw an increase in root development and root hairs after being inoculated with PGP Actinobacteria. The biomass of the plant increased significantly, according to research by Sadeghi et al. (2012) on Streptomyces isolates with the capacity to synthesise auxins and siderophores in saline soil. On wheat shoots produced in both normal and saline soil, their analysis revealed an increase in the concentration of N, P, Fe, and Mn. Improved root development encourages an increase in root surface area, which facilitates the uptake of nutrients and water (Dimkpa et al. 2009). Actinobacterial strains have been known to encourage Rhizobia to produce root nodules (Tokala et al. 2002).When S. lydicus WYEC108 is colonised by pea plants, the nodules grow larger and more robust, which helps them absorb soil nutrients (Tokala et al. 2002). Streptomyces, Actinoplanes, and Micromonospora were only a few of the Actinobacteria species that had an impact on Frankia species' ability to produce nodules (Solans 2007). Culture filtrates have also been found to improve nodulation in addition to Actinobacteria's direct role in the promotion of nodule development (Solans 2007).A lack of phosphorus, an essential nutrient for plants, has a direct impact on crop output. The majority of the phosphorus in soil is insoluble.Enhancement of Indirect Plant Growth in Grain Legumes: Actinobacteria's function 25 metal complexes that reduce the amount of phosphorus that is accessible to plants (Hamdali et al. 2008). These complexes can be solubilized by helpful bacteria, releasing free phosphates (Rodr'guez and Fraga 1999). Many Actinobacterial species have been shown to solubilize phosphorus (Hamdali et al. 2008; Oliveira et al. 2009; Franco-Correaa et al. 2010). The growth of bean plants (*Phaseolus vulgaris* L. roots)'s and shoots was particularly enhanced in greenhouse research by El-Tarabily et al. (2008) utilising the phosphate-solubilizing isolate Micromonosporaendolithica as opposed to the nonphosphate-solubilizing M. olivasterospora. The positive isolates allowed the bean plants to receive phosphorus (El-Tarabily et al. 2008). In order to combat stress caused by nutrient deprivation, the capacity of microorganisms to solubilize critical components and make them accessible for plant use, coupled with stimulation of nodulation, can be considered a significant trait.

8.2.2 Indirect Mechanism

8.2.2.1 Nitrogen Fixation

Nitrogen application controls plant growth and development in large part, however inorganic nitrogen application has a high manufacturing cost. Thus, the free or symbiotic fixation of nitrogen from the atmosphere by heterotrophic bacteria is crucial for agriculture. Free-living organisms can get reasonably close to a plant's roots, but they can't go inside the plant or form a symbiosis with it. These microbes are able to fix nitrogen from theatmosphere into the soil, which is mostly utilised by plants and not for their own advantage. Examples of naturally occurring bacteria that fix nitrogen include Azospirillum, Azotobacter, Burkholderia, Herbaspirillum, and Bacillus (Vessey 2003). The first report on the use of Azotobacter in agricultural studies was made in 1902. To fix nitrogen and produce phytohormones including indole acetic acid, gibberellic acid, cytokinins, and vitamins, Azotobacter and Azospirillum are utilised (Rodelas et al. 1999). Yet, it was shown that Azotobacteraceae is the most representative genus employed on various crops for their ability to fix nitrogen, accounting for 20 and 30 kg per hectare per year (Mayea, 1998; Bhattacharya and Chaudhuri, 1993). The majority of nodulating legumes can fix atmospheric N2 through a symbiotic relationship with rhizobium bacteroids and the nitrogenase enzyme, and they account for 65% of the nitrogen used in agriculture (Matiru and Dakora 2004). In Rhizobia (Rhizobium, Mesorhizobium, Bradyrhizobium, Azorhizobium, Allorhizobium, and Sinorhizobium), a flavonoid molecule released from the legume host induces the expression of "nod genes," which produce lipo-chitooligosaccharide (LCO) signals that influence the cell division for nodule formation in roots (Dakora 1995, 2003; Lhuissier et al. 2001; Matiru and Dakora 2004). As a result, this nodulation-forming legume crop undergoes a process known as biological nitrogen fixation (BNF), which turns atmospheric elemental nitrogen (N_2) into ammonia (NH_3) (Shiferaw et al. 2004). Yet, host-microsymbiont compatibility and the physicochemical characteristics of the soil affect how effective BNF is (Tisdale et al. 1990; Zafar-ul-Hye et al. 2007). Due to biological nitrogen fixation in the soil, all non-legume crops (cereals) benefit from intercropping as well (Snapp et al. 1998).

8.2.2.2 Phosphate Solubilization

Both the organic and inorganic forms of phosphorus, which are insoluble by nature, stay in the soil. Tricalcium phosphate, dicalcium phosphate, hydroxyl apatite, and rock phosphate are examples of the insoluble substances that PGPR (Pseudomonas, Bacillus, Rhizobium, Burkholderia, Achromobacter, Agrobacterium, Microccocus, Aerobacter, Flavobacterium, and Erwinia) can transform into accessible form. According to Rodriguez and Fraga (1999), Igual et al. (2001), Goldstein and Krishnaraj (2007), Goldstein (2007), Delvasto et al. (2008), organic acid generation makes this conversion possible in acidic media. The production of organic acids like gluconic acid secreted by *Pseudomonas* sp., Erwinia herbicola, Pseudomonas cepacia and Burkholderiacepacia (Rodríguez and Fraga 1999) and 2-ketogluconic acid present in Rhizobium leguminosarum, Rhizobium meliloti (Halder and Chakrabarty 1993), Bacillus firmus (Banik and Dey 1982) responsible for mineral phosphate solubilization.

8.2.2.3 Siderophore-Production

Iron (Fe), which functions as a cofactor of many enzymes necessary to critical physiological activities like respiration, photosynthesis, and nitrogen fixation, is the most important micronutrient for plants. Fe+, an oxidised form of iron that interacts to produce insoluble oxides and hydroxides inaccessible to plants or microbes, is the most common chemical form of iron. As a result, plants and even bacteria have evolved alternative mechanisms for iron nutrition or absorption, such as the synthesis of organic acids, which can bind the iron complex (Fe+) and act as a chelating agent. Another process is the creation of siderophores, which have low molecular weights and include functional groups capable of reversibly binding iron. Pseudomonas fluorescens is the most prevalent genus of bacteria, and bacteria in the rhizosphere release these substances to boost their competitive potential in relation to antibiotic action and increased iron nutrition for the plant (Glick 1995).

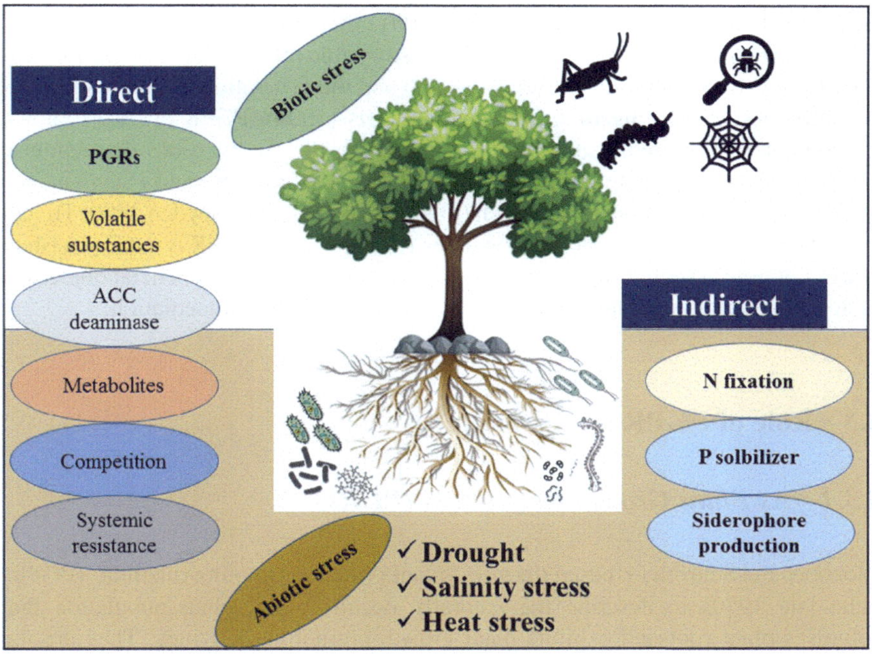

Plates: Mechanism of PGPR on crop plant

8.2.2.4 Plant Microbe Interaction

The endosphere, phyllosphere, and rhizosphere are the three levels where plants and bacteria interact. Rhizosphere refers to the area where a root and dirt come into touch. It is essentially a cloud of microorganisms that surrounds the roots of plants and is essential to their life and development. Lorenz Hiltner first used the term "rhizosphere" in 1904. For the external root surface and the closely adherent soil and debris, Clark proposed the word "rhizoplane". It has been extensively investigated how root exudates affect the growth of soil microorganisms surrounding and inside roots (Hartmann et al. 2008) and the interactions between rhizosphere colonists, plant hosts, and soil microbes (Dennis et al. 2010; Friesen et al. 2011; Berendsen et al. 2012). The microbial community in the rhizosphere is different from that in the soil in terms of both quantity and quality. According to the proposed theory, mycorrhizae envelop the majority of plant roots. As a result, the term "mycorrhizosphere" should be used instead of "rhizosphere" (Shrivastava et al. 2014). Sugars and amino acids found in plant exudates are abundant sources of nourishment and energy. Combinatorial effects of root-root, root-microbe, and root-insect interactions result in plant root interaction in the rhizosphere. According to studies using molecular methods, there are over 4000 different microbial species present in every gramme of soil (Montesinos 2003). Filamentous actinomycetes are one of the most significant communities in the rhizosphere microbiome (Benizri et al. 2001). According to Cardoso and Freitas (1992), rhizosphere microbial colonies have a dynamic relationship with the biogeochemical cycling of nutrients (C, P, N, and S) and the generation of phytohormones or antibiotics. According to Andrews and Harris (2000), PGPR are well known for colonising plant roots and promoting plant growth. As microbes that colonise the rhizosphere of plants, *Azospirillum* sp., *Bacillus subtilis* sp., and *Pseudomonas* sp. have all received extensive research (Steenhoudt and Vanderleyden 2000; Trivedi et al. 2005). Although the colonisation of the rhizosphere by PGPR is not a consistent process, it is an essential stage in the use of microorganisms for beneficial objectives such biofertilization, phytostimulation, biocontrol, and phytoremediation.

8.3 Role of PGPR on Agriculture

8.3.1 As Plant Growth Regulators (PGRs)

Kloepper and Schroth first used the term "plant growth-promoting rhizobia" (PGPR) inthe late 1970s to describe the naturally occurring microbial population that colonises plant rootsand exhibits plant growth-promoting activities. They are the potential tools for sustainableagriculture and trend for the future. The plant roots basically provide niche for the proliferation of soil bacteria that thrive on root lysates and root exudates. Rhizosphere has a bacterial population density that is one hundred times greater than that of bulk soil, and up to fifteen percent of a plant's roots may be covered by micro-colonies of different bacterial strains (Van loon 2007). Plant growth is stimulated by many bacteria, including those of the genera Pseudomonas, Bacillus, Rhizobium, Azospirillum, Arthrobacter, Burkholderia, Serratia, Flavobacterium, Enterobacter, Erwinia, Acinetobacter, Alcaligenes, etc. (Egamberdiyeva 2007). Bacillus, Pseudomonas, Burkholderia, and Azotobacter are examples of free-living or non-symbiotic rhizobacteria that exist outside plant cells and are therefore classified as extracellular PGPR (ePGPR) (Babalola and Akindolire 2011). In fact, they generate metabolites like plant growth regulators that aid plants in expanding and absorbing nutrients.These bacteria are a significant component of many plants' rhizospheres. Additionally, bacterial inoculants may boost seedling emergence, increase plant development rate and germination percentage, and protect plants from disease (Kozdrój et al. 2004).

Characteristics of Ideal PGPR

1. Augmentation of plant growth
2. Environment friendly.
3. Compatible with different rhizobacteria.
4. High rhizosphere proficiency
5. Superior capacity to compete as a saprophyte
6. Ease of large-scale production

7. Effective against a wide variety of infectious organisms
8. Resistant to saltwater, drought high temperatures, oxidising chemicals, and exposure to ultraviolet radiation.
9. a reliable and efficient pest control provider.

8.3.2 Application of PGPR in Agriculture

Biotechnological methods that are beneficial to the environment provide alternatives to chemical fertilisers (Dobbelaere et al. 2003). Agricultural practises are negatively impacted by the excessive usage and increasing cost of chemical fertilisers; consequently, PGPB is being investigated as an alternate method for minimising the use of chemicals in agriculture (Welbaum et al. 2004). Globally, the use of plant growth promoting rhizobacteria (PGPR) for sustainable agriculture has expanded dramatically over the last few of decades. Treatment with PGPR has enhanced the percentage of germination, seedling vigour, emergence, shoot growth, root growth, total plant biomass, seed weight, early blooming, grain, forage, and fruit yields, etc. In response to PGPR inoculation, significant increases in the growth and yield of economically significant crops have been consistently documented (Araujo, 2008). The use of PGPR with the aim of improving nutrient availability for plants is an important practice in agriculture production.Based on their method of action, PGPR are categorised into two general categories: (1) PGPR that directly impact plant development, seed emergence, or crop yields; and (2) biocontrol PGPR that indirectly promote plant growth (Mallik Freitas et al. 2007). The direct process by which PGPR promotes plant development includes improved nitrogen fixation, synthesis of phytohormones, nutrient solubilization, and nutrient absorption.

8.3.3 P-Solubilizer

Phosphorus (P) is a key macronutrient recognised for plant biological and physiological growth such as cellular structure, promotion of root development, seed generation, and quicker crop maturity. The majority of phosphorus in soil is present in organic form. The plant uses merely 0.1% of the soil in certain cases (Beever and Burns 1981). However, a considerable fraction of phosphates stay in inaccessible form. When applied as fertiliser, a large quantity of P interacts with reactive cations in acidic soil and Ca^{2+} in calcareous or normal soils through precipitation reaction (Richardson et al. 2011). Organic matter, according to Richardson (2011), should also exist as the main reservoir (30–80%) of immobilised P, primarily in the form of orthophosphoric acid, phytin, inositol phosphates, phospholipids, nucleic acids, and so on, which are degraded by the activity of enzymes such as acid phosphatase, phytase, and nucleases. PSB bacteria performed the enzymatic breakdown of organic forms of phosphorus. Phosphorus solubilizing bacteria (PSB) play a key function in phosphorus nutrition. They could increase the availability of accumulated phosphate through solubilization while simultaneously improving the efficiency of biological nitrogen fixation and increasing the availability of Fe, Zn, and other trace elements through the production of plant growth promoting substances (Kucey et al. 1989) Phosphate solubilizing bacteria (PSB) have been used as a biofertilizer since the 1950s (Marchner 1995). These microorganisms secrete various types of organic acids, such as carboxylic acids, which reduce the pH in the rhizosphere (He and Zhu 1998) and thus dissociate the bound forms of phosphate, such as $Ca_3(PO_4)_2$, in calcareous soils. As a result, using biofertilizer helps to reduce the amount of expensive phosphatic fertiliser in the soil. They can also help to reduce the negative effects of pathogens on crop yield, act as biocontrol agents against expensive pesticides, and protect the soil and environment (Ojuederie and Babalola 2017). Microorganisms with phosphate solubilizing ability were found to synthesise a diverse range of metabolites with biocidal and fungicidal properties. Using of phosphorous solubilizing microorganisms PSM has an impact on plant growth and development in both direct and indirect ways. PSM suppresses pathogen growth in plants by producing antimicrobial compounds such as phenolic compounds, PAL and flavonoids, antibiotics, cyanide hydrogen, indoleacetic acid (IAA), siderophores, and several lytic enzymes (Köhl et al. 2019). Burkholderiacepacia, Citrobacter freundii, Enterobacter aerogenes, Klebsiella pneumonia, Proteus vulgaris and Acinetobacter lwoffii Alternaria alternate, Macrophominaphaseolina, Fusarium oxysporum and Sclerotium rolfsii were also posse's biocontrol capacity (Zhang et al. 2016) by exerting antagonistic activities against pathogens by synthesizing antimicrobial peptides, secreting lytic enzymes and inducing systemic resistance against pathogenic microbes (Ullah et al. 2018). The insertion or over-expression of phosphatesolubilizing genes in rhizospheric bacteria is an unique method for increasing their capacity to function as agricultural biofertilizers. *Chryseobacterium* sp. strain ISE14 is a phosphatesolubilizing endophytic bacterium that promotes plant development and has anti-Phytophthora blight and anthracnose activity on pepper. Significant agronomic characteristics of these PSMs include the production of exopolysaccharide (EPS) (Fatima et al. 2022), various phytohormones, 1-aminocyclopropane-1 carboxylate (ACC) deaminase, volatile compounds, increased nutrient uptake, induction of osmolyte accumulation, upregulation or downregulation of stress responsive genes, or changes in root morphology that exaggerate the improvement in soil properties, assist in plant growth, development and biocontrol activities (Meena et al. 2017). *Enterobacter* sp. may improve RP solubilization by up to 17.5%, suggesting that inoculation with PSB and amendment with RP might be a potential alternative strategy to act as source of P fertiliser and maintaining better nutrient availability in soils (Park et al. 2011). The results demonstrated that the integrated effect of PGPR and PSB in conjunction with chemical fertilisers had a significant impact on soil fertility improvement, plant growth, and biomass development (Abbas et al. 2013).

8.3.4 Plant Stress Resistance

Agriculture productivity suffers greatly from abiotic stress, which differs based on the soil's quality and plant characteristics like disease susceptibility, abscission, etc. (Nadeem et al. 2010). According to Pishchik et al. (2002)'s research, cadmium contamination on barley plants can have an impact on how well the PGPR functions. In addition, Nautiyal et al. (2008) reported that Bacillus lentimorbus strains improved the antioxidant activity of spinach, carrots, and lettuce, and that their growth also increased. Under stressful circumstances, it was also discovered that the leaf water status had improved (Naveed et al. 2014). Sarma and Saikia (2014) showed that Pseudomonas aeruginosa was found to promote the growth of mung bean (Vigna radiate) plants in drought-like conditions. In their research, Naveed et al. (2014) found that plants with PGPR inoculation transpired at a higher rate than plants without PGPR inoculation. According to Marulanda et al. (2010), Bacillus megaterium sprayed to the roots of the maize plant improved water absorption under stressful circumstances. Additionally, Pantoeaagglomerans improved water absorption in maize roots, according to research by Gond et al. (2015). Azospirillumbrasilense was applied to the jojoba plant by Gonzalez et al. (2015) when it was under salt stress. Fasciglione et al. (2015) observed similar results, noting that applying Azospirillum to lettuce improved its growth. According to Viscardi et al. (2016), Azotobacter chroococcum promotes tomato plant development in salinity and drought. According to research, Serratia marcescens improves wheat plant growth when exposed to salinity stress (Singh and Jha, 2016). According to research by Barnawal et al. (2017), the PGPR strains of Arthrobacter protophormiae and Dietzianatronolimnaea grow wheat more quickly in saline settings whereas Bacillus subtilis grows wheat more quickly in dry conditions. Citrus macrophylla (alemow) plant development was improved by two plant growth-promoting rhizobacteria, *Pseudomonas putida* and *Novosphingobium* sp. under salt stress, according to Vives-Peris et al. (2018). Chinese cabbages have also been shown to grow more quickly in hot and dry environments when treated with Bacillus aryabhattai and B. siamensis (Da Jeong Shin et al. 2019).

8.3.5 Nutrient Acquisition

By making nutrients more readily available in the root zone, PGPR improves nutrient uptake (Choudhary et al. 2011). As atmospheric nitrogen becomes a limiting factor for plant growth (Lloret and Martnez-Romero 2005; Raymond et al. 2004), fixation of atmospheric nitrogen by prokaryotes (such as Azospirillum) is once again a significant phenomenon. Rhizobacteria that encourage plant growth, including Kocuriaturfanensis strain 2M4, solubilize phosphate (Goswami et al. 2014; Wani et al. 2007), increasing the amount of phosphate ions that are available to plants in the soil. According to Yadav et al. (2014), Pseudomonas fluorescens and Pseudomonas putida improve rice's ability to absorb nutrients. Pseudomonas putida was also reported to enhance the uptake of phosphorous, nitrogen, and potassium in chickpeas (Israr et al. 2016). Xiaohui et al. (2017) demonstrated that Bacillus pumilus enhance the nitrogen and phosphorous uptake in tomato.

8.3.6 Crop Growth, Yield and Productivity

Plant growth stimulation by rhizobacteria may occur directly or indirectly at many stages of the plant life cycle of plant. Direct growth promotion includes nitrogen fixation (Mallik et al. 1997), phosphate solubilization (Alori et al. 2017), phytohormone production (Egamberdieva et al. 2017) or enhancement in the availability of minerals (Korir et al. 2017). The nitrogen fixation process is mediated by an oxygen-sensitive enzyme complex that transforms atmospheric nitrogen into an ammonical form (biologically fixed nitrogen) that is either made accessible to plants or liberated into the soil. Phosphate solubilizers make Phosphorous accessible to plants from fixed fforms that are already being present in the soil. Plant growth is also influenced by the synthesis of plant hormones such as auxins, gibberellins, and cytokinins. The production of siderophores by PGPR provides the plant with sufficient iron in iron-limited soils. Additional beneficial effects of PGPR on plant development include osmotic adjustment, stomatal modulation, root shape modification, and so on under abiotic stress situations (El-Esawi et al. 2018). Indirect growth stimulation of PGPR is associated with the secteation of antagonistic chemicals such as phenazine, diacetylphloroglucinol (DAPG), hydrogen cyanide (HCN), 2–3 butanediol, acetoin, and siderophores (Compant et al. 2005) to prevent deleterious effects of phytopathogens (Sharma et al. 2016). The lytic enzymes generated by these PGPRs, chitinase and glucanase, may breakdown the cell wall of fungal infections, generating systemic resistance throughout the plant system (Kloepper et al. 2006). Yet, the mechanisms in which PGPRs directly influence plant development may differ among species or be strain specific.

Rhizobial symbiosis is very important in agriculture because it improves soil fertility and crop yield via synergistic and antagonistic interactions with other bacteria in the soil environment. Yet, the functional responsibilities of PGPR are heavily reliant on their survival and expansion in the field (del Carmen Rivera-Cruz et al. 2008). It is generally recognised that organic matter is a crucial limiting factor for microbial development, thus it is not surprising that earlier research revealed that using soil (Shahzad et al. 2014). The use of vermicompost and microbial agents into tomato and spinash crop cultivation is a viable substitute for traditional chemical fertilisation procedures (Song et al. 2015). More explicit engineering of PGPR strains is required to enhance their ability to suppress infections or reduce the sysnthesis of plant stress hormones (Ryan et al. 2008). In the field, for example, Azospirillum-lipoferum transplanted onto field grown maize seed induced a statistically significant difference in the diversity of the indigenous rhizobial community (Baudoin et al. 2009). P. fluorescens strain CHA0 in conjunction with chitin has the potential to boost banana plant growth, leaf nutrient content, and yield in perennial cropping systems (Kavino et al. 2010). development and commercialization of rhizobacterial strains T19, T29, S7, and A26 are feasible due to their capacity to increase maize yield when applied as seed treatments in the field. Half-fertilization with individual inoculation with isolates AC32 (*Herbaspirillum* sp.), AG15 (*Burkholderia* sp.), CA21 (*Pseudoacidovorax* sp.), and UR51 (*Azospirillum* sp.) resulted in optimal rice growth and production (Souza et al. 2013). The combination application of PGPR and silicon fertiliser may improve growth, yield, and biochemical activity P. vulgaris bacteria in saline soils (Kumar et al. 2020). Plants infected with PGPR and 33 kg/ha result in greater root dry weight and nutrient content (N, P, K) of sweet potato shoot and root. Plants infected with PGPR and 33 kg/ha result in increased root dry weight and nutrient content (N, P, K) of sweet potato shoot and root.

8.4 Role of PGPR in Soil Fertility and Sustainability

8.4.1 Organic Matter (OM) Decomposition

Decomposition of organic matter means decomposition of component it is made of e.g., cellulose (20–50%), hemicellulose (10–30%), protein (1–8%), carbohydrates (5%), oils and fats (2%) and lignin ad polyphenols (2%). Few of these compounds are easily decomposable e.g., carbohydrates and protein while few components e.g., lignin and polyphenols are resistant in nature and can't be decomposed easily. PGPR secrets extracellular hydrolytic enzymes e.g., Endoglucanases, exoglucanases, β-glucosidases, endoxylanase, ecetylxylan esterase, α-glucuronidase and β-xylosidase, lignin peroxidases etc. in order to catalyze organic matter decomposition (Berg and McClaugherty 2008). Beside these enzymes there are list of some additional hydrolytic enzymes also (e.g., cellulases, pectianses, proteases, catalases and chitinases) which along with decomposition of organic matter also capable of providing protection measures against plant pathogens e.g. Chitinase enzyme. Chitinase enzymes act on chitin and help in inhibiting fungal pathogens like Botrytis cinerea (Sadeghi et al. 2017). Likewise cellulolytic soil microbes are also multi-functions: decompose cellulose present in plant residues (Dadwal et al. 2019) and produce biofuelsas well from biomass (Siqueira et al. 2020) etc.

8.4.2 Nutrient Transformation and Availability

PGPR have multiple genes and countless pathway to synthesis enzymes/protein to mineralize nutrients locked in different form (Khatoon et al. 2020). PGPR are also involved in biofixation and biosolubilization which act as nutrient fertilization activity to make it available for plant use (Chaparro et al. 2012). PGPR in association with plant root synthesis the biomolecules responsible for increasing availability of different nutrients (Kumar and Verma 2019). There is various mechanism used by PGPR for nutrient transformation. In the process of decomposition nutrients locked in organic matter e.g., nitrogen, phosphorus, sulphurare released and utilized by plants. Many of the PGPR e.g., nitrogen fixer (Acetobacter diazotrophicus), Ammonium production (*Achromobacterxylosoxidans*), phosphate solubilizer (*Azospirilliumbrasiliensis*), potassium solubilizer (*Bacillus circulans*), Zinc solubilizer (*Bacillus filamentous, Bacillus sphericus*) are responsible for utilizing part of nutrients which are present in environment (soil and atmosphere but can't be utilized). e.g., Diazotrophic PGPR fix atmospheric nitrogen and convert dinitrogen to ammonical form likewise phosphate solubilizing bacteria release phosphorus fixed by Al, Fe, Ca etc. Such PGPR can easily be used as biofertilizer (del Carmen Rivera-Cruz et al. 2008; Khatoon et al. 2020). Various PGPR (Pseudomonas, Bacillus, Azotobacter etc.) has been used in vegetables e.g. lettuce and soyabean and found to increase nutrient content in soil (Alori and Babalola 2018).

8.4.3 Soil Bioremediations

It is the technique of cleaning, managing, and recovering technique of polluted environment through microbial activity. It helps in detoxifying or degrading contaminants present in the soil, wastewater, or industrial sludge by biological means. Continuous use of cyanide fertilizers and pesticides in agriculture resulting to increased nitrile concentration in soil. Nitrile in soil act as pollutant and casing danger to ground water and food chain. Group of PGPR has been reported that show nitrilase activity and control nitrile concentration in nitrile polluted soil (Vaishnav et al. 2022). Polychlorinated biphenyls (PCBs), total petroleum hydrocarbons (TPHs),polycyclic aromatic hydrocarbon (PCAH), and creosote are the organic pollutants, widely used in the industrial process and are most degradation-resistant. In addition, heavy metals like Cd, Cr, Cu, Hg, Pb, Ni and Zn are the primary inorganic contaminants (Zhuang et al. 2007). Soil bacteria e.g., *Azotobacter chroococcum, Bacillus megaterium, Bacillus mucilaginosus, Bacillus subtilis, Pseudomonas fluorescens*, act as hyperaccumulator (Wu et al. 2006; Zaidi et al. 2006), of heavy metals and sequester in their own body and reducing mobility of these heavy metals by changing pH or redox potential of soil. However, these heavy metals reduce microbial proliferation and their biomass accumulation in soil to offer bioremediation of soil. Some other limitation of bioremediation technologies through PGPr are: chlorinated contaminants and high molecular weight PAHs do not easily undergo microbial degradation; bioremediation of few chemicals may increase toxicity and mobility of transformed pollutant than the parent compound (Zhuang et al. 2007).

8.4.4 Drought and Nutrient Stress/Deficiency

A drought or nutrient stress are abiotic stress encountered by the plants throughout their life cycle due to inadequate water availability, which results to lowering of crop development and yield by negatively affecting the morphological, physiological, and biochemical attributes of the plants (Osakabe et al. 2014; Seleiman et al. 2021). Large number of PGPR has been reported to secrete phytohormones, such as auxin, cytokinin, gibberellin, abscisic acid (ABA), ethylene, salicylic acid, and jasmonic acid, for supporting plant growth. Gowtham and co-workers (2021) reported list of PGPR that support plant growth and development by releasing various plant hormones e.g., *Bacillus megatherium* (IAA), *Bacillus subtilis* (cytokinin), *Pseudomonas putida* (gibberellin), *B. licheniformis* and *P. fluorescens* (ABA) etc. Beside secreting plant hormones some soil microbes are also reported to secrete exopolysaccharides that enable soil particles to form soil aggregates which can retain and restore soil moisture for prolong to mitigate effect of drought. These polysaccharides also act as hydrophilic biofilms on root surface, to protect the roots and, allowing for a slower rate of water release in drought stress (Ansari et al. 2021; Naseem et al. 2018) (Table 8.2).

8.5 Future Prospects

World's rapid increasing population may be upto 10 billion in near future which may lead to a huge pressure on agricultural land. Overburdened soils, with all the pressure of producing higher to feed the earth's people, mostly terminated as degraded land. Thought of getting high yield with high dose of fertilizer/input application most of the time results into a less yield producing piece of land. Application of enormous quantity of chemicals to the soil for decades after decades has left it almost barren. Not only soil, environment has also deteriorated because of that. Era o green revolution has awarded the society with many boons like HYVs, hybrids etc. but has taken away many more things from us. It did teach us that over dependability on inorganics led us nowhere but towards an extremely dangerous situation. Agriculture is not only a profession for the people in India; it is a source of generating major portion of the country's employment. So faulty practices may endanger the sustainability of the system but here is always some way to get rid of this. Among the various ways out found in last decades, biofertilizers came out as a good option to manage and sustain the system. They are natural, not at all harmful for soil, water and environment. PGPR or plant growth promoting rhizobacteria are safe to environments, cost effective and a promising long term tool for achieving sustainability in agriculture.

Worldwide different experiments show the amazing responses of different crop species towards these tiny little microscopic creatures. Their mutual interaction is really surprising and going to enlighten a completely new vast arena of crop production with PGPR in future. A lot more studies to be done in future in this field.

Table 8.2 Various PGPR strain according to crop

Groups	Crops	PGPRs	Influence	References
Cereals	Rice	*Azospirillumlipoferum*	Biostimulation by Gibberellin production	Cassan et al. (2001)
		Bacillus amyloliquefaciens	Biostimulation by Gibberellin production	Shahzad et al. (2016)
		Bacillus firmus	Bioprotection by Siderophore production	Chaiharn et al. (2009)
		Rhizobium leguminosarum	Biostimulation by Gibberellin production	Yanni et al. (2001)
	Wheat	*R. leguminosarum and Pseudomonas* sp.	Improve phosphorus uptake and hence yield	Afzal and Asghari (2008)
		Azospirillumbrasilense	Improve growth attributes	Dobbelaere et al. (2002)
	Maize	*P. putida strains R-168 and DSM-291*	Improves seed germination, seedling growth and yield	Nezarat and Gholami (2009)
		P. putida strain R-168	Enhance seed, germination, growth parameters	Gholami et al. (2009)
		Azospirillumbrasilense	Improve growth attributes	Dobbelaere et al. (2002)
		P. alcaligenes PsA15 and B. polymyxa	Enhance nutrient uptake	Egamberdiyeva (2007)
		Azospirillumbrasilense	Phosphate solubilization	Lucy et al. (2004)
	Pearl millet	*P. fluorescens strain*	Growth and yield	Niranjan et al. (2003)
Legumes	Chickpea	*Pseudomonas and Azotobacter*	Stimulates growth and yield	Rokhzadi et al. (2008)
		Bacillus subtilis	Biostimulation (by IAA and Cytokinin production)	Ait-Kaki et al. (2014)

(continued)

Table 8.2 (continued)

Groups	Crops	PGPRs	Influence	References
	Soybean	*Bradyrhizobium japonicum*	Biofertilization (Nitrogen fixation), Biostimulation (Gibberellin production)	Kaneko et al. (2002)
		Proteus vulgaris	Siderophore production	Rani et al. (2009)
		Bacillus cereus	Acquired and induced systemic resistance	Ait-Kaki et al. (2014)
Vegetbles	Tomato	*P. putida strain*	Growth stimulation	Gravel et al. (2007)
Fruit	Banana	*P. fluorescens strain R-93, P. fluorescens DSM 50,090,*	Increase growth, leaf nutrient contents and yield	Kavino et al. (2010)
Sugarcane	Sugarcane	*Azospirillumdiazotrophicus*	Nitrogen fixation	Steenhoudt and Vanderleyden (2000)

Few things are there to be taken care of at field as well as in laboratory level. The most important is to explore the group of micro organisms by knowing their species diversity and their interaction pattern. A details study of available data upto molecular level is required. That may flourish some new paradigm in the field of PGPR. Another important aspect is response and reaction of microbiomes to the abiotic stresses. Detailed study of individual species and their behavior with different crops would be needed for practical implement in farmers' field. Researches must also be done there in a large scale along with laboratory. That may increase the chances of social acceptance. The newer technologies like nanotechnology, rhizo-engineering and metaproteonomics may have crucial role in making new formulations. This is another very important area to explore because of its future role in determining the acceptability of newly made microbial formulation. Low temperature plant growth and development may be explored by experimenting on ice-nucleating plant growth promoting rhizobacteria.

While inventing formulation under laboratory is the first step of this task, the second and most important part is to launch a pocket friendly, eco friendly, safe to handle formulation in the market. Shelf life of the product is also should be longer. Commercialization is the next important step. For this actual study must be done in the grass root level for acquiring the proper knowledge regarding existing farmers practices for different crop, market scenario and survey, transport chain, market linkage etc. Some risk factors may also be analyzed before commercialization. Like, there should not be any chances of negative interaction with the used micro organism to the other crops. So that recommendation and use of crop specific strains must be assured. Chances of any kind of antagonistic activity of used microbes in formulation towards any fungi should also be taken care of before selection. For good promotion and popularization of these formulations among farmers, sufficient knowledge sharing platforms also should be created because knowledge about the positive possible outcomes of the formulation is necessary. Lack of that may result into improper interpretation and unsatisfactory results. Awareness programmes, trainings interaction sessions with the input dealers and the direct users can fill up this gap.

8.6 Challenges

People are working a lot on PGPR, a huge number of research works are being carried out in different parts of the world. PGPR is making its own domain as agricultural input slow but steadily. Farmers are also started to understand the benefits of organic biofertilizer approach over the inorganic one. But worldwide large scale spreading of use of PGPR still has to overcome many challenges and problems.

Firstly, a laboratory set up is required for initial testing of these bioinoculants to observe how good they are at promoting the growth and development. If performance is not satisfactory, elimination may be done at laboratory level only (Cardinale et al, 2015).

Selecting a particular species of microbes to develop biofertilizer is also a big challenge everytime. Behavior of the strain (s) to the crop species is very important. Higher selectivity of the strain is not acceptable; there must be wider host range. Experimental results in the field level are always dissimilar to the laboratory results. Because in open fields, many more factors are active other than the stains inoculated like other soil micro organisms, soil type and environmental conditions. Therefore selection of strains should always include multiple testing in field condition in different location (Meena 2020). As these are the growth promoting strains working in the roots zone o the plant, there must be a good interaction with plant root also by making a colony. In addition to that availability of soil nutrient also should be increased (Vessey 2003; Vejan et al. 2016). Also matter of interaction with non target plant species in field condition should be solved before release. The behaviour of the PGPR changes with different host plant, soil and place. Crop rotation history of the field has also have effect on the PGPR behavior (Amara et al. 2015). The bio control potential shown by the gram negative bacteria also should be explored keeping in mind the point of making their formulations with a longer shelf life. Formulation with shorter shelf life has a challenge to be sold before the expiry date (Goswami et al. 2016; Zandi and Basu 2016; Arora et al. 2011). Living cells present in the inoculants is very much delicate and requires proper storage and transport. It the product is deteriorated because of the shorter life cycle that will cause a monetary loss to the seller. The next important challenge is registration and patent filing. As the registration process is quite complex, costly and having different rules and regulation which varies countrywise (Meena et al. 2020; Timmusk et al. 2017). Also lack of proper policy regarding this makes it more complicated (Baker et al. 2018). Lack of trained and skilled person to handle the entire laboratory and field work including manufacture and quality control, handling of the delicate instruments are again another challenge (Mahajan and Gupta 2009). Not only this, all this work requires a supply of huge fund to run smooth system for the journey from laboratory to field, which is also not sufficiently available. This breaks the supply chain and sometimes produced material is completely wasted because of the shorter life cycle (Mahajan and Gupta 2009). Adaptability and efficiency of the PGPR formulations in different farm condition with changing soil, water, microclimate etc. is another big challenge to overcome (Kamilovva et al. 2015; Barea 2015). As the products are very delicate, using them in the farmers' field needs delicate handling, proper carrier, less residual effect of the previously used synthetics etc. Residual effects may hamper or delay the colonization process of the microbes. Amount of presence of salt, heavy metal, lack of important nutrient etc. in the soil may reduce the activity of microbes. Plant microbe interaction is affected by all these factors. Because of this, farmers' rely and satisfaction on PGPR formulation is not up to the mark (Mahanty et al. 2017). Safety of the user is also another challenge to overcome. PGPR Microbial formulation should not cause any harm to the person handling or applying this. Any chances of health hazard should be avoided.

8.7 Conclusive Remarks

Use of PGPR in place of conventional inorganic fertilizer, will help plant growth as well as ultimate production. But implementation at field level in large scale is still a far way to go! For that proper infrastructure is required which includes well equipped laboratories, trained persons and controlled field experiments with all the facilities. Not only this, a elaborated homework is needed for the market, demand, commercial production, safety, possible hazards etc. before launching any PPR formulation to the market.

References

Abbas Z, Zia MA, Ali S, Abbas Z, Waheed A, Bahadur A, Sultan T (2013) Integrated effect of plant growth promoting rhizobacteria, phosphate solubilizing bacteria and chemical fertilizers on growth of maize original research article. Int J Agric Crop Sci 6(13):913

Ahemad M, Khan MS (2009) Effect of insecticide-tolerant and plant growth-promoting Mesorhizobium on the performance of chickpea grown in insecticide stressed alluvial soils. J Crop Sci Biotechnol 12:217–226

Alori ET, Glick BR, Babalola OO (2017) Microbial phosphorus solubilization and its potential for use in sustainable agriculture. Front Microbiol 8:971

Alori ET, Babalola OO (2018) Microbial inoculants for improving crop quality and human health in Africa. Front Microbiol 9:2213. https://doi.org/10.3389/fmicb.2018.02213

Amara U, Khalid R, Hayat R (2015) Soil bacteria and phytohormones for sustainable crop production. In: Maheshwari DK (ed) Bacterial metabolites in sustainable agroecosystem. Springer, Cham, Switzerland, pp 87–103

Andrews JH, Harris RF (2000) The ecology and biogeography of microorganisms on plant surfaces. Annu Rev Phytopathol 38:145–180. https://doi.org/10.1146/annurev.phyto.38.1.145

Ansari FA, Jabeen M, Ahmad I (2021) Pseudomonas azotoformans FAP5, a novel biofilm-forming PGPR strain, alleviates drought stress in wheat plant. Int J Environ Sci Technol 18:3855–3870

Araujo FFD (2008) Seed inoculation with Bacillus subtilis, formulated with oyster meal and growth of corn, soybean and cotton. Ciência e Agrotecnologia 32:456–462

Armada E, Portela G, Roldan A, Azcon R (2014) Combined use of beneficial soil microorganism and agrowaste residue to cope with plant water limitation under semiarid conditions.Geoderma 232:640–648

Arora NK, Khare E, Maheshwari DK (2011) Plant growth promoting rhizobacteria: constraints in bioformulation, commercialization, and future strategies. In: Plant growth and health promoting bacteria, pp 97–116

Babalola OO, Akindolire AM (2011) Identification of native rhizobacteria peculiar to selected food crops in Mmabatho municipality of South Africa. Biol Agric Hortic 27(3–4):294–309

Backer R, Rokem JS, Ilangumaran G, Lamont J, Praslickova D, Ricci E, Subramanian S, Smith DL (2018) Plant growth promoting rhizobacteria: context, mechanisms of action, and roadmap to commercialization of biostimulants for sustainable agriculture. Front Plant Sci 9:1473

Barea JM (2015) Future challenges and perspectives for applying microbial biotechnology in sustainable agriculture based on a better understanding of plant-microbiome interactions. J Soil Sci Plant Nutr 15(2):261–282

Bashan Y, de-Bashan LE, Prabhu S, Hernandez JP (2014) Advances in plant growth-promoting bacterial inoculant technology: formulations and practical perspectives (1998–2013). Plant Soil 378:1–33

Baudoin E, Nazaret S, Mougel C, Ranjard L, Moënne-Loccoz Y (2009) Impact of inoculation with the phytostimulatory PGPR Azospirillumlipoferum CRT1 on the genetic structure of the rhizobacterial community of field-grown maize. Soil Biol Biochem 41(2):409–413

Beever RE, Burns DJW (1981) Phosphorus uptake, storage and utilization by fungi. In: Advances in botanical research, vol 8. Academic Press, pp 127–219

Benizri E, Baudoin E, Guckert A (2001) Root colonization by inoculated plant growth promoting rhizobacteria. Biocontrol Sci Tech 11:557–574

Berendsen RL, Pieterse CMJ, Bakker PAHM (2012) The rhizosphere microbiome and plant health. Trends Plant Sci 17:478–486

Berg B, McClaugherty C (2008) Plant litter: decomposition, humus formation, carbon sequestration (No. 04, QH541. 5. S6, B4 2008.). Springer, Berlin

Butterbach-Bahl K, Baggs EM, Dannenmann M, Kiese R, Zechmeister-Boltenstern S (2013) Nitrous oxide emissions from soils: how well do we understand the processes and their controls? Philos Trans R Soc B: Biol Sci 368

Calvo P, Nelson L, Kloepper JW (2014) Agricultural uses of plant biostimulants. Plant Soil 383:3–41

Cardinale M, Ratering S, Suarez C, Montoya AMZ, Geissler-Plaum R, Schnell S (2015) Paradox of plant growth promotion potential of rhizobacteria and their actual promotion effect on growth of barley (Hordeum vulgare L.) under salt stress. Microbiol Res 181:22–32

Cardoso EJBN, Freitas SS (1992) A rizosfera. In: Cardoso EJBN, Tsai SM, Neves PCP (eds) Microbiologia do solo. Sociedade Brasileira de Ciencia do Solo, Campinas, pp 41–57

Chaparro JM, Sheflin AM, Manter DK, Vivanco JM (2012) Manipulating the soil microbiome to increase soil health and plant fertility. Biol Fertil Soils 48(5):489–499. https://doi.org/10.1007/s00374-012-0691-

Choudhary D, Sharma K, Gaur R (2011) Biotechnological perspectives of microbes in agro-ecosystems. Biotech Lett 33:1905–1910

Compant S, Duffy B, Nowak J, Clément C, Barka EA (2005) Use of plant growth-promoting bacteria for biocontrol of plant diseases: principles, mechanisms of action, and future prospects. Appl Environ Microbiol 71(9):4951–4959

Dadwal A, Sharma S, Satyanarayana T (2019) Diversity in cellulose-degrading microbes and their cellulases: role in ecosystem sustainability and potential applications. Microbial diversity in ecosystem sustainability and biotechnological applications. Soil & Agroecosystems 2:375–402

Del Carmen Rivera-Cruz M, Narcía AT, Ballona GC, Kohler J, Caravaca F, Roldan A (2008) Poultry manure and banana waste are effective biofertilizer carriers for promoting plant growth and soil sustainability in banana crops. Soil Biol Biochem 40(12):3092–3095

Dennis PG, Miller AJ, Hirsch PR (2010) Are root exudates more important than other sources of rhizodeposits in structuring rhizosphere bacterial communities? Fed Eur Microbiol Societies 72:313–327

Dey R, Pal K, Bhatt D, Chauhan S (2004) Growth promotion and yield enhancement of peanut (Arachis hypogaea L.) by application of plant growth promoting rhizobacteria. Microbiol Res 159:371–394

Dobbelaere S, Vanderleyden J, Okon Y (2003) Plant growth-promoting effects of diazotrophs in the rhizosphere. Crit Rev Plant Sci 22(2):107–149

Egamberdieva D, Wirth SJ, Alqarawi AA, Abd_Allah EF, Hashem A (2017). Phytohormones and beneficial microbes: essential components for plants to balance stress and fitness. Front Microbiol 8:2104

Egamberdiyeva D (2007) The effect of plant growth promoting bacteria on growth and nutrient uptake of maize in two different soils. Appl Soil Ecol 36(2–3):184–189

El-Esawi MA, Alaraidh IA, Alsahli AA, Alamri SA, Ali HM, Alayafi AA (2018) Bacillus firmus (SW5) augments salt tolerance in soybean (Glycine max L.) by modulating root system architecture, antioxidant defense systems and stress-responsive genes expression. Plant Physiol Biochem 132:375–384

Fatima F, Ahmad MM, Verma SR, Pathak N (2022) Relevance of phosphate solubilizing microbes in sustainable crop production: a review. Int J Environ Sci Technol 19(9):9283–9296

Friesen ML, Porter SS, Stark SC, Wettberg EJ, Sachs JL, Martinez-Romero E (2011) Microbially mediated plant functional traits. Annu Rev Ecol Evol Syst 42:23–46

Galloway JN, Townsend AR, Erisman JW, Bekunda M, Cai Z, Freney JR et al (2008) Transformation of the nitrogen cycle: recent trends, questions, and potential solutions. Science 320:889–892

Glick BR (1995) The enhancement of plant growth by free-living bacteria. Can J Microbiol 41:109–117

Glick BR (2012) Plant growth-promoting bacteria: mechanisms and applications. Scientifica 2012:1–15

Goswami D, Thakker JN, Dhandhukia PC (2016) Portraying mechanics of plant growth promoting rhizobacteria (PGPR): a review. Cogent Food & Agriculture 2(1):1127500

Gowtham HG, Duraivadivel P, Ayusman S, Sayani D, Gholap SL, Niranjana SR, Hariprasad P (2021) ABA analogue produced by Bacillus marisflavi modulates the physiological response of Brassica juncea L. under drought stress. Appl Soil Ecol 159:103845

Hartmann A, Rothballer M, Schmid M (2008) Lorenz Hiltner, a pioneer in rhizosphere microbial ecology and soil bacteriology research. Plant Soil 312:7–14

He Z, Zhu JUN (1998) Microbial utilization and transformation of phosphate adsorbed by variable charge minerals. Soil Biol Biochem 30(7):917–923

Hiltner L (1904) Iiber neuereErfahrungen und unterbesondererBerucksichtingung der Grundunging und Brache. Arb. Deut. Landw. Ges. 98:59–78

Joshi KK, Kumar V, Dubey RC, Maheshwari DK, Bajpai VK, Kang SC (2006) Effect of chemical fertilizer-adaptive variants, Pseudomonas aeruginosa GRC2 and Azotobacter chroococcum AC1, on Macrophominaphaseolina causing charcoal rot of Brassica juncea. Korean J Environ Agric Stud 25:228–235

Kavino M, Harish S, Kumar N, Saravanakumar D, Samiyappan R (2010) Effect of chitinolytic PGPR on growth, yield and physiological attributes of banana (Musa spp.) under field conditions. Appl Soil Ecol 45(2):71–77

Khalid A, Arshad M, Zahir ZA (2004) Screening plant growth-promoting rhizobacteria for improving growth and yield of wheat. J Appl Microbiol 96(3):473–480

Khatoon Z, Huang S, Rafique M, Fakhar A, Kamran MA, Santoyo G (2020) Unlocking the potential of plant growth-promoting rhizobacteria on soil health and the sustainability of agricultural systems. J Environ Manage 273:111118

Köhl J, Kolnaar R, Ravensberg WJ (2019) Mode of action of microbial biological control agents against plant diseases: relevance beyond efficacy. Front Plant Sci 845

Korir H, Mungai NW, Thuita M, Hamba Y, Masso C (2017) Co-inoculation effect of rhizobia and plant growth promoting rhizobacteria on common bean growth in a low phosphorus soil. Front Plant Sci 8:141

Kour D, Rana KL, Yadav AN, Yadav N, Kumar M, Kumar V et al (2020) Microbial biofertilizers: bioresources and eco-friendly technologies for agricultural and environmental sustainability. Biocatal Agric Biotechnol 23:101487. https://doi.org/10.1016/j.bcab.2019.101487

Kozdrój J, Trevors JT, Van Elsas JD (2004) Influence of introduced potential biocontrol agents on maize seedling growth and bacterial community structure in the rhizosphere. Soil Biol Biochem 36(11):1775–1784

Kucey RMN, Janzen HH, Leggett ME (1989) Microbially mediated increases in plant-available phosphorus. Adv Agron 42:199–228

Kumar A, Verma JP (2019) The role of microbes to improve crop productivity and soil health. In: Ecological wisdom inspired restoration engineering, pp 249–265. https://doi.org/10.1007/978-981-13-0149-0_14

Kumar V, Kumar P, Khan A (2020) Optimization of PGPR and silicon fertilization using response surface methodology for enhanced growth, yield and biochemical parameters of French bean (Phaseolus vulgaris L.) under saline stress. Biocatal Agric Biotechnol 23:101463

Lucas Garcia JA, Probanza A, Ramos B, Barriuso J, Gutierrez Manero FJ (2004) Effects of inoculation with plant growth promoting rhizobacteria (PGPRs) and Sinorhizobium fredii on biological nitrogen fixation, nodulation and growth of Glycine max cv. Osumi. Plant and Soil 267:143–153

Mahajan A, Gupta RD (eds) (2009) Integrated nutrient management (INM) in a sustainable rice—wheat cropping system. Springer, Netherlands, Dordrecht

Mahanty T, Bhattacharjee S, Goswami M, Bhattacharyya P, Das B, Ghosh A, Tribedi P (2017) Biofertilizers: a potential approach for sustainable agriculture development. Environ Sci Pollut Res 24:3315–3335

Maheshwari D, Dubey R, Aeron A, Kumar B, Kumar S, Tewari S et al (2012) Integrated approach for disease management and growth enhancement of Sesamum indicum L. utilizing Azotobacter chroococcum TRA2 and chemical fertilizer. J Microbiol Biotechnol 28:3015–3024

Mallik Freitas ADSD, Vieira CL, Santos CEDR, Stamford NP, Lyra MDCCPD (2007) Characterization of isolated rhizobia from Pachyrhyzuserosus L. cultivated in saline soil of the State of Pernambuco, Brazil. Bragantia 66:497–504

Marchner H (1995) Mineral nutrition of higher plants, 2nd edn. Academic Press, London

Martinez-Viveros O, Jorquera M, Crowley D, Gajardo G, Mora M (2010) Mechanisms and practical considerations involved in plant growth promotion by rhizobacteria. J Soil Sci Plant Nutr 10:293–319

Meena KK, Sorty AM, Bitla UM, Choudhary K, Gupta P, Pareek A, Minhas PS (2017) Abiotic stress responses and microbe-mediated mitigation in plants: the omics strategies. Front Plant Sci 8:172

Meena M, Swapnil P, Divyanshu K, Kumar S, Tripathi YN, Zehra A, Upadhyay RS (2020) PGPR-mediated induction of systemic resistance and physiochemical alterations in plants against the pathogens: current perspectives. J Basic Microbiol 60(10):828–861

Montesinos E (2003) Plant-associated microorganisms: a view from the scope of microbiology. Int Microbiol 6:221–223. https://doi.org/10.1007/s10123-003-0141-0

Nakkeeran S, Fernando WD, Siddiqui ZA (2005) Plant growth promoting rhizobacteria formulations and its scope in commercialization for the management of pests and diseases. In: Siddiqui ZA (ed) PGPR: biocontrol and biofertilization. Springer Netherlands, Dordrecht, pp 257–296. https://doi.org/10.1007/1-4020-4152-7_10
Naseem H, Ahsan M, Shahid MA, Khan N (2018) Exopolysaccharides producing rhizobacteria and their role in plant growth and drought tolerance. J Basic Microbiol 58(12):1009–1022
Osakabe Y, Osakabe K, Shinozaki K, Tran LSP (2014) Response of plants to water stress. Front Plant Sci 5:86
Park JH, Bolan N, Megharaj M, Naidu R (2011) Concomitant rock phosphate dissolution and lead immobilization by phosphate solubilizing bacteria (Enterobacter sp.). J Environ Manage 92(4):1115–1120
Ryan RP, Germaine K, Franks A, Ryan DJ, Dowling DN (2008) Bacterial endophytes: recent developments and applications. FEMS Microbiol Lett 278(1):1–9
Sadeghi A, Koobaz P, Azimi H, Karimi E, Akbari AR (2017) Plant growth promotion and suppression of Phytophthora drechsleri damping-off in cucumber by cellulase-producing Streptomyces. Biocontrol 62:805–819
Sagar S, Dwivedi A, Yadav S, Tripathi M, Kaistha SD (2012) Hexavalent chromium reduction and plant growth promotion by Staphylococcus arlettae strain Cr11. Chemosphere 86:847–852
Seleiman MF, Al-Suhaibani N, Ali N, Akmal M, Alotaibi M, Refay Y, Battaglia ML (2021) Drought stress impacts on plants and different approaches to alleviate its adverse effects. Plants 10(2):259
Shahzad SM, Khalid A, Arif MS, Riaz M, Ashraf M, Iqbal Z, Yasmeen T (2014) Co-inoculation integrated with P-enriched compost improved nodulation and growth of Chickpea (Cicer arietinum L.) under irrigated and rainfed farming systems. Biol Fertil Soils 50:1–12
Sharma M, Mishra V, Rau N, Sharma RS (2016) Increased iron-stress resilience of maize through inoculation of siderophore-producing Arthrobacter globiformis from mine. J Basic Microbiol 56(7):719–735
Shrivastava S, Prasad R, Varma A (2014) Anatomy of roots from eyes of microbiologists. In: Asun M, Ajit V (eds) Root engineering, vol 40. Soil biology series. Springer, Berlin, pp 3–22
Siqueira JGW, Rodrigues C, de Souza Vandenberghe LP, Woiciechowski AL, Soccol CR (2020) Current advances in on-site cellulase production and application on lignocellulosic biomass conversion to biofuels: a review. Biomass Bioenerg 132:105419
Song X, Liu M, Wu D, Griffiths BS, Jiao J, Li H, Hu F (2015) Interaction matters: synergy between vermicompost and PGPR agents improves soil quality, crop quality and crop yield in the field. Appl Soil Ecol 89:25–34
Steenhoudt O, Vanderleyden J (2000) Azospirillum, a free-living nitrogen-fixing bacterium closely associated with grasses: genetic, biochemical and ecological aspects. FEMS Microbiol Rev 24:487–506
Timmusk S, Behers L, Muthoni J, Muraya A, Aronsson AC (2017) Perspectives and challenges of microbial application for crop improvement. Front Plant Sci 8:49
Trivedi P, Pandey A, Palni LMS (2005) Carrier-based preparations of plant growth-promoting bacterial inoculants suitable for use in cooler regions. J Microbiol Biotechnol 21:941–945. https://doi.org/10.1007/s11274-004-6820-y
Ullah A, Mushtaq H, Ali U, Hakim AE, Mubeen S, Chaudhary HJ (2018) Screening, isolation, biochemical and plant growth promoting characterization of endophytic bacteria. Microbiol Curr Res 2(2):24–30
Vaishnav A, Kumar R, Singh HB, Sarma BK (2022) Extending the benefits of PGPR to bioremediation of nitrile pollution in crop lands for enhancing crop productivity. Sci Tot Environ:154170
Van Loon LC (2007) Plant responses to plant growth-promoting rhizobacteria. Eur J Plant Pathol 119:243–254
Vejan P, Abdullah R, Khadiran T, Ismail S, Nasrulhaq Boyce A (2016) Role of plant growth promoting rhizobacteria in agricultural sustainability—a review. Molecules 21(5):573
Vessey JK (2003) Plant growth promoting rhizobacteria as biofertilizers. Plant Soil 255:571–586

Welbaum GE, Sturz AV, Dong Z, Nowak J (2004) Managing soil microorganisms to improve productivity of agro-ecosystems. Crit Rev Plant Sci 23(2):175–193

Wu SC, Cheung KC, Luo YM, Wong MH (2006) Effects of inoculation of plant growth-promoting rhizobacteria on metal uptake by Brassica juncea. Environ Pollut 140(1):124–135

Youssef M, Eissa M (2014) Biofertilizers and their role in management of plant parasitic nematodes. A review. Indian Journal of Biotechnology and Pharmaceutical Research 5:1–6

Zahir ZA, Abbas SA, Khalid M, Arshad M (2000) Substrate dependent microbially derived plant hormones for improving growth of maize seedlings. Pak J Biol Sci 3(2):289–291

Zandi P, Basu SK (2016) Role of plant growth-promoting rhizobacteria (PGPR) as biofertilizers in stabilizing agricultural ecosystems. In: Organic farming for sustainable agriculture, pp 71–87

Zaidi S, Usmani S, Singh BR, Musarrat J (2006) Significance of Bacillus subtilis strain SJ-101 as a bioinoculant for concurrent plant growth promotion and nickel accumulation in Brassica juncea. Chemosphere 64(6):991–997

Zhang F, Ge H, Zhang F, Guo N, Wang Y, Chen L, Li C (2016) Biocontrol potential of Trichoderma harzianum isolate T-aloe against Sclerotinia sclerotiorum in soybean. Plant Physiol Biochem 100:64–74

Zhuang X, Chen J, Shim H, Bai Z (2007) New advances in plant growth-promoting rhizobacteria for bioremediation. Environ Int 33(3):406–413

Chapter 9
From Pollution to Prosperity: The Role of PGPRs in Bioremediation

Krishnendu Sen, Priyanka Patra, Suman Mallick, Sk Saruk Islam, Subrata Dutta, and Sujoy Midya

Keywords Bioaugmentation · Environmental pollutant · Ex situ remediation · Heavy metal · In situ remediation · Microbial enzymes · PGPR · Phytoremediation

9.1 Introduction

Global industrialization has led to environmental, freshwater and topsoil pollution (Bala et al. 2022). Environmental pollution is the presence of noxious materials and toxins in the environment, which might be detrimental to ecosystems, biodiversity, and the wellbeing of people. Environmental issues are raised by a variety of industrial processes, transportation, agriculture, and incorrect waste disposal. The environment is harmed by pollutants such as polythenes, petroleum, and trace metals. HMs are contaminants that naturally occur in the outer surface of Earth and are challenging to breakdown (Bala et al. 2022; Pushkar et al. 2021). HMs could get into the environment on account of high-level exposures. Many of these contaminants can cause mutations in both humans and their surroundings. The brain, liver, and kidney store

K. Sen (✉) · S. Dutta
Department of Plant Pathology, Bidhan Chandra Krishi Viswavidyalaya, Mohanpur, West Bengal 741252, India
e-mail: krishnendusen10@gmail.com

K. Sen
Department of Microbiology, Vidyasagar University, Midnapore, West Bengal 721102, India

P. Patra · S. Mallick
Department of Microbiology, Bankura Sammilani College, Bankura, West Bengal 722102, India

S. S. Islam · S. Midya
Department of Zoology, Raja Narendra Lal Khan Women's College (Autonomous), Midnapore, West Bengal 721102, India

P. P. Adhikary et al. (eds.), *Soil, Water Pollution and Mitigation Strategies*, Environmental Science and Engineering, https://doi.org/10.1007/978-3-031-63296-9_7

HMs that have been absorbed. Additional consequences for animals include cancer, damage to the nervous system, development retardation, and even loss of life. By blocking dietary sources, growth of plants, and physiological metabolic procedures, the quality and amount of food produced are both reduced by HMs in the soil (Bala et al. 2022; Ajona and Vasanthi 2021). Additionally, the agricultural area has significantly decreased as a result of urbanization and industrialization, it has prompted new inspiration for scientists to create viable methods to boost crop output from the previously \ declining farmed region (Saeed et al. 2021). A biological process known as bioremediation includes reducing the severity of pollutants through bioreduction of complex hazardous materials into simpler and less toxic molecules by the help of microorganisms (Pushkar et al. 2021). This technique has gained popularity due to its potential to remediate contaminated sites without causing further environmental damage. By using enzymes or other metabolic pathways dependent on their development and metabolism, microbes break down the organic material in pollution and also possess the ability to degrade heavy metal (HM), dyes, hydrocarbon, polychlorinated biphenyls etc. (Ajona and Vasanthi 2021). These processes help crops to grow again on remediated soil. In order to bioremediate contaminated soils, microorganisms like PGPR bacteria and plants utilize a various type of mechanisms (Chibuike and Obiora 2014), and according to certain suggestions, they play a remarkable and pivotal role in reducing the toxicity in various polluted soils (Cetin et al. 2011). Utilizing PGPR strains with a variety of traits, such as resistant property of metal/ability to reduce, and the capability to promote the growth of plants through a variety of mechanisms in contaminated soils, is thought to be crucial for the successful outcome of the bioremediation program. This chapter synthesizes and critically evaluates the PGPR based bioremediation, including the choice of microbial strains, the types and sources of pollutants, and the ways in which PGPR improve remediation effectiveness.

The function of PGPR in the phytoremediation process is also discussed in this chapter. In the process of phytoremediation, contaminants and toxins are removed from the environment through the utilization of plants. It entails utilizing different plant species to metabolize, degrade, or absorb pollutant substances from soil, water, and air. Both inorganic and organic pollutant substances, such as pesticides, HMs, petroleum hydrocarbons, and other harmful chemicals, may be present (Wu et al. 2017). The process of phytoremediation can be adopted to clean up soils contamination by removing organic compounds and HMs from the soil both in-situ and ex-situ (Rada et al. 2019). Through the activation of soil-based microorganisms, the in-situ remediation process breaks down and removes pollutants. Additionally, this approach can be used on active manufacturing facilities and on soil that is immediately below structures. In order to promote microbial decomposition, Ex-situ bioremediation procedures include excavating polluted soil or pumping groundwater (Sasikumar and Papinazath 2003). PGPR-assisted phytoremediation is a promising and sustainable approach to remediate contaminated environments.

In order to effectively change various microbial enzymes into other relatively harmless compounds, the pollutant substrate must be able to recognize them under the optimal circumstances (temperature/concentration//contact time/pH). It is seen to be novel, cost-efficient, and promising to remove persistent organic contaminants

(POPs) using microbial enzymes (Tahri et al. 2013). Cell wall degrading enzymes are secreted by various types of biocontrol strains of PGPR, i.e., chitinase β-1,3-glucanase, protease, cellulase, and chitinase, having potentially inhibitory effect on fungal pathogens (Saeed et al. 2021). Enzymes from rhizobacteria that degrade cell walls have an impact on the target pathogen's ability to maintain their structural integrity (Saeed et al. 2021). By generating important enzymes including chitinase, ACC-deaminase, and rhizobitoxine exopolysaccharides, the PGPR promote plant growth in adverse environmental situations (Chitara et al. 2021). A bioremediation method used In-situ to treat contaminated soil is known as bioaugmentation. This approach involves treating the polluted soil with a microbial culture, which has enormous potential for soil remediation via a variety of biological mechanisms. The favourable environmental conditions are fully responsible for the microbial activity (Zaidi et al. 2012). By introducing or increasing the population of particular microorganisms in a contaminated environment, bioaugmentation procedures can be beneficial in increasing the efficacy of bioremediation. The type of pollutants, the environment, and the microorganisms metabolic capacities all influence the bioaugmentation technique that is used.

This chapter is also highlighted by a significant factor in bioremediation is the PGPR's phytotoxic impact-based bioremediation and the biohazards associated with PGPR. It is significant to remember that some PGPR strains can also exhibit phytotoxic effect on plants. The creation of excessive amounts of growth regulators for plants, the excretion of extracellular enzymes, or the production of secondary metabolites that may be poisonous to plants are only a few of the causes of phytotoxicity. It has been shown that PGPR helps plants better tolerate abiotic stresses like salinity (Bhat et al. 2020) and drought (Leontidou et al. 2020).

This study attempts to comprehend the interconnection between microorganisms and plants, including those mediated by signal exchange, and to evaluate the contribution of helpful microbes, particularly PGPR, as a sustainable method of increasing agricultural yield. The negative impacts related to the overuse of chemical inputs (pesticides and fertilizers) could be decreased by including PGPRs members in agricultural systems as a sustainable strategy to disease management and nutritional supplementation (Antar et al. 2021). Additionally, PGPRs members have been used as a successful technique to reduce certain biotic and abiotic stresses that may have an impact on crop growth and productivity (Khan et al. 2020), and it has a potential implication in bioremediation for promoting sustainable agriculture practices.

9.2 Environmental Contaminants: Types and Sources

Environmental contamination has largely been caused by industrialization, a strong reliance on petroleum, and the use of pesticides. The widespread use of chemicals and fossil fuels as well as the rapid rise of industrialization have caused dangerous pollutants to be released into the air, soil, and waterways. Millions of people worldwide are impacted by the major environmental problem of soil and groundwater pollution.

The quality of water has deteriorated on accounts of human activities like mining and the eventual elimination of hazardous metal wastes from steel factory, battery manufacturing, and energy generating, raising significant environmental issues (Bala et al. 2022). Pollutants such as petroleum, and trace metals, polythenes harm the ecosystem. HM substances are contaminants that occur naturally on the Earth's surface and are difficult to degrade. At high amounts of exposure, HMs can be discharged into the environment. They are far more dangerous once they are in the environment (Masindi et al. 2021).

Soil and groundwater contaminants might be chemical, physical, or biological in nature. Chemical contaminants include pesticides, HMs, petroleum-based products, and solvents, to name a few. Physical pollutants include debris, silt, and other solid substances that can have an impact on the quality of soil and groundwater. Pathogens, bacteria, and viruses are examples of biological pollutants that can impair the health of humans and animals (Mulla et al. 2019). Due to the continually growing industrial sector's pollution,atmospheric deposition, pesticides trace metal waste, leaded fuel petrochemical spills, mine tailings, high concentrations of metalloids coal combustion residues, and sewage sludge can contaminate soil (Phulpoto et al. 2021). Different types of environment pollutants are listed in Fig. 9.1.

Soil is nature's most amazing gift to human society. Several pollutants, also termed as soil contaminants, contaminate the soil, and this event is known as soil pollution, which occurs when the presence of a chemical or other material in the soil at a higher

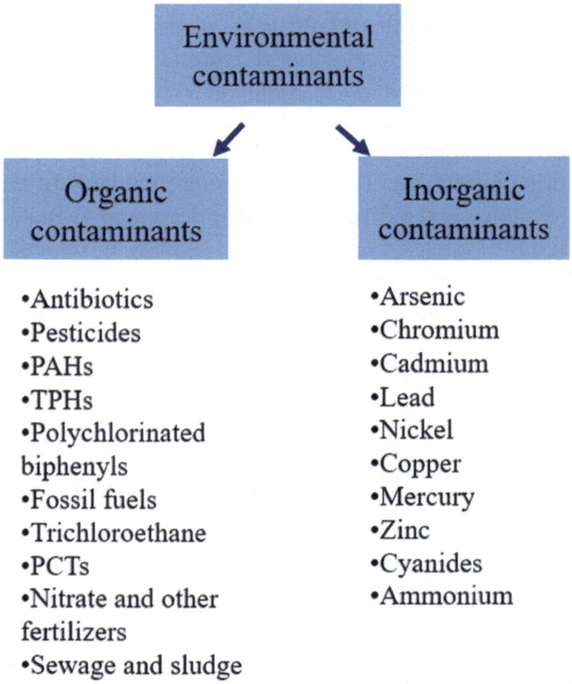

Fig. 9.1 Various types of environmental contaminants

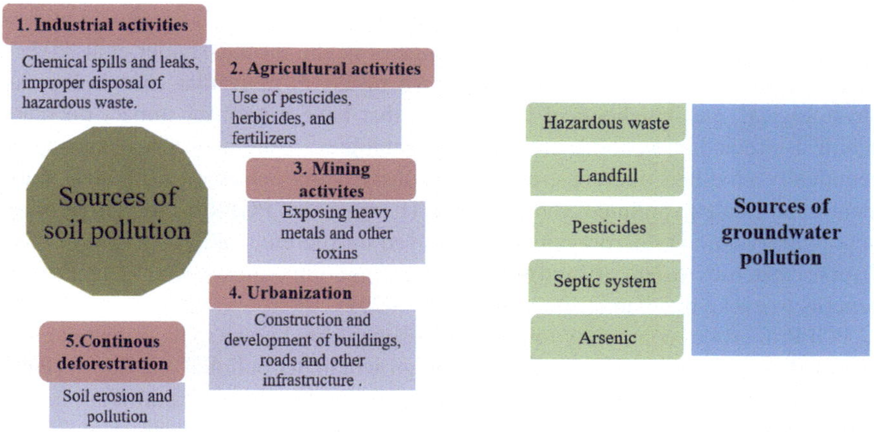

Fig. 9.2 Sources of soil and groundwater pollution

concentration than normal has a harmful influence on species that are not intended targets, because soil pollution is frequently difficult to measure directly, it is classified as a hidden risk (Rodrguez-Eugenio et al. 2018). Incomplete combustion of various compounds, transport, spray residue left over from using pesticides, atmospheric deposition from smelting (Zhang et al. 2012), and other factors can also pollute soil. The sources of soil pollution and groundwater pollution are listed in Fig. 9.2.

These toxins can cause mutations in both people and the environment. HMs are taken up by the brain, kidney, and liver through absorption. Other negative consequences on animals include cancer, dangerous to the neurological system, development retardation, and even death (Briffa et al. 2020). Children are more likely to come into touch with polluted soil through play and other activities, making them more susceptible to the impacts of soil and groundwater pollution. For instance, lead exposure might result kids learning problems and developmental delays. In addition to affecting the safety of drinking water, contaminated groundwater can also cause kidney damage and a host of other health issues in adult's body.

9.3 PGPR and Bioremediation: General Overview

Increasing rate of trace metals and metalloids can pollute environmental soil as a outcome of accelerating industry-based pollution, leaded fuel, petrochemicals, metal waste, mine waste, chemical pesticides, sewage waste, atmospheric deposition, and combustion of coal residues (Mulla et al. 2019). The most frequent inorganic contaminants detected in industrial wastes include arsenic, copper, chromium, cadmium, lead, mercury, nickel and zinc (Haider et al. 2021). Organic pollutants include pesticides like bentazon and atrazine, antibiotics like tetracyclines, sulfonamides, macrolides, total petroleum hydrocarbons (TPHs), quinolones and polycyclic aromatic hydrocarbons (PAHs) are allied with a varied range of health issues (Haider et al. 2021). As Some pollutants cause plants to undergo a variety of molecular, morphological, physiological, and biochemical fluctuations that have a harmful impact on plant health. Bioremediation and remediation technologies, using naturally occurring or genetically modified microbes, are continuously being enhanced to remove trace metals and clean up polluted environments. The usages of PGPR bacteria is growing in popularity among microorganisms employed for bioremediation because of their diverse capacities to detoxify and breakdown toxins compounds as well as their major benefits to evoke the plant growth (Ali et al. 2020).

PGPR microbes commonly found in rhizosphere, are known to have the potential to remediate polluted soil and enhance plant growth under polluted conditions (Patel et al. 2016). PGPR microbes can evoke plant growth through varied numerous mechanisms such as siderophore production, biological N_2 fixation (Yadegari et al. 2010), phosphate solubilization, signal interference, 1-aminocyclopropane-1-carboxylate deaminase (ACC) production (Gontia-Mishra et al. 2017), phytohormone production, and quorum sensing (Podile et al. 2014). Additionally, PGPR bacteria can exhibit biocontrol activity (Islam et al. 2022), produce volatile organic compounds (VOCs), promote plant–microbe symbioses association (Bhattacharyya and Jha 2012), induce systemic resistance, biotransform hazardous metals into less toxic forms (Shukla et al. 2009), sequester metal ions, and detoxify contaminated environments and uptake and release metals (Fig. 9.1). PGPR can also improve crop yields by influencing plant development mechanisms and enhancing plant health, soil fertility, and pollution cleanup by producing various key enzymes i.e.,chitinase, cellulase, exopolysaccharides and ACC-deaminase. The use of formulated biofertilizers like PGPR can promote soil health and productivity by altering the composition and structure of soil microbial population and influencing afforestation. Thus, PGPR has been identified as a critical factor in the efficiency of bioremediation programs and plant growth promotion (Fig. 9.3).

9.3.1 In-Situ Remediation

When we talk about "in-situ" bioremediation, we we refer to cleaning up pollution on the spot without moving contaminated materials elsewhere (Banerjee et al. 2016). Natural bio-attenuation and improved bioremediation are two types of in-situ approaches (Smets and Pritchard 2003) (Fig. 9.4).

9.3.2 Natural Bio-attenuation

It is a biological degradation process which solely depends on the metabolism of local microbes to knock out harmful chemicals, devoid of artificial phases utilized

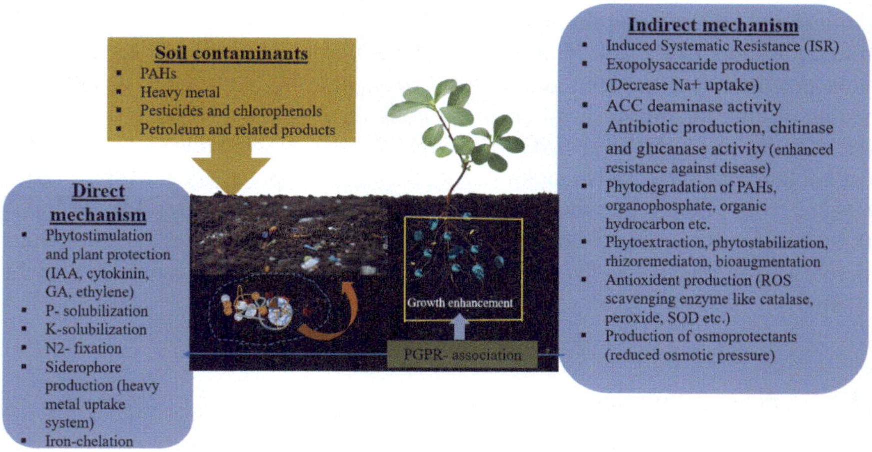

Fig. 9.3 Role of PGPRs in contaminated soil and their mechanism of action in bioremediation

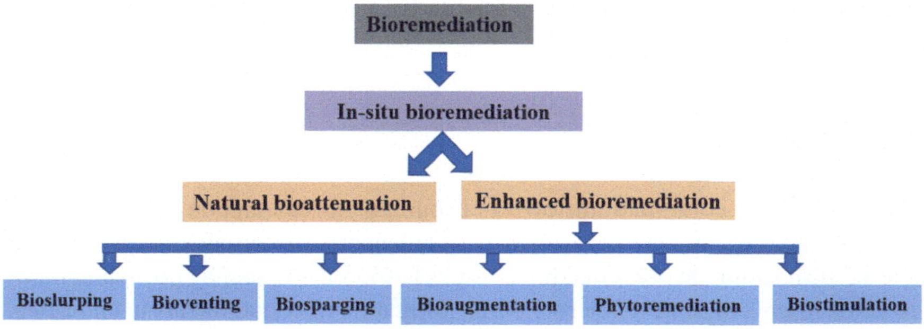

Fig. 9.4 Diverse technique under the In-situ bioremediation

to increase biodegradation process (Sharma 2019). Due to the lack of external influences, this procedure is the most affordable in-situ bioremediation technology. Native bacteria degrade pollutants very effectively because they modified according ti to the site's environment, and that evolution has established a microbe-hydrocarbon interaction, capacity to degrade hydrocarbon is common among microbial community (Kumar et al. 2018a, b).

9.3.2.1 Biostimulation

Native bacteria are induced to proliferate in biostimulation by adding growth stimulants such as nitrogen, phosphorus. Native remediation is sometimes effective. Under normal conditions, microbes cannot exist; thus, they are encouraged by adding nutrients, oxygen, or other oxidising chemicals. Stimulating compounds are often administered underground through narrow wells (Zeneli et al. 2019). Carbon availability, nutrients like P and N, amount of oxygen available, soil pH, temperature, redox potential, and the kind and quantity of organic impurity itself all influence the rate of microbial turnover of chemical pollutants. Water soluble nutrients (e.g., NH_4NO_3, KNO_3, K_2HPO_4, $NaNO_3$, and $MgNH_4PO_4$), delayed release (e.g., IBDU, customblen, max-bac), and oleophilic (e.g., S200, F1, Inipol EAP22, MM80) fertilizers are given to accelerate microbial decomposition (Nikoloulou and Kalogerakis 2008). The fundamental advantage of this approach is that it involves autochthonous microorganisms that are highly adapted and disseminated in the environment (Speight 2018).

9.3.2.2 Bioaugmentation

The autochthonous microbiome of the contaminated soil is boosted by the inclusion of formerly selected native or genetically engineered microbes to facilitate oil breakdown in bioaugmentation. When the local microorganisms are unable to breakdown the contaminants, this approach is particularly successful (Wozniak-Karczewska et al. 2019a, b).

9.3.2.3 Bioventing

This method uses controlled airflow to encourage bioremediation by transporting oxygen to the unsaturated (vadose) zone by enhancing the activities of indigenous bacteria. Bioventing involves making changes to bioremediation by providing moisture and nutrients to achieve microbial transformation of contaminants to aninoffensive condition (Boudh et al. 2019). Humidity and nutrients are typically provided to accomplish pollution alteration into contaminants that are less harmful to the soil and to recover by cleanup. This technique was used successfully in the cleanup of oil-contaminated soils (Gholami-Shabani et al. 2018).

9.3.2.4 Bioslurping

According to Azubuike et al. (2016), the remediation of soil and groundwater can be achieved through a technology which combines vacuum-enhanced pumping, bioventing, and soil vapor extraction. This method indirectly supplied oxygen and evoke pollutant biodegradation. It is effective in removing light non-aqueous phase liquids (LNAPLs), low-density water-insoluble pollutants, from the water table in unsaturated, capillary, and saturated zones. Which also clean up polluted soil with semi-volatile and volatile organic compounds. The technology uses a "slurp" to extract liquids and soil gas from the free product layer, similar to a straw from a vessel. Bioventing is employed to promote aerobic breakdown of contaminants in air and water after they have been removed from the surface. Various study suggest that bioventing is an effective way to biodegrade toxins in the air and water (Kumar et al. 2018a, b). Miller (1996) described that, the bioslurping tube extracts vapours from the unsaturated area when the fluid level in the well drops due to vacuum extraction of LNAPLs. However, the vacuum pump is not effective at sucking LNAPLs at larger depths, so the bioslurping system should only be installed if the contaminants are not in excess of 7 m below ground. The primary limitation of bioslurping is that excessive soil moisture affects air permeability, the rate of oxygen transfer, and soil microbial activity. Moreover, bioslurping is not suitable for treating low-permeability soils, in accordance to Godheja et al. (2019).

9.3.2.5 Biosparging

Air injection into the soil is a common practice during biosparging, a remediation technique used to enhance the biodegradation of pollutants by microbes. In contrast to bioventing, which delivers air into the unsaturated zone, biosparging introduces air into the saturated zone to promote the volatilization of contaminants. The efficiency of biosparging relies on soil permeability, which affects the availability of impurities to microbes and their biodegradability (Godheja et al. 2019). Biosparing is similar to in-situ air sparging (IAS) in operation, but biosparging focuses on promoting biodegradation over pollutant volatilization. Neither technique is mutually exclusive, and they may work in tandem to remove pollutants. Biosparging is commonly used to treat petroleum-contaminated aquifers, especially those containing kerosene and diesel, which are highly biodegradable, such as the BTEX group and naphtalenes (Kao et al. 2008). Anaerobic bacteria can break down BTEX, mineral oils, and naphtalenes. Injection filters supply oxygen directly to the polluted soil and below-groundwater, promoting the growth of aerobic microorganisms and enhancing their capacity to digest pollutants.

9.4 Phytoremediation

Phytoremediation, an eco-friendly practice that utilizes plants in contaminated areas to induce a range of biological, microbiological, physical, biochemical and chemical communications that decrease the toxicity of pollutants (Godheja et al. 2019). Depending on the type of contaminant present, phytoremediation employs a variety of methods to treat the soil. Sedimental contaminants such as HMs and radioactive wastes are primarily mined, altered, and sequestered, while organic pollutants are primarily eradicated through biodegradation, rhizo-degradation, vaporization and stabilization (Kuiper et al. 2004a, b).

There are several methods of phytoremediation, each with its unique mechanism for treating contaminants in soil. Phytoaccumulation is a process in which the plant removes pollutants from soil and accumulates it in some portion of plant, such as the leaf, root, or stem (Sidhu et al. 2018). Phytovolatilization involves the plant drawing the contaminant out of the soil and transforming it into a volatile compound that is subsequently released into the atmosphere. Phytostabilization involves immobilization of contaminants in soil, which prevents erosive processes and allows for interaction with lignin and humus (Shackira and Puthur 2019). Phytostimulation occurs when the plant roots increase the precipitation, absorption, and removal of inorganic and organic contaminants from groundwater, wastewater, and surface water (Awa and Hadibarata 2020). Rhizo-degradation is another process in which bacteria, fungi, and yeasts in the rhizosphere have induced 10–100 times metabolic activity of the soil, letting them to digest pollutants alongside natural organic nutrients (Cristaldi et al. 2020). Phytoremediation is considered a cost-effective and less invasive method of cleaning up contaminated sites than traditional cleanup methods. Contaminants recovered from the soil are accumulated in the shoots and roots of plants, which are then collected for biomass processing, including mining, composting, and burning (Ashraf et al. 2019).

9.4.1 Phytoremediation: Using Plants to Remove Pollutants

According to Chaudhry et al. (1998), the growing industrialization of society causes widespread environmental contamination with permanent inorganic and organic wastes. Soil contaminants, such as radionuclides, HMs, nitrate, inorganic acids, phosphate, organic chemicals, wastes, weapons, chemical fertilizers, pesticides, drugs, acid rain, and radioactive fallout, are of concern (Sparks 1995). An important part of the environment's pollution is being played by HMs. HMs may be found in soil because of many human activities. HM pollution cannot be readily eliminated or degraded. They enter our bodies through food, soil, air and water in extremely minute amounts (Duruibe et al. 2007). A few trace amounts of various metals are required to keep the human body's metabolism running smoothly (Garbisu and Alkorta 2003).

According to Alkorta et al. (2004), HMs can become toxic when existing in high quantities. HMs such as mercury, copper, arsenic lead, and zinc are highly poisonous when entersin biological systems (Sharma et al. 2015). Jaishankar et al. (2014) suggested that HMs bind to protein at specific site while displacing the inherent metals from the protein and causes deformation. While HMs have played a role in enhancing human welfare throughout history, their toxicity poses a serious risk to both humans and aquatic life (Etim 2012). Therefore, the removal of HMs from polluted sites is of utmost importance (Naushad et al. 2017), and regulatory organizations have set limits on the acceptable levels of these metals in water and land sources.

Various physicochemical techniques, such as solidification, cremation, vapor extraction, enzymatic degradation, ion exchange adsorption, thermal desorption, and soil leaching, have been developed to remediate contaminated sites (Naushad et al. 2015; Karthik et al. 2016). However, these techniques are often expensive and can cause secondary air or groundwater pollution, impair soil fertility, and render lands unsuitable for cultivation (Sivarajasekar et al. 2017). Phytoremediation, on the other hand, is a plant-based technique that uses natural or genetically modified plant species to restore polluted water sources and landscape (Parmar and Singh 2015). Phytoremediation is a new, cost-effective, and eco-friendly remediation technique that can be used in situ. To optimize phytoremediation, a multiprocess phytoremediation system (MPPS) has been developed. The process includes a combination of microbiological and mechanical stages to promote biomass accumulation in the soil, particularly plant belowground parts, and accelerate remediation kinetics (Huang et al. 2005). Plants can manage toxins without harming the topsoil, preserving its fertility and improving it with the addition of organic matter.

9.4.2 *Mechanisms of Phytoremediation*

9.4.2.1 Phytoaccumulation

Plants intake pollutants from contaminated areas, together with other essential micro and macro-nutrients and water essential for growth, in a process called as phytoaccumulation. The Pollutants that are absorbed accumulate in the plant's shoots, leaves, and other parts rather than being removed. This approach is commonly used for metals and radio-waste (Rashid et al. 2014). Phytoremediation is an efficient procedure for the cleanup of contaminated soils. One such method is the use of a mix of microbiological, mechanical, and plant growth phases, known as the Microbial Phytoremediation Procedure (MPPS) to promote biomass accumulation in the soil, particularly plant below ground parts, and therefore accelerate remediation kinetics (Huang et al. 2005). Land farming and PGPR inoculation are the methods used in this procedure. Croprhizobateria provide various aids to their hosts, the most important of which is HM protection. Heterotrophs, members of the β-proteobacteria exhibit the highest levels of HM tolerance. The *Alcaligenes eutrophus* CH34 species have two endogenous megaplasmids covering heavy metal resistance. The bacterial plasmid pMOL28 (180 kb) encodes HM resistance to nickel, cobalt, mercury, chromate, and thallium. Another plasmid pMOL30 (240 kb) has resistance to cadmium, cobalt and zinc (Taghavi et al. 1997). Most plant-associated bacteria can create HM chelator siderophores, which solubilize of inaccessible HM forms to accessible forms, boosting phytoaccumulation efficiency (Rajkumar et al. 2010). *Streptomyces tendae* F4 siderophores greatly increased Cd absorption by sunflower plants (Dimkpa et al. 2009). *Pseudomonas aeruginosa* produced pyoverdine and pyochelin which increase the bioavailability of Pb and Cr in maize rhizosphere (Braud et al. 2009). Siderophore generation by bacteria is influenced by a variety of parameters, with iron availability, pH, soil nutritional type, status, and concentration of HMs, among others. As a result, increasing HM concentrations function as triggers for bacteria to manufacture more siderophore (Braud et al. 2009). Organic acids with low molecular weight produced by plant–microbe communication show a significant function in increasing trace element and metal bioavailability in soil. They produced stable compounds with HMs, although their stability is influenced by variables such as carboxyl groups, HM form, and soil pH (Ryan et al. 2001). The bacterial strains *Gluconobacterdia zotrophicus* and *Pseudomonas aeruginosa*, which are engaged in the generation of gluconic acid and the solubilization of Zn, are primarily responsible for the solubilization and mobilization of insoluble ZnO, $Zn_3(PO_4)_2$, and $ZnCO_3$ (Saravanan et al. 2007). By generating citric and malic acids, ericoid mycorrhizal fungus (*O. maius*) and soil fungi (*B. caledonica*) can improve Zn solubility from pyromorphite, insoluble $Zn_3(PO_4)_2$, and ZnO (Fomina et al. 2005).

9.4.2.2 Phytovolatilization

Phytovolatilization is a process that involves the intake of impurities by the plants and release into the atmosphere in a less harmful form (Moreno et al. 2004). Plants absorb organic and inorganic contaminants with water, which they then transfer onto their leaves, where they volatilize into the atmosphere (Mueller et al. 1999). The process by which pollutants are absorbed by plants and then translocated before finally volatilizing is known as direct phytovolatilization. Indirect phytovolatilization, on the other hand, refers to the increase in subsurface volatile pollutant flux caused by plant root activity. Phytovolatilization is most effective for pollutants that can easily be converted into vapor and discharged into the atmosphere, such as volatile and semivolatile organic chemicals and certain HMs (Ali et al. 2013).

Mercury is the first metal to be eliminated by phytovolatilization (Henry 2000). Toxic metals such as selenium, arsenic, and mercury can be transformed into volatile forms, such as mercuric oxide and dimethyl selenide and then evaporated or volatilized into the atmosphere (Sakakibara et al. 2010). Aquatic plant *T. latifolia* L. is used for phytovolatilization of selenium-polluted soil (LeDuc and Terry 2005). *Nicotiana tabacum* L. and *Arabidopsis thaliana* L. and have been genetically modified with bacterial organomercurial lyase and mercuric reductase to absorb CH_3-mercury and elemental Hg (II) from soil and release volatile Hg (0) into the atmosphere (Pillon-Smith and Pilon 2000). Plant–microbe communications have also been used to help the volatilization of selenium from soils (Karlson and Frankenberger 1989). *Pteris vittata* L. (Chinese Brake fern) has been identified as an arsenic (As) hyperaccumulator that may also collect a significant quantity of Se. Catalase, ascorbate peroxidase, and peroxidase are anti-oxidative enzymes that contribute to Se hyperaccumulation (Feng and Wei 2012).

Brassica juncea was known as a plant accomplished of removing Se from soils (Bauelos and Meek 1990). Organochlorines (OCs), 1,2,4-trichlorobenzene (TCB), 1,4-dichlorobenzene (DCB) and hexachlorocyclohexane (HCH) have prolong persistency in the nature. Their absorption is primarily determined by their hydrophobicity, solubility, and volatility (San Miguel et al. 2013). Other VOCs including 1,4-dioxane, have also been studied, and it was discovered that phytovolatilization successfully eliminated dioxane (2.5 g/L) (Ferro et al. 2013). *Brassica juncea* may emit up to 40 g Se/ha/d as different gaseous chemicals (Padmavathiamma and Li 2007). It is worth noting that phytovolatilization has the potential to expose humans to harmful substances (Marr et al. 2006).

9.4.2.3 Phytostabilization

The method known as phytostabilization involves the process by which plants absorb pollutants in the soil or groundwater through absorption, adsorption onto the roots, or the creation of insoluble compounds. This mechanism restricts the mobility of pollutants, preventing their migration into groundwater or the atmosphere, thus reducing their bioavailability and restricting their flow into the food chain (Soudek et al. 2012). The plant-associated microbiota employs specific mechanisms to immobilize or inhibit metal adsorption in plants, such as forming a permeability barrier, active ejection of metal outside the cell, sticking it to extracellular polymers, and detoxification. Soil micro-flora play a crucial role in phytostabilization and phytoremediation by enhancing the immobilization properties and accumulation rate of HMs (Ma et al. 2011). Anionic functional groups like amide and amine groups, hydroxyl, sulfonate, carboxyl, and sulfhydryl interact with HMs, leading to metal ion immobilization and restriction of metal ion entrance into the root system (Pulsawat et al. 2003). The Variovorax strain isolated from mine tailings is effective for the bio-augmentation process (Tamburini et al. 2017). Soil supplements such as clay manures, minerals, biosolids, and biological inoculants increase HM immobilization and pesticide retention, thereby playing a crucial role in phytostabilization (Ashraf et al. 2019; Yuan et al. 2018). Inoculating green chilli with *Cellulosimicrobium cellulans* reduces Cr absorption by 37% in the shoot and 56% in the root when grown in Cr(VI) contaminated soils, as the soil's microbial conversion of mobile and hazardous Cr(VI) to nontoxic and stationary Cr(III) prevents absorption (Chatterjee et al. 2009). Similarly, *S. maltophilia* isolated from the rhizospheric soil of *A. bisulcatus* converts soluble and toxic Se(IV) to insoluble and inaccessible Se(0), lowering Se absorption by plants (Di Gregorio et al. 2005).

9.4.2.4 Phytostimulation

Phytostimulation refers to the degradation of organic pollutants in plant rhizosphere, which is facilitated by rhizospheric bacteria (Abhilash and Singh 2010). In the rhizosphere, the synergistic activities of plant roots and microbes break down organic impurities and enhance the phytoremediation of HMs and organic pollutants (Chawla et al. 2013). Plants promote the growth of microorganisms in the rhizosphere by providing a surface for attachment, secreting root exudates, and altering the properties of contaminants in the soil. Root exudates, which contain vitamins, sugars, organic acids, alcohols, and inorganic compounds, play a crucial role in this process. Some root exudates act as surfactants, promoting the desorption of hydrophobic organic pollutants and enhancing their bioavailability. Other root exudates can bind HM ions, affecting their solubility, mobility, and bioavailability in soils. It was observed that DDT desorption increased with an increase in root exudates (Luo et al. 2006).

PGPRs, Arbuscular Mycorrhizal Fungi (AMF), and decompose bacteria are becoming increasingly important for soil remediation. PGPR produce phytohormones, siderophores, and ammonia, which enable them to fix nitrogen from the air and convert insoluble phosphorus into a soluble form that can be used by plants. In addition, PGPR excrete metabolites such as exo-polysaccharides, metal chelating agents, biosurfactants and antibiotic compounds that help reduce plant stress caused by biotic and abiotic factors (Rajtor and Piotrowska-Seget 2016). Furthermore, rhizo-micro-biome can degrade organic contaminants through their own degradation processes, which involve the development of enzymes such as hydrolases, laccases, catalases and dehalogenases in the rhizosphere. These enzymes play a crucial role in the sequestration, degradation, mineralization, stabilization, and detoxification of organic pollutants and HMs (Hong et al. 2015). Some metabolites and chemicals, such as siderophores, organic acids, exo-polysaccharides, vitamins, biosurfactants and phenols and, not only promote plant development but also aid in the degradation of organic contaminants and the modification of the nature of HMs, either activating or deactivating them (Manoj et al. 2020).

9.4.2.5 Phytodegradation

Phytodegradation, which involves the absorption and breakdown of pollutants inside the plant or in the surrounding environment, is a method which uses plants and their associated microbes to transform harmful organic impurities such as 2,4,6-trinitrotoluene (TNT), herbicides, pesticides, polychlorinated biphenyls and insecticides into non-toxic forms (Lee 2013). This method is particularly effective for removing hydrophobic organic contaminants such as ethyl benzene, benzene, toluene and xylene at shallow depths, short-chain aliphatic hydrocarbons, and chlorinated solvents (Zayed and Terry 2003). Peroxidase, Nitroreductase, dehalogenase, laccase, and nitrilase are among the enzymes that contribute to phytodegradation. The aquatic plant *M. aquaticum*, for example, produces the nitroreductase enzyme, which decreased TNT levels in the soil from 10 to 15 ppm within one week. However, because metals are not biodegradable, this method is limited to organic contaminants (Fig. 9.5).

9.5 Bioaugmentation: Enhancing Remediation Efficiency

The idea of bioaugmentation has received a lot of interest. It is based on the direct introduction into the contaminated site of particular microbial species that are trained in the biodegradation of particular chemicals (Ławniczak et al. 2020). The method of bioaugmentation contains the inclusion of microorganisms that can break down resistant compounds in a contaminated environment (Nzila et al. 2016). Associating this approach to physico-chemical methods, it is inexpensive and more ecofriendly when the pollutant cannot be degraded by the native microorganism (Wozniak-Karczewska et al. 2019a, b).

By introducing native, allochthonous, wide-form, genetically modified microbial strains or microbial consortia with favored catabolic ability to break down resistant

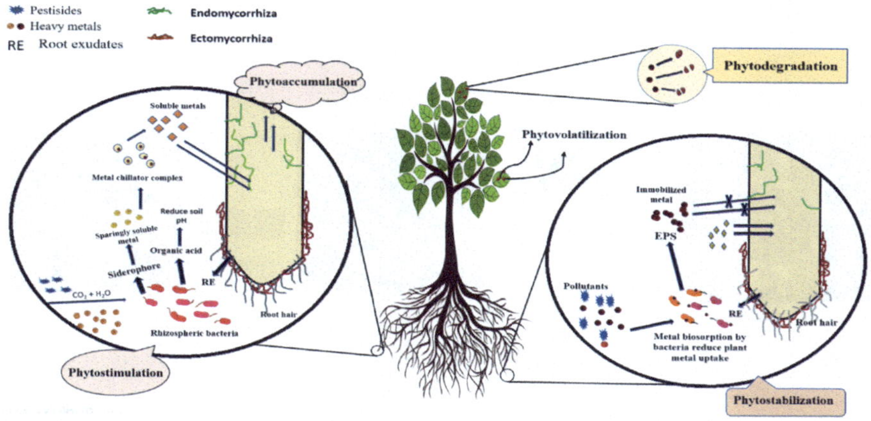

Fig. 9.5 Different processes of phytoremediation

chemicals in such habitats, bioaugmentation intends to enhance the biodegradative capabilities of polluted locations (El Fantroussi and Agathos 2005). Numerous substances, including chlorinated solvents, nitrophenols, oil, methyl tert-butyl ether, polychlorinated biphenyls, pentachlorophenol, polycyclic aromatic hydrocarbons, and a number of pesticides, including atrazine, dicamba, and carbofuran, have been shown to degrade more quickly when subjected to bioaugmentation with microorganisms. Microorganisms are chosen based on two key factors: their capacity to break down certain contaminants and their resistance to and ability to thrive in a variety of settings. Many microorganisms, including bacteria, fungus, yeast, actinomycetes, and algae, are able to endure in a variety of settings and can even remove HMs from contaminated places. Bacteria can be isolated from specific polluted soils, and then after being cultured under controlled conditions, pure strains can be prepared and returned to the original soil. This strategy with the indigenous microorganism is called the reinoculation of soil. The second option is to choose suitable microbiota from locations with analogous pollutants that are prevalent in soils and are then give in toto clean up by these chosen microbiomes (Fig. 9.6).

The majority of bioaugmentation experiments used gram-negative bacteria from the genera *Pseudomonas*, *Sphingobium*, *Flavobacterium*, *Achromobacter* and *Alcaligenes* (Mallavarapu and Leee 2020). G(+) bacteria from the genera *Bacillus*, *Mycobacterium*, and *Rhodococcus* also need to receive further attention (Radwan et al. 2019). As for the fungi themselves, species from the genera *Absidia*, *Aspergillus*, *Achremonium*, *Penicillium*, *Verticillium* and *Mucor* (Jayaprakash et al. 2019). Pure *Pseudomonas putida* ZWL73 culture was applied to soil that had been polluted with 4-chloronitobenzene (4CNB), and this increased the breakdown of 4CNB in soil micro-environment (Niu et al. 2009).

Thus, bioaugmentation includes the inoculation of individual cells as well as groups of cells, as well as the introduction of rhizospheric bacteria as well as those

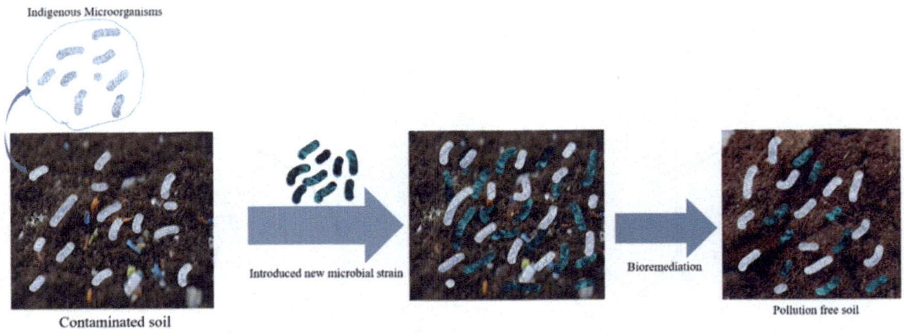

Fig. 9.6 Process of bioremediation by bioaugmentation

that do not. When plant-associated rhizospheric bacteria are implicated in remediation problems, the process is called bioaugmentation technique of rhizoremediation (Kuiper et al. 2004a, b).

9.6 Ex-Situ Remediation: Handling Pollutant Off-Site

Many different applications for bioremediation are discussed here, including some of the most popular ones. The most prevalent technique is Ex-situ bioremediation, but In-situ bioremediation can also result in it (Bala et al. 2022). Better environmental conditions can be controlled using ex-situ procedures, increasing the comparative biodegradation rate to in-situ treatment methods. Additionally, the process is generally more uniform and faster because it is possible to homogenize the polluted ground.

To remediate contaminated material, this method involves extracting the contaminated material from the site and transporting it to a separate facility for degradation and treatment. When selecting ex-situ bioremediation techniques, various factors, including operational costs, depth and extent of pollution, type of pollutant, location, and geological characteristics of the polluted site, are taken into consideration. Nonetheless, because of the need for excavation, site restoration, and treatment, these methods are typically more costly (Fig. 9.7).

9.6.1 Slurry Phase Bioremediation

Slurry-phase bioremediation is a technique which uses a bioreactor to mix contaminated soil, sludge, or sediment with water and other chemicals. This creates three phases—liquid, solid, and gas—and maintains blending conditions to enhance the biodegradation rate of pollutants that are both water-soluble and connected to the

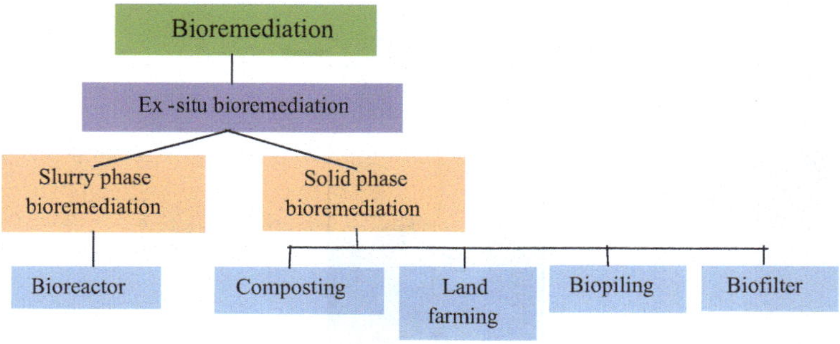

Fig. 9.7 Diverse techniques under the ex-situ bioremediation

soil. Additionally, the method involves the use of indigenous organisms' biomass to aid in the process (Kumar et al. 2018a, b).

9.6.1.1 Sludgebioreactor

"Bioreactor", refers to any machinery or enclosed space that facilitates a bioactive system to turn basic materials into specific products through a series of biological processes (Bala et al. 2022). Compared to ex-situ bioremediation techniques, the use of a bioreactor in treating contaminated soil has several advantages. Sludge bioreactors can safely and effectively remediate hydrocarbon pollutants by creating a three-phase solid, liquid, and gaseous system through mixing the sludge with various tools to retain contaminants in a containment container where the resulting biofilm promotes the breakdown of pollutants and enhances biomass levels (Salehi and Hakimghiasi 2017). The successful bioremediation process utilizing bioreactors is greatly enhanced by accurately controlling parameters such as pH, temperature, agitation, substrate concentration, aeration and inoculum concentration, which can significantly reduce the remediation duration. This technique can eliminate a variety of pollutants including pesticides, HMs, and volatile organic chemicals like BTEX (Balseiro-Romero et al. 2019). The volume of a bioreactor can vary from a few liters to several cubic meters, typically taking a cylindrical shape and being constructed of strong glass or stainless steel. The reactor can receive the contaminated material in either a dry form or suspension, and the remediation process can be either aerobic or anaerobic. The use of bioreactors is considered one of the best methods for remediating polluted soil since it allows for the precise management of working parameters, resulting in an augmented microbial biodegradation potentiality (Dzionek et al. 2016) (Fig. 9.8).

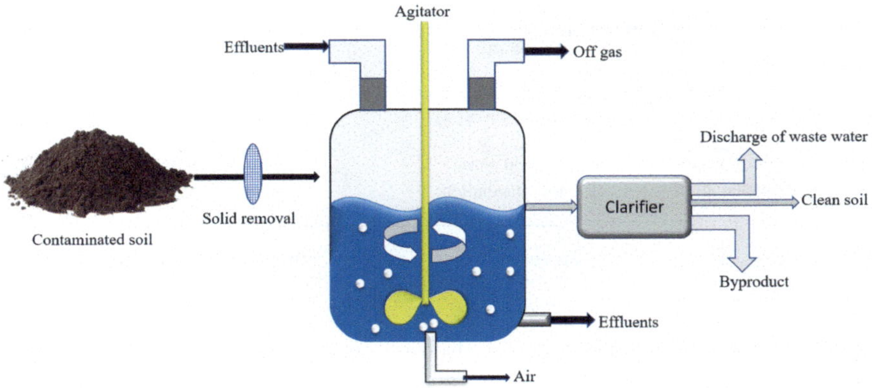

Fig. 9.8 Bioremediation process by sludge bioreactor

9.6.2 Solid Phase Bioremediation

Solid phase bioremediation, a long-term, cost-effective way of reducing environmental contamination caused by many organic and inorganic contaminants (like-HMs, hydrocarbons, pesticides, and new pollutants). Microorganisms or their enzymes are used to transform or breakdown pollutants in solid matrices such as soil, sediment, and sludge. For the detoxification and disintegration of the dangerous toxic pollutants, solid-phase soil treatments include composting, land farming, soil biopiles and biofilters (Kumar et al. 2018a, b).

9.6.2.1 Composting

Organic waste is decomposed by thermophilic microbes during the ex-situ aerobic composting process to create a humic amendment known as compost, which is use as a soil fertilizer (Chadar et al. 2018). In the process of composting, the contaminated soil is excavated and mixed with an organic material (such as wood chips, animal manure, or vegetal wastes, etc.). The presence of these organic components promotes the growth of a miscellaneous microbial communal (Bacteria, algae, fungi etc.), which uses its enzymatic activity to convert the organic matter into compost (Kumar et al. 2018a, b). When the wastes are combined with soil or injected with finished compost, soil microbes such as bacteria, fungus, actinomycetes, and protozoa become familiar. There are three types of microorganisms that carry out the composting process: Psychrophiles, mesophiles, and thermophiles are listed in that order. High oxygen and nutrient availability, as well as a temperature between 40 and 60 °C Celsius, are necessary for achieving high biodegradation rates by earthworm; soil insects, such as, mites, ants, sow bug, beetles, springtails, and nematodes escalating their bioavailability for the soil microbiota (like, *Actinobifidachromogena*, *Therinomnonosporacurvata*, *Streptomyces rectus* (Zeng et al. 2011; Buyksonmez et al.

2000). These microbes are introduced to soil when the waste or finished compost is applied in the ground.

Composting is commonly used to recycle organic waste, but it is also used to bioremediate polluted soil. Microbial activities capable of biodegrading toxic organic substances while providing metal bioavailability are present in such a process (Aguelmous et al. 2019). In this process, bacteria break down organic substances to provide energy for their metabolic processes and nutrients (N, P, and K) for their long-term survival. C and N are the most significant of the numerous components that are necessary for microbial breakdown. The C:N ratio used in the composting model is around 30:1. As the composting process continues, the C:N ratio steadily decreases from around 30:1 to 10–15:1 to produce the completed product (Kumar et al. 2018a, b).

This method has been shown to be efficient in the degradation of a varieties of organic soil contaminants, including pesticides, chlorophenols, explosives, and petroleum hydrocarbons (Sayara and Sánchez 2020).

9.6.2.2 Land Farming

Land farming is an eco-friendly bioremediation technique because of its low operational costs and lack of specific equipment, that utilizes the power of microorganisms to clean up contaminated soils and restore them to their natural state. In land farming, contaminated soil\sediment, or sludge is distributed across a predetermined area so that microorganisms (bacteria, actinomycetes, fungi, protozoa, and nematodas) can use biodegradation, bioaccumulation, and other processes to break down the toxins.

In this method, sediment, contaminated soil and sludge is dug up, spread out, and then repeatedly tilled to aerate the mixture until contaminants are broken down by increasing aerobic microbial activities in the soils as a result of aeration and/or the addition of moisture, minerals, and nutrients (Kumar et al. 2018a, b). It has been demonstrated that the land farming strategy is effective in reducing the concentrations of all petroleum product components typically found on underground storage tank sites.

9.6.2.3 Biopiling

Biopiling (known as biocells, bioheaps or biomounds) is an ex-situ bio-remediation technique which has been widely employed to treat a variety of petrochemical contaminants in the soils and sediments (Germaine et al. 2015). Biopiling involves piling up polluted soils or dried sediments while fostering aerobic microbial populations' biodegradation activities by fostering optimal or nearly optimal growth conditions within the pile. The biopile's adaptableness allows for a reduction in remediation time by increasing microbial potentiality and pollutant availability while also enhancing biodegradation efficacy. Bioremediation is enhanced when warm air is supplied into the biopile system to give both air and heat (Bala et al. 2022).

The biopile method of bioremediation makes use of crucial bio-available organic carbon (BOC) (Bala et al. 2022). Total petroleum hydrocarbon (TPH) has been removed from petroleum-contaminated soil utilising mesophilic conditions (30–40 °C) and a low aeration rate employing *alpha, beta,* and *gamma proteobacteria* (Naeem and Qazi 2020). Depending on the source of carbon, temperature, various bacterial strains, including *R. rhodochrous, P. alcaligens, R. erythropolis*, and some *Actinobacter* sp., have been discovered to breakdown petroleum (Czarny et al. 2020). Additionally, it has been revealed that certain genera, such as *Corynebacteria*, *Nocardia*, *Aeromonas*, *Rhodococcus*, *Bacillus*, *Mycobacterium*, *Alcaligenes*, *Flavobacterium*, *Micrococcus*, and *Pseudomonas* are capable of biodegrading aromatic hydrocarbons with an efficiency of 60–70% (Alegbeleye et al. 2017). The sub-Antarctic region's diesel-contaminated soil has also been treated using biopile devices. Within a year, the biopile system was able to remove 93% of the total petroleum hydrocarbon (TPH).

9.6.2.4 Biofilters

The biofilter, a type of passive, natural, or designed device that makes use of microorganisms to turn harmful compounds into less toxic or benign forms, is one often used oldest technology in bioremediation. To reduce the negative effects of pollutants on both human health and the environment, biofilters are frequently used in a variety of situations, such as industrial sites, wastewater treatment facilities, landfills, and agricultural operations.

Utilizing microbial stripping columns as part of the biofilters process, air pollution is treated. In most cases, the bacteria convert the poisonous compounds into a non-toxic component, such as carbon dioxide (CO_2), water (H_2O), and salts (Chitara et al. 2021).

9.7 Microbial Enzymes: Degrading Environmental Pollutants

The process of bioremediation uses microbial enzymes to convert dangerous contaminants into less dangerous ones (Bala et al. 2022). The enzymatic metabolic pathways of microorganisms accelerate the biochemical activities that break down contaminants. It belongs to the lyase, transferase, hydrolase, and oxidoreductase family. The ability of many enzymes to break down a variety of substrates is enabled by their non-specific and specific substrate affinities. For bioremediation to be effective, there needs to be enzymatic activity on the contaminants (Bala et al. 2022).

Some fungal pathogens' hyphal development is significantly inhibited by the cell wall-degrading enzymes released by biocontrol strains of PGPR bacteria, such as cellulase, 1,3-glucanase, chitinase, and protease. Chitin, a soluble linear polymer of

1,4-N-acetylglucosamine and a significant element of fungal cell walls, is broken down by the enzymes chitinase and -1,3-glucanase (Saeed et al. 2021; Midya and Islam 2022). For example, *Bacillus cepacia* generates 1,3-glucanase, an enzyme that degrades the cell walls of soilborne fungal pathogens i.e., *R. solani, S. rolfsii, and P.* In order to stop the activity of phytopathogens, a number of bacteria create enzymes that hydrolyze chitin, hemicelluloses, cellulose, and proteins. Chitinase is produced by *Streptomyces* sp., *S. marcescens, S. plymuthica* C48, *Paenibacillus* sp., and *P. stutzeri* and it breaks down the mycelia of many fungal phytopathogens (Saeed et al. 2021).

Microbial enzymes (such as amylases, lipases, cellulases, esterases, nitrilases, peroxidases, proteases, cutinase, etc.,) can be used to treat biofilm deposits, oil-contaminated soils, manage waste generated throughout the food processing, break down plastics and insecticides, and more (Bhandari et al. 2021). Hydrolytic enzymes have a wide range of potential applications in the biomedical sciences, chemical industries, and as feed additives. The enzyme amidases, esterases, and proteases can break down amide, ester, and peptide bonds respectively, resulting in the production of less harmful byproducts. In order to convert pollutants like diazinon, carbaryl or parathion, carbofuran, and coumaphos by the process of hydrolysis, microbial hydrolase from *Nocardia, Flavobacterium, Pseudomonas, Achromobacter*, and *B. cereus* has been utilized effectively (Czarny et al. 2020).

Additionally, protease enzymes are utilized in bioremediation of marine crustacean wastes used in the chitin removal process following deproteinization (Bhandari et al. 2021). Shrimp wastes are deproteinized to a 75% degree by the alkaline protease generated by *B. licheniformis* MP1 (Jellouli et al. 2011). On a 7-day fermentation, the bacteria *S. marcescens* FS-3 deproteinized up to 84% of the proteins in the crab shell (Jo et al. 2008). Lipases can also speed up the bioremediation of greasy effluents, which are released from a variety of locations and contain oils, fats, and protein. To stop oil spills (including those involving aromatic hydrocarbons, n-alkanes, and PAHs), lipase from the bacteria *Acinetobacter* sp., *Rhodococcus* sp., and *Mycobacterium* sp., has been utilized (Casas-Godoy et al. 2012). To reduce the amount of phosphate-based compounds in detergent compositions, detergent businesses use the crude lipase from the *B. subtilis* strain (Saraswat et al. 2017).

Animal horns, nails, poultry excrement, moulting and shedding of appendages, as well as animal mortality, are resistant to breakdown because they include insoluble keratin protein. In addition to being foul-smelling, they also contribute to environmental damage. Keratinase, a protease enzyme, may break down keratin proteins and be used in the bioremediation of poultry wastes by breaking down and recycling keratinous wastes into usable byproducts (Bhandari et al. 2021). Chicken feather biodegradation has been significantly accelerated by the protease enzyme keratinase from *S. maltophilia* KB13 (Bhange et al. 2016).

Other bacterial species, including *Rhizobium* sp., *Pseudomonas* sp., *Ancylobacter aquaticus* strain UV5, *Pseudomonas umsongensis* YCIT1612, and *Pseudomonas* sp., have also produced enzymes that can break down diverse halogenated substrates. The list of other microbial enzymes, their mechanisms along with their functions, is mentioned in Table 9.1.

Table 9.1 Function of various type of microbial enzyme and their application in bioremediation

Sl no	*Microorganism*	Substrate	Enzymes	Function	References
1	*Pseudomonas putida F6*	Syringaldazine (SGZ)	Laccase	Degradation of synthetic dyes	Villa et al. (2007)
2	*Sphingobacterium* sp. *Strain S2*	Chromogenic PNPL	Lipase	Degradation of PLA	Satti (2018)
3	*Bacillus subtilis*	Feather culture medium	Serine protease	Degradation of casine and feather	Suh and Lee (2001)
4	*Streptomyces thermoviolaceus*	Muscles, hair, collagen, nail, feather	Protease	Hydrolyse fibrin muscles, collagen, nail and hair	Chitte et al. (1999)
5	*Bacillus megaterium*	1-MCDD, 2,3-DCCD, 2,3,7-triCDD	Cytochrome P450	Hydrolyzation of PCCDs	Sulistyaningdyah et al. (2004)
6	*S. rhizophila*	Polyvinyl alcohol, and vinyl alcohol	Dehydrogenase	Polyvinyl alcohol degradation	Wei et al. (2017)
7	*Bacillus subtilis*	Olive oil	Lipase	Removal of oil or grease stain from detergent	Saraswat et al. (2017)
8	*Bacillus* sp. *and Geobacillus*	Diethylethanolamine (DEAE)	Amylase	Starch liquification	Nigam (2013)
9	*Pseudomonas* sp *TL*	2-Chloropropionoate	Dehalogenase	Degradation of halogen acid	Liu et al. (1994)
10	*Pichia pastoris*	Pesticides, polyurethanes, eromatic and aliphatic polyesters	Esterase	Degradation of organophosphorus harbicides, pesticides like chemicals	Mousavi et al (2021)
11	*Rhodococcusrhodochrous*	Herbicides, polymers, and plastics	Nitrilase	Degradation of harbicides, polymers, plastic	Nigam et al. 2017
12	*Armoracia rusticana*	H_2O_2	Horseradish peroxidase (HRP)	Oxidation of H_2O_2	Köller et al. 2000
13	*Trametesvillosa*	2,4-DCP (2,4-dichlorophenol)	Laccase	Degradation of 2,4-DCP	Dave and Das 2021

9.8 Future Direction and Challenges for PGPR-Based Bioremediation

PGPR are a promising tool for natural bioremediation process, as they can enhance plant growth and remove contaminants from soil. However, the future direction and challenges for PGPR-based bioremediation are still being studied and debated in the scientific community.

One challenge is the selection of the appropriate PGPR strain for a specific contaminant, as different strains may have varying levels of efficacy in removing different pollutants (Gulzar and Mazumder 2022). Additionally, the effectiveness of PGPR-based bioremediation may be limited by environmental constraints such as pH, temperature and each soil types (Saeed et al. 2021). Therefore, further research is needed to optimize the use of PGPR for bioremediation in various environments. Another challenge is the potential ecological risks associated with the use of genetically modified PGPR strains for bioremediation. While genetically modified PGPR strains may have improved bioremediation capabilities, their release into the environment may pose risks to non-target organisms and ecosystems. Thus, more research is needed to assess the potential risks and benefits of using genetically modified PGPR (GM-PGPR) strains for bioremediation. The mechanism involve in PGPR-based bioremediation is still less understood mechanisms. While it is known that PGPR can enhance plant growth and stimulate the degradation of contaminants, the precise mechanisms are not well understood. Further investigation and studyarerequired to elucidate the biochemical and molecular mechanisms underlying PGPR-mediated bioremediation. Moreover, the scale-up and commercialization of PGPR-based bioremediation is a challenge, as it requires large-scale production and delivery of PGPR to contaminated sites. The advancement of efficient and cost-effective methods for PGPR production and delivery is necessary for the widespread adoption of PGPR-based bioremediation. The long-term effects of PGPR-based bioremediation on soil quality and microbial diversity are not well understood. While PGPR-based bioremediation has shown promising results in removing contaminants, it is vital to assess the potential impacts on soil ecology and microbial diversity in the long term.

Furthermore, PGPR-based bioremediation is a promising technique for environmental remediation, but several challenges must be addressed before it can become a reliable and widely implemented tool. Further research is needed to understand the mechanisms involved in PGPR-mediated bioremediation, optimize PGPR production and delivery, and evaluate the long-term effects on soil ecology and microbial diversity.

Author Contribution All authors are equally contributed to prepared the manuscript. All authors read and approved the final manuscript.

Funding No funding was received to assist with the preparation of this manuscript.

Ethical Approval Not applicable.

Conflict of Interest The authors declare that they have no conflict of interest.

References

Abhilash PC, Singh N (2010) Withania somnifera Dunal-mediated dissipation of lindane from simulated soil: implications for rhizoremediation of contaminated soil. J Soils Sed 10:272–282

Aguelmous A, El Fels L, Souabi S, Zamama M, Hafifidi M (2019) The fate of total petroleum hydrocarbons during oily sludge composting: a critical review. Rev Environ Sci Bio/technol 18:473–493

Ajona M, Vasanthi P (2021) Bioremediation of petroleum contaminated soils–a review. Mater Today Proc 45:7117–7122

Alegbeleye OO, Opeolu BO, Jackson V (2017) Bioremediation of polycyclic aromatic hydrocarbon (PAH) compounds:(acenaphthene and fluorene) in water using indigenous bacterial species isolated from the Diep and Plankenburg rivers, Western Cape, South Africa. Braz J Microbiol 48:314–325

Ali H, Khan E, Sajad MA (2013) Phytoremediation of heavy metals—concepts and applications. Chemosphere 91:869–881

Ali MH, Sattar MT, Khan MI, Naveed M, Rafifique M, Alamri S, Siddiqui MH (2020) Enhanced growth of Mungbean and remediation of petroleum hydrocarbons by enterobacter sp. MN17 and biochar addition in diesel contaminated soil. Appl Sci 10:8548

Alkorta I, Hernandez-Allica J, Becerril JM, Amezaga I, Albizu I, Garbisu C (2004) Recent findings on the phytoremediation of soils contaminated with environmentally toxic heavy metals and metalloids such as zinc, cadmium, lead, and arsenic. Rev Environ Sci Biotech 3:71–90

Antar M, Lyu D, Nazari M, Shah A, Zhou X, Smith DL (2021) Biomass for a sustainable bioeconomy: an overview of world biomass production and utilization. Renew Sustain Energy Rev 139:110691

Ashraf S, Ali Q, Zahir ZA, Ashraf S, Asghar HN (2019) Phytoremediation: environmentally sustainable way for reclamation of heavy metal polluted soils. Ecotoxicol Environ Saf 174:714–727

Awa SH, Hadibarata T (2020) Removal of heavy metals in contaminated soil by phytoremediation mechanism: a review. Water Air Soil Pollut 231:e47

Azubuike CC, Chikere CB, Okpokwasili GC (2016) Bioremediation techniques–classifification based on site of application: principles, advantages, limitations and prospects. World J Microbiol Biotechnol 32:e180

Bala S, Garg D, Thirumalesh BV, Sharma M, Sridhar K, Inbaraj BS, Tripathi M (2022) Recent strategies for bioremediation of emerging pollutants: a review for a green and sustainable environment. Toxics 10:484

Balseiro-Romero M, Monterroso C, Kidd PS, Lu-Chau TA, Gkorezis P, Vangronsveld J, Casares JJ (2019) Modelling the ex-situ bioremediation of diesel-contaminated soil in a slurry bioreactor using a hydrocarbon-degrading inoculant. J Environ Manage 246:840–848

Banerjee A, Roy A, Dutta S, Mondal S (2016) Bioremediation of hydrocarbon—a review. Int J Adv Res 4:1303–1313

Bauelos GS, Meek DW (1990) Accumulation of selenium in plants grown on selenium-treated soil. J Environ Qual 19:772

Bhandari S, Poudel DK, Marahatha R, Dawadi S, Khadayat K, Phuyal S, Parajuli N (2021) Microbial enzymes used in bioremediation. J Chem 2021:1–17

Bhange K, Chaturvedi V, Bhatt R (2016) Feather degradation potential of Stenotrophomonas maltophilia KB13 and feather protein hydrolysate (FPH) mediated reduction of hexavalent chromium. 3 Biotech 6(1):42

Bhat MA, Kumar V, Bhat MA, Wani IA, Dar FL, Farooq I, Bhatti F, Koser R, Rahman S, Jan AT (2020) Mechanistic insights of the interaction of plant growth-promoting rhizobacteria (PGPR) with plant roots toward enhancing plant productivity by alleviating salinity stress. Front Microbiol 11:1952

Bhattacharyya PN, Jha DK (2012) Plant growth-promoting rhizobacteria (PGPR): emergence in agriculture. World J Microbiol Biotechnol 28(4):1327–1350

Boudh S, Singh JS, Chaturvedi P (2019) Microbial resources mediated bioremediation of persistent organic pollutants. In: New and future developments in microbial biotechnology and bioengineering. Elsevier, pp 283–294

Braud A, Jézéquel K, Bazot S, Lebeau T (2009) Enhanced phytoextraction of an agricultural Cr, Hg- and Pb-contaminated soil by bioaugmentation with siderophore producing bacteria. Chemosphere 74:280–286

Briffa J, Sinagra E, Blundell R (2020) Heavy metal pollution in the environment and their toxicological effects on humans. Heliyon 8;6(9):e04691

Buyuksonmez F, Rynk R, Hess TG, Bechinski E (2000) Occurrence, degradation and fate of pesticides during composting. Part II: occurrence and fate of pesticides in compost and composting systems. Compost Sci Util 8:61–81

Casas-Godoy S, Duquesne FB, Sandoval G, Marty A (2012) Lipases: an overview. In: Lipases and phospholipases, methods in molecular biology. Humana Press, Totowa, NJ, USA

Cetin SC, Karaca A, Kizilkaya R, Turgay OC (2011) Role of plant growth promoting bacteria and fungi in heavy metal detoxification. Detoxification Heavy Met 369–388

Chadar SN, Chadar K, Singh M (2018) Composting as an eco-friendly method to recycle organic waste. Prog Petrochem Sci 2(5):252–254

Chatterjee S, Sau GB, Mukherjee SK (2009) Plant growth promotion by a hexavalent chromium reducing bacterial strain, cellulosimicrobium cellulans KUCr3. World J Microbiol Biotechnol 25:1829–1836

Chaudhry TM, Hayes WJ, Khan AG, Khoo CS (1998) Phytoremediation focusing on accumulator plants that remediate metal contaminated soils. Aust J Ecotoxicol 4:37–51

Chawla N, Sunita S, Kamlesh K, Kumar R (2013) Bioremediation: an emerging technology for remediation of pesticides. Res J Chem Environ 17:88–105

Chibuike GU, Obiora SC (2014) Heavy metal polluted soils: effect on plants and bioremediation methods. Appl Environ Soil Sci 752708

Chitara MK, Chauhan S, Singh RP (2021) In: © Panpatte DG, Jhala YK (eds) Microbial rejuvenation of polluted environment. Microorganisms for sustainability, p 25

Chitte RR, Nalawade VK, Dey S (1999) Keratinolytic activity from the broth of a feather-degrading thermophilic *streptomyces thermoviolaceus* strain SD8. Lett Appl Microbiol 28(2):131–136

Cristaldi A, Oliveri Conti G, Cosentino SL, Mauromicale G, Copat C, Grasso A, Zuccarello P, Fiore M, Restuccia C, Ferrante M (2020) Phytoremediation potential of Arundo donax (Giant Reed) in contaminated soil by heavy metals. Environ Res 185:e109427

Czarny J, Staninska-Pięta J, Piotrowska-Cyplik A, Juzwa W, Wolniewicz A, Marecik R, Lawniczak L, Chrzanowski L (2020) Acinetobacter sp. as the key player in diesel oil degrading community exposed to PAHs and heavy metals. J Hazard Mater 383:121168

Dave S, Das J (2021) Role of microbial enzymes for biodegradation and bioremediation of environmental pollutants: challenges and future prospects. In: Saxena G, Kumar V, Shah MP (ed) Bioremediation for environmental sustainability. Elsevier, pp 325–346

Di Gregorio S, Lampis S, Vallini G (2005) Selenite precipitation by a rhizospheric strain of Stenotrophomonas sp isolated from the root system of Astragalus bisulcatus: a biotechnological perspective. Environ Int 31:233–241

Dimkpa CO, Merten D, Svatos A, Buchel G, Kothe E (2009) Siderophores mediate reduced and increased uptake of cadmium by Streptomyces tendae F4 and sunflower (Helianthus annuus), respectively. J Appl Microbiol 107:1687–1696

Duruibe JO, Ogwuegbu MOC, Egwurugwu JN (2007) Heavy metal pollution and human biotoxic effects. Intl J Phys Sci 2(5):112–118

Dzionek A, Wojcieszynska D, Guzik U (2016) Natural carriers in bioremediation: a review. Electron J Biotechnol 23:28–36

El Fantroussi S, Agathos S (2005) El Fantroussi S, Agathos SN. Is bioaugmentation a feasible strategy for pollutant removal and site remediation? Curr Opin Microbiol 8:268–275

Etim EE (2012) Phytoremediation and its mechanisms: a review. Int J Environ Bioenergy 2:120–136

Feng RW, Wei CY (2012) Antioxidative mechanisms on selenium accumulation in Pteris vittata L., a potential selenium phytoremediation plant. Plant Soil Environ 58(3):105–110

Ferro AM, Kennedy J, LaRue JC (2013) Phytoremediation of 1,4-dioxane-containing recovered groundwater. Int J Phytorem 15(10):911–923. https://doi.org/10.1080/15226514.2012.687018

Fomina MA, Alexander IJ, Colpaert JV, Gadd GM (2005) Solubilization of toxic metal minerals and metal tolerance of mycorrhizal fungi. Soil Biol Biochem 37:851–866

Garbisu C, Alkorta I (2003) Basic concepts on heavy metal soil bioremediation. Eur J Min Proc Environ Protect 3:58–66

Germaine KJ, Byrne J, Liu X, Keohane J, Culhane J, Lally RD, Kiwanuka S, Ryan D, Dowling DN (2015) Ecopiling: a combined phytoremediation and passive biopiling system for remediating hydrocarbon impacted soils at field scale. Front Plant Sci 5(5):756

Gholami-Shabani M, Gholami-Shabani Z, Shams-Ghahfarokhi M, Razzaghi-Abyaneh M (2018) Application of nanotechnology in mycoremediation: current status and future prospects. Fungal Nanobionics Principles Appl 89–116

Godheja J, Modi DR, Kolla V, Pereira AM, Bajpai R, Mishra M, Sharma SV, Sinha K, Shekhar SK (2019) Environmental remediation: microbial and nonmicrobial prospects. In: Microbial interventions in agriculture and environment, vol 1. Springer, Singapore, pp 379–409

Gontia-Mishra I, Sapre S, Kachare S, Tiwari S (2017) Molecular diversity of 1-aminocyclopropane-1-carboxylate (ACC) deaminase producing PGPR from wheat (Triticumaestivum L.) rhizosphere. Plant Soil 414(1–2):213–227

Gulzar ABM, Mazumder PB (2022) Helping plants to deal with heavy metal stress: the role of nanotechnology and plant growth promoting rhizobacteria in the process of phytoremediation. Environ Sci Pollut Res 29(27): 40319–40341

Haider FU, Ejaz M, Cheema SA, Khan MI, Zhao B, Cai L, Salim MA, Naveed M, Khan N, Núñez-Delgado A et al (2021) Phytotoxicity of petroleum hydrocarbons: sources, impacts and remediation strategies. Environ Res 197:111031

Henry JR (2000) In an overview of phytoremediation of lead and mercury. NNEMS Report, Washington, DC, pp 3–9

Hong Y, Liao D, Chen J, Khan S, Su J, Li H (2015) A comprehensive study of the impact of polycyclic aromatic hydrocarbons (PAHs) contamination on salt marsh plants Spartina alterniflflora: implication for plant-microbe interactions in phytoremediation. Environ Sci Pollut Res 22:7071–7081

Huang XD, El-Alawi Y, Gurska J, Glick BR, Greenberg BM (2005) A multi-process phytoremediation system for decontamination of persistent total petroleum hydrocarbons (TPHs) from soils. Microchem J 81:139–147

Islam SS, Sen K, Dutta S, Midya S (2022) Performance of rice-lentil cropping under different tillage influencing soil suppressiveness: a short-term approach. Soil health and environmental sustainability: application of geospatial technology. Springer International Publishing, Cham, pp 551–577

Jaishankar M, Tseten T, Anbalagan N et al (2014) Toxicity, mechanism and health efects of some heavy metals. Interdiscip Toxicol 7:60–72

Jayaprakash K, Govarthanan M, Mythili R, Selvankumar T, Chang YC (2019) Bioaugmentation and biostimulation remediation technologies for heavy metal lead contaminant. Microb Biodegradation Xenobiotic Compdounds 24

Jellouli K, Ghorbel-Bellaaj O, Ayed HB, Manni L, Agrebi R, Nasri M (2011) Alkaline-protease from Bacillus licheniformis MP1: purification, characterization and potential application as a detergent additive and for shrimp waste deproteinization. Process Biochem 46(6):1248–1256

Jo GH, Jung WJ, Kuk JH, Oh KT, Kim YJ, Park RD (2008) Screening of protease-producing Serratia marcescens FS-3 and its application to deproteinization of crab shell waste for chitin extraction. Carbohyd Polym 74(3):504–508

Kao CM, Chen CY, Chen SC, Chien HY, Chen YL (2008) Application of in situ biosparging to remediate a petroleum- hydrocarbon spill site: field and microbial evaluation. Chemosphere 70:1492–1499

Karlson U, Frankenberger WT (1989) Accelerated rates of selenium volatilization from California soils. Soil Sci Soc Am J 53:749–753

Karthik V, Saravanan K, Sivarajasekar N, Suriyanarayanan N (2016) Bioremediation of dye bearing efuents using microbial biomass. Ecol Environ Conserv 22:423–434

Khan N, Bano A, Ali S, Babar MA (2020) Crosstalk amongst phytohormonesfrom planta and PGPR under biotic and abiotic stresses. Plant Growth Regul 90:189–203

Köller G, Möder M, Czihal K (2000) Peroxidative degradation of selected PCB: a mechanistic study. Chemosphere 41(12):1827–1834

Kuiper I, Lagendijk EL, Bloemberg GV, Lugtenberg BJJ (2004a) Rhizoremediation: a beneficial plant-microbe interaction. Mol Plant Microbe Interact 7:6–15

Kuiper I, Lagendijk EL, Bloemberg GV, Lugtenberg BJJ (2004b) Rhizoremediation: a benefificial plant-microbe interaction. Mol Plant Microbe Interact 17:6–15

Kumar V, Shahi SK, Singh S (2018a) In: Singh J et al (eds) Microbial bioprospecting for sustainable development. © Springer Nature Singapore Pte Ltd.

Kumar V, Shahi SK, Singh S (2018b) Bioremediation: an eco-sustainable approach for restoration of contaminated sites. In: Microbial bioprospecting for sustainable development. Springer, Singapore, pp 115–136

Ławniczak Ł, Wozniak-Karczewska M, Loibner AP, Heipieper HJ, Chrzanowski Ł (2020) Microbial degradation of hydrocarbons—basic principles for bioremediation: a review. Molecules 25(4):856

LeDuc DL, Terry N (2005) Phytoremediation of toxic trace elements in soil and water. J Ind Microbiol Biotechnol 32:514–520

Lee SB (2013) Toxin binding receptors and the mode of action of Bacillus thuringiensis subsp. israelensis cry toxins. Ph.D. Thesis, University of California, Riverside

Leontidou K, Genitsaris S, Papadopoulou A et al (2020) Plant growth promoting rhizobacteria isolated from halophytes and drought-tolerant plants: genomic characterisation and exploration of phyto-beneficial traits. Sci Rep 10:14857

Liu JQ, Kurihara T, Hasan AK et al (1994) Purification and characterization of thermostable and nonthermostable 2- haloacid dehalogenases with different stereospecificities from Pseudomonas sp. strain YL. Appl Environ Microbiol 60(7):2389–2393

Luo L, Zhang S, Shan X-Q, Zhu Y-G (2006) Oxalate and root exudates enhance the desorption of p, p′-DDT from soils. Chemosphere 63:1273–1279

Ma Y, Prasad MNV, Rajkumar M, Freitas H (2011) Plant growth promoting rhizobacteria and endophytes accelerate phytoremediation of metalliferous soils. Biotechnol Adv 29:248–258

Mallavarapu M, Leee YB (2020) Biostimulation and bioaugmentation: modern strategies for the successful bioremediation of contaminated environments. The handbook of environmental remediation: classic and modern techniques, p 299.

Manoj SR, Karthik C, Kadirvelu K, Arulselvi PI, Shanmugasundaram T, Bruno B, Rajkumar M (2020) Understanding the molecular mechanisms for the enhanced phytoremediation of heavy metals through plant growth promoting rhizobacteria: a review. J Environ Manage 254:109779

Marr LC, Booth EC, Andersen RG, Widdowson MA, Novak JT (2006) Direct volatilization of naphthalene to the atmosphere at a phytoremediation site. Environ Sci Technol 40:5560–5566

Masindi V, Mkhonza P, Tekere M (2021) Sources of heavy metals pollution. In: Inamuddin, Ahamed MI, Lichtfouse E, Altalhi T (eds) Remediation of heavy metals. Environmental chemistry for a sustainable world, vol 70. Springer, Cham

Midya S, Islam SS (2022) Assessment of chitinolytic bacteria isolated from zooplankton of freshwater ecosystem. River health and ecology in south Asia: pollution, restoration, and conservation, pp 81–102

Miller RR (1996) Ground-water remediation technologies analysis center prepared for.: technology overview report. Ground-Water Remediation Technologies Analysis Center, Pittsburgh, PA, USA

Moreno FN, Anderson CWN, Stewart RB et al (2004a) Mercury phyto extraction and phyto-volatilisation from hg-contaminated artisanal mine sites. Phytoremediat Mercur Mine Wastes 147–159

Mousavi SM, Hashemi SA, Iman Moezzi SM, Ravan N, Gholami A, Lai CW, Chiang WH, Omidifar N, Yousefi K, Behbudi G (2021) Recent advances in enzymes for the bioremediation of pollutants. Biochem Res Int 22(2021):5599204

Mueller B, Rock S, Gowswami D, Ensley D (1999) Phytoremediation Decision Tree. Prepared by—Interstate Technology and Regulatory Cooperation Work Group, pp 1–36

Mulla SI, Bharagava RN, Belhaj D, Saratale GD, Kumar A, Rajasekar A, Tallur PN, Edalli VA, Hu A, Yu CP (2019) Microbial degradation of phenolic compounds. In: Kumar A, Sharma S (eds) Microbes and enzymes in soil health and bioremediation; microorganisms for sustainability, vol 16. Springer, Singapore

Naeem U, Qazi MA (2020) Leading edges in bioremediation technologies for removal of petroleum hydrocarbons. Environ Sci Pollut Res 27:27370–27382

Naushad M, Mittal A, Rathore M, Gupta V (2015) Ion-exchange kinetic studies for Cd (II), Co (II), Cu (II), and Pb (II) metalions over a composite cation exchanger. Des Water Treat 54:2883–2890

Naushad M, Ahamad T, Al-Maswari BM et al (2017) Nickel ferrite bearing nitrogen-doped mesoporous carbon as efcient adsorbent for the removal of highly toxic metal ion from aqueous medium. Chem Eng J 330:1351–1360

Nigam P (2013) Microbial enzymes with special characteristics for biotechnological applications. Biomolecules 3(4):597–611

Nigam VK, Arfi T, Kumar V, Shukla P (2017) Bioengineering of nitrilases towards its use as green catalyst: applications and perspectives. Indian J Microbiol 57(2):131–138

Nikolopoulou M, Kalogerakis N (2008) Enhanced bioremediation of crude oil utilizing lipophilic fertilizers combined with biosurfactants and molasses. Mar Pollut Bull 56:1855–1861

Niu G-L, Zhang J-J, Zhao S, Liu H, Boon N, Zhou N-Y (2009) Bioaugmentation of a 4-chloronitrobenzene contaminated soil with Pseudomonas putida ZWL73. Environ Pollut 157:763–771

Nzila A, Razzak SA, Zhu J (2016) Bioaugmentation: an emerging strategy of industrial wastewater treatment for reuse and discharge. Int J Environ Res Public Health 13(9):846

Padmavathiamma PK, Li LY (2007) Phytoremediation technology: hyper-accumulation metals in plants. Water Air Soil Pollut 184:105–126

Parmar S, Singh V (2015) Phytoremediation approaches for heavymetal pollution: a review. J Plant Sci Res 2:1–8

Patel PR, Shaikh SS, Sayyed RZ (2016) Dynamism of PGPR in bioremediation and plant growth promotion in heavy metal contaminated soil. Indian J Exp Biol 54:286

Phulpoto AH, Maitlo MA, Kanhar NA (2021) Culture dependent to culture independent approaches for the bioremediation of paints: a review. Int J Environ Sci Technol 18:241–262

Pillon-Smith EA, Pilon M (2000) Breeding mercury breathing plants for environmental clean-up. Trends Plants Sci 5:235–236

Pulsawat W, Leksawasdi N, Rogers PL, Foster LJR (2003) Anions effects on biosorption of Mn(II) by extracellular polymeric substance (EPS) from *Rhizobium etli*. Biotechnol Lett 25:1267–1270

Pushkar B, Sevak P, Parab S, Nilkanth N (2021) Chromium pollution and its bioremediation mechanisms in bacteria: a review. J Environ Manage 287:112279

Rada EC, Andreottola G, Istrate IA, Viotti P, Conti F, Magaril ER (2019) Remediation of soil polluted by organic compounds through chemical oxidation and phytoremediation combined with DCT. Int J Environ Res Public Health 16(17):3179

Radwan SS, Al-Mailem DM, Kansour MK (2019) Bioaugmentation failed to enhance oil bioremediation in three soil samples from three different continents. Sci Rep 9(1): 19508

Rajkumar M, Ae N, Prasad MNV, Freitas H (2010) Potential of siderophore-producing bacteria for improving heavy metal phytoextraction. Trends Biotechnol 28:142–149

Rajtor M, Piotrowska-Seget Z (2016) Prospects for arbuscular mycorrhizal fungi (AMF) to assist in phytoremediation of soil hydrocarbon contaminants. Chemosphere 162:105–116

Rashid A, Mahmood T, Mehmood F et al (2014) Phytoaccumulation, competitive adsorption and evaluation of chelators-metal interaction in lettuce plant. Environ Eng Manage J 13:2583–2592

Rodríguez-Eugenio N, McLaughlin M, Pennock D (2018) Soil pollution: a hidden reality. FAO, Rome, p 142

Ryan PR, Delhaize E, Jones DL (2001) Function and mechanism of organic anion exudation from plant roots. Annu Rev Plant Physiol Plant Mol Biol 52:527–560

Saeed Q, Xiukang W, Haider FU, Kucerik J, Mumtaz MZ, Holatko J, Naseem M, Kintl A, Ejaz M, Naveed M, Brtnicky M, Mustafa A (2021) Rhizosphere bacteria in plant growth promotion, biocontrol, and bioremediation of contaminated sites: a comprehensive review of effects and mechanisms. Int J Mol Sci 22(19):10529

Sakakibara M, Watanabe A, Inoue M et al (2010) Phytoextraction and phytovolatilization of arsenic from As-contaminated soils by Pteris vittata. In: Proceedings of the annual international conference on soils, sediments, water and energy, p 26

Salehi MA, Hakimghiasi N (2017) Hydrodynamics and mass transfer inthree-phase airlift reactors for activated carbon and sludge fifiltration. Adv Environ Technol 2:179–184

San Miguel A, Ravanel P, Raveton M (2013) A comparative study on the uptake and translocation of organochlorines by Phragmites australis. J Hazard Mat 244:60–69

Saraswat R, Verma V, Sistla S, Bhushan I (2017) Evaluation of alkali and thermotolerant lipase from an indigenous isolated Bacillus strain for detergent formulation. Electron J Biotechnol 30:33–38

Saravanan VS, Madhaiyan M, Thangaraju M (2007) Solubilization of zinc compounds by the diazotrophic, plant growth promoting bacterium Gluconacetobacter diazotrophicus. Chemosphere 66:1794–1798

Sasikumar CS, Papinazath T (2003) Environmental management:- bioremediation of polluted environment. In: Bunch MJ, Suresh VM, Kumaran TV (eds) Proceedings of the third international conference on environment and health, Chennai, India, 15–17 December, 2003. Department of Geography, University of Madras and Faculty of Environmental Studies, York University, Chennai, pp 465–469

Satti SM (2018) Statistical optimization of lipase production from sphingobacterium sp. strain S2 and evaluation of enzymatic depolymerization of poly (lactic acid) at mesophilic temperature. Polym Degrad Stab 160:1–13

Sayara T, Sánchez A (2020) Bioremediation of PAH-contaminated soils: process enhancement through composting/compost. Appl Sci 10:3684

Shackira AM, Puthur JT (2019) Phytostabilization of heavy metals: understanding of principles and practices. In: Plant-metal interactions. Springer International Publishing, Cham, Switzerland, pp 263–282

Sharma J (2019) Advantages and limitations of in situ methods of bioremediation. Recent Adv Biol Med 5:1

Sharma G, Pathania D, Naushad M (2015) Preparation, characterization, and ion exchange behavior of nanocomposite polyaniline zirconium (IV) selenotungstophosphate for the separation of toxic metal ions. Ionics 21:1045–1055

Shukla OP, Rai UN, Dubey S (2009) Involvement and interaction of microbial communities in the transformation and stabilization of chromium during the composting of tannery efflfluent treated biomass of Vallisneria spiralis L. Bioresour Technol 100:2198–2203

Sidhu GPS, Bali AS, Singh HP, Batish DR, Kohli RK (2018) Ethylenediamine disuccinic acid enhanced phytoextraction of nickel from contaminated soils using Coronopusdidymus (L.) Sm. Chemosphere 205:234–243

Sivarajasekar N, Paramasivan T, Subashini R, Kandasamy S, Prakashmaran J (2017) Central composite design optimization of fuoride removal by spirogyra biomass. Asian J Microbiol Biotechnol Environ Sci 19:130–137

Smets BF, Pritchard PH (2003) Elucidating the microbial component of natural attenuation. Curr Opin Biotechnol 14:283–288

Soudek P, Sarka Petrova S, Vanek T (2012) Phytostabilization or accumulation of heavy metals by using of energy crop Sorghum sp. 3rd international conference on biology, environment and chemistry IPCBEE. IACSIT Press, Singapore

Sparks DL (1995) Environmental soil chemistry. Academic Press, San Diego, CA

Speight JG (2018) Environmental organic chemistry for engineers, vol 1. Elsevier Ltd., Butterworth-Heinemann, UK

Suh HJ, Lee HK (2001) Characterization of a keratinolytic serine protease from Bacillus subtilis KS-1. J Protein Chem 20(2):165–169

Sulistyaningdyah WT, Ogawa J, Li Q-S et al (2004) Metabolism of polychlorinated dibenzo-p-dioxins by cytochrome P450 BM-3 and its mutant. Biotech Lett 26(24):1857–1860

Taghavi S, Mergeay M, Nies D, van der Lelie D (1997) Alcaligenes eutrophus as a model system for bacterial interactions with heavy metals in the environment. Res Microbiol 148:53551

Tahri N, Bahafid W, Sayel H, El Ghachtouli N (2013) Biodegradation: involved microorganisms and genetically engineered microorganisms. Biodegradation Life Sci 41

Tamburini E, Sergi S, Serreli L et al (2017) Bioaugmentation-assisted phytostabilisation of abandoned mine sites in south west Sardinia. Bull Environ Contam Toxicol 98:310–316

Villa AM, Doyle EM, Brooks S (2007) Biochemical characterisation of the coexisting tyrosinase and laccase in the soil bacterium Pseudomonas putida F6. Enzyme Microb Technol 40(5):1435–1441

Wozniak-Karczewska M, Lisiecki P, Białas W, Owsianiak M, Piotrowska-Cyplik A, Wolko L, Ławniczak L, Heipieper HJ, Gutierrez T, Chrzanowski L (2019a) Effect of bioaugmentation on long-term biodegradation of diesel/biodiesel blends in soil microcosms. Sci Total Environ 671:948–958

Wozniak-Karczewska M, Lisiecki P, Białas W, Owsianiak M, Piotrowska-Cyplik A, Wolko L, Ławniczak L, Heipieper HJ, Gutierrez T, Chrzanowski L (2019b) Effffect of bioaugmentation on long-term biodegradation of diesel/biodiesel blends in soil microcosms. Sci Total Environ 671:948–958

Wu Y, Yang J, Tang J, Kerr P, Wong PK (2017) The remediation of extremely acidic and moderate pH soil leachates containing Cu (II) and Cd (II) by native periphytic biofilm. J Clean Prod 162:846–855

Yadegari M, Rahmani HA, Noormohammadi G, Ayneband A (2010) Plant growth promoting rhizobacteria increase growth, yield and nitrogen fifixation in Phaseolus vulgaris. J Plant Nutr 33(12):1733–1743

Yuan X, Xiong T, Wang H, Wu Z, Jiang L, Zeng G, Li Y (2018) Immobilization of heavy metals in two contaminated soils using a modifified magnesium silicate stabilizer. Environ Sci Pollut Res Int 25:32562–32571

Zaidi A, Wani PA, Khan MS (2012) Bioremediation: a natural method for the management of polluted environment. Toxicity of heavy metals to legumes and bioremediation. Springer, Vienna, pp 101–114

Zayed AM, Terry N (2003) Chromium in the environment: factors afecting biological remediation. Plant Soil 249:139–156

Zeneli A, Kastanaki E, Simantiraki F, Gidarakos E (2019) Monitoring the biodegradation of TPH and PAHs in refinery solid waste by biostimulation and bioaugmentation. J Environ Chem Eng 7:e103054

Zeng G, Yu Z, Chen Y, Zhang J, Li H, Yu M, Zhao M (2011) Response of compost maturity and microbial community composition to pentachlorophenol (PCP)-contaminated soil during composting. Biores Technol 02:5905–5911

Zhang X, Yang L, Li Y, Li H, Wang W, Ye B (2012) Impacts of lead/zinc mining and smelting on the environment and human health in China. Environ Monit Assess 184(4):2261–2273

Chapter 10
Zooplankton Attached Bacteria: Potentiality Towards Antifungal and PGPR Properties

Sk Saruk Islam, Sumana Mahato, Suchetana Karmakar, Saheli Maiti, Krishnendu Sen, Sk Md Abu Imam Saadi, and Sujoy Midya

Keywords Aquatic ecosystem · PGPR · Secondary metabolites · Surface bacteria · Zooplankton

10.1 Introduction

Aquatic ecosystem provides a habitat for aquatic organisms. Aquatic organisms can be of two types microorganisms like zooplankton, phytoplankton, bacteria etc. and microorganisms like macrophytes, fishes and other large organisms (Midya ad Islam 2022). Zooplankton plays an important intermediate role in food chain between phytoplankton and small fish and helps in transferring the energy to higher trophic level, they also help in regulating the self-purification capacity of the waterbody (Midya et al. 2018a). Environmental heterogeneity effects the zooplankton population and create a certain distribution pattern that varies with different habitats, temperature, salinity, conductivity has immense effect on zooplankton diversity and

S. S. Islam · S. Mahato · S. Karmakar · S. Maiti · S. Midya (✉)
Department of Zoology, Raja Narendra Lal Khan Women's College (Autonomous), Midnapore, West Bengal 721102, India
e-mail: sujoy.midya@gmail.com

K. Sen
Department of Plant Pathology, Bidhan Chandra Krishi Viswavidyalaya, Mohanpur, West Bengal 741252, India

Department of Microbiology, Vidyasagar University, Midnapore, West Bengal 721102, India

S. M. A. I. Saadi
Department of Biological Sciences, Aliah University, IIA/27, New Town, Kolkata, West Bengal 700160, India

P. P. Adhikary et al. (eds.), *Soil, Water Pollution and Mitigation Strategies*, Environmental Science and Engineering, https://doi.org/10.1007/978-3-031-63296-9_21

abundance (Midya et al. 2018b). Climatic condition of West Bengal is extremely diverse and create considerable effect on zooplankton diversity.

In an aquatic system bacterial population can be seen as planktonic form (free floating) or can form biofilm that remains attached with the different part of zooplankton (Flemming et al. 2021). Bacterial attachment can be seen on the exoskeleton as well as on the gut lining of copepods which act as a favorable growth surface for the zooplankton associated bacteria (Midya et al. 2019). This interactive association must help the bacteria to have a protective shelter and can provide organic carbon comes from chitinous appendages of copepods (Midya ad Islam 2022). This process of bacterial attachment can also be observed in non- crustaceans' zooplankton like Appendicularia's (Flood 1991), jellyfish (Schuett and Doepke 2009) and rotifers (Selim 2001). These bacterial population can have some antifungal properties. Fungal inhibition can be useful in so many ways, it may have a medicinal prospective, can also be used in prevention of food spoilage and bio-preservation (Ibrahim et al. 2021).

Fungi are the group of heterotrophic eukaryotic organisms that consume food from environment (Sudharsan et al. 2023). The frequent growth of fungus can result in damage of foods which causes considerable economical loss and also causing diseases in a population. Bacteria and fungi become colonized in the food products with high nutrient value and physiologically adopted in the new ecological environment (Castiglione et al. 2021). But some bacteria that grows in food products having many antifungal components which helps to kill the fungal population like lactic acid bacteria (LAB). Scientists research that *Lactobacillus acidophilus* can produce antifungal product against *Candida albicans* (Guillo 1958; Ibrahim et al. 2021), *Leuconostoc citrovorum* secrets some active components which kill four types of yeast colonies *S. fragilis, S. cerevisiae, T. glutinis and M. lipolytica* (Marth and Hussong 1963). Even also *Aspergillus parasiticus* is also inhibited by *Streptococcus lactis* bacteria (Moller et al. 2021).

Microorganisms are involved in the agriculture in almost all areas. Microorganisms provides various nutrient content, mineral, antibiotics and growth substances to the plant species and it evoke the growth of plant (Khan et al. 2022). There are few microorganisms, which are used as biofertilizers such as *Rhizobium, Azobactor*, Blue-green algae etc., are promote the growth of plants (Rehmaan et al. 2022). Microorganisms can promote certain nutrients availability such as, phosphorus, nitrogen, water directly or indirectly which support to produces plant hormone that promote the plant growth.Till now the exact beneficial role of plankton appendage bacterial species are unknown. So, present research aimed to identify the unknown roles of plankton associate bacterial species as a bio-controlled agent and a plant growth promoting activator.

10.2 Materials and Methods

10.2.1 Study Site

The current experiment was conducted in and around a few wetlands in West Bengal, India, as well as freshwater aquatic environments. The samples were gathered from six study locations, including Medinipur (lat- 22.41402 and long- 87.323168), Mecheda (lat- 22.40.5004 and long- 87.898614), Gope Palace (lat- 22.429564 and long- 87.290623), Khargapur (lat- 22.347953 and long- 87.276495), Jhargram (lat- 22.432885 and long- 87.001219) and Tamluk (lat- 22.278367 and long- 87.91933). The samples are taken in 2022, primarily in the early monsoon season, which is the best time for Rotifera breeding. These ponds are primarily utilised for home purposes, though occasionally the edges of the ponds are used to grow vegetables. The majority of the research site in Kharagpur is located in an industrial region, and compared to other study sites, substantially more chemicals are found here. It has an impact on the formation and growth of planktons.

10.2.2 Zooplankton Diversity Study

Scientifically, planktons, primarily zooplanktons, were gathered from the pond's surface and subsurface area as well as from nearby wetlands. The plankton is gathered using a Nylonbolt plankton net that is 50 m in diameter. Zooplanktons are the group with the lowest taxonomic rank, according to accepted literature (Midya et al. 2018a; 2018b). Under the phase contrast microscope, the zooplankton was quantified using a Sedgwick-rafter cell counter. The cubet holds a solution of formalin that is used to collect planktons. The numbering of the zooplanktons is done according to the amount of litter. Individuals/ml = A * (n/v)/L is the formula used to compute the Sedgewick rafter cell (Midya et al. 2018a; 2018b).

10.2.3 Isolation and Screening of Surface Attachment Bacteria

The planktons that were gathered and their fragments that were kept in 5% formalin. They are homogenised using a sterilised glass rod and incubated in Alkaline Peptone Water for 18–24 h at a nearly suitable temperature of 37 °C (Midya et al. 2019). The APW medium is used to isolate bacteria, which are then distributed onto nutrient agar media and cultured for 72 h at 30 °C. Then, using the streaking technique, the specific bacterial colonies from them were extracted and screened on another fresh nutrient agar, where they were incubated for three days at 37 °C. Then conduct further study using these focused bacterial colonies (Midya et al. 2019; Midya and Islam 2022).

10.2.4 Effect of pH, Temperature and Salt Concentration on Isolates

Effect of pH: The culture medium's nature, whether acidic or alkaline, was determined by pH. Alkalinity is managed by adding the right amount of NaOH, whereas acidity is maintained by supplying concentrated H_2SO_4. The culture media is divided into five petri dishes with pH values of 1, 3, 5, 7, and 9. The nine bacteria isolates, S1–S9, were then added to nutritional agar media and cultured there for 24 h at 37 °C. The results were then recorded (Midya et al. 2020).

Effect of temperature: In incubator, the bacteria are cultured at various temperatures. The native bacterial isolates will grow between 24 and 48 h at temperatures of 10, 25, 37, 50, and 60 degrees Celsius, and the findings will be recorded (Midya et al. 2020).

Effect of salt concentration: The chosen bacterial isolates are kept in separate incubations with varying salt concentrations (1%, 3%, 5%, 7%, 9%) on nutrient agar media. Five Petri dishes are filled with various salt concentrations of culture media. The nine previously chosen bacterial isolates are all present in each petridish. Results can be seen within two to three days (Midya et al. 2020).

10.2.5 Enzymatic Activity of Isolated Antagonist Microorganisms

10.2.5.1 Amylase Activity

Amylase, which is composed of amylose and amylopectine. Maltose, a disaccharide comprised of two glucose molecules, is created when starch and amylase react. Amylase has a pH range of 5.0–10.5 with 7.0 being the ideal pH = 7.Starch agar medium was used to conducted the amylase activity test. The test bacterial cultures were spot injected onto the plates, and they were then incubated at 28 °C temp., for up to a week. The plate was saturated for five minutes with iodine solution to detect amylase activity. After that, the iodine solution was drained out. The presence of clear halo zones around the growth shows amylase activity if the bacterium hydrolyzed amylase and the growth was surrounded by a clear zone (Islam et al. 2022).

10.2.5.2 Cellulase Activity

Cellulase, an enzyme that aids in the breakdown of certain polysaccharides and cellulose, is mostly produced by bacteria, fungi, and protozoans. It facilitates the conversion of cellulose molecules into simple sugars, such as beta-glucose. In cellulose chain hydrolyzed by cellulase enzyme, β-1,4 glycosidic bond is present. Cellulose agar medium was used for the cellulase activity test. The culture media conical is then sterilised in an autoclave. The sterilised solution is then put into three sterile petri dishes after being autoclaved. Three bacterial isolates are present in each petri dish, and after three days, the growth of nine different bacterial isolates has been noted. The Congo red solution (1 mg/ml) was flooded onto the plate for 15 min in order to identify Cellulase activity. It was then time to pour off the Congo red solution. After 15 min of flooding with 1(M) NaCl solution, the plates were further de-stained. If the bacterium hydrolyzed cellulose, there would have been obvious orange zones surrounding the growth, which would have shown cellulase activity (Islam et al. 2022).

10.2.5.3 Pectinase Activity

Pectinase enzyme is mostly produced by microorganisms like yeast, bacteria, and fungus. A polysaccharide and pectine are this enzyme's most frequent substrates. Pectine is mostly found in plant cell walls, which are hydrolyzed by the enzyme to be digested. Pectinase has a molecular mass between 43.5 and 47 kDa. Pectinase's stable pH ranges from 3.5 to 9.5, with 4.0 being the ideal number. Pectinase agar medium was used for the pectinase activity test. For sterilisation, this specific conical is placed in an autoclave. Three petri dishes that are likewise autoclave-sterilized are then filled with the liquid-sterilized media. Three petri dishes each contained three of the nine chosen bacterial isolates, total, which were separated into 9. The growth is seen three days later. The plate was then submerged in 1% tri-methyl ammonium bromide for 48 h and drained off to detect pectinase activity. The presence of clear halo zones around the growth shows pectinase activity if the bacterium hydrolyzed pectinase and the growth was surrounded by a clear zone (Islam et al. 2022).

10.2.5.4 Lipase Activity

Lipase is utilised to break down fat molecules, primarily triglycerides and glycerol. The pH range for stable solutions is 7–9, although the ideal pH for lipase activity is 8, and the ideal temperature for lipase activity is 37 °C. Lipase agar medium was used for the lipase activity test. By using an autoclave, the conical and three petri dishes are sterilised. Then, 3 petri plates are filled with the solution, and 9 bacterial isolates are cultivated there. Three days later, the outcome is seen. Around bacterial colonies that produce the esterase enzyme, opaque zones form for the purpose of detecting lipase activity. The development of opaque zones around bacterial colonies shows the presence of lipase activity if the bacterium hydrolyzed lipase (Islam et al. 2022).

10.2.6 Secondary Metabolites Production of Native Isolates

10.2.6.1 *Production of Ammonia (NH_3)*

Testing for ammonia generation was done using peptone water medium. The cultures were cultivated in tubes filled with peptone water, which were then incubated at 30 °C for four days. Each tube received 1 ml of Nessler's reagent following incubation. The presence of a faint yellow hue suggests a low concentration of ammonia, while a deep yellow to brownish colour indicates a high synthesis of ammonia (Islam et al. 2022).

10.2.6.2 Production of HCN

With the use of King's B broth and picric acid solution, an HCN production test was conducted. 4.4 g/l of glycine was added. On top of glycine-added King's B broth, the bacteria were injected. The conical flask's top was assembled using sterile filter paper sandwiched between glass and cotton and saturated with picric acid solution. For 48 h, the flasks were incubated at 28 °C. As a faint (+), moderate (++), or strong (+++) reaction, the filter paper's colour changed from yellow to light brown, brown, or reddish-brown, respectively. According to, colour changes signal the synthesis of HCN (Islam et al. 2022).

10.2.6.3 Production of Salicylic Acid

King's broth medium was used to produce salicylic acid. Bacterial isolates were cultured in a 250 ml conical flask with 50 ml of the King's broth medium at room temperature (28 ± 2 °C) for 48 h on a rotary shaker. Following the centrifugation of cells at 10,000 rpm for 10 min, 4 ml of the cell-free culture filtrate was adjusted to a pH of 2.0 by adding 1N HCl, and salicylic acid (SA) was extracted using an equal volume of chloroform. 5 ml of 2 M $FeCl_3$ and 4 ml of distilled water were added to the pooled chloroform extracts. Using a spectrophotometer, the purple iron-SA complex's absorbance, which had evolved in the aqueous phase, was measured at 527 nm. SA was dissolved in King's B broth to create a standard curve (Islam et al. 2022). According to Meyer et al. (1992), the concentration of SA in the culture filtrate was expressed as mg ml^{-1}.

10.2.6.4 Production of IAA

Each isolate's culture was inoculated in a loopful into 25 ml of sterile Nutrient broth that had been added L-Tryptophan (0.1 g/l) before being cultured for 24 h at 28 °C on a rotary shaker. After that, cultures were centrifuged for 15 min at 10,000 rpm. O-phosphoric acid was added in two drops to two millilitres of supernatant. The aliquot was combined with four millilitres of the Salkowski reagent. The resulting sample was incubated for 25 min at room temperature, and then, using a spectrophotometer, the absorbance was measured at 535 nm (Islam et al. 2022). The standard graph of IAA (Hi-Media), which was acquired in the range of 10–120 g/ml, was used to measure the concentration of IAA produced by cultures.

10.2.7 Seed Germination Assay Supplemented with Isolated Microorganisms

10.2.7.1 Seed Surface Sterilization

70% ethyl alcohol sterilised the seed's surface. Lentil seed was first dissolved for two minutes in 70% ethyl alcohol. Then, give the seed two or three washes in sterile distilled water. When sterile seed's excess surface water is soaked in sterilised Whatman filter paper, the seed is prepared for germination.

10.2.7.2 Seed Germination

The isolated microbes received a minimally modified version of the ESTA procedure as reported by Elliott and Fredrickson (1985) to be bio-assessed for their capacity to stimulate and/or inhibit seedling growth. Young cultures were spun at 10,000 revolutions per minute for 6 min and the supernatant was discarded in order to cultivate seeds for germination tests in nutrient broth. The pellet was then dissolved in sterile distilled water, followed by a 30-min soak of the surface-sterilized seed in the dissolved pellet. Extra water was taken out of the seed and put on water agar (0.6% agar) media. On control plates, seeds that had only been sterilised water treatments were put. After five days of incubation at room temperature, the plant's root and shoot lengths were measured, along with the germination percentage and vigour index (Islam et al. 2022).

10.2.8 In Vitro Antagonistic Activity of the Isolates Against Soil Borne Fungal Pathogen

The Dual Culture plate method was used to measure the antagonistic activity of newly isolated bacteria (Kerr 1999). A potato dextrose agar (PDA) medium was used to grow the soil-borne fungus pathogen. The pathogens were inserted as a 4 mm disc from the 3-day old culture at the centre of each petriplate, and the bacteria were streaked linearly leaving 1 cm from the margin on PDA media on both sides (Ahmad et al. 2008; Islam et al. 2022). For seven days, an experiment was conducted in a plant growth chamber at 27°. In solid media, fungal inhibition was assessed after 7 days.

10.2.9 Statistical Analysis

On a pooled dataset, a univariate description of the variables was performed based on the calculation of sample statistics like mean, SD, maximum and minimum values. The relationship between species abundance and diversity index was calculated. Utilising the software XLSTAT, the relative abundance indices were frequently calculated, PCA analysis was done in SPSSv-25.0.

10.3 Result and Discussion

10.3.1 Abiotic Characteristics

Among the study sites, there was a significant variation in the wetlands, ponds, and industrial belts due to ecological factors and climate variability. The average pH value falls in between the pH-6.30 to pH-6.90, making them mildly acidic, which is good for aquatic ecosystems. The temperature of the water sample is close to 30–35 °C. Site 2 and Site 3 samples are mostly polluted as a result of heavy local human interaction and industrial drainage, respectively, because the concentration of dissolved oxygen is so low.

10.3.2 Zooplankton and Bacterial Diversity in the Study Sites

During our research, we discovered two phyla (Arthropoda, Rotifer) and four orders (Cyclopoida, Calanoida, Podocopida, and Anomopoda). Discontinuous distribution was found in the study sites according to the zooplankton species composition distribution patterns. The calanoida population is absent from study sites SS1 and SS5, whereas SS3 has the highest diversity of calanoida in SS3. Among the six study sites the diversity index i.e., abundance (Ab) and relative abundance (RAb) was highest in rotifera population (Ab-0.31 and RAb-31.32) than podocopida (Ab-0.23 and RAb-23.39), cyclopoida (Ab-0.21 and RAb-21.13), anomopoda (Ab-0.17 and RAb-17.36) and calanoida (Ab-0.06 and RAb-6.79) (Fig. 10.1). Intriguingly, the dynamics of culturable microbiological parameters changed according to their abundance i.e., the culturable total bacterial CFU/ml was higher in SS5 (6.35 ± 0.52) than SS2 (6.27 ± 0.63), SS6 (5.34 ± 0.24), SS1 (5.34 ± 0.45), SS4 (4.34 ± 0.24) and

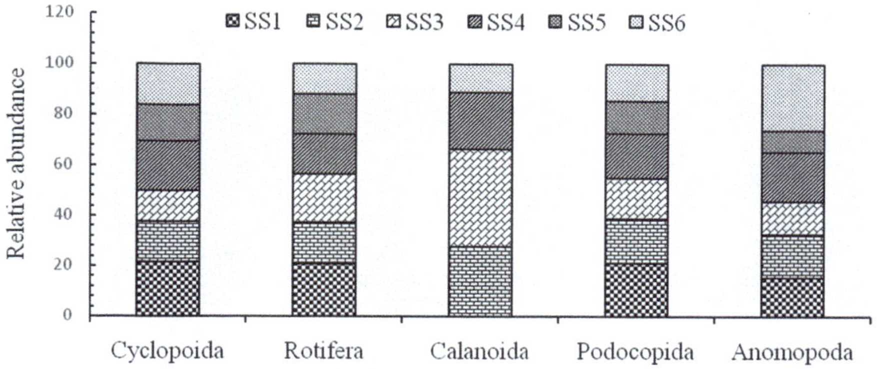

Fig. 10.1 Study site wise diversity index of zooplankton population

SS3 (4.31 ± 0.38). So, the present study reveals that the microbial load was higher in case of SS5 than SS2, SS6, SS1, SS4 and SS3.

10.3.3 *Effect of Edaphic Factors on Bacterial Growth*

Studies on how bacteria grow in relation to pH show that recently isolated bacteria thrive best at a pH-7. According to reports (Rosier et al. 2020; Narsing Rao et al. 2021), the majority of bacteria grow at neutral or slightly basic pH. In a comparison of recently isolated bacterial stain SSI2, SSI6, SSI8, and SSI9 at various pH concentrations, pH-9 was shown to be optimal. The advantageous pH range for these bacterial isolates are (pH3-pH9) for SSI2, and (pH7-pH9) for SSI6, SSI8, and SSI9. All of the newly discovered bacterial isolates SSI1, SSI2, SSI3, SSI4, SSI5, SSI6, SSI7, SSI8, and SSI9 thrive best at pH-7. The bacteria SSI1, SSI3, and SSI9 can grow between pHs of 5–7, but SSI4, SSI5, and SSI7 can only thrive at pH 7. For the majority of isolates, pH values between 1 and 9 can impede bacterial growth by changing the structural arrangement of the enzyme and protein (Narsing Rao et al. 2021).

The examination of bacterial growth revealed that some of the nine native isolates are halotolerant, or able to endure high salt concentrations, depending on salinity (Kushner 2020). These five bacterial isolates, designated SSI2, SSI4, SSI5, SSI6, and SSI8, are expanding synchronously at 9% salt concentration. However, native isolates SSI3 and SSI7 formed colonies with salt concentrations ranging from 3 to 9%, respectively. All nine native isolates produced minimal findings at salt concentrations of 1% and 3%. With the exception of SSI7 isolates, SSI1, SSI2, SSI4, SSI5, SSI6, SSI7, and SSI8 these isolates grow relatively greater salt concentration under 5% as well as in 7%. Some isolates, including SSI2, SSI4, SSI5, SSI6, and SSI8, had their best growth when exposed to 9% salt concentration, while the remainder died as a result of this treatment (Rodriguez-Valera 2020).

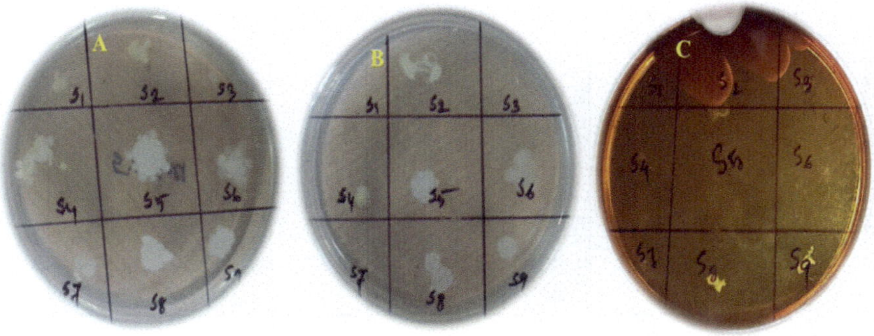

Fig. 10.2 Effect of temperature (**A**), salt concentration (**B**) and pH (**C**) on native isolates

The investigation of bacterial proliferation at particular temperatures yielded noteworthy findings. The very low and very high temperatures of 10 °C and 60 °C had no results. All nine of the native bacterial isolates grow more quickly between 25 and 37 °C. A few bacterial growths, SSI1, SSI2, SSI3, and SSI6, could be seen as the temperature comparatively rises in 50 °C only (Midya et al. 2020). From the study above, it can be inferred that SSI1 might exhibit its best growth at 50 °C, 7% concentration of salt, and pH 7. Under 50 °C and 9% salt concertation at pH 9, S2 grew at its maximum rate. SSI6 would be most active at pH 9 and 9% concentrations of salt at the ideal temperature of 50 C. SSI3 would exhibit optimal development at pH 7 and 50 °C temperature at 3% salt concentration. At pH 7 and 9% concentrations of salt, SSI4 and SSI5 are most active at 37 °C. In the case of SSI7, SSI8, and SSI9, the highest growth was observed at 37 °C, however the pH and concentration of salt were not the same (Donhauser et al. 2020) (Fig. 10.2).

10.3.4 Secondary Metabolites Activity of Native Bacterial Isolates

In the current study, the active synthesis of secondary metabolites by all of the plankton-associated bacterial isolates was examined in vitro condition. Native bacterial isolates can indirectly stimulate plant development by shielding plants against plant diseases in addition to directly promoting plant growth. Their capacity to produced phytohormones such IAA, HCN, salicylic acid (SA), and NH_3 activity was examined. IAA, HCN, and NH_3 were discovered to be present in every isolate. Out of these nine isolates, SSI5 (110.47 μg/ml), SSI1 (106.47 μg/ml), and SSI6 (103.80 μg/ml) generate the largest amounts of IAA plant hormones. Isolate SSI5 (13.02 μg/ml) produced the most salicylic acid, followed by SSI1 (11.77 μg/ml) and SSI6 (10.94 μg/ml). Isolates SSI1, SSI4, and SSI5 produced the most NH_3 and HCN, respectively (Table 10.1).

Table 10.1 Biochemical quantification observed from the different native bacteria isolates

Isolates	IAA (μg/ml)	SA (μg/ml)	NH_3	HCN
SSI1	106.47	11.77	+++	+++
SSI2	70.13	2.42	+	+
SSI3	81.47	3.96	++	+
SSI4	99.80	9.69	+++	++
SSI5	110.47	13.02	+++	+++
SSI6	103.80	10.94	++	++
SSI7	95.47	8.33	++	+
SSI8	79.13	3.23	+	++
SSI9	94.47	8.02	++	++

10.3.5 Enzymatic Efficacy of Native Isolates

All nine selected bacterial isolates were grown in amylase, pectinase, cellulose and lipase agar medium. The study of enzymatic activity shows that in a favourable condition only SSI1 and SSI5 could produce all possible enzymatic activity where as other isolates could not. SSI5 showed maximum activity in amylase, where all isolates showing amylase activity except SSI8. Maximum CMC (Carboxymethylated cellulose) activity showed in SSI2, where except SSI4, SSI7, and SSI9 all native isolates showing the CMC activity. SSI1 isolate showing maximum pectinase hydrolyzed activity, except SSI3, SSI4, SSI6 and SSI8 all isolates are showing the pectinase activity. Maximum lipase activity showed by the isolate SSI7, where all isolates are showed the lipase activity exceptSSI2 and SSI9 isolates (Fig. 10.3).

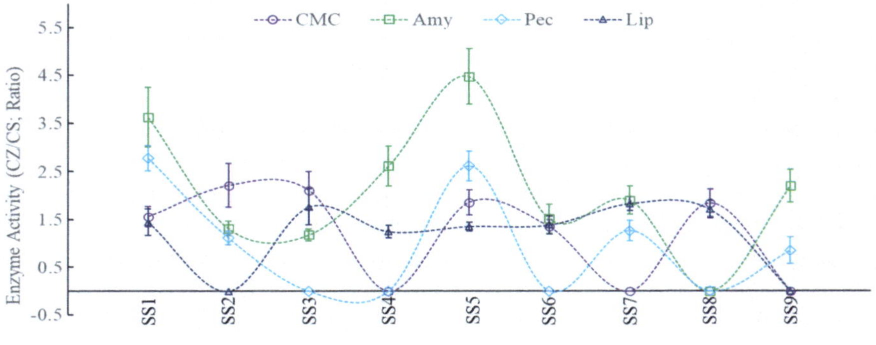

Fig. 10.3 Enzymatic activity (CZ/CS, ration) of different isolates

10.3.6 Plant Growth Promotion and Antifungal Activity of Different Potential Native Bacterial Isolates

After five days of seed bacterization, the germination rate, root length, and shoot length of seedlings were measured for the seeds of cauliflower, cabbage, ladies' finger, tomato, and mung bean. Analysis of the obtained results showed that in all treatments over control, groundnut demonstrated the best performance in enhancing plant growth activity related to seed germinability, root and shoot length, as well as vigour index. When compared to the other soil bacterial isolates, isolate SSI1 and SSI5 among the nine bacterial isolates was revealed to be the best plant growth stimulator. The data revealed that SSI1 and SSI5bacterized seeds supplied the highest percentage of seed germination, followed by the others bacterial isolates. The isolate SSI1 (1558.27; Mung bean) and SSI5 (1636.57; Ladies finger) had the highest computed vigour index values based on germination percentage, root length, and shoot length, respectively (Table 10.2).When compared to the other soil bacterial isolates, isolate SSI1 and SSI5 among the nine bacterial isolates was revealed to be the best plant growth stimulator. According to the findings, seeds bacterized with different isolates had showing the varied germination percentage against the different plant species, such as SSI1 and SSI5 (GI-93.33%) for cauliflower, SSI4, SSI5, and SSI9 (GI-96.67%) for cabbage, SSI4 and SSI7 (GI-96.67%) for tomato, SSI5 (GI-96.67%) for ladies finger and SSI1 (GI-96.67%) for mung bean (Table 10.2).When compared to the other soil bacterial isolates, isolate SSI1 and SSI5 among the nine bacterial isolates was revealed to be the best plant growth stimulator. Computed vigour index based on germination %, root length, and shoot length are differ for every experimented plant species, such as SSI1 has the highest vigour index against the plant species cauliflower (VI-478.80), and mung bean (VI-1558.27) where SSI5 showing the highest vigour index to the plant species cabbage (VI-828.43), tomato (VI-1046.27) and ladies finger (VI-1636.57) (Table 10.2). Yeole and Dube reported similar results, finding that bacterizing seeds with soil bacterial isolates increased the germination rate, root length, and shoot length of cotton, groundnut, chilli, and soybean. According to studies, field crops like wheat, sugar beets, and potatoes grew and produced more when their seeds were bacterized with fluorescent *Pseudomonas* and *Bacillus sp.* For determination of fungal growth inhibition, newly isolated bacterial species were applied to the lab isolated fungal species. It was found that the fungal growth was inhibited through isolation time depended manner. Visual observation of fungal growth inhibition also revealed that the isolates SSI5 imparts better antifungal activity in compare to other bacterial strains that is SSI1, SSI6, SSI4, SSI8, SSI9, whereas SSI2, SSI3 and SSI7 have no fungal inhibition property.

Table 10.2 Observed plant growth promotion (PGPR) activity of different potential native bacterial isolates against the selected seeds

	Isolates	Control	SSI1	SSI2	SSI3	SSI4	SSI5	SSI6	SSI7	SSI8	SSI9	Sem (±)	CD (<0.05)
Cauliflower	GI%	56.67	93.33	43.33	46.67	83.33	93.33	76.67	83.33	46.67	83.33	6.36	12.90
	Y	1.40	1.86	1.57	2.22	1.02	2.34	1.11	0.98	1.27	1.64	0.15	0.31
	Z	0.89	3.27	2.21	1.22	1.61	2.26	0.85	0.92	0.88	0.96	0.26	0.53
	VI*	129.77^{9}	478.80^{1}	163.80^{5}	160.53^{6}	219.17^{3}	429.33^{2}	150.27^{8}	158.33^{7}	100.33^{10}	216.67^{4}	40.64	82.42
Cabbage	GI%	66.67	86.67	43.33	43.33	96.67	96.67	86.67	93.33	46.67	96.67	7.37	14.95
	Y	2.22	4.22	2.53	4.36	3.75	4.65	4.23	3.68	2.67	2.32	0.30	0.60
	Z	1.88	4.39	2.28	4.32	3.46	3.92	3.88	4.26	3.54	3.64	0.27	0.54
	VI*	273.33^{9}	746.20^{2}	208.43^{10}	376.13^{7}	696.97^{5}	828.43^{1}	702.87^{4}	741.07^{3}	289.80^{8}	576.13^{6}	73.71	149.49
Tomato	GI%	56.67	83.33	53.33	43.33	96.67	93.33	83.33	96.67	33.33	93.33	7.65	15.51
	Y	4.22	5.89	4.82	5.67	5.92	7.65	5.12	6.59	5.64	6.85	0.32	0.65
	Z	2.20	2.89	1.95	4.69	3.84	3.56	3.33	4.23	2.46	3.84	0.29	0.58
	VI*	363.80^{8}	731.67^{5}	361.07^{9}	448.93^{7}	943.47^{4}	1046.27^{1}	704.17^{6}	1045.93^{2}	270.00^{10}	997.73^{3}	97.96	198.67
Ladies finger	GI%	53.33	86.67	26.67	33.33	73.33	96.67	83.33	86.67	26.67	86.67	8.75	17.75
	Y	3.22	10.23	6.06	4.56	8.12	8.33	5.61	3.84	3.95	5.22	0.73	1.47
	Z	2.45	8.26	2.68	3.12	6.00	6.80	4.26	3.26	2.24	4.62	0.74	1.50
	VI*	302.40^{7}	1602.47^{2}	233.07^{9}	256.00^{8}	1035.47^{3}	1636.57^{1}	822.50^{5}	615.33^{6}	165.07^{10}	852.80^{4}	172.63	350.09
Mung bean	GI%	56.67	96.67	33.33	46.67	93.33	86.67	80.00	93.33	30.00	93.33	8.42	17.07
	Y	4.82	8.64	5.63	9.67	2.88	2.28	5.14	4.66	6.12	7.36	0.74	1.49
	Z	3.22	7.48	5.97	4.59	3.48	4.68	6.90	8.41	6.94	5.68	0.55	1.11
	VI*	455.60^{8}	1558.27^{1}	386.67^{10}	665.47^{5}	593.60^{7}	603.20^{6}	963.20^{4}	1219.86^{2}	391.80^{9}	1217.06^{3}	129.40	262.44

Germination% (GI%), Shoot Length (Y), Root Length (Z), Vigour Index (VI), *Rank.

10.4 Discussion

Plant growth promoting rhizosphere bacteria have to eliminate or inhibit or reduces the mycelial growth of soil borne fungal pathogens. Different bacterial species such as *Trichoderma* sp., *Ulocladium* sp., *Psedumonas*sp., *Bacillus* sp., *Azotobacter* sp., *Burkholderia* sp., etc., have the potentiality towards the plant growth and antifungal activity as well as (Elad et al. 2000; Kohl and Molhoek 2001). In present investigation, zooplankton surface attached bacteria have the potent activity to reduces the growth of soil borne fungal pathogen, in invitro condition. Different enzymatic activity (such as pectinase, cellulase and amylase) are showing the positive correlation to with the fungal inhibition i.e., fungal growth inhibition highly reduces with increasing the enzyme activity. PCA analysis (2-D plot) was created according with the first 2-components (PCA_1 and PCA_2), the eigen value and percentage of variance coverage for PCA_1 is 5.172 and 57.462, and for PCA_2 is 1.52 and 16.866, respectively (Fig. 10.4). From rotated component matrix analysis, it was concluded that the all-enzymatic assay and plant secondary metabolites are positively corelated with the fungal inhibition properties of the native isolates. Biochemical parameters of native potential biocontrol isolates showed that fungal inhibition or antagonistic potentiality (0.690) was closely related to the amylase (0.932), pectinase (0.853), HCN (0.881) and cellulase (0.552) activity. Whereas, the others secondary metabolites i.e., SA (0.906), IAA (0.893) and NH_3 (0.853) are closely related and positively corelated with the fungal inhibition (Fig. 10.4). Various study also supported that, increasing rate of enzymatic activity reduces the soil borne fungal growth in vivo and in vitro condition (Orozco-Mosqueda et al. 2023; Urgiles-Gomez et al. 2021).

Climate changes hampered the soil profile and diminish the potential PGPRs microbial population, because of this the crop growth yield loss increases simultaneously. To enhance the crop yield required potent active PGPRs bacterial strain (Cao et al. 2023; Ajijah et al. 2023). So, here we implicated the plankton associate microbes to evoke the plant growth in vitro condition, various biochemical assay showed the positive correlation with plant growth. PCA analysis (2-D plot) was created according with the first 2-components (PCA_1 and PCA_2), the eigen value and percentage of variance coverage for PCA_1 is 8.232 and 63.327, and for PCA_2 is 1.785 and 13.730, respectively (Fig. 10.5). From rotated component matrix analysis, it was concluded that the all-enzymatic assay and plant secondary metabolites are positively corelated with the plant growth promoting activity of the native isolates and evoke the plant growth. Biochemical parameters and secondary metabolites activity of the native isolates showed the positive correlation with the vigour index, such as vigour index of cabbage (0.914), tomato (0.731) and mung bean (0.664) plant are closely related to the IAA (0.951), SA (0.899), NH_3 (0.899) and lipase (0.541) activity. Whereas, the others enzyme and secondary metabolites i.e., HCN (0.808), amylase (0.863) and pectinase (0.814) are closely related to the vigour index properties of cauliflower (0.867) and ladies finger (0.981) plant species. Various previous study also suggests that plant hormone like IAA, SA and other metabolites such as HCN and NH_3 production activity directly effect on plant growth (El-Saadony

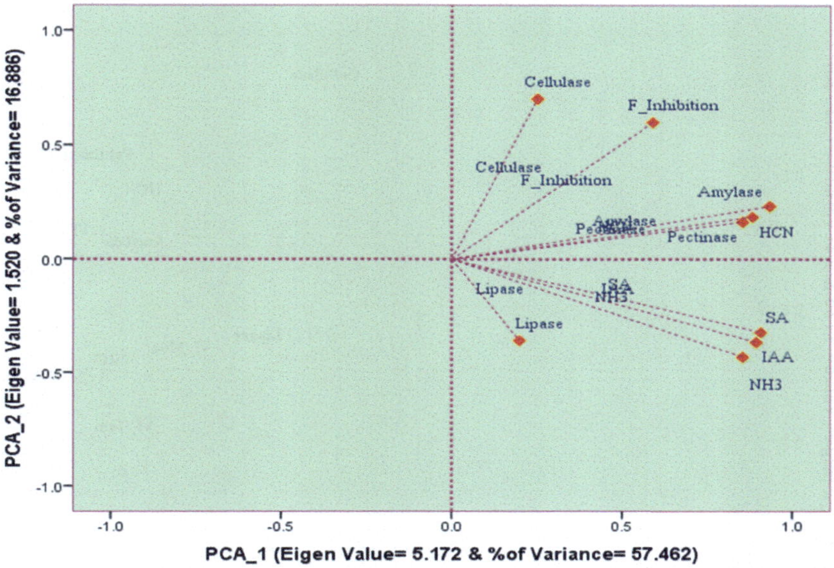

Fig. 10.4 PCA analysis of among all the biochemical parameters with the fungal inhibition properties

et al. 2023; Fadiji et al. 2022). An increasing rate of enzymatic activity evoke the plant growth, here the biochemical activity like IAA, SA, HCN and NH_3 production are showing mostly in case of isolates SSI1 and SSI5. The isolates SSI1 and SSI5 actively promote the plant growth and also showed the maximum vigour index value on treated plant species in terms of their root and shoot length properties.

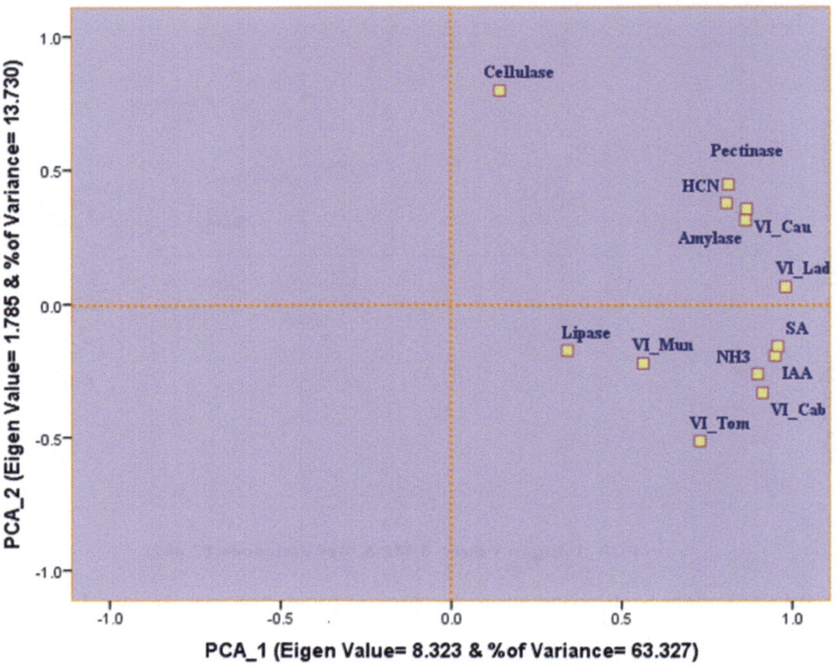

Fig. 10.5 PCA analysis of among all the biochemical parameters with the vigour index properties of the selected plant species

10.5 Conclusion

It is clear from the study above that among the nine bacterial strains, SSI1 and SSI5 are two potent isolates with substantial plant growth promotion (PGPR) activities. SSI1 and SSI5 isolates were actively working against cauliflower, mung bean and cabbage, tomato, ladiesfinger, respectively, in comparison to the control plant seed. IAA, SA, HCN, and NH3 are only a few examples of the many plants' growth-promoting substances that each isolate actively creates. SSI5 potent isolates among the nine bacterial strains that exhibit antifungal activity, according to the abovementioned study. The bacterial strain SSI1 and SSI5 can be regarded as a potent one for its industrial usage due to its wide range of temperature, pH, salt concentration, antifungal activity and plant growth promotion phenomena.

Acknowledgements The authors acknowledge Raja N. L. Khan Women's College (Autonomous) for supporting all sorts of library and laboratory facilities.

Author Contribution All authors are equally contributed to prepared the manuscript. All authors read and approved the final manuscript.

Funding No funding was received to assist with the preparation of this manuscript.

Ethical Approval Not applicable.

Conflict of Interest The authors declare that they have no conflict of interest.

References

Ahmad F, Ahmad I, Khan MS (2008) Screening of free-living rhizospheric bacteria for their multiple plant growth promoting activities. Mol Res 163:173–181

Ajijah N, Fiodor A, Pandey AK, Rana A, Pranaw K (2023) Plant growth-promoting bacteria (PGPB) with biofilm-forming ability: a multifaceted agent for sustainable agriculture. Diversity 15(1):112

Cao M, Narayanan M, Shi X, Chen X, Li Z, Ma Y (2023) Optimistic contributions of plant growth-promoting bacteria for sustainable agriculture and climate stress alleviation. Environ Res 217:114924

Castiglione AM, Mannino G, Contartese V, Bertea CM, Ertani A (2021) Microbial biostimulants as response to modern agriculture needs: composition, role and application of these innovative products. Plants 10(8):1533

Donhauser J, Niklaus PA, Rousk J, Larose C, Frey B (2020) Temperatures beyond the community optimum promote the dominance of heat-adapted, fast growing and stress resistant bacteria in alpine soils. Soil Biol Biochem 148:107873

Elad Y (2000) Biological control of foliar pathogens by means of Trichoderma harzianum and potential modes of action. Crop Prot 19:709–714

Elliott FL, Fredrickson KJ (1985) Colonization of winter wheat roots by inhibitory rhizobacteria. Soil Sci Soc Am J

El-Saadony MT, Saad AM, Soliman SM, Salem HM, Ahmed AI, Mahmood M, AbuQamar SF (2023) Plant growth-promoting microorganisms as biocontrol agents of plant diseases: mechanisms, challenges and future perspectives. In: Plant root interaction with associated microbiomes to improve plant resiliency and crop biodiversity, volume II, vol 16648714, p 193

Fadiji AE, Santoyo G, Yadav AN, Babalola OO (2022) Efforts towards overcoming drought stress in crops: revisiting the mechanisms employed by plant growth-promoting bacteria. Front Microbiol 13

Flemming HC, Baveye P, Neu TR, Stoodley P, Szewzyk U, Wingender J, Wuertz S (2021) Who put the film in biofilm? The migration of a term from wastewater engineering to medicine and beyond. Npj Biofilms Microbiomes 7(1):10

Flood PR (1991) Architecture of andand water circulation and flow rate in the house of the planktonic tunicate Oikopleuralabradoriensis. Mar Biol 111:95–111

Guillo N (1958) Élaboration par Lactobacillus acidophilus d'un produitactifcontre Candida albicans. Ann Inst Pasteur 95:194–207

Ibrahim SA, Ayivi RD, Zimmerman T, Siddiqui SA, Altemimi AB, Fidan H, Bakhshayesh RV (2021) Lactic acid bacteria as antimicrobial agents: food safety and microbial food spoilage prevention. Foods 10(12):3131

Islam SS, Sen K, Dutta S, Midya S (2022) Performance of rice-lentil cropping under different tillage influencing soil suppressiveness: a short term approach. Soil health and environmental sustainability: application of geospatial technology. Springer International Publishing, Cham, pp 551–577

Kerr RJ (1999) Bacterial inhibition of fungal growth and pathogenicity. Microb Ecol Health Dis 11(3):129–142

Khan NF, Rasool A, Mansoor S, Saleem S, Baba TR, Haq SM, Popesc SM (2022) Potential applications of Rhizobacteria as eco-friendly biological control, plant growth promotion and soil metal bioremediation. Sustain Crop Prod Recent Adv 104–170

Kohl J, Molhoek WML (2001) Effect of water potential on conidial germination and antagonism of Ulocladiumatrum against Botrytis cinerea. Phytopathology 91:485–491

Kuddus M, Saima, et al (2013) Isolation of novel chitinolytic bacteria and protection optimization of extracellular chitinase. J Genet Eng Biotechnol 11(1):39–46

Kushner, D. J. (2020). Growth and nutrition of halophilic bacteria. In The biology of halophilic bacteria (pp. 87–103). CRC Press.

Marth EH, Hussong RV (1963) Effect of skim milks cultured with different strains of Leuconostoc citrovorum on growth of some bacteia and yeasts 46:1033–1037

Meyer JM, Ajelvandre P, Georges C (1992) Iron metabolism in pseudomonas: salicylic acid, a siderophore of pseudomonas fluorescens CHA0. Bio Factors 4:23–27

Midya S, Ganguly RK, Islam SS, Chakraborty SK (2019) Attachment and antimicrobial susceptibility of bacterial associates of zooplanktonic copepod: lesson for environmental safety. Ecotoxicol Environ Saf 2019(180):749–755

Midya S, Islam SS (2022) Assessment of chitinolytic bacteria isolated from zooplankton of freshwater ecosystem. In: Patra BC, Shit PK, Bhunia GS, Bhattacharya M (eds) River health and ecology in South Asia. Springer, Cham, pp 81–102

Midya S, Bhattacharya S, Islam SS, Ganguly RK, Chakraborty SK (2018a) Observation on freshwater zooplankton and hydrophytes composition in different wetlands of Paschim Medinipur, West Bengal (India). Int J Zool Stud 3(2):05–09

Midya S, Islam SS, Jana A (2020) Enumeration, isolation and identification of crustacean surface attachment bacteria. Int Res J Basic Appl Sci 2020(5):23–32

Midya S, Islam SS, Paul R, Chakraborty SS (2018b) Ecological gradients determining the zooplankton-macrophyte interaction and diversity in brackish water wetlands of midnapore (East), West Bengal, India. Int J Adv Sci Res Manage 3:118–125

Moller CODA, Freire L, Rosim RE, Margalho LP, Balthazar CF, Franco LT, Oliveira CAFD (2021) Effect of lactic acid bacteria strains on the growth and aflatoxin production potential of Aspergillus parasiticus, and their ability to bind aflatoxin B1, ochratoxin A, and zearalenone in vitro. Front Microbiol 12:655386

Narsing Rao MP, Dong ZY, Luo ZH, Li MM, Liu BB, Guo SX, Li WJ (2021) Physicochemical and microbial diversity analyses of indian hot springs. Front Microbiol 12:627200

Orozco-Mosqueda MDC, Kumar A, Fadiji AE, Babalola OO, Puopolo G, Santoyo G (2023) Agroecological management of the grey mould fungus Botrytis cinerea by plant growth-promoting bacteria. Plants 12(3):637

Rehmaan IU, Jan B, Khan NF, Islam T, Rehman S, Sheergojri IA, Wani AH (2022) Nitrogen biofertilizers: role in sustainable agriculture. In: Advances in plant nitrogen metabolism. CRC Press, pp 170–184

Rodriguez-Valera, F. (2020). Introduction to saline environments. In The biology of halophilic bacteria (pp. 1–23). CRC Press.

Rosier BT, Moya-Gonzalvez EM, Corell-Escuin P, Mira A (2020) Isolation and characterization of nitrate-reducing bacteria as potential probiotics for oral and systemic health. Front Microbiol 11:555465

Schuett, C., Doepke, H. Endobiotic bacteria and their pathogenic potential in cnidarian tentacles. Helgol Mar Res. 2009.

Selim G (2001) Ectosymbiotic bacteria on ciliated cells of a rotifer. Tissue Cell 33:258–261

Sudharsan MS, Rajagopal K, Banu N (2023) An insight into fungi in forest ecosystems. Plant mycobiome: diversity, interactions and uses. Springer International Publishing, Cham, pp 291–318

Urgiles-Gomez N, Avila-Salem ME, Lojan P, Encalada M, Hurtado L, Araujo S, Cornejo P (2021) Plant growth-promoting microorganisms in coffee production: from isolation to field application. Agronomy 11(8):1531

Chapter 11
Nematode Community Profiling as Indicators of Soil Quality: Current Scenario and Future Prospective

Himani Sharma and Ashok Kumar Chaubey

Keywords *Caenorhabditis elegans* · Enrichment index · Fertilizer use efficiency · Maturity index · Nematodes · Soil food web

11.1 Introduction

The soil health is the base of fulfilling the growing population's demands for basic necessities within a restricted range of available landmass. Furthermore, it becomes challenging with varying climatic condition which requires multifaceted ways in order to handle the situation (Lal 2011; Assefa et al. 2020; Fagodiya et al. 2016; Jankowski et al. 2018; Millar et al. 2010; Pimentel and Giampietro 1994; Robertson et al. 2012). The soil heath found to be dependent upon various factors such as microorganism, physical and chemical characteristics, nutrient availability and water holding capacities (Lal 2011; Anon 2016; Doran and Zeiss 2000; Sánchez-Moreno 2018). The present activities i.e., anthropogenic actions lead to the soil degradation pathways which in turn deteriorates soil health (Háněl 2004). There are various ways for the evaluation of the soil degradation by exploiting either bioindicators or chemicals, which also focuses on the microorganism associated with the soil (Edwards 2002; Nahmani et al. 2007; Özdemir et al. 2010; Skubała and Zaleski 2011). Previous researches revealed the important relationship between the soil fauna and their responses to different disturbances which highlights the present conditions of the soil and its future outcomes also (Korthals et al. 2000). Soil comprises of various microorganisms which are directly or indirectly depend on the soil for their

H. Sharma (✉) · A. K. Chaubey
Nematology Laboratory, Department of Zoology, Chaudhary Charan Singh University, Meerut 250004, India
e-mail: sharmahimani1995hariom@gmail.com

P. P. Adhikary et al. (eds.), *Soil, Water Pollution and Mitigation Strategies*, Environmental Science and Engineering, https://doi.org/10.1007/978-3-031-63296-9_8

fundamental processes, however, nematodes found to be interesting organism as slight disturbances in the soil affect them (Sohlenius 1980; Freckman and Baldwin 1990). The researchers exploited this characteristic of nematodes to assess the degradation factors, chemical concentration from the 1990s (Korthals et al. 1996a; Nagy et al. 2004; Li et al. 2011). The interdependency of the soil biota found to be based on the carbon and other carbohydrate molecules. However, this dependency was interrupted by various chemicals, disturbances, climatic conditions, indiscriminate pesticides and many more. These disturbances weren't easy to assess as soil is not a homogeneous compound. The nemic fauna inhabited in the soil known to occupied an important position the food web. Furthermore, their highly appreciated characteristic of omnipresent except Antarctica helps in the *in-situ* evaluation of changes affecting the soil conditions. There are different categories of nematodes based on the trophic groups and their association with higher organism. Among all the categories, free living nematodes found to be most suitable indicators for different ecosystems due to their quick responses to slight disturbances (Balsamo et al. 2012; Semprucci et al. 2015). Several diversity and community indices were implemented for the assessment of Šalamún et al. (2014), Caixeta et al. (2016), Park et al. (2011), Rodríguez Martín et al. (2014), Gutiérrez et al. (2016). Likewise, they are used for the estimation of changes occurred in the agroecosystem (Sánchez-Moreno et al. 2018; Zhong et al. 2017).

11.2 Nematode's Importance in Soil Health

The planet Earth is divided into different geographical realms and each realm is a hotspot for diverse endemic flora and fauna. Among this diverse flora and fauna, nematodes are one of the most fascinating animals on earth after insects, with many free living and parasitic forms. They have been evolved in such a way that all the possible habitats of the ecosystem *viz.*, marine, soil and aquatic, are populated with these nematodes' species. Diversity can also be seen in their shape and size as they occur in variable shapes such as melon shape (*Meloidogyne*), elongated (roundworm) etc. as well as their size ranges from microscopic plant parasitic nematode (*Paratylenchus* 0.18–0.6 mm) to several animal nematodes (*Placentonema gigantissima* 8.4 m long) meters. The general body structure of nematodes is elongated, cylindrical and tapered at both ends as suggested by their name 'nema' which means thread like. The basic body design consists of an inner alimentary canal which is surrounded by an outer tube of body wall which is commonly referred as 'tube within a tube' structure. The alimentary canal basically comprises of gonads and digestive tract. Body wall consists of variety of muscles *viz.*, ventral and dorsal longitudinal muscles which regulates nematode locomotion as well as these muscles are activated by the dorsal and ventral nerves. Unlined, pressurized, fluid filled cavity which is present between the tubes is known as 'pseudocoelom' formed from the blastula surrounding the gut cavity. Its primary function is to provide the turgor pressure for the entire body and subsequently aids the transfer of oxygen, nutrients and metabolic products. Exoskeleton of nematodes, also known as cuticle, is made up of collagens, glycoproteins, insoluble proteins and lipids. It's a noteworthy characteristic in case of nematodes that all the conceivable habitats are occupied by this amazing group (Sánchez-Moreno 2018; Glavatska et al. 2017; Ferris et al. 2001; Bongers and Ferris 1999; Van der Hoogen et al. 2019). Based on their feeding habits, nematodes have been characterized into five different trophic groups. These trophic groups are bacteriovores, fungivores, herbivores, predators and omnivores. The bacteriovores feed on the bacteria available in the soil. Fungivores takes fungi as their food source. Herbivores are also known as plant parasitic nematodes found associated with the roots of economically important plants. Predatory nematodes take other small nematodes as their food source. Omnivores feed on plants and other microorganisms (Van der Hoogen et al. 2019). The nematodes can be categorized as colonizer-persister (cp) on the basis of similarity in the environment circumstances and feeding habits along with commonality in anatomy and physiology. Nematodes which are categorized in the same cp class showed similarity in reactions to various parameters especially disturbances (Bongers 1999; Bongers and Ferris 1999). There colonizer-persister scale was given to the nematodes based on population characteristics. The r and k strategies adopted by the nematodes were exploited to assign the specific taxa to a cp1-cp5 scale. The cp1 taxa comprise mainly bacteriovores as they possess small size, short life cycle, and high fecundity. The cp2 taxa comprises of bacterial and fungal feeders which showed high tolerance level with a high generation time. On the contrary, cp3 taxa consist of bacteriovores which are sensitive to changing conditions. The cp4 taxa consist of omnivores and found to be very sensitive to adverse environment. The cp5 taxa found to be most sensitive even to slight changes in the environment and consist of omnivores and predators (Bongers 1999; Bongers and Ferris 1999; Martinez 1992). The colonizer are always ranges from 1 to 3 and present in the disturbed, changing environment as they opportunistic, however, persisters occupy the undisturbed areas (Bongers 1999; Bongers and Ferris 1999). To maintain soil conditions, it's very important that all the nematode groups ranging from cp1-cp5 present in the soils (Doran and Zeiss 2000; Gebremikael et al. 2015; Sánchez-Moreno 2018; Van der Hoogen et al. 2019).

11.2.1 Challenges in Maintaining Stable Soil Ecosystem

To attain a sustainable soil ecosystem is very difficult due to the presence of challenging situations. These are as follows:

(1) Soil health is not an isolated component, it's a multidisciplinary subject which requires in depth knowledge of the system and its functions, related microorganism, interconnection, etc. (Bulluck et al. 2002; Gibson et al. 2014; Schorpp and Schrader 2016; Toosi et al. 2017).

(2) The situation becomes more challenging when both plant parasitic nematodes and bacteriovores and other valuable fauna exist together in the same system. It's become a tricky situation where we have to restrict the growth of harmful fauna and increase the beneficial nematode population in relation to services provided from the system (Ferguson 2015; Melakeberhan 2006, 2010; Melakeberhan and Avendaño 2008).

(3) The situation becomes more complicated when we don't have any standard measures which describes the healthy soil differs from a disturbed soil (Ferguson 2015; Riga et al. 2003; Glavatska et al. 2017).

These challenges need to be overcome in order to maintain a healthy soil system which supports life and provide services. This can be achieved from the alignment of all the related components so that desired services shouldn't be disturbed. The bio-component plays an essential role from the soil perspective as it regulates all the nutrient cycle in the soil food web analysis (Doran and Zeiss 2000; Sánchez-Moreno 2018). The other step includes the microorganism present in the soil food web analysis, among all some of them work as bioindicators of soil status (Kokalis-Burelle et al. 2002; Jian et al. 2020; Wander et al. 2019; Fine et al. 2017; Kihara et al. 2019; Stewart et al. 2020). The final step will be more focused on the results/solutions that can be implemented on agricultural systems irrespective of the sustainable actions. For developing sustainable measures, fertilizer efficiency model based on pathogenic and beneficial community must be exploited (Melakeberhan 2006, 2010; Melakeberhan and Avendaño 2008). The integrated soil food web analysis and fertilizer efficiency model will be helpful in maintaining the soil health with the sustainable measures.

11.3 Nematode as Bioindicators

There are various characteristics possessed by the nematodes which make them a suitable indicator. The exoskeleton of nematodes, also known as cuticle, comprises of many proteins, lipids and other molecule which acts as signaling molecules and reacts with a variety of reactants such as pollution reactants etc. and their reaction with these also helpful in understanding the ecosystem health (Saly and Ragala 1984; Wasilewska 1979, 1989). Furthermore, there are nematode populations which have the ability to deal with the adverse conditions by undergoing some stages helping them in survival such as an hydrobiosis, cryptobiosis and many more. Anhydrobiosis is the stage at which nematodes lose all the water in the extreme dry conditions and in the presence of water, again, regain their shape and metabolism also (Cooper and Van Gundy 1971; Crowe and Clegg 1973; Ellenby 1969). Besides this, there are nematode populations also which doesn't possess these stages and they are extremely sensitive to slight differences in the environment. The nematode group which showed highest sensitivity is Dorylaimidae (Bongers 1999). In addition to this, there are some populations which exploit the HSP (heat shock proteins) for facing the adversities (Hashmi et al. 1997). Heat shock proteins are present in bacteria and humans for combating the stress environment. It's been reported in the studies that stress condition triggers the production of HSP in nematodes also and the conditions are heat and various toxins molecules (Kammenga et al. 1998). The stress conditions such as extreme heat or high concentrations of metal, ions or other molecule lead to the production of HSP in the nematode population. It could be beneficial for the environmentalist to assess the soil by evaluating the levels of HSP in the nematodes (Guven et al. 1994, 1999; Kammenga et al. 2000). The history of nematodes using as indicator is not new; but dated back in 1970. The literature also describes the potential use of nematodes for determining the adverse effects of approximately 400 molecules (Samoiloff 1987). The marine nematology has a special place in this area as free-living nematodes are very popular due to their high abundance. There are some limitations with free living nematode regarding their size as it becomes very difficult for the taxonomist to identify these on morphological basis. Still, with great expertise, it is possible to combat this limitation. Despite of this, they are important for the bioindicator purposes as they are easy to collect and their presence in every layer of the marine ecosystem is highly appreciated (Ridall and Ingels 2021). It's been found that these nematodes could be as abundant as 5 million/m^2 as a result of which approximately more than half percent of biomass found to be dependent on these nematodes (Soetaert et al. 2009; Semprucci and Balsamo 2014). In the mid nineteenth century, the nematodes along with other invertebrates were used for the assessment of marine ecosystem (Raffaelli and Mason 1981). Previous studies also suggested the use of copepods for the assessment of soil health. The copepods showed more sensitivity towards the changing conditions or disturbances than free living nematodes. But there were some limitations with the copepods. The copepod generation found to be increasing with the increased sand particle diameter which caused a statistic constraint in estimating the damage done by the pollutants (Coull et al. 1981). In addition to this, copepod found to be consumer of the nemic population leading to a disturbed ratio used for environmental health monitoring (Lehman and Reid 1993). Later, in the 1980, the researchers started focusing on the nematode profiling for the evaluation of changing environment (Neher 2001a, b). One of the major contributions was the development of maturity index (MI) for the assessment of nematode community (Bongers 1990; Yeates 1970, 1984). The MI was first developed for the soil nematodes, but it was applied to marine and freshwater ecosystem nematodes also (Bongers et al. 1991). The MI exploited the colonizer-persister scale of the nematode to determine the level of disturbances present in the soils (Bongers and Ferris 1999). The values of MI played major role in revealing the soil health. It's been observed that the value of soils presents in less disturbance environmental conditions found to be 64 (Fig. 11.1). On the contrary, the soils facing dynamic change either naturally or man-made showed a value of ≤ 2. The man-made practices in agriculture also reduces the MI value as fertilizers increment in the soils leads to enhanced activity of the bacteriovores (Ferris et al. 1996; Yeates and Bird 1994). The MI is only calculated for the free-living nematodes which are bacteriovores, omnivores, predators and fungivores. However, there are herbivores nematodes which derived nutrients from the roots of the plants, trees etc. The population of herbivores found to be dependent upon the plant on which they feed. Though, the generation time of the herbivores is higher in comparison to bacteriovores, but their recovery and generation time both becomes faster in presence of their host (Ettema 1998). For the herbivores, plant parasitic index (PPI) is calculated which showed a negative correlation with MI (Ettema 1998).

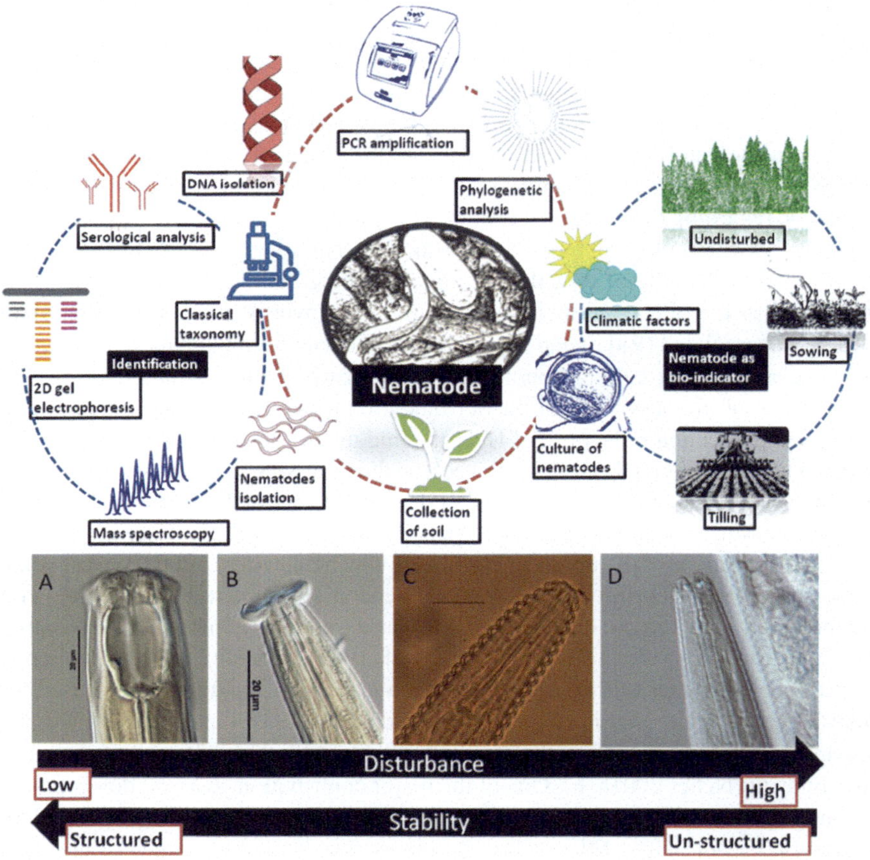

Fig. 11.1 Nematode as bio-indicators of the soil quality. The figure depicting the process of nematode collection, identification techniques, molecular amplification, phylogenetic analysis, culturing, indicators of climatic factors and disturbances like tilling, sowing etc. A = *Aporcelaimellus* (Omnivores, cp 5), B = *Discolaimus* (Predators, cp 4), C = *Hemicriconemoides* (Herbivores, cp 3), D = *Zeldia* (Bacteriovores, cp 2). The disturbance increasing from high cp value to low. The soil quality decreasing from high cp value to low

11.3.1 Indices as Enrichment Indicator of Soil

The indiscriminate use of fertilizers in the agricultural field leads to the enrichment of the soils with nitrogen and phosphorus. These minerals and nutrients will support the growth of colonizers as their number increases with the enrichment of the soils. The addition of fertilizers or organic manures in the beginning favours the growth of cp1 and then cp2. To find the tilling system effects or the timing of tilling in the agricultural soils, MI values have been utilized (Yeates and Bird 1994). If we are concerning about the levels of disturbances occurred during the tilling and other such processes, then, MI and PPI values were compared. The trend revealed that high level of tilling showed high PPI values but decreased MI values (Freckman and Ettema 1993). The MI value found to be affected by the various microorganisms' activity. The soil health was determined by various parameters such as maturity index (MI), plant parasitic index (PPI), enrichment index (EI), structural index (SI) and variety of diversity indices. The agricultural land soil's quality was checked by maturity index and plant parasitic index ratio. The ratio of PPI and MI describes the condition of soil health. If the ratio of PPI and MI found to be high, then, soil comprises of excess of nitrogen, phosphorus and ammonium which favours plant parasitic nematodes. In short, the soil has less nutrients which supports plant or any other crop growth, however, it consists of enough supplements for the harmful nematode growth (Freckman and Ettema 1993).

Previous studies and researches describe the fidelity of MI and various diversity indices for the evaluation of soil health for a specified place and for the comparison of soil health status in different areas (Neher and Campbell 1996; Neher et al. 1995). The diversity indices were exploited for the analyzing the soil status for supporting the micro and macro fauna. Though, they are useful for soil analysis, still, they possess some limitations as they don't show sensitivity towards taxon (Neher 2001a, b). These indices were based on the number of taxa present in a specific region. For example, regions with more alien species diversity index in comparison to region with more inhabited species showed similar diversity indices. On the contrary, Jaccard and Sørensen indices were used for differentiating these regions where other diversity indices gave similar values (Jongman et a1. 1995; Topham et al. 1991). It's very wrong if we understood diversity is just counting number of taxa. The diversity indices like Shannon diversity index and Simpson diversity index always assess the abundance of a particular taxon, and the value of taxon in relation to others. For diversity, it's very important to consider, the richness, evenness along with relative abundance. Furthermore, the Camargo index mostly exploited for understanding the disturbances in marine ecosystems. The communities affected by the disturbances were recorded in this index more accurately than Shannon diversity index (Camargo 1992). The MacArthur indices used for estimating the different number of taxon of a specified region (Camargo 1992). The Shannon index has been utilized for finding the endemic taxon or rarest taxa of given region. The statistical analysis of these indices is not easy but applying Hill's family numbers makes it possible for interpreting the data. The statistics always demands for a concluding statement whether the studied regions have similar taxa or some regions have higher number or this region has lower diversity than this. To make this possible, Hill's diversity numbers were used. The N_0 represents the total number of taxa, N_1 represents the abundance of a particular taxa and N_2 represent the highest abundance of a particular taxa (Ludwig and Reynolds 1988).

The diversity indices also help in determining the abundance of taxa in a particular soil type. For example, the agricultural soils support the large number microorganism, however, when diversity indices were calculated for these soils and they found to be higher, then, it could be interpreted on the presence or absence of specific feeding groups. As it's already mentioned, there are five feeding groups of nematodes, the abundance of fungal feeders, predators and omnivores nematodes showed positive correlation with diversity index. On the other hand, the bacterial feeders and PPN showed negative correlation with diversity indices (Wasilewska 1979). Diversity indices totally dependent upon the taxa recovered from the soils. It's very important for the taxonomist to precisely identify the taxa for correct placement in taxonomy and feeding groups. But there are some challenges in the feeding group categorization faced by the taxonomist (Yeates et al. 1993). There are some groups such as *Aphelenchus* which is considered as plant parasitic nematode; however, this group also feeds on fungus also (Sohlenius et al. 1977). Similar condition observed with predatory nematodes as there are some predators which also feed on bacteria (Russell 1986). This is problematic in the feeding group classification. There are some cases in which nematodes feed on fungus, bacteria or any other but there juvenile or reproductive stages showed differential feeding preferences. Adding on to other drawbacks, the isolation of soil nematodes were done by adopted the methods of Cobb 1918 and Baermann's funnel method (Goodey 1957; Southey 1986). The originally described Cobb sieving method has some limitations and to overcome this modified sieving method is practiced now-a-days. The modified Cobb sieving method comprises of various sieves of different pore/aperture sizes which were placed in increasing order to collect the maximum amount and variety of nematodes. On the contrary, the nematode recovered from one type of sieve found deficient in some feeding groups (McSorley and Walter 1991; Neher et al. 1995).

11.4 Role of Nematode in Pollution Assessment

The MI showed high sensitivity towards pollutant concentration. If the concentration of pollutants found higher in the soil, then, the nematodes of cp1-2 showed dominance in the community. The dominant taxa i.e., cp1-2 leads to decrement of MI values as cp4-5 nematode growth favored in stable environment (Odum 1985) (Fig. 11.1). The cp4-5 nematode taxa known to be regulators of controlling the plant parasitic nematodes action and several other soil related diseases. The cp4-5 scale comprises of predatory and omnivores nematode which showed high sensitivity towards slight changes. They showed an inverse correlation with the pollution, as pollution increases their population starts decreasing. The studies also revealed that high level of pollution also restricts the growth of opportunistic nematodes (Korthals et al. 1996a, b). MI is not only an indicator of pollution but also gave a picture of metal concentration in the aquatic ecosystem and land degradation (Bongers et al. 1991; Popovici and Korthals 1995; Coleman et al. 1998; Korthals et al. 1998). The MI in the agro-ecosystem doesn't provide satisfactory data for the metal disturbances as opportunistic nematodes already showed dominance in this ecosystem. The cp1-2 nematode taxa played an important role in determining the values of MI in different ecosystems. They also revealed the soil status whether it is highly enriched, susceptible to diseases, supporting plant growth or not, balance between decomposers and producers etc. The cp1-2 nematode which are already recognized as opportunistic or colonizers or tolerant to stress can be further classified into enrichment based opportunistic and general nematode (Ettema and Bongers 1993; Venette and Ferris 1998; Laasli et al. 2022). This classification is helpful in understanding the scenario of agro-ecosystems. The increased population of enrichment based opportunistic nematode supply the information of enriched soils. Likewise, if the population of general nematode starts increasing, then, it's an alarming condition as the enrichment decreases and stress condition prevailing in the system (Korthals et al. 1996a, b; Bongers et al. 1995).

Nematodes also play role in assessment of marine sediments pollution mostly occurred by the polycyclic aromatic hydrocarbons. The communities grow on the marine sediment consist of nematode communities which showed high sensitivity towards hydrocarbons (Egres et al. 2019). To understand the relationship of the nematodes in the presence of hydrocarbons, various studies were carried out by the scientist to get a clear picture of this. Furthermore, a trend has been observed in the nematode communities which describes that some populations are sensitive to this pollutant and their growth restricted leading to reduced population, however, some species withstand the conditions and some species population increases (Balsamo et al. 2012; Monteiro et al. 2018). There are various species which showed high tolerance towards these hydrocarbons' pollution viz., Rhabditis, *Calamicrolaimus honestus* etc. (Elarbaoui et al. 2015; Soto et al. 2017; Allouche et al. 2020). There are some events recorded in the past which signify the use of nematodes as bioindicator of increasing and decreasing pollution. The increasing pollution in the marine sediments changes the nema communities from higher to lower cp values. However, after surviving the high concentration of PAH in the marine sediments, the nema communities increased with an increase of the cp values which demarcate the suitability of nematode as eco-indicators also (Soto et al. 2017; Stark et al. 2017).

11.4.1 Assessing Heavy Metal Concentration Using Nematodes

The concentration of heavy metal found to be responsible for inviting various diseases and affect the microorganism population dependent upon the soil irrespective of the land and water. The major sources of these high concentration occurred from the anthropogenic activities like wastes from industries etc. The literature also provides role of nematodes in assessing the metal ion concentration in the soil. The high level of Cr affects the 1 species of nema communities out of 4 (Boufahja et al. 2011). Likewise, increased concentration of Hg favored the nema population in the studied area (Nasira et al. 2010). To understand more deeply, various studies had been carried out on the heavy metal concentration and the outcomes describes the two types of meiobenthos population in the sediment—one population which showed sensitivity and their population decreases in the presence of high metal concentration and other population found to be more of resistant one which is not affected by the high or low levels of metals (Bastami et al. 2017; Stark et al. 2020). Previous studies also supported the information which revealed that nematode groups showed basic trends of decreasing species and population in the presence of metal ion concentration (Nasira et al. 2010; Sedano et al. 2014; Bastami et al. 2017).

11.5 Responses of Nematodes to Various Disturbances

There is some nematode genus which showed either resistance or sensitivity towards specific metal ions. To find out the specific genus affected from the disturbances, a literature survey was carried out and here are some outcomes which were found consistent with the previous studies like Ridall and Ingels (2021). The *Pomponema* genus found to be sensitive towards hydrocarbons, heavy metals specifically Chromium, Copper, and Nickel; diesel organic enrichment and chemical pollution. Other genera i.e., *Halalaimus* showed sensitivity towards hypoxia, organic enrichment, sewage and mariculture. *Setosabaeria* genus found to be sensitive for the oil spills, sewage and mariculture. *Paracanthonchus* genus showed sensitivity towards hydrocarbons and oil spills. The *Oncholaimellus* genus showed sensitivity towards various disturbances like hydrocarbons, oil spills, heavy metal cobalt and resistance to high concentration of Chromium, Copper, and Nickel, harbours and marine deposition. The Belbolla genus is affected by harbours concentration and neutral to all the described disturbances. The *Paralongi cyatholaimus* genus showed sensitivity towards heavy metal copper and resistant to harbours and showed neutral responses to other disturbances. The *Richtersia* genus found sensitive for the organic enrichment, sewage, mariculture and deposition disturbances. The *Desmoscolex* genus known to be affected by the high disturbances of mariculture and hypoxia, whereas, the population found to be resistant towards harbor disturbances. The *Xyala* population showed high sensitivity towards hydrocarbons, heavy metals specifically Chromium, Copper, and Nickel and diesel. It showed resistance to zinc concentrations in the soil and gave neutral responses to other disturbances. *Leptonemella* genus known to be affected by the high disturbances created from high hydrocarbons, metals i.e., chromium, nickel, copper and resistance to diesel concentration. *Chaetonema* genus was affected by the diesel and harbor disturbances whereas *Elzalia* genus showed sensitivity towards mariculture. The high concentration of the lead heavy metal known to affected the *Calomicrolaimus* genus. The hypoxia disturbances affected the *Tricoma, Anticoma, Chromadorita, Aponema, Oxystomina, Daptonema* where one genus showed resistance known as *Diplolaimella.* The heavy metal disturbances mainly, chromium, nickel and copper were signified by the changing population of these nematode genera i.e., *Trichotheristus, Ditlevsenella, Gammanema, Viscosia* and *Theristus.* The *Oncholaimellus* genus showed resistance towards this disturbance. Furthermore, the heavy metal barium and cadmium also affected the *Litodis* genus only. If the population found to be increasing or decreasing, there might be chances that heavy metal concentration is high or low, but there are some challenges in this case, as the population of *Litodis* nematode known to be shifting from higher to lower sides from other disturbances also such as copper and lead heavy metal pollution. *Caenorhabditis* genus showed sensitivity towards high metal concentration of zinc, lead and nickel. The sewage pollution found to affect the populations of *Halalaimus, Richtersia* and *Bolbolaimus* which showed an inverse relationship with the pollution. The physical disturbances were traced from the changing population of Enoplidae family, Cyatholaimidae family, *Desmodora, Prochomadorella, Sprinia* and *Ptycholaimellus.* The heavy metal Cobalt pollution found to affect the *Oncholaimus, Neochromadora* and *Oncholaimellus* population.

11.6 Soil Food Web Analysis

The system of soil food web act as pool of several of nutrients, defines the structural part of the soil, keeps a check on pests' population. The structural and functional attributes of the soil food web system found sensitive to changes in the environment. It's very important to understand the soil food web system for the conservation, restoration purposes. The food web system can be analyzed by evaluating the function and structure. However, this is not an easy job as it requires whole microorganism community identification and assessing each and every taxon. Furthermore, the isolation of the microbial communities is a tedious task which demands precise handling and expertise (Ferris et al. 2001). To overcome these challenges, some alternatives are available i.e., biochemical reaction of the microorganism to look for the characteristic biomolecules (Bossio and Scow 1995; Pankhurst 1997). The other alternatives include respiratory rates, organic carbon, available phosphorus, nitrogen and other micronutrient of the soils (Griffiths et al. 2001; Gunapala et al. 1998). These alternatives have some drawbacks as they don't provide any information on the soil functional aspects and sustainable levels. To avoid this complicated evaluation of the soil food web system, guilds methods can be used as better alternatives. Guilds can be defined as the group of species which showed similarity in the responses to various environment circumstances. There are some guilds which can be assessed for the structural evaluation (Ferris et al. 1999; Ritz and Trudgill 1999; Wardle et al. 1995). Guilds comprises of nematode community which are easily assessed for the structural and functional aspects of the soil food web system. There is some limitation with the guild method as changes occurring at species level were not analyzed by this system (Polis 1995; Wall and Moore 1999). Hence, nematode trophic diversity and colonizer-persister scale were also introduced for the analysis of soil food web function (Yeates et al. 1993).

11.6.1 Role of Soil Food Web in the Assessment of Soil Conditions

Soil food web model estimate the soil health by assessing the abundance, population dynamics and different factors of nematode group especially free living which was finally demonstrated in four quadrants (Ferris et al. 2001). This model also considered the cp groups as their increase or decrease population reflects the changing soil conditions (Ferris et al. 2001). The plot of this model drawn between enrichment index and structural index to show the nematode population growth based on the availability of resources and stress conditions. Both the indices were calculated separately to avoid errors and the calculation for the indices are as follows

$$\text{Enrichment Index (EI)} = (e/e + b) \times 100$$

$$\text{Structure Index (SI)} = (s/s + b) \times 100$$

where, $s = Ba_n + Pr_n + Fu_n + Om_n$ ($n = 3$–5); $b = (Ba_2 + Fu_2) * W_2$; $e = Ba_1 * W_1 + Fu_2 * W_2$ (Ba-bacteriovorous, Pr-Predatory, Fu-fungivorous and Om-omnivorous nematodes).

The upper left quadrant comprises of high enrichment index and low structural index represents the soil consists of appreciable nitrogen concentration which favoured the growth of bacteriovores showing high resistivity towards changes. The colonizer-persister 1 and 2 showed dominance in this quadrant. The upper right quadrant consists of high values of both the indices i.e., enrichment and structural index represents the stable conditions of soil favored the growth of predators and omnivores nematodes showing high sensitivity to slight changes. The colonizer-persister 4 and 5 showed dominance in this quadrant. The lower right quadrant comprises of low enrichment index and high structural index represents the soil with moderate nitrogen concentration favored the growth of fungivores which showed sensitivity to disturbances. The lower left quadrant comprises of low enrichment index and low structural index represents the soil with lower nitrogen concentration favored fungivores showing resistance to changes. Different categories of nematodes based on feeding and colonizer-persister scale also drives the nutrient cycling which plays an important role in maintenance and restoration of soil. Therefore, upper quadrants (left and right) possessed low carbon to nitrogen ratio whereas lower left quadrant consists of high carbon to nitrogen ratio and carbon to nitrogen ratio of lower right quadrant exists between these. This model also depicts the structured and degraded soil ecosystems based on the population growth of nemic fauna and the movement,

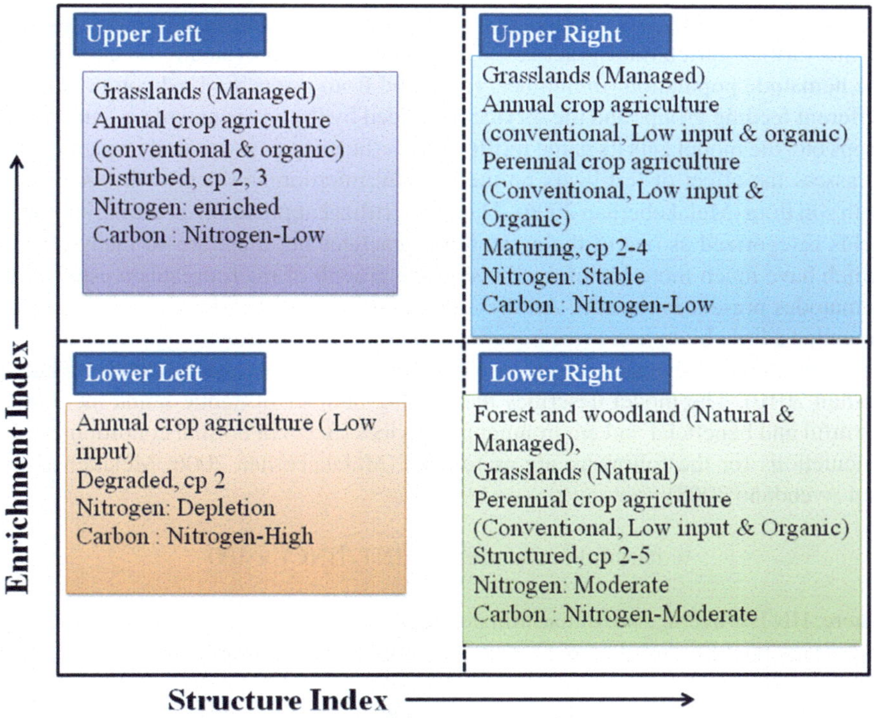

Fig. 11.2 Soil food web analysis model based on the nematode indices for the assessment of disturbances. The biplot of enrichment and structure index demonstrating the effect of disturbances, minerals on the nematode population (Ferris et al. 2001)

exchange of nutrients in the system (Fig. 11.2). Based on these, upper left quadrant found to be affected by the disturbances whereas lower left quadrant represents the soil degradation. On the contrary, upper right and lower right quadrant found to be more stable and regulatory in nature. The degradation of the soils invites a lot of pests; however, the regulatory systems restrict the population of various harmful diseases (Fig. 11.2).

11.7 Fertilizer Use Efficiency Analysis

The fertilizer efficiency model (Melakeberhan 2006, 2010; Melakeberhan and Avendaño 2008) focused on both parasitic and free-living forms of nematodes. The plot between the nematodes and the yield or nutrients reveals the nutrient uptake load and soil health. There are different agricultural practices carried out in the agricultural fields which affects the ecosystem services (Sánchez-Moreno 2018; Fixen et al. 2011; Adesemoye and Kloepper 2009; Chagas et al. 2015; Nissen and Wander 2003).

The fertilizer use efficiency analysis used to evaluate the changes in the services due to the various agricultural practices. This model found to evaluate the changes in the nematode population (in number) recovered from a specific land comprising of different feeding groups and the services provided by the environment i.e., nutrients, crops etc. the model gets its name fertilizer use efficiency as it was first implemented to assess the effect of fertilizers on the harmful microorganism growth associated with soil flora (Melakeberhan 2006). Though, fertilizer application on the agricultural fields categorized as one of the agricultural practices but there are so many others which have much more harmful effects on the growth of microorganism especially nematodes present in the soil. All these points were considered and the model was amended a little bit to assess the beneficial microorganism growth with respect variable conditions of the agroecosystem (Melakeberhan and Avendaño 2008; Melakeberhan 2010). The model describes the varying number of nemic fauna including harmful and beneficial and environmental services (in %) in normal conditions. The calculations for the following are as follows (Melakeberhan 2006; Melakeberhan and Avendaño 2008).

$$\%\ \text{harmful nematodes} = (\text{HNT}/\text{HNC}) \times 100,$$

where, HNT = average no. of harmful nemic fauna in a disturbed environment; HNC = average no. of harmful nemic fauna in a undisturbed environment.

$$\%\ \text{beneficial nematodes} = (\text{BNT}/\text{BNC}) \times 100,$$

where, BNT = average no. of beneficial nemic fauna in a disturbed environment; BNC = average no. of beneficial nemic fauna in a undisturbed environment

$$\%\ \text{ecosystem service parameter} = (\text{ECT}/\text{ECC}) \times 100$$

where, ECT = ecosystem services in a disturbed environment; BNC = ecosystem services in a undisturbed environment.

11.7.1 Role of Fertilizer Use Efficiency in the Assessment of Soil Conditions

The model describes the conditions of soil health with the help of quadrants by interpreting the effects of nutrients/pollutant concentration on the nemic population. The graph is plotted between ecosystem services and the number of nematode populations where, y-axis was governed by the services and percentage of nematodes on the x-axis. Comparing to the soil food web analysis model, this model also compares the harmful and beneficial nematode number which was affected by different changes; however, the harmful and beneficial nematodes were assessed separately in the quadrants. According to the fertilizer use efficiency model, there are eight quadrants depicting the soil conditions with the help of nematode abundance in relation the various services provided by the environment. The quadrant I-IV depicts the role of plant parasitic nematodes in relation the services whereas the V-VIII quadrant describes the abundance of beneficiary nemic population in relation to services. The graph describes the structured to disrupted conditions of the soil on the basis of nematode population found to be increasing or decreasing in relation to environmental activities/services (Fig. 11.3). The appreciable and desired conditions existed in the I quadrant where environmental activities increases and leads to reduction in the population of plant parasitic nematodes which are harmful for the plants and many agricultural crops like sugarcane, okra, eggplant, lady finger etc. The quadrant VI also depicts the ideal conditions where population of beneficial nematodes becomes abundant due to the enhancement of environmental services. The quadrant II and V depicts the scenarios where increment in the population of harmful nematodes and reduction in population occurred in relation to environmental activities. These cases should be considered for providing necessary treatments, so that system regains its original health and equilibrium between the harmful and beneficial nematodes maintained (Fig. 11.3). There are some other cases which required attention i.e., quadrant III and VIII where plant parasitic and beneficial nematodes population increased and environmental services decreased (Fig. 11.3). Here, in these scenarios, practices should be done in favour of environment so that services increase in both the cases. There are some cases also present where the population of plant parasitic rising indiscriminately and this leads to loss of flora and environment which also affects the population of beneficial nemic fauna. In these cases, treatments should be applied for the regulation of plant parasitic nematodes growth and to sustain the environment.

11.8 Future Perspectives

There are two models viz., soil food web analysis and fertilizer use efficiency, both of which are used for nematode community profiling; however, they are not restricted to just profiling, but they have other applications also. The soil health is not related to one field; it's a multidisciplinary topic which should be addressed with the help of integrated practices. There are some barriers viz., agricultural practices, different soil category, climatic conditions and very less soil health standards (quantitative), in attaining the soil health sustainably. All these barriers make it difficult for the researchers, practitioners and related authorities to manage soil health. To understand what are the ongoing conditions in an agricultural field will be depicted from the soil food web analysis model which revealed the ideal and the worse cases. Likewise, the fertilizer use efficiency model takes all the prescribed measures of sustainable soil health into consideration which helps the researchers, agriculturalists and other to adopt specific measures. These measures will be applied to a specific location or

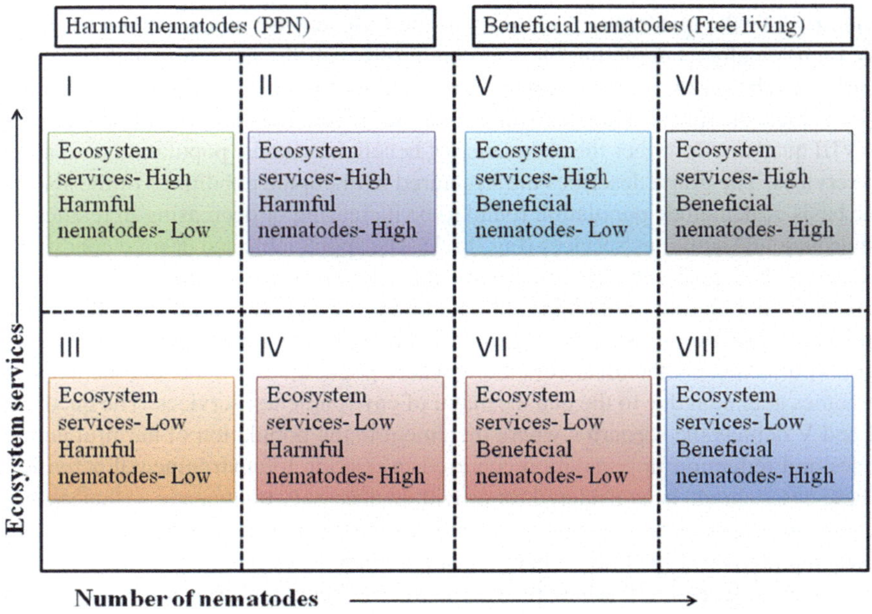

Fig. 11.3 Fertilizer use efficiency model for the assessment of ecosystem services based on nematode population. The I-IV quadrants represent the relation between harmful nematodes and ecosystem services (Melakeberhan 2006); V-VIII quadrants establish the relation between ecosystem services and beneficial nematodes (Melakeberhan and Avendaño 2008). The IV and VII quadrants present the worst case

can be exploited for everywhere whenever soil health degrades also described by the fertilizer use efficiency model.

These models serve as a great source of acknowledging the present conditions and to manage these conditions by specific measures. Now-a-days, these models were potentially used in the integrated management programs to sustain soil health which also increases the funding, policies and practices. Furthermore, to enhance these integrated management practices it is highly required to pass these models functioning, theories to the other disciplines. To understand this, let's take one example, if a specific field was evaluated by the nematologist, he/she will be focused on the nematodes, same field evaluated by the soil scientist, the results will be based on soil properties, and to manage soil health, both perspectives are required. This is the reason it is that scientist shouldn't be focused on one field only and for better management practices interdisciplinary concepts should be used.

The graphs from soil food web and fertilizer efficiency model is very easy to understand, however, the information comes out from these is potential for sustainable soil health programs. The soil food web analysis provides the information on a scale of 0–100. On the contrary, the fertilizer use efficiency model has a range from 0 to no higher limit. The soil conditions described from both models can be exploited by researchers to reduce those factors which influences the result in a hypothesis analysis.

The assessment of soil conditions can also be done in lab environment with the help of *Caenorhabditis elegans* (Tejeda-Benitez and Olivero-Verbel 2016; Chowdhary et al. 2020). This is a free-living nematode used to be model organism for understanding various diseases effects on humans and for the testing of potential drugs for a specific disease, however, it's not used for the soil assessment in previous studies in laboratory environment. *Caenorhabditis elegans* possess various characteristics which make it suitable for the soil assessment, these are easy to culture, short lifecycle, high fecundity, whole genome availability and quick reactions in presence of a certain environmental factor (Bouyanfif et al. 2019). They are exploited for the assessment of toxicants and harmful contaminants in the soil which directly affects the growth or fecundity of the nematode (Höss et al. 2009). The metal pollution also assessed by these nematodes in the laboratories from some gram of the soil (Kim et al. 2015). The effect of heavy metal was evaluated by the number of juveniles recovered from a specific metal, ultimately, developing a trend and finally leads to the classification of the heavy metal based on their toxic levels. The number of juvenile's decreases when the levels of clay, silt increase. On the contrary, the % sand found to be directly proportional to the juveniles number. Hence, it can be suggested that *Caenorhabditis elegans* should be exploited for primary assessment of the soil health.

References

Adesemoye AO, Kloepper JW (2009) Plant-microbe interactions in enhanced fertilizer-use efficiency-mini-review. Appl Microbio Biotechnol 85:1–12

Allouche M, Nasri A, Harrath AH, Mansour L, Beyrem H, Boufahja F (2020) Migratory behavior of free-living marine nematodes surrounded by sediments experimentally contaminated by mixtures of polycyclic aromatic hydrocarbons. J King Saud Univ 32:1339–1345. https://doi.org/10.1016/j.jksus.2019.11.025

Anon (2016) Soil health. NRCS. Available online http://www.nrcs.usda.gov/wps/portal/nr.cs/main/soils/health/

Assefa F, Elias E, Soromessa T, Ayele G (2020) Effect of changes in land-use management practices on soil physiochemical properties in Kabe Watershed, Ethiopia. Air Soil Water Res 13:1–16

Balsamo M, Semprucci F, Frontalini F, Coccioni R (2012) Meiofauna as a tool for marine ecosystem biomonitoring. In: Cruzado A (ed) Marine ecosystem. INTECH, London

Bastami KD, Taheri M, Foshtomi MY, Haghparast S, Hamzehpour A, Bagheri H et al (2017) Nematode community structure in relation to metals in the southern of Caspian Sea. Acta Oceanol Sin 36:79–86. https://doi.org/10.1007/s13131-017-1051-x

Bongers T (1990) The maturity index: an ecological measure of environmental disturbance based on nematode species composition. Oecologia 83:14–19

Bongers T (1999) The Maturity Index, the evolution of nematode life history traits, adaptive radiation and cp-scaling. Plant Soil 212:13–22

Bongers T, Ferris H (1999) Nematode community structure as a bioindicator in environmental monitoring. Trends Evol Ecol 14:224–228

Bongers T, Alkemade R, Yeates GW (1991) Interpretation of disturbance-induced maturity decrease in marine nematode assemblages by means of the maturity index. Mar Ecol Prog Ser 76:135–142
Bongers T, de Goede RGM, Korthals GW, Yeates GW (1995) Proposed changes of c-p classification for nematodes. Russ J Nematol 3:61–62
Bossio DA, Scow KM (1995) Impact of carbon and flooding on the metabolic diversity of microbial communities in soils. Appl Environ Microbiol 61:4043–4050
Boufahja F, Hedfi A, Amorri J, Aïssa P, Beyrem H, Mahmoudi E (2011) An assessment of the impact of chromium-amended sediment on a marine nematode assemblage using microcosm bioassays. Biol Trace Elem Res 142:242–255. https://doi.org/10.1007/s12011-010-8762-6
Bouyanfif A, Jayarathne S, Koboziev I, Moustaid-Moussa N (2019) The nematode *Caenorhabditis elegans* as a model organism to study metabolic effects of ω-3 polyunsaturated fatty acids in obesity. Advan Nutrit 10(1):165–178
Bulluck LR III, Barker KR, Ristaino JB (2002) Influences of organic and synthetic soil fertility amendments on nematode trophic groups and community dynamics under tomatoes. Appl Soil Ecol 21:233–250
Caixeta LB, Pereira TJ, Castañeda NE, Cares JE (2016) Nematode communities as indicators of the status of a soil ecosystem influenced by mining practices in Brazil. Nematology 18:265–276
Camargo JA (1992) New diversity index for assessing structural alterations in aquatic communities. Bull Environ Contam Toxicol 48:428–434
Chagas WFT, Emrich EB, Guelfi DR, Caputo ALC, Faquin V (2015) Productive characteristics, nutrition and agronomic efficiency of polymer-coated MAP in lettuce crops. Cienc Agron 46:266–276
Chowdhary P, Sammi SR, Pandey R, Kaithwas G, Raj A, Singh J, Bharagava RN (2020) Bacterial degradation of distillery wastewater pollutants and their metabolites characterization and its toxicity evaluation by using Caenorhabditis elegans as terrestrial test models. Chemosphere 127689. https://doi.org/10.1016/j.chemosphere.2020.1
Cobb NA (1918) Estimating the nematode population of soil, vol 1. United States, Department of Agri Tech Circu, pp 1–48
Coleman DC, Hendrix PF, Odum EP (1998) Ecosystem health: an overview. In: Soil chemistry and ecosystem health, SSSA Special Publ., vol 52, pp 1–20
Cooper AF Jr, Van Gundy SD (1971) Senescence, quiescence and cryptobiosis. In: Zuckerman BM, Mai WF, Rohde RA (eds) Plant parasitic nematodes, vol 2. Academic Press, New York, pp 297–318
Coull BC, Hicks GRF, Wells JBJ (1981) Nematode/copepod ratios for monitoring pollution: a rebuttal. Mar Pollut Bull 12:378–381. https://doi.org/10.1016/0025-326x(81)90408-2
Crowe JH, Clegg JS eds (1973) Anbydrobiosis. Dowden, Hutchinson, and Ross, Inc., Stroudsburg, Pennsylvania, p 477
Doran JW, Zeiss MR (2000) Soil health and sustainability: managing the biotic component of soil quality. Appl Soil Ecol 15:3–11
Edwards CA (2002) Assessing the effects of environmental pollutants on soil organisms, communities, processes and ecosystems. Eur J Soil Biol 38:225–231
Egres AG, Hatje V, Miranda DA, Gallucci F, Barros F (2019) Functional response of tropical estuarine benthic assemblages to perturbation by polycyclic aromatic hydrocarbons. Ecol Indic 96:229–240. https://doi.org/10.1016/j.ecolind.2018.08.062
Elarbaoui S, Richard M, Boufahja F, Mahmoudi E, Thomas-Guyon H (2015) Effect of crude oil exposure and dispersant application on meiofauna: an intertidal mesocosm experiment. Environ Sci Process Impacts 17:997–1004. https://doi.org/10.1039/c5em00051c
Ellenby D (1969) Dormancy and survival in nematodes. SOCE Xp Biol Symp 23:83–97
Ettema CH (1998) Soil nematode diversity: species coexistence and ecosystem function. J Nematol 30:159–169
Ettema CH, Bongers T (1993) Characterization of nematode colonization and succession in disturbed soil using the maturity index. Biol Fertil Soils 16:79–85

Fagodiya RK, Pathak H, Kumar A, Bhatia A, Jain N (2016) Global temperature change potential of N use in agriculture: a 50 year assessment. Sci Rep 7:44928 [PubMed]

Ferguson RB (2015) Groundwater quality and nitrogen use efficiency in Nebraska's central platte river valley. J Environ Qual 44:449–459

Ferris H, Venette RC, Lau SS (1996) Dynamics of nematode communities in tomatoes grown in conventional and organic farming systems, and their impact on soil fertility. Appl Soil Ecol 3:161–175

Ferris H, Bongers T, De Goede RGM (1999) Nematode faunal indicators of soil food web condition. J Nematol 31:534–535

Ferris H, Bongers T, de Goede RGM (2001) A framework for soil food web diagnostics: extension of the nematode faunal analysis concept. Appl Soil Ecol 18:13–29

Fine AK, van Es HM, Schindelbeck RR (2017) Statistics, scoring functions, and regional analysis of a comprehensive soil health database. Soil Sci Soc Am J 81:589

Fixen P, Brentrup F, Bruulsema TW, Garcia F, Norton R, Zingore S (2011) Nutrient/fertilizer use efficiency: measurements, current situation and trends. In: Drechsel P, Heffer P, Magen H, Mikkelsen R, Wichelns D (eds) Managing water and fertilizer for sustainable agricultural intensification, 1st edn. International Fertilizer Industry Association, International Water Management Institute, International Plant Nutrition Institute, and International Potash Institute, Paris, France, pp 8–38

Freckman DW, Baldwin JG (1990) Nematoda. In: Dindal DL (ed) Soil biology guide. Wiley, New York, pp 155–200

Freckman DW, Ettema CH (1993) Assessing nematode communities in agroecosystems of varying human intervention. Agr Ecosyst Environ 45:239–261

Gebremikael MT, Steel H, Bert W, Maenhout P, Sleutel S, De Neve S (2015) Quantifying the contribution of entire free-living nematode communities to carbon mineralization under contrasting C and N availability. PLoS ONE 10:e0136244

Gibson J, Shokralla S, Porter TM, King I, van Konynenburg S, Janzen DH, Hallwachs W, Hajibabaei M (2014) Simultaneous assessment of the macrobiome and microbiome in a bulk sample of tropical arthropods through DNA metasystematics. Proc Natl Acad Sci USA 111:8007–8012

Glavatska O, Muller K, Boutenschoen O, Schmalwasser A, Kandeler E, Scheu S, Totsche KU, Ruess L (2017) Disentangling the root- and detritus-based food chain in the micro-food webs of an arable soil by plant removal. PLoS ONE 13:e0180264

Goodey JB (1957) Laboratory methods for work with plant and soil nematodes. Technical Bulletin No. 2 Ministry of Agriculture, Fisheries and Food, London, p 47

Griffiths BS, Bonkowski M, Roy J, Ritz K (2001) Functional stability, substrate utilization and biological indicators of soils following environmental impacts. Appl Soil Ecol 16:49–61

Gunapala N, Venette RC, Ferris H, Scow KM (1998) Effects of soil management history on the rate of organic matter decomposition. Soil Biol Biochem 30:1917–1927

Gutiérrez C, Fernández C, Escuer M, Campos-Herrera R, Rodríguez MEB, Carbonell G, Martín JAR (2016) Effect of soil properties, heavy metals and emerging contaminants in the soil nematodes diversity. Environ Pollut 213:184–194

Guven K, Duce JA, Depomerai DI (1994) Evaluation of a stress-inducible transgenic nematode strain for rapid aquatic toxicity testing. Aquatic Toxicol 29:119–137

Guven K, Power RS, Avramides S, Allender R, de Pomerai DI (1999) The toxicity of dithiocarbamate fungicides to soil nematodes, assessed using a stress-inducible transgenic strain of Caenorhabditis elegans. J Bioch Mol Toxicol 13:324–333

Háněl (2004) Colonization of chemical factory wastes by soil nematodes. Pedobiologia 48:373–381

Hashmi G, Hashmi S, Selvan S, Grewal P, Gaugler R (1997) Polymorphism in heat shock protein gene (hsp70) in entomopathogenic nematodes (Rhabditida). J Thermal Biol 22:143–149

Höss S, Jänsch S, Moser T, Junker T, Römbke J (2009) Assessing the toxicity of contaminated soils using the nematode Caenorhabditis elegans as test organism. Ecotoxicol Environ Saf 72(7):1811–1818

Jankowski K, Neill C, Davidson EA, Macedo MN, Costa C, Galford GL, Santos LM, Lefebvre P, Nunes D, Cerri CE et al (2018) Deep soils modify environmental consequences of increased nitrogen fertilizer use in intensifying Amazon agriculture. Sci Rep 8:13478

Jian J, Du X, Stewart RD (2020) A database for global soil health assessment. Nat Sci Data 7:16 [PubMed]

Jongman RHG, ter Braak CJF, van Tongeren OFR (1995) Data analysis in community and landscape ecology. Cambridge University Press, Cambridge, UK

Kammenga JE, Arts MSJ, Oude-Breuil WJM (1998) HSP60 as a potential biomarker of toxic stress in the nematode *Plectus acuminatus*. Arch Environ Contami Toxicol 34:253–258

Kammenga JE, Dallinger R, Donker MH, Kohler HR, Simonsen V, Triebskorn R, Weeks JM (2000) Biomarkers in terrestrial invertebrates for ecotoxicological soil risk assessment. Rev Environ Contami Toxicol 164:93–147

Kihara J, Bolo P, Kinyua M, Nyawira SS, Sommer R (2019) Soil health and ecosystem services: lessons from sub-Saharan Africa. Geoderma 370:141342

Kim SW, Moon J, An YJ (2015) A highly efficient nonchemical method for isolating live nematodes (Caenorhabditis elegans) from soil during toxicity assays. Environ Toxicol Chem 34(1):208–213

Kokalis-Burelle N, Mahaffee WF, Rodriguez-Kabana R, Klopper JW, Brown KL (2002) Effects of switchgrass (Panicum virgatum) rotations with peanut (Arachis hypogaea L.) on nematode populations and soil microflora. J Nematol 34:98–105

Korthals GW, Alexiev AD, Lexmond TM, Kammenga JE, Bongers T (1996a) Long-term effects of copper and pH on the nematode community in an agroecosystem. Environ Toxicol Chem 15:979–985

Korthals GW, van de Ende A, van Megen H, Lexmond TM, Kammenga JE, Bongers T (1996b) Short-term effects of cadmium, copper, nickel and zinc on soil nematodes from different feeding and life-history strategy groups. Appl Soil Ecol 4:107–117

Korthals GW, Popovici J, Van Megen HHB (1998) Soil nematodes in heathland around a zinc-smelter near Budel, The Netherlands. In: De Goede RGM, Bongers T (eds) Nematode communities of northern temperate grassland ecosystems. Giessen, Germany, pp 155–158

Korthals GW, Bongers M, Fokkema A, Dueck TA, Lexmond TM (2000) Joint toxicity of copper and zinc to terrestrial nematode community in an acid sandy soil. Ecotoxicology 9:219–228

Laasli SE, Mokrini F, Lahlali R, Wuletaw T, Paulitz T, Dababat AA (2022) Biodiversity of nematode communities associated with wheat (Triticum aestivum L.) in southern Morocco and their contribution as soil health bioindicators. Diversity 14(3):194

Lal R (2011) Soil health and climate change: an overview. In: Singh BP, Cowie AL, Chan KY (eds) Soil health and climate change. Springer, Berlin/Heidelberg, Germany; Dordrech, The Netherlands; London, UK; New York, NY, USA, pp 3–24

Lehman PS, Reid JW (1993) Phyllognathopus viguieri (Crustacea:Harpacticoidea), a predaceous copepod of phytoparasitic, entomopathogenic, and free-living nematodes. Soil Crop Sci Soc Fla Proc 52:78–82

Li Q, Zhong S, Li F, Lou Y, Liang W (2011) Nematode community structure as bioindicator of soil heavy metal pollution along an urban-rural gradient in southern Shenyang, China. Int J Environ Pollut 45:297–309. https://doi.org/10.1504/IJEP.2011.040276

Ludwig JA, Reynolds JF (1988) Statistical ecology: a primer on methods and computing. Wiley, New York, NY

Martinez ND (1992) Constant connectance in community food webs. Am Nat 139(6):1208–1218

McSorley R, Walter DE (1991) Comparison of soil extraction methods for nematodes and microarthropods. Agri Ecosy Environ 34:201–207

Melakeberhan H (2006) Fertiliser use efficiency of soybean cultivars infected with Meloidogyne incognita and Pratylenchus penetrans. Nematology 8:129–137

Melakeberhan H (2010) Cross-disciplinary efficiency assessment of agronomic and soil amendment practices designed to suppress biotic yield-limiting factors. J Nematol 42:73–74

Melakeberhan H, Avendaño MF (2008) Spatio-temporal consideration of soil conditions and site-specific management of nematodes. Precis Agric 9:341–354

Millar N, Robertson GP, Grace PR, Gehl RJ, Hoben JP (2010) Nitrogen fertilizer management for nitrous oxide (N2O) mitigation in intensive corn (Maize) production: an emissions reduction protocol for US Midwest agriculture. Mitig Adapt Strateg Glob Change 15:185–204

Monteiro L, Traunspurger W, Roeleveld K, Lynen F, Moens T (2018) Direct toxicity of the water-soluble fractions of a crude and a diesel-motor oil on the survival of free-living nematodes. Ecol Indic 93:13–23. https://doi.org/10.1016/j.ecolind.2018.04.066

Nagy P, Bakonyi G, Bongers T, Kadar I, Fabian M, Kiss I (2004) Effects of microelements on soil nematode assemblages seven years after contaminating an agricultural field. Sci Total Environ 320(2–3):131–143

Nahmani J, Hodson ME, Black S (2007) Effects of metals on life cycle parameters of the earthworm Eisenia fetida exposed to field-contaminated, metal-polluted soils. Environ Pollut 149:44–58. https://doi.org/10.1016/j.envpol.2006.12.018

Nasira K, Shahina F, Kamran M (2010) Response of free living marine nematode community to heavy metal contamination along the coastal areas of Sindh and Baluchistan, Pakistan. Pak J Nematol 28:263–278

Neher DA (2001a) Nematode communities as ecological indicators of agroecosystem health. In: Gliessman SR (ed) Agroecosystem sustainability: developing practical strategies. CRC/Lewis Press, Boca Raton, FL, pp 105–120

Neher DA (2001b) Role of nematodes in soil health and their use as indicators. J Nematol 33(4):161

Neher DA, Campbell CL (1996) Sampling for regional monitoring of nematode communities in agricultural soils. J Nematol 28:196–208

Neher DA, Peck SL, Rawlings JO, Campbell CL (1995) Measures of nematode community structure for an agroecosystem monitoring program and sources of variability among and within agricultural fields. Plant Soil 170:167–181

Nissen TM, Wander MM (2003) Management and soil-quality effects on fertilizer-use efficiency and leaching. Soil Sci Soc Am J 67:1524–1532

Odum PE (1985) Trends expected in stressed ecosystems. Bioscience 35:419–422

Özdemir G, Limoncu MH, Yapar S (2010) The antibacterial effect of heavy metal and cetylpyridinium-exchanged montmorillonites. Appl Clay Sci 48:319–323. https://doi.org/10.1016/j.clay.2010.01.001

Pankhurst CE (1997) Biodiversity of soil organisms as an indicator of soil health. In: Pankhurst CE, Doube BM, Gupta VVSR (eds) Biological indicators of soil health. CAB International, Wallingford, UK, pp 297–324

Park BY, Lee JK, Ro HM, Kim YH (2011) Effects of heavy metal contamination from an abandoned mine on nematode community structure as an indicator of soil ecosystem health. Appl Soil Ecol 51:17–24

Pimentel D, Giampietro M (1994) Food, land, population and the US economy. Available online http://www.dieoff.com/page40.htm

Polis GA (1995) Complex food webs. In: Patten BC, Jørgensen SE (eds) Complex ecology. Prentice-Hall, Englewood Cliffs, NJ, pp 513–548

Popovici I, Korthals G (1995) Soil nematodes used in the detection of habitat disturbance due to industrial pollution. Studia Univ Babes Royal Biol XXXVIII (1–2): 37–41

Raffaelli DG, Mason CF (1981) Pollution monitoring with meiofauna, using the ratio of nematodes to copepods. Mar Pollut Bull 12:158–163. https://doi.org/10.1016/0025-326x(81)90227-7

Ridall A, Ingels J (2021) Suitability of free-living marine nematodes as bioindicators: status and future considerations. Front Mar Sci 8:685327

Riga E, Mojtahedi H, Ingham RE, McGuire AM (2003) Green manure amendments and management of root knot nematodes on potato in the Pacific Northwest of USA. Nematol Monogr Perspect 2:151–158

Ritz K, Trudgill DL (1999) Utility of nematode community analysis as an integrated measure of the functional state of soils: perspectives and challenges. Plant Soil 212:1–11

Robertson GP, Bruulsema TD, Gehl RJ, Kanter D, Mauzerall DL, Rotz CA, Williams CO (2012) Nitrogen-climate interactions in US agriculture. Biogeochemistry 14:41–70

Rodríguez Martín JA, Gutiérrez C, Escuer M, García-González MT, Campos-Herrera R, Águila N (2014) Effect of mine tailing on the spatial variability of soil nematodes from lead pollution in La Union (Spain). Sci Total Environ 473:518–529

Russell CC (1986) The feeding habits of a species of Mesodorylaimus. J Nematol 18:641

Šalamún P, Kucanová E, Brázová T, Miklisová D, Renčo M, Hanzelová V (2014) Diversity and food web structure of nematode communities under high soil salinity and alkaline pH. Ecotoxicology 23:1367–1376

Saly A, Ragala P (1984) Free-living nematodes-bioindicators of the effects of chemization on the soil fauna. Sbornık Uvtiz Ochrana Rostlin 20:15–21

Samoiloff MR (1987) Nematodes as indicators of toxic environmental contaminants. In: Veech JA, Dickson DW (eds) Vistas on nematology: a commemoration of the 25th annual meeting of the Society of Nematologists. Society of Nematologists, Hyattsville, MD, pp 433–439

Sánchez-Moreno S (2018) Biodiversity and soil health: the role of the soil food web in soil fertility and suppressiveness to soil-borne diseases. Acta Hortic 1196:95–104

Sánchez-Moreno S, Cano M, López-Pérez A, Benayas JMR (2018) Microfaunal soil food webs in Mediterranean semi-arid agroecosystems. Does organic management improve soil health? Appl Soil Ecol 125:138–147

Schorpp Q, Schrader S (2016) Earthworm functional groups respond to the perennial energy cropping system of the cup plant (Silphium perforliatum). Biomass Bioenergy 87:61–68

Sedano F, Marquina D, Espinosa Torre F (2014) Usefulness of meiofauna at high taxonomic levels as a tool to assess harbor quality status. Marina del Este Harbor (Granada, Spain) as a case study. Rev Ciencias Mar Y Costeras 6:103–113

Semprucci F, Balsamo M (2014) Free-living marine nematodes as bioindicators: past, present and future perspectives. Trends Environ Sci 6:17–36

Semprucci F, Losi V, Moreno M (2015) A review of Italian research on freeliving marine nematodes and the future perspectives on their use as Ecological Indicators (EcoInds). Mediterr Mar Sci 16:352. https://doi.org/10.12681/mms.1072

Skubała P, Zaleski T (2011) Heavy metal sensitivity and bioconcentration in oribatid mites (Acari, Oribatida): gradient study in meadow ecosystems. Sci Total Environ 414:364–372

Soetaert K, Franco MA, Lampadariou N, Muthumbi A, Steyaert M, Vandepitte L et al (2009) Factors affecting nematode biomass, length and width from the shelf to the deep sea. Mar Ecol Prog Ser 392:123–132. https://doi.org/10.3354/meps08202

Sohlenius B (1980) Abundance, biomass and contribution to energy flow by soil nematodes in terrestrial ecosystems. Oikos 34:186–194

Sohlenius B, Persson H, Magnusson C (1977) Distribution of root and soil nematodes in a young Scots pine stand in central Sweden. Ecol Bull (stockholm) 25:340–347

Soto LA, Salcedo DL, Arvizu K, Botello AV (2017) Interannual patterns of the large free-living nematode assemblages in the Mexican Exclusive Economic Zone, NW Gulf of Mexico after the Deepwater Horizon oil spill. Ecol Indic 79:371–381. https://doi.org/10.1016/j.ecolind.2017.03.058

Southey JF (1986) Laboratory methods for work with plant and soil nematodes. Ministery of Agriculture, Fisheries and Food, Reference Book 402, London

Stark JS, Mohammad M, McMinn A, Ingels J (2017) The effects of hydrocarbons on meiofauna in marine sediments in Antarctica. J Exp Mar Bio Ecol 496:56–73. https://doi.org/10.1016/j.jembe.2017.07.009

Stark JS, Mohammad M, McMinn A, Ingels J (2020) Diversity, abundance, spatial variation, and human impacts in marine meiobenthic nematode and copepod communities at Casey Station, East Antarctica. Front Mar Sci 7:480. https://doi.org/10.3389/fmars.2020.00480

Stewart ZP, Pierzynski GM, Middendorf BJ, Prasad PVV (2020) Approaches to improve soil fertility in sub-Saharan Africa. J Exp Bot 71:632–641

Tejeda-Benitez L, Olivero-Verbel J (2016) *Caenorhabditis elegans*, a biological model for research in toxicology. Rev Environ Contam Toxicol 237:1–35

Toosi ER, Kravchenko AN, Quigley MM, Mao J, Rivers ML (2017) Effects of management and pore characteristics on organic matter composition of macroaggregates, evidence from X-ray μ-tomography, FTIR and 13C-NMR. Eur J Soil Sci 68:200–211

Topham PB, Boag B, McNicol JW (1991) An assessment of some measures of association between species based on presence/absence and applied to plant-parasitic nematode data. Nematologica 37:470–480

Van der Hoogen J, Geisen S, Routh D, Ferris H, Traunspurger W, Wardle DA, De Goede RG, Adams BJ, Ahmad W, Andriuzzi WS et al (2019) Soil nematode abundance and functional group composition at a global scale. Nature 572:194–198

Venette RC, Ferris H (1998) Influence of bacterial type and density on population growth of bacterial–feeding nematodes. Soil Biol Biochem 30:949–960

Wall DH, Moore JC (1999) Interactions underground, soil biodiversity, mutualism, and ecosystem processes. Bioscience 49:109–117

Wander MM, Cihacek LJ, Coyne M, Drijber RA, Grossman JM, Gutknecht JLM, Horwath WR, Jagandamma S, Olk DC, Ruark M et al (2019) Developments in agricultural soil quality and health: reflections by the research committee on soil organicmatter management. Front Environ Sci 7:1–9

Wardle DA, Yeates GW, Watson RN, Nicholson KS (1995) The detritus food web and the diversity of soil fauna as indicators of disturbance regimes in agroecosystems. Plant Soil 170:35–43

Wasilewska L (1979) The structure and function of soil nematode communities in natural ecosystems and agrocenoses. Polish Ecol Stud 5:97–145

Wasilewska L (1989) Impact of human activities on nematodes. In: Clarholm C, Bergstro¨m L (eds) Ecology of arable land. Kluwer Academic, Dordrecht, The Netherlands, pp 123–132

Yeates GW (1970) The diversity of soil nematode faunas. Pedobiologia 10:104–107

Yeates GW (1984) Variation in soil nematode diversity under pasture with soil and year. Soil Biol Biochem 16:95–102

Yeates GW, Bird AF (1994) Some observations on the influence of agricultural practices on the nematode faunae of some South Australian soils. Fund Appl Nematol 17(2):133–145

Yeates GW, Bongers T, de Goede RGM, Freckman DW, Georgieva SS (1993) Feeding habits in soil nematode families and genera—an outline for soil ecologists. J Nematol 25:315–331

Zhong S, Zeng HC, Jin ZQ (2017) Influences of different tillage and residue management systems on soil nematode community composition and diversity in the tropics. Soil Biol Biochem 107:234–243

Chapter 12
An Overview on the Impact of Heavy Metals in Faunal Community of Soil and Its Controlling Approaches

Namita Joshi and Garima

Keywords Bioremediation · Faunal communities · Heavy metals · Pollution · Soil

12.1 Introduction

Those metals which have at least five times more density and atomic weight than water is considered heavy metals. There are many sources of heavy metals like medical industries, vehicle exhaust, incomplete combustion of fossil fuel, volcanic eruption, mining, construction, technology, agriculture, domestic, and industrial waste which contaminate soil, water, and air (Yedjou et al. 2012). Mostly anthropogenic activity leads to heavy metal pollution. Soil contamination by metals is quite a common environmental disorder, nowadays. Soil is the habitat of a large number of micro and macro-organisms. Different soil faunal communities like bacteria, fungi, protozoans, and plants have the diverse impact of heavy metals. Increased soil pollution not only affects the soil faunal population, but also have a major threat to human society (Hou et al. 2017), as this pass-through bioaccumulation and bio-magnification along with the food chain (Weissmannová and Pavlovský 2017). How heavy metals affect living organisms, mainly depends on the concentration and type of heavy metals. Heavy metals not only alter the soil's physiochemical properties but also alter the nutrient dynamics of soil. Organisms that reside in contaminated soil also get affected by cellular, individual, and population levels. Soil contamination also has an indirect effect on humans, as the bioaccumulation of heavy metals in plants, reaches animals through the food chain, and causes severe toxicity to humans (Wang et al. 2007).

N. Joshi · Garima (✉)
Department of Environmental Science, Kanya Gurukula Campus (Haridwar) of Gurukula Kangri (Deemed to Be University), Haridwar, Uttarakhand 249404, India
e-mail: garimasaini39@gmail.com

P. P. Adhikary et al. (eds.), *Soil, Water Pollution and Mitigation Strategies*,
Environmental Science and Engineering, https://doi.org/10.1007/978-3-031-63296-9_9

The main focus of this book chapter is to analyze how heavy metal affects (1) soil structure, and soil faunal community including (2) protozoan (3) bacteria (4) fungal (5) plant and (6) earthworm community, along with the different methods used by an organism to resist metal toxicity.

12.2 Soil

The loose surface material that covers the land is soil, a major substituent for the earth's ecosystem. It is composed of many organic, inorganic, gaseous, liquid, minerals, and organisms including micro and major fauna and flora. The pedosphere is used for soil covering the earth's body. It mainly has four functions, i.e., provides a medium for the growth of plants, acts as a purifier and reservoir of water, and is home to various organisms and earth's atmosphere modifier. The process of soil formation is called paedogenesis (Van Breemen and Buurman 2002).

12.3 Soil Types

The U.S. Department of Agriculture (USDA) system classified soil into four different types based on particle size, i.e., clay, silt, sand, and gravel. Particle sizes equal to or less than 0.002 mm come under clay, particle sizes between 0.002 and 0.05 come under silt, sand contains particles between 0.05 and 2 mm, and more than 2 mm come under gravel. Out of them, the relative mass percentage of sand, silt, and clay define the soil texture (Yolcubal et al. 2004). Sandy soil is dry, warm, and light, slightly less alkaline, and has a low amount of nutrients. Plants and shrubs which are suitable for sandy soil are called Psammophytes. Crops like potatoes, parsnips, and carrots favor sandy soil for growth (Zhao et al. 2014). Clay soil is heavy soil, and holds a high amount of water. Crops like cabbage, cauliflower, pea, bean, brussels sprouts, kale, and daikon radish are found to show good growth in clay soil (Nonnecke 1989). Silt is light soil with a high fertility rating, it is rich in potash. Crops like hellebore, sage, tomatoes, daffodils, ferns, butterfly bushes, and roses are suitable for this soil. Loam soil is a mixture of clay, silt, and sand. This soil type is quite fertile with a good drainage capacity. Typical loamy soil has 50% water and pore space and the other 50% is formed by the mixture of sand, clay, and silt particles, in which approximately, 20% clay, 40% silt, and 40% is sand composition (Libohova et al. 2018).

12.4 Properties of Soil

Soil has many physical and chemical properties. Soil comprises many physical properties like soil structure, porosity, texture, temperature, color, consistency and resistivity, soil moisture, soil gas, etc. Chemical properties like cation and anion exchange capacity determine the exchangeable cation in between soil water and plant roots, and the soil's ability to eliminate anions. Reactivity (pH) helps in the determination of soil whether it is alkaline or acidic. Soil organic matter determines the amount of organic matter present in the soil. The main components of organic matter in soil are an animal, plant, and microbial constituents including both living and non-living components. Various ions and minerals present in soil and climatic conditions help in the determination of soil type and its properties (Baldock and Skjemstad 2000).

12.5 Heavy Metals

Metals are essential for organisms' metabolic reactions; they are considered heavy metals when the concentration of it above their threshold/permissible limit. When the concentration of it is high, they cause dysregulation of metabolic reactions and organ damage, etc. Those metals which have a higher density and atomic weight than water come under heavy metals. Their multiple application use in industries, medical, domestic, medical, and technological use let their wide distribution in the environment (Yedjou et al. 2012).

12.6 Sources of Heavy Metals

Metals are naturally and anthropogenically released in the soil in many forms like free metal ions, precipitated or insoluble forms, soluble forms, and in complex with organic matter (Baldrian 2010). Figure 12.1 shows the various natural and anthropogenic source of heavy metals that persists in the environment, wearing and tearing of electric waste emission of industries, pharmaceutical companies, application of fertilizers, fertilizers, sewage sludge, combustion residue, herbicides, pesticides, wastewater irrigation, mining, parent rock, volcanic eruption, acid rain, forest fire like many other activities are responsible for making soil the major sink of heavy metals. E-waste also becomes a prominent contamination source of soil, nowadays. It has different physical and chemical properties than municipal and industrial waste. Cu, Cd, Pb, Sb, Hg, and Ni, are vulnerably found in e-waste. These metals have a negative impact on soil properties as the degradation of these metals is not as simple as the organic matter. When an element accumulation occurs more than its natural level it becomes contaminants. Many heavy metals like zinc, vanadium, nickel, manganese, molybdenum, iron, cobalt and copper, etc. are absorbed and used

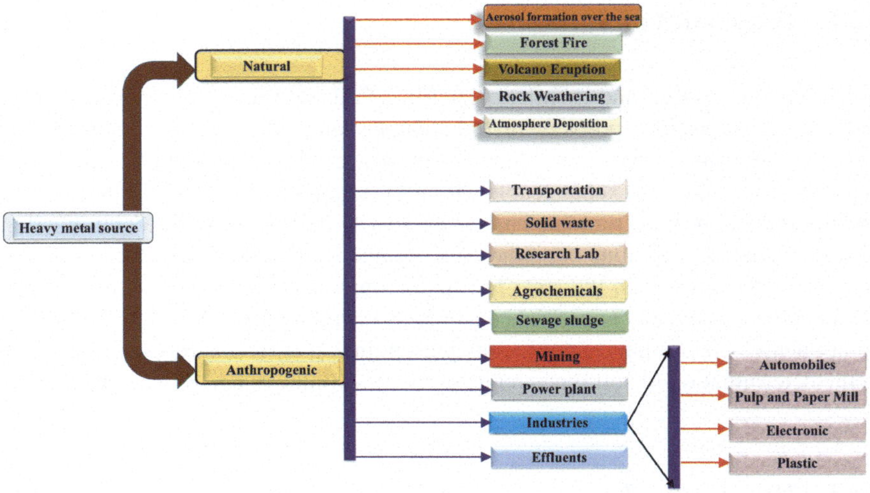

Fig. 12.1 Showing the various natural and anthropogenic sources of heavy metals

by plants for their metabolic processes. However, exceeding amounts of these metals can lead to harmful impacts on plants' metabolic activities. Lead, cadmium, copper, nickel, chromium, and zinc are the major heavy metals that are generally found in higher amounts in contaminated soil (Singh and Kalamdhad 2011). They are a potential threat to human health and the environment. The main sources of heavy metals are given in Table 12.1.

12.7 How It Affects the Soil Properties

Several studies indicate that the major factor which is affected by heavy metals is the pH. A positive correlation has been seen in the water holding capacity, and moisture content with heavy metals (Castaldi et al. 2004). Immobilization of metals can change the colloidal charge and soil density, soil's relative surface area, ligand complexation degree, amount of anion and cation, hydrous oxides of metals and their solubility (Giller et al. 1998; Nannipieri et al. 1997). Biotic components of soil, i.e. its microbiota are found more sensitive towards contamination, like the microbial community of soil which plays an important role in maintenance and prevention also affected by heavy metal contamination. Heavy metals can decrease microorganisms' activity, growth speed, and reproduction capacity (Simon 1999). Major sources of heavy metals are represented in Fig. 12.2.

The impact on soil microfauna; heavy metals directly affect the soil nutrient dynamics and decomposition rate of organic matter (Anderson and Ineson 1984). Abiotic factors like nutrient dynamics, nutrient mineralization, and soil moisture are also affected by heavy metals. Metal pollution, affects litter decomposition, nitrogen

Table 12.1 Source of heavy metals

S. No.	Metal	Source	References
1.	Pb	Coal burning exhaust, industrial fumes, domestic waste, mining, traffic exhaust, gasoline vehicle, battery production, thermal power plant, water–gas treatment system, scrap battery recovery facilities, iron steel industries, atmospheric deposition, sewage sludge, brass corrosion	Cheng et al. (2018), Sun et al. (2013), Yaylalı-Abanuz (2011)
2.	Cd	Mining, traffic exhaust, Industrial dust, agriculture and slag, agronomic practices, industrial effluents	Cai et al. (2015), Dai et al. (2013)
3.	Hg	Mining, traffic exhaust, industrial effluent, agricultural activities, coal-burning exhaust, vehicle exhaust, domestic waste, industrial fumes, fungicides	Cai et al. (2015), Wang et al. (2022a; b)
4.	As	Agronomic practices, industrial effluents, phosphate fertilizers	Cai et al. (2015), Jayasumana et al. (2015)
5.	Cr	Tannery waste, leather manufacturing waste, electroplating sludge, Cr pigment, municipal sludge waste, sewage sludge, phosphate fertilizer, paper sludge	Yaylalı-Abanuz (2011)
6.	Cu	Fertilizers, agricultural and municipal waste, spray, corrosion of construction material, industrial emission, urban road dust, atmospheric deposition, sewage sludge, livestock manure, nitrogen fertilizers, irrigation water, paper sludge, mining, refining of copper ores, electroplating industries, misc., milling, refining melting plant, brass corrosion	Shrivastava (2009), Yaylalı-Abanuz (2011)
7.	Se	Manure, sewage sludge, mining, coal burning, waste irrigation, selenium fertilizers, irrigation in combination with high evaporation rates, oil refining, insecticide production	He et al. (2018), Huang et al. (2009a), Winkel et al. (2012)
8.	Ni	Paper sludge, phosphate fertilizers, industrial waste materials, sewage sludge, lime, fertilizer, Leather factory and tannery, metal processing factories, coal combustion	Bhalerao et al. (2015). Yahaya (2011)
9.	Zn	Agricultural practice, Nonferric metal industries, atmospheric deposition, sewage sludge, livestock manure, nitrogen fertilizers, phosphate fertilizer, irrigation water, paper sludge, acid mine drainage, Zn smelter	Liu et al. (2020), Yaylalı-Abanuz (2011)
10.	Co	Leather factories and tanneries, manure, metal processing factories, coal combustion, mining	Khan et al. (2019), Křibek et al. (2014)
11.	Fe	Metallurgical and industrial dust	Lecoanet et al. (2001)

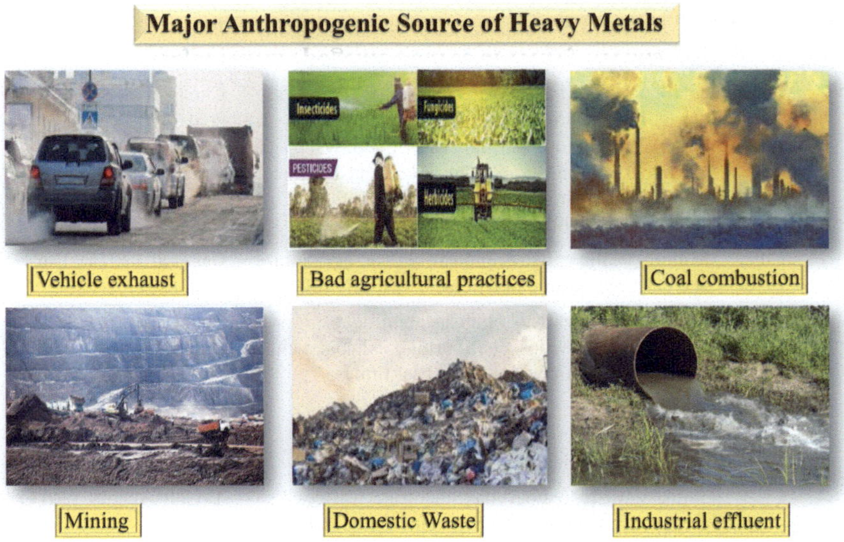

Fig. 12.2 Showing the major anthropogenic sources of heavy metals

mineralization, and soil respiration. Contaminated soil, decrease the nitrification rate and respiration rate. Increased mineralization of nitrogen also seems in metal-polluted soil, a higher concentration of NH^{4+}-N was observed in contaminated soil than in control soil. Soil processes as influenced by heavy metals and the composition of soil fauna. Soil processes as influenced by heavy metals and the composition of soil fauna (Komulainen and Mikola 1995).

The effect of acute toxicity on microbes is quite unpredictable in short-term assay experiments and has little resemblance of results with the long-term effect of metal toxicity experiments (Renella et al. 2002). Metal toxicity generally decreases microbial biomass, as most of the energy utilization occurs in the compensating of stress produced due to metal toxicity. A decrease in Microbial biomass was found to be related to the microbial community (Abaye et al. 2005).

12.8 Impact on Protozoans

Protozoans also belong to the faunal community of the soil. Terrestrial biotopes contain 400 different species of ciliates, 200 species of testate, around 260 species of testate amoeba, and about 60 species of naked amoeba (Foissner 1996). Fewer species were found at higher altitudes and latitudes, and most of the protozoan species were found at the uppermost 10 cm layer of soil. Soil is considered the place of the biosphere where the transformation of various biogeochemical cycles, occur and this transformation is performed by the activity of various soil faunal communities.

Protozoan species also regulate the metabolic pathways and decomposition rate and by doing so they also facilitate the carbon and nitrogen cycle (Swift and Anderson 1993). Harsh conditions like intense cultivation, erosion, acidic deposition, desiccation, burning, and pollutants of heavy metals or nitrogenous waste have a negative impact on soil microorganisms. The toxicity of heavy metals hit different protozoan species. Depending upon the type and concentration of heavy metals. Heavy metals present in sewage sludge had found severe impact on the protozoan community by altering the richness and density of species. The order of metal toxicity was found accordingly, i.e., Cd, Pb, and Cu were found more toxic than Zn and Cr. Out of three toxic metals, Cd is the most toxic, copper slightly less than that, and Zn the least toxic among the three. From Zn and Cr, Cr is the found least toxic to protozoan communities (Madoni et al. 1996). Laboratory studies performed on common ciliated protozoans, i.e., *Cyrtolophosis elongate*, *Colpoda inflate* and *Colpoda steinii* indicated that Cd and Cr were found more toxic than Zn. It was also seen that the strains of colpodea ciliates have diversity related to heavy metal resistance. The main mechanism found for metal resistance was bioaccumulation (Díaz et al. 2006).

12.9 Impact on Soil Bacterial Diversity

Change in soil microbial population is considered one of the characterizations which shows the impact of soil heavy metals contamination. Beyond a certain level, these metals become hazardous to living organisms. Interference in biological processes, enzymatic activities, cellular disruption, DNA damage, and stressful environment are some of the effects caused by these metals in organism's bodies, which directly impact their biodiversity and biomass (Bruins et al. 2000; Göhre and Paszkowski 2006; Rathnayake 2010). Soil microbes are considered one of the main faunal communities in the soil environment. Bacteria, fungi, and Protista-like microorganisms participate in the degradation of these metals or their transformation into non-toxic compounds. This microorganism reduces these metals' mobility and bioavailability by using methods like bio-precipitation and bio-sorption. Some studies proved that the harmful effect on microbial species is diminished by the coexistence of plant communities (Yang et al. 2007). Various environmental factors like soil type, species of microbiota, and source of contamination also affect the impact (Vig et al. 2003). Heavy metal pollutes soil and reduces the different types of aromatic compounds in soil used by microbes for their growth and development thus heavy metals impair microbial community structure (Reber 1992). Long-termexposure to metal pollution leads to adaptation in microbes through changes in community structure and composition (Li et al. 2017).

12.10 Impact on Fungal Diversity

Fungal in symbiotic association with plants like Mycorrhiza helps in the uptake of heavy metals by increasing the adsorptive surface area (Alloway 1995). Symbiotic organisms are found more sensitive toward Cd, Pb, and Hg (Ahmad et al. 2005a, b). Mycelial fungi have various methods like toxic soil domain avoidance or cell wall mobilization techniques to reduce metal toxicity. The toxicity of metal reducesnutrient availability and leads to a reduction in the growth of fungi and morphological changes in the whole fungal community (Baldrian 2003a; b). Cellular changes like mitochondrial fragmentation, tubularity, and vacuolar mobility, production of protamines are seen in the ectomycorrhizal fungus *Paxillusinvolutus*. Some metals also induce cellular oxidative stress, which can damage the lipid and DNA content in *Boletus edulis* (Collin-Hansen et al. 2005). Expression of enzymes like superoxide dismutase, laccase, aconitase, and metallothionein is elevated in *Paxillusinvolutus* due to Cd stress (Jacob et al. 2018). In mycorrhizal and saprotrophic fungi metal contamination mostly affect the life stages like germination and spore formation rather than mycelial growth. Some studies report more stain-specific sensitivity than species-specific in micromycetes and basidiomycetes fungi (Baldrian 2003a; b). Fungal survival in a contaminated zone is more sensitive, i.e., depending on the type of species (Gadd 1994). Microbial biomass is found less in metal-contaminated sites, which decreases the activity of many enzymes like β-glucosidase, phosphomonoesterase, endoglucanase, and N-acetylglucosaminiase (Kuperman and Carreiro 1997). Species which can tolerate a high amount of toxicity level, generally have a low population in the uncontaminated area (Gadd 2007a, b).

12.11 Impact on Plant Community

Cu, Fe, Mn, Mo, Ni, and Zn are considered essential beneficial heavy metals for plants as they directly play role in cellular mechanisms, important for plant survival and growth. However higher concentrations than the threshold value of these metals can cause toxicity in plants (Arif et al. 2016). These metals then affect the plants' metabolism either directly or indirectly. Direct methods generally involve the introduction of oxidative stress leads to destruction in cellular components, and cellular integrities and affect enzymatic activities. The indirect effect may include the change in cations at cation exchange sites (Vangronsveld and Clijsters 1994). Toxic amounts of Hg and Pb inhibit the enzymatic activity of δ-aminolaevulinic acid (ALA) in *Pennisetum typhoideum* seedlings (Prassad and Prassad 1987). Cd inhibits the activity of the protochlorophyllide reductase enzyme which is important in the production of chlorophyll (StiborovÁ 1988). Cd also inhibits photosystem II in the *Lycopersicum* plant (Singh and Prasad 2017). The same effect is also shown by Cd and Hg in *Euglena* (Filippis et al. 1981). For the proper functioning of the rubisco enzyme metal group SH interaction is very crucial (Lorimer 1981) but increased levels of Cd,

Pb, Zn, and Cu hindered this interaction and affect the activity of the rubisco enzyme (StiborovÁ 1988). Treatment of *Phaseolus vulgaris* seedlings with Zn impairs the activity of the carboxylase enzyme, which decreases the C/O ratio (Van Assche and Clijsters 1986). Brown algae *Luminaria saccharina* rubisco enzyme is also inhibited by Cd by indirect inhibition of de novo protein synthesis (Kremer and Markham 1982). Cadmium stress can lead to an increase in mean germination time and reduce the growth of wheat seedlings, however, the effect varies according to different cultivars (Ahmad et al. 2012). A higher concentration of Zn can decrease the germination rate, biochemical contents and growth and development of *Cyamopsis tetragonoloba* seedlings (Manivasagaperumal et al. 2011). An experiment performed on a pea plant described that the effect of zinc on plants varies with the developmental stage of the plant, amount, and type of Zn treatment (Doncheva et al. 2001). Studies show that Mn accumulation is found more in roots than shoots, and affects the growth of the broad bean (*Vicia faba* L.) plant in a concentration-dependent manner. An increased amount of Guaiacol peroxidase (GPOD) activity is also seen in Mn-treated plants which indicates that to reduce cellular damage the cell enhances its activity (Arya and Roy 2011). Cu is an essential component of electron transport chain. Industries, mining, and various human activity leads to increasing concentration of Cu in the ecosystem. Excess Cu harms the plants and can cause leaf chlorosis and retard growth (Asati et al. 2016). The toxicity of Cu causes metabolic stress which damage and disturbs macromolecules and molecular pathways respectively (Lin et al. 2020). In chlorosis gayana (Rhodes grass), copper toxicity reduces the growth of roots (Itanna and Coulman 2003). Biomass reduction, plant mortality, and reduction in seed production are observed in black bindweed (*Polygonum convolvulus*), and root malformation is observed in beans (*Phaseolus vulgaris*) due to excess accumulation of copper (Kjær and Elmegaard 1996). Zinc is an essential micronutrient and its deficiency can cause chlorosis and affect the crucial metabolic metabolism of the plant. Zn and Cd toxicity leads to oxidative stress causing a reduction in development, growth, and metabolic activities in *Brassica juncea* and *Phaseolus vulgaris* (Prasad et al. 1999; Sarret et al. 2001). Excess Cd is initiated with inhibition of growth, chlorosis, and root tip browning, and terminates to death if the toxicity becomes severe (Foy et al. 1978). Hg causes high oxidative stress in plants like alfalfa plant causing peroxidation of lipids and altering the enzymatic activities and metal-chelating components in roots (Zhou et al. 2009). Oxidative stress caused by mercury can lead to ROS production and disrupt the integrity of the mitochondrial membrane, and cellular metabolism. It was observed that high hg levels, bioaccumulate in roots and shoots, and reduce tiller and panicle formation, plant height, and grain production in rice (*Oryza sativa*) (Du et al. 2005). Stress-induced due to chromium affect photophosphorylation, CO_2 fixation, electron transport chain and various enzymatic activities. Chromium has a different effect on different pigments like disorganizing the chloroplast which automatically lowers the photosynthesis rate, and affects the production rate of chlorophyll and anthocyanine (Nikolaou et al. 2022). The formation of some components like glutathione and ascorbic acid increases due to overdue stress caused by chromium. Plant strength of nutrient acquisition decrease in *lycopersicon esculentum* (tomato) (Goupil et al. 2009), and a reduction in plant biomass and seed germination in *Allium cepa* (onion) is also observed (Nematshahi et al. 2012). Lead is considered one of the most toxic elements. It not only affects the morphology but also disturbed the physiological function, like, toxicity affecting photosynthesis and in *Spartina alterniflora* it inhibits the seeds' germination (Mrozek and Funicelli 1982). Cobalt when crosses its threshold limit then can become harmful for plants, its higher concentration not only stimulates lipid peroxidation, metabolic pathways destruction, and ROS generation, but an excessive amount can become so lethal that a plant could die because of its toxicity (Karuppanapandian and Kim 2013). Smaltite, erythrite, and cobaltite are naturally occurring cobalt sources. Co toxicity can leadto less biomass production and reduced shoot growth in tomatoes (*Lycopersicon esculentum* L.), oilseed rape (*Brassica napus* L.), and barley (*Hordeum vulgare* L.) (Li et al. 2009). In cauliflower leaves excess of Co limits the production of protein, iron, and chlorophyll and restricts the functioning of the catalase enzyme translocation of other elements like Cu, Zn, S, P, and Mn is also affected but Co in Cauliflower (Sinha et al. 2015). A negative correlation between plant sugar, protein, amino acids, antioxidant enzymes, plant nutrients, chlorophyll, the total surface area of leaves, and the length of shoot and root is also seen in radishes (*Raphanus sativus*) (Manesh et al. 2018). Ni toxicity can cause necrosis and chlorosis in plants. Impairments in, lipid content, cell membrane and plant nutrition, and H-ATPase activity in shoots of *Oryza sativa* also occur. Water uptake reduction in monocot and dicot plant is also an indication of Ni toxicity (Ramzani et al. 2016), Iron competes with Mn for root absorption. Many cells have redox systems constituted with Fe like leghaemoglobin, catalase, cytochromes, peroxidase, aconitase, ferredoxin, and superoxide dismutase (Römheld and Marschner 1991), so its deficiency can cause an effect various cellular functioning. Iron toxicity generally reduces the photosynthesis rate and generation of reactive oxygen species which further disrupts cellular membrane, cell components, proteins, and DNA (Bhattacharjee 2010). Mn is easily absorbed and transmitted by roots, its toxicity can cause chlorosis, 'crinkle leaf', and necrosis on leaves, stems, and petioles (Adams and Wear 1957). Lowering pH below 6 may be one of the indicators of deficiency of Mn. In beans (*Vicia faba*) and spearmint (*Mentha spicata*) Mn toxicity can cause chlorosis, limit the growth of shoot, root and reduce carotenoid and chlorophyll a content (Ng et al. 2018) chlorophyll a and b content, O_2 evolution in photosystem II, lowers in *Pisum sativum*, which reduce photosynthetic rate (Gangwar et al. 2010). The chlorophyll content is also affected in tomatoes (*Lycopersicon esculentum*) (Kleiber and Grajek 2015). Arsenic toxicity reduces dry weight, seedling height, leaf area, and seed germination rate in rice (*Oryza sativa*) (Panaullah et al. 2009). In tomatoes (*Lycopersicon esculentum*) and canola (*Brassica napus*) reduction in leaf weight, fruit yield, growth along with chlorosis and wilting occur (Tlustoš et al. 2006; Farooq et al. 2015).

12.12 Impact on Earthworms

Earthworms are one of the most crucial components of the earth, it not only decomposes matterbut also participated in various nutrient cycle and helps in maintaining soil structure. The presence of earthworms also stimulates the growth of other microorganisms. Different physiological, chemical, and climatic conditions have an impact on soil structure and earthworm diversity composition. Due to their behavioral pattern of feeding habits and burrowing, they are found more susceptible to soil contamination. Long-term exposure to Pb, Zn, Cu, and Cd impacts the reproductive rate of *Metaphire californica* by damaging the seminal vesicle and by increasing sperm deformity (Huang et al. 2023). Exposure of *Eisenia fetida* to Cd, Cu, Pb, and Zn showed a decrease in cocoon production and an increase in mortality rate (Spurgeon et al. 1994). Cd toxicity also leads to a decrease in body weight (Elyamine et al. 2018). An inverse relation between earthworm diversity and heavy metal ion concentration was also found in some studies. Species like *Aporrectodea rosea*, *Allolobophora chlorotica*, and *Aporrectodea caliginosa* are found to be more sensitive to Zn metal contamination (Spurgeon and Hopkin 1996). Most of the heavy metals get bio-accumulated inside the earthworm, but it is not universal for all the metal ions. By atomic absorption spectrometry, it was found that Cd accumulated in seminal vesicles of earthworms however Ni ion did not bioaccumulate even after long-term exposure (Nakashima et al. 2008). Bioaccumulation of metals along who metal concentration also depends on soil pH and cation exchange capacity (Morgan and Morga 1988). Copper toxicity was found more sensitive to cocoon production than earthworm body weight change and their survival. Cocoon production about Cu toxicity was also found to be related to organic carbon content and cation exchange capacity of the soil (Duan et al. 2016). The bioavailability of Pb in soil reduces the soil pH. Soil pH in turn found inversely related to the availability of Pb in soil solution, bioaccumulation of Pb in earthworms, and mortality rate of earthworms (Bradham et al. 2006). Artificial soil tests determined that Cu, Zn, and Ni are more highly toxic to *Eisenia fetida* than Cd and Pb. All three have almost equal toxicity effects, including physiological stress which leads to weight loss in earthworms.

12.13 Remediation of Soil Contaminated with Heavy Metals

Various methods can be used to treat the contamination of heavy metals like reduction, ion exchange, membrane filtration, adsorption, precipitation, and chemical coagulation. Most of these methods can produce secondary pollutants (Kumar et al. 2019) and are unable to remove heavy metal contamination below 100 mg/L. along with that these are costly and eco-friendly up to a certain limit (Dixit et al. 2015). So, it becomes necessary to use a more naturally stable and cost-effective method, which is bioremediation. In bioremediation, in which organisms like plants, bacteria, fungus are used to treat heavy metal-contaminated sources (Kumar and Dwivedi 2021).

1. **Phytoremediation**: It is a technique by which plants are used to purify the contaminated environment. It is considered the safest, cheap, safe, and most conventional technique to reduce contamination from soil, water, sediments, and other contaminated zones (Zavoda et al. 2001; Pandey et al. 2022). Phytoremediation can be of various types (Nedjimi 2021).

    1.1. **Phytoextraction**: It is also known as phytoabsorption, Phyto sequestration, and phytoaccumulation, as the name indicates in this, plants absorb the contamination from the soil by roots and translocate to shoots (Abdel-Shafy and Mansour 2018). It involves the extraction of heavy metals, like As, Zn, Cd, and Ni (McGrath and Zhao 2003) by uptake into the harvestable part of the plant (Suman et al. 2018).

    1.2. **Phytostabilization**: it is a type of phytoremediation in which the metal ion mobility in the soil reduces to minimize heavy metal toxicity (Suman et al. 2018). It reduces off-sitecontamination by immobilization of heavy metals in the rhizosphere or by accumulation in roots, thus it reduces mobility inside the vadosa zone (Bolan et al. 2011).

    1.3. **Phytodegradation**: As the name indicates, this process involves the direct degradation of organic contaminants by the enzymatic or metabolic activity of roots and plant tissue. Roots take the contaminant and convert it into a less toxic substance.

    1.4. **Rhizofiltration**: Also known as a biogeochemical filter. It is a type of phytoremediation in which roots separated toxic chemicals by adsorption and absorption. Contaminated water passed through the mass of roots to remove a low level of contaminants (Rao et al. 2022).

    1.5. **Phytovolatilization**: It includes the transformation of contaminants into volatile forms. Contaminants are absorbed by the roots from the soil and transported through the xylem. These contaminants are converted into less toxic volatile forms and released into the atmosphere (Awa and Hadibarata 2020).

2. **Mycoremediation**: When the fungus is used as a bioremediation reagent for the treat heavy metal contamination, the process through which heavy metal contamination is diminished is termed mycoremediation. For mycoremediation different species of fungus-like*Penicillium*, *Trichoderma*, *Fusarium* and *Aspergillus* can be used in dead biomass form or growing form (Kumar and Dwivedi 2021). Heavy metals like Pb, Mn, Zn, Co, Cu, Ni, Cr, and Cd can be treated by species of *Aspergillus* genus fungi (Kumar and Dwivedi 2021). Fungus shows tolerance against heavy metals through adsorption, bioreduction, biosynthesis, or biotransformation (Sriharsha et al. 2017). Dead biomass removes the contaminants by bioadsorption (Shi et al. 2018). However different factors like pH, de of biomass, temperature, and concentration of heavy metals are some factors that can alter

the rate of mycoremediation, as every fungal species has different characteristics (Kumar and Dwivedi 2021).

3. **Vermiremidiation**: It is one of the emerging techniques of bioremediation, in which earthworm species are used. Soil contaminated with inorganic and organic substances is removed from soil by bioaccumulation in earthworm tissue or by adsorption on vermicast. Earthworms' species like *Eisenia fetida* (Javed and Hashmi 2021), *Alma millsoni*, *Eudrilus eugeniae* and *Libyodrilus violaceus* (Owagboriaye et al. 2020), *Drawidamodesta* and *Lampitomauritii* (Rajadurai et al. 2022), *Amynthas robustus* (Lin et al. 2021) can be used for vermiremediation of heavy metal. Earthworms not only bioaccumulate contaminants but also make the soil environment more favorablefor microbes. By increasing the microbial population, earthworms also stimulate microbial remediation (Dabke 2013).
4. **Micro remediation**: When microbes like bacteria are used for the remediation then it is termed as micro-remediation (Ahmed 2018). Certain bacteria, such as Nostoc and Anabaena, which fix atmospheric nitrogen in the form of ammonium ions, have a symbiotic connection with the roots of some higher plants (Pandey and Kamboj 2022). Bacteria have the ability to sense the heavy metals in their surroundings. Active efflux of heavy metals, intracellular and extracellular sequestration, metal ion reduction, and extracellular barriers are the five mechanisms used by bacteria for heavy metal resistance (Ianeva 2009). Thirteen resistance, nodulation, and cell division (RND), family proteins are found as a part of Protein complexes present in the trans envelope of *Cupriavidusmetallidurans* CH34, used in the exportation of heavy metal ions from periplasm to the outer side of the cell (von Rozycki and Nies 2009). The presence of heavy metals also enhances the activity of the class1 integrons gene family, which along with lateral gene transfer also helps in adaptation to stress (Rosewarne et al. 2010). A normal level of metal toxicity is resisted by the bacteria without hindering bacterial growth. In many bacteria, the genes which control metal resistance are found on the plasmid, but it does not mean that the non-plasmid bacterial species cannot resist the heavy metal. For i.e., Pseudomonas sp. is on plasmid bacteria but can resist the lead concentration up to 1000 μg/ml lead (Trevors et al. 1985).

12.14 Conclusion

There are numerous sources of heavy metals that contaminate land, water, and air, including medical industries, automobile emissions, incomplete combustion of fossil fuels, volcanic eruptions, mining, construction, technology, agriculture, residential, and industrial waste. Metals are necessary for an organism's metabolic processes, and when their concentration exceeds their permitted limit or threshold, they are referred to as contaminants. When their concentration is high, they disrupt metabolic processes and become harmful to organisms. Their extensive spread in the environment is due to their numerous applications in various industries, home settings, and the medical and technological fields. A wide variety of micro and macro-organisms live in the soil. The effects of heavy metals on various soil faunal communities, including bacteria, fungi, protozoans, and plants, vary. According to Hou et al. (2017), increased soil pollution possesses a serious threat to human society in addition to its effects on the soil faunal population since it passes down the food chain and undergoes bioaccumulation and bio-magnification. Treatment options for heavy metal pollution include chemical coagulation, ion exchange, membrane filtration, adsorption, reduction, and ion exchange. Most of these methods can produce secondary pollutants. Phytoremediation, Mycoremediation, Vermiremidiation, Micro remediation, and various other methods are currently ongoing for the treatment of heavy metal contamination.

12.15 Future Perspective of This Study

This book chapter is written to check the current heavy metal generation in various sectors from natural and anthropogenic activities. Heavy metals are mostly generated from anthropogenic sources due to unscientific agricultural farming, industrial effluents, bad construction, mining excessive use of fossil fuel, or incomplete combustion of petrol products. These all are the major source of heavy metals and soil contaminants. For a better and healthy future, heavy metal pollution must be mitigated from the environment, especially from the soil and water bodies, as both are very crucial for sustainable life. From the soil and water bodies heavy metals directly bioaccumulate in the food chain and through it, these metals or contaminants reach mankind and impose adverse impacts on human health. There is an urgent need to adopt such natural technologies which can reduce heavy metal contamination without polluting the environment. Today environmentalists and microbiologists are focusing to find the best and cheapest way to diminish heavy metal pollution and its associated problems.

Acknowledgements The author thanks the Department of Environmental Science, Kanya Gurukula Campus, Gurukula Kangri (Deemed to be University), Haridwar, Uttarakhand, for providing the lab facility.

Conflict of Interest The author declares that there is no conflict of interest.

References

Abaye DA, Lawlor K, Hirsch PR, Brookes PC (2005) Changes in the microbial community of an arable soil caused by long-term metal contamination. Eur J Soil Sci 56(1):93–102

Abdel-Shafy HI, Mansour MS (2018) Phytoremediation for the elimination of metals, pesticides, PAHs, and other pollutants from wastewater and soil. Phytobiont and ecosystem restitution. Springer, Singapore, pp 101–136

Adams F, Wear JI (1957) Manganese toxicity and soil acidity in relation to crinkle leaf of cotton. Soil Sci Soc Am J 21(3):305–308
Ahmad I, Hayat S, Ahmad A, Inam A (2005) Effect of heavy metal on survival of certain groups of indigenous soil microbial population
Ahmad I, Zafar S, Ahmad F (2005) Heavy metal biosorption potential of Aspergillus and Rhizopus sp. isolated from wastewater-treated soil
Ahmad I, Akhtar MJ, Zahir ZA, Jamil A (2012) Effect of cadmium on seed germination and seedling growth of four wheat (*Triticum aestivum* L.) cultivars. Pak J Bot 44(5):1569–1574
Ahmed A (2018) Micro-remediation of chromium-contaminated soils. PeerJ 6:e6076
Alloway BJ (1995) Soil processes and the behavior of metals. Heavy Met Soils 13:3488
Anderson JM, Ineson P (1984) Interactions between microorganisms and soil invertebrates in nutrient flux pathways of forest ecosystems [Soil macrofauna]. In *Symposium Series-British Mycological Society,* pp 59–88
Arif N, Yadav V, Singh S, Singh S, Ahmad P, Mishra RK, Chauhan DK (2016) Influence of high and low levels of plant-beneficial heavy metal ions on plant growth and development. Front Environ Sci 4:69
Arya SK, Roy BK (2011) Manganese induced changes in growth, chlorophyll content and antioxidants activity in seedlings of broad bean (*Vicia faba* L.). J Environ Biol 32(6):707–711
Asati A, Pichhode M, Nikhil K (2016) Effect of heavy metals on plants: an overview. Int J Appl Innovation Eng Manage 5(3):56–66
Awa SH, Hadibarata T (2020) Removal of heavy metals in contaminated soil by phytoremediation mechanism: a review. Water Air Soil Pollut 231(2):1–11
Baldock JA, Skjemstad JO (2000) Role of the soil matrix and minerals in protecting natural organic materials against biological attack. Org Geochem 31(7–8):697–710
Baldrian P (2003a) Interactions of heavy metals with white-rot fungi. Enzym Microbial Technol 32(1):78–91
Baldrian P (2003b) Interactions of heavy metals with white-rot fungi. Enzyme Microb Technol 32(1):78–91
Baldrian P (2010) Effect of heavy metals on saprotrophic soil fungi. Soil heavy metals. Springer, Berlin, Heidelberg, pp 263–279
Bhalerao SA, Sharma AS, Poojari AC (2015) Toxicity of nickel in plants. Int J Pure Appl Biosci 3(2):345–355
Bhattacharjee S (2010) Sites of generation and physicochemical basis of formation of reactive oxygen species in plant cell. React Oxygen Species Antioxid High Plants 1:1–30
Bolan NS, Park JH, Robinson B, Naidu R, Huh KY (2011) Phytostabilization: a green approach to contaminant containment. Adv Agron 112:145–204
Bradham KD, Dayton EA, Basta NT, Schroder J, Payton M, Lanno RP (2006) Effect of soil properties on lead bioavailability and toxicity to earthworms. Environ Toxicol Chem Int J 25(3):769–775
Bruins MR, Kapil S, Oehme FW (2000) Microbial resistance to metals in the environment. Ecotoxicol Environ Saf 45(3):198–207
Cai L, Xu Z, Bao P, He M, Dou L, Chen L, Zhu YG (2015) Multivariate and geostatistical analyses of the spatial distribution and source of arsenic and heavy metals in the agricultural soils in Shunde, Southeast China. J Geochem Explor 148:189–195
Castaldi P, Garau G, Melis P (2004) Influence of compost from sea weeds on heavy metal dynamics in the soil-plant system. Fresenius Environ Bull 13(11):1322–1328
Cheng X, Danek T, Drozdova J, Huang Q, Qi W, Zou L, Xiang Y (2018) Soil heavy metal pollution and risk assessment associated with the Zn-Pb mining region in Yunnan, Southwest China. Environ Monit Assess 190:1–16
Collin-Hansen C, Andersen RA, Steinnes E (2005) Damage to DNA and lipids in Boletus edulis exposed to heavy metals. Mycol Res 109(12):1386–1396
Dabke SV (2013) Vermi-remediation of heavy metal-contaminated soil. J Health Pollut 3(4):4–10

Dai Z, Feng X, Zhang C, Wang J, Jiang T, Xiao H, Qiu G (2013) Assessing anthropogenic sources of mercury in the soil in the Wanshan Hg mining area, Guizhou, China. Environ Sci Pollut Res 20:7560–7569

De Filippis LF, Hampp R, Ziegler H (1981) The effects of sublethal concentrations of zinc, cadmium, and mercury on Euglena. Arch Microbiol 128(4):407–411

Díaz S, Martín-González A, Gutiérrez JC (2006) Evaluation of heavy metal acute toxicity and bioaccumulation in soil ciliated protozoa. Environ Int 32(6):711–717

Dixit R, Malaviya D, Pandiyan K, Singh UB, Sahu A, Shukla R, Singh BP, Rai JP, Sharma PK, Lade H (2015) Bioremediation of heavy metals from soil and aquatic environment: an overview of principles and criteria of fundamental processes. Sustainability 7(2):2189–2212

Doncheva S, Stoynova Z, Velikova V (2001) Influence of succinate on zinc toxicity of pea plants. J Plant Nutr 24(6):789–804

Du X, Zhu YG, Liu WJ, Zhao XS (2005) Uptake of mercury (Hg) by seedlings of rice (Oryza sativa L.) grown in solution culture and interactions with arsenate uptake. Environ Exp Bot 54(1):1–7

Duan X, Xu M, Zhou Y, Yan Z, Du Y, Zhang L, Li F (2016) Effects of soil properties on copper toxicity to earthworm Eisenia fetida in 15 Chinese soils. Chemosphere 145:185–192

Elyamine AM, Afzal J, Rana MS, Imran M, Cai M, Hu C (2018) Phenanthrene mitigates cadmium toxicity in earthworms Eisenia fetida (epigeic specie) and Aporrectodea caliginosa (endogeic specie) in soil. Intern Environ Res Public Health 15(11):2384

Farooq MA, Li L, Ali B, Gill RA, Wang J, Ali S, Zhou W (2015) Oxidative injury and antioxidant enzymes regulation in arsenic-exposed seedlings of four Brassica napus L. cultivars. Environ Sci Pollut Res 22(14):10699–10712

Foissner W (1996) Soil protozoan diversity: the state of art. Mitt Deut Bodenk Gesell 81:219–220

Foy CD, Chaney RT, White MC (1978) The physiology of metal toxicity in plants. Annu Rev Plant Physiol 29(1):511–566

Gadd GM (1994) Interactions of fungi with toxic metals. The genus Aspergillus. Springer, Boston, MA, pp 361–374

Gadd GM (2007a) Geomycology: biogeochemical transformations of rocks, minerals, metals, and radionuclides by fungi, bioweathering and bioremediation. Mycol Res 111(1):3–49

Gadd GM (2007b) Transformation and mobilization of metals, metalloids, and radionuclides by microorganisms. *Biophysico-chemical processes of heavy metals and metalloids in soil environments*. Wiley, Hoboken, pp 53–96

Gangwar S, Singh VP, Prasad SM, Maurya JN (2010) Modulation of manganese toxicity in *Pisum sativum* L. seedlings by kinetin. Sci Hortic 126(4):467–474

Giller KE, Witter E, Mcgrath SP (1998) Toxicity of heavy metals to microorganisms and microbial processes in agricultural soils: a review. Soil Biol Biochem 30(10–11):1389–1414

Göhre V, Paszkowski U (2006) Contribution of the arbuscular mycorrhizal symbiosis to heavy metal phytoremediation. Planta 223(6):1115–1122

Goupil P, Souguir D, Ferjani E, Faure O, Hitmi A, Ledoigt G (2009) Expression of stress-related genes in tomato plants exposed to arsenic and chromium in nutrient solution. J Plant Physiol 166(13):1446–1452

He Y, Xiang Y, Zhou Y, Yang Y, Zhang J, Huang H, Tang L (2018) Selenium contamination, consequences and remediation techniques in water and soils: a review. Environ Res 164:288–301

Hou D, O'Connor D, Nathanail P, Tian L, Ma Y (2017) Integrated GIS and multivariate statistical analysis for regional scale assessment of heavy metal soil contamination: a critical review. Environ Pollut 231:1188–1200

Huang S, Ming HUA, Jinshun FENG, Zhong X, Yang JIN, Baiwan ZHU, Hua L (2009a) Assessment of selenium pollution in agricultural soils in the Xuzhou District, Northwest Jiangsu, China. J Environ Sci 21(4):481–487

Huang C, Shen Z, Li L, Yue S, Jia L, Wang K, Qiao Y (2023) Reproductive damage and compensation of wild earthworm *Metaphire californica* from contaminated fields with long-term heavy metal exposure. Chemosphere 311:137027

Ianeva OD (2009) Mechanisms of bacteria resistance to heavy metals. Mikrobiolohichnyizhurnal (Kiev, Ukraine: 1993) 71(6):54–65
Itanna F, Coulman B (2003) Phyto-extraction of copper, iron, manganese, and zinc from environmentally contaminated sites in Ethiopia, with three grass species. Commun Soil Sci Plant Anal 34(1–2):111–124
Jacob JM, Karthik C, Saratale RG, Kumar SS, Prabakar D, Kadirvelu K, Pugazhendhi A (2018) Biological approaches to tackle heavy metal pollution: a survey of literature. J Environ Manage 217:56–70
Javed F, Hashmi I (2021) Vermiremediation–remediation of soil contaminated with oil using earthworm (Eisenia fetida). Soil Sedim Contam Int J 30(6):639–662
Jayasumana C, Fonseka S, Fernando A, Jayalath K, Amarasinghe M, Siribaddana S, Paranagama P et al (2015) Phosphate fertilizer is a main source of arsenic in areas affected with chronic kidney disease of unknown etiology in Sri Lanka. SpringerPlus 4:1–8
Karuppanapandian T, Kim W (2013) Cobalt-induced oxidative stress causes growth inhibition associated with enhanced lipid peroxidation and activates antioxidant responses in Indian mustard (Brassica juncea L.) leaves. Acta Physiologiae Plant 35(8):2429–2443
Khan ZI, Arshad N, Ahmad K, Nadeem M, Ashfaq A, Wajid K, Ugulu I et al (2019) Toxicological potential of cobalt in forage for ruminants grown in polluted soil: a health risk assessment from trace metal pollution for livestock. Environ Sci Pollut Res 26:15381–15389
Kjær C, Elmegaard N (1996) Effects of copper sulfate on black bindweed (Polygonum convolvulusL). Ecotoxicol Environ Saf 33(2):110–117
Kleiber T, Grajek M (2015) Tomato reaction on excessive manganese nutrition. Bul J Agric Sci 21(1):118–125
Komulainen M, Mikola J (1995) Soil processes as influenced by heavy metals and the composition of soil fauna. J Appl Ecol 234–241
Kremer BP, Markham JW (1982) Primary metabolic effects of cadmium in the brown alga, Laminaria saccharina. Z Pflanzenphysiol 108(2):125–130
Kříbek B, Majer V, Knésl I, Nyambe I, Mihaljevič M, Ettler V, Sracek O (2014) Concentrations of arsenic, copper, cobalt, lead, and zinc in cassava (*Manihot esculenta* Crantz) growing on uncontaminated and contaminated soils of the Zambian Copperbelt. J Afr Earth Sci 99:713–723
Kumar V, Dwivedi SK (2021) Mycoremediation of heavy metals: processes, mechanisms, and affecting factors. Environ Sci Pollut Res 28(9):10375–10412
Kumar V, Singh S, Singh G, Dwivedi SK (2019) Exploring the Cadmium Tolerance and Removal Capability of a Filamentous Fungus *Fusarium solani*. Gemicrobiol J 36(9):782–791
Kuperman RG, Carreiro MM (1997) Soil heavy metal concentrations, microbial biomass and enzyme activities in a contaminated grassland ecosystem. Soil Biol Biochem 29(2):179–190
Lecoanet H, Leveque F, Ambrosi JP (2001) Magnetic properties of salt-marsh soils contaminated by iron industry emissions (southeast France). J Appl Geophys 48(2):67–81
Li HF, Gray C, Mico C, Zhao FJ, McGrath SP (2009) Phytotoxicity and bioavailability of cobalt to plants in a range of soils. Chemosphere 75(7):979–986
Li X, Meng D, Li J, Yin H, Liu H, Liu X, Yan M (2017) Response of soil microbial communities and microbial interactions to long-term heavy metal contamination. Environ Pollut 231:908–917
Libohova Z, Seybold C, Wysocki D, Wills S, Schoeneberger P, Williams C, Owens PR (2018) Reevaluating the effects of soil organic matter and other properties on available water-holding capacity using the National Cooperative Soil Survey Characterization Database. J Soil Water Conserv 73(4):411–421
Lin S, Qiao N, Chen H, Tang Z, Han Q, Mehmood K, Zhang H et al (2020) Integration of transcriptomic and metabolomic data reveals metabolic pathway alteration in mouse spermatogonia with the effect of copper exposure. Chemosphere 256:126974
Lin Z, Zhen Z, Luo S, Ren L, Chen Y, Wu W, Zhang D (2021) Effects of two ecological earthworm species on tetracycline degradation performance, pathway and bacterial community structure in laterite soil. J Hazard Mater 412:125212

Liu Y, Gao T, Xia Y, Wang Z, Liu C, Li S, Lv Y (2020) Using Zn isotopes to trace Zn sources and migration pathways in paddy soils around the mining area. Environ Pollut 267:115616

Lorimer GH (1981) Ribulosebisphosphate carboxylase: amino acid sequence of a peptide bearing the activator carbon dioxide. Biochem Mar 3, 20(5):1236–40. https://doi.org/10.1021/bi00508a028. PMID: 6784749

Madoni P, Davoli D, Gorbi G, Vescovi L (1996) Toxic effect of heavy metals on the activated sludge protozoan community. Water Res 30(1):135–141

Manesh RR, Grassi G, Bergami E, Marques-Santos LF, Faleri C, Liberatori G, Corsi I (2018) Co-exposure to titanium dioxide nanoparticles does not affect cadmium toxicity in radish seeds (Raphanus sativus). Ecotoxicol Environ Saf 148:359–366

Manivasagaperumal R, Balamurugan S, Thiyagarajan G, Sekar J (2011) Effect of zinc on germination, seedling growth, and biochemical content of cluster bean (Cyamopsis tetragonoloba (L.) Taub). Curr Bot 2(5):11–15

McGrath SP, Zhao FJ (2003) Phytoextraction of metals and metalloids from contaminated soils. Curr Opin Biotechnol 14(3):277–282

Morgan JE, Morgan AJ (1988) Earthworms as biological monitors of cadmium, copper, lead, and zinc in metalliferous soils. Environ Pollut 54(2):123–138

Mrozek E Jr, Funicelli NA (1982) Effect of zinc and lead on germination of Spartina alterniflora Loisel seeds at various salinities. Environ Exp Bot 22(1):23–32

Nakashima T, Okada T, Asahi J, Yamashita A, Kawai K, Kasai H, Hirano T (2008) 8-Hydroxydeoxyguanosine generated in the earthworm Eisenia fetida grown in metal-containing soil. Mutat Res/Genet Toxicol Environ Mutagenesis 654(2):138–144

Nannipieri P, Badalucco L, Landi L, Pietramellara G (1997) Measurement in assessing the risk of chemicals to the soil ecosystem. Ecotoxicol Responses Biomark Risk Assess (1):507

Nedjimi B (2021) Phytoremediation: a sustainable environmental technology for heavy metals decontamination. SN Appl Sci 3(3):1–19

Nematshahi N, Lahouti M, Ganjeali A (2012) Accumulation of chromium and its effect on growth of (Allium cepa cv. Hybrid). Eur J Exp Biol 2(4):969–974

Ng KT, Herrero P, Hatt B, Farrelly M, McCarthy D (2018) Biofilters for urban agriculture: metal uptake of vegetables irrigated with stormwater. Ecol Eng 122:177–186

Nikolaou KE, Chatzistathis T, Theocharis S, Argiriou A, Koundouras S, Zioziou E (2022) Effects of chromium toxicity on physiological performance and nutrient uptake in two Grapevine Cultivars (Vitis vinifera L.) Growing on own roots or grafted onto different rootstocks. Horticulturae 8(6):493

Nonnecke IL (1989) Vegetable production. Springer Science & Business Media

Owagboriaye F, Dedeke G, Bamidele J, Aladesida A, Isibor P, Feyisola R, Adeleke M (2020) Biochemical response and vermiremediation assessment of three earthworm species (*Alma millsoni*, *Eudrilus eugeniae* and *Libyodrilus violaceus*) in soil contaminated with a glyphosate-based herbicide. Ecol Ind 108:105678

Panaullah GM, Alam T, Hossain MB, Loeppert RH, Lauren JG, Meisner CA, Duxbury JM et al (2009) Arsenic toxicity to rice (*Oryza sativa* L.) in Bangladesh. Plant Soil 317(1):31–39

Pandey N, Kamboj N (2022) Consequences of soil quality on crop yield: an update. Indian J Ecol 49(4):1406–1416

Pandey N, Kamboj N, Sharma AK, Kumar A (2022) An overview of recent advancements in the irrigation, fertilization, and technological revolutions of agriculture. Environ Pollut Nat Resource Manage 167–184

Prasad DDK, Prasad ARK (1987) Effect of lead and mercury on chlorophyll synthesis in mung bean seedlings. Phytochemistry 26(4):881–883

Prasad KVSK, Saradhi PP, Sharmila P (1999) Concerted action of antioxidant enzymes and curtailed growth under zinc toxicity in Brassica juncea. Environ Exp Bot 42(1):1–10

Rajadurai M, Karmegam N, Kannan S, Yuvaraj A, Thangaraj R (2022) Vermiremediation of engine oil-contaminated soil employing indigenous earthworms, Drawidamodesta, and Lampitomauritii. J Environ Manage 301:113849

Ramzani PMA, Khan WUD, Iqbal M, Kausar S, Ali S, Rizwan M, Virk ZA (2016) Effect of different amendments on rice (Oryza sativa L.) growth, yield, nutrient uptake, and grain quality in Ni-contaminated soil. Environ Sci Pollut Res 23(18):18585–18595

Rao TS, Panigrahi S, Velraj P (2022) Transport and disposal of radioactive wastes in the nuclear industry. In Microbial biodegradation and bioremediation. Elsevier, pp 419–440

Rathnayake IVN (2010) Development of whole-cell microbial biosensors to assess the heavy metal bioavailability in the soil environment (Doctoral dissertation)

Reber HH (1992) Simultaneous estimates of the diversity and the degradative capability of heavy-metal-affected soil bacterial communities. Biol Fertil Soils 13(3):181–186

Renella G, Chaudri AM, Brookes PC (2002) Fresh additions of heavy metals do not model long-term effects on microbial biomass and activity. Soil Biol Biochem 34(1):121–124

Römheld V, Marschner H (1991) Function of micronutrients in plants. Micronutrients in agriculture, 4:297–328

Rosewarne CP, Pettigrove V, Stokes HW, Parsons YM (2010) Class 1 integrons in benthic bacterial communities: abundance, association with Tn 402-like transposition modules and evidence for co-selection with heavy-metal resistance. FEMS Microbiol Ecol 72(1):35–46

Sarret G, Vangronsveld J, Manceau A, Musso M, D'Haen J, Menthonnex JJ, Hazemann JL (2001) Accumulation forms of Zn and Pb in Phaseolus vulgaris in the presence and absence of EDTA. Environ Sci Technol 35(13):2854–2859

Shi L, Xue J, Liu B, Dong P, Wen Z, Shen Z, Chen Y (2018) Hydrogen ions and organic acids secreted by ectomycorrhizal fungi, Pisolithus sp1, are involved in the efficient removal of hexavalent chromium from wastewater. Ecotoxicol Environ Saf 161:430–436

Shrivastava AK (2009) A review on copper pollution and its removal from water bodies by pollution control technologies. Indian J Environ Prot 29(6):552–560

Simon T (1999) The effect of increasing rates of nickel and arsenic on the growth of radish and soil microflora. RostlinnaVyroba-UZPI (Czech Republic)

Singh J, Kalamdhad AS (2011) Effects of heavy metals on soil, plants, human health, and aquatic life. Int J Res Chem Environ 1(2):15–21

Singh S, Prasad SM (2017) Effects of 28-homobrassinoloid on key physiological attributes of Solanum lycopersicum seedlings under cadmium stress: photosynthesis and nitrogen metabolism. Plant Growth Regul 82:161–173

Sinha P, Shukla AK, Sharma YK (2015) Amelioration of heavy-metal toxicity in cauliflower by application of salicylic acid. Commun Soil Sci Plant Anal 46(10):1309–1319

Spurgeon DJ, Hopkin SP (1996) The effects of metal contamination on earthworm populations around a smelting works: quantifying species effects. Appl Soil Ecol 4(2):147–160

Spurgeon DJ, Hopkin SP, Jones DT (1994) Effects of cadmium, copper, lead and zinc on growth, reproduction and survival of the earthworm Eisenia fetida (Savigny): assessing the environmental impact of point-source metal contamination in terrestrial ecosystems. Environ Pollut 84(2):123–130

Sriharsha DV, Lokesh KR, Savitha J (2017) Immobilized fungi on Luffa cylindrica: an effective biosorbent for the removal of lead. J Taiwan Inst Chem Eng 80:589–595

StiborovÁ M (1988) Cd2+ ions affect the quaternary structure of ribulose-1, 5-bisphosphate carboxylase from barley leaves. Biochem Physiol Pflanz 183(5):371–378

Suman J, Uhlik O, Viktorova J, Macek T (2018) Phytoextraction of heavy metals: a promising tool for clean-up of the polluted environment? Front Plant Sci 9:1476

Sun C, Liu J, Wang Y, Sun L, Yu H (2013) Multivariate and geostatistical analyses of the spatial distribution and sources of heavy metals in agricultural soil in Dehui, Northeast China. Chemosphere 92(5):517–523

Swift MJ, Anderson JM (1993) Biodiversity and ecosystem function in agricultural systems. In: Schulze ED, Mooney HA (eds) Biodiversity and ecosystem function. Ecological Studies 99, Springer, Heidelberg, pp 15±41

Tlustoš P, Száková J, Pavlíková D, Balík J (2006) The response of tomato (Lycopersicon esculentum) to different concentrations of inorganic and organic compounds of arsenic. Biologia 61(1):91–96

Trevors JT, Oddie KM, Belliveau BH (1985) Metal resistance in bacteria. FEMS Microbiol Rev 1(1):39–54

Van Assche F, Clijsters H (1986) Inhibition of photosynthesis in Phaseolus vulgaris by treatment with toxic concentration of zinc: effect on ribulose-1, 5-bisphosphate carboxylase/oxygenase. J Plant Physiol 125(3–4):355–360

Van Breemen N, Buurman P (2002) Soil formation. Springer Science & Business Media

Vangronsveld J, Clijsters H (1994) Toxic effects of metals. Plants and the chemical elements: biochemistry, uptake, tolerance and toxicity, pp 149–177

Vig K, Megharaj M, Sethunathan N, Naidu R (2003) Bioavailability and toxicity of cadmium to microorganisms and their activities in soil: a review. Adv Environ Res 8(1):121–135

von Rozycki T, Nies DH (2009) Cupriavidusmetallidurans: evolution of a metal-resistant bacterium. Antonie Van Leeuwenhoek 96(2):115–139

Wang Y, Shi J, Wang H, Lin Q, Chen X, Chen Y (2007) The influence of soil heavy metals pollution on soil microbial biomass, enzyme activity, and community composition near a copper smelter. Ecotoxicol Environ Saf 67(1):75–81

Wang X, Sun Y, Guo H, Wang H (2022a) The genome of the mesopolyploid crop species *Brassica rapa*. Nat Genet 43(10):1035–1039

Wang X, Sun Y, Guo H, Wang H (2022b) Analysis of soil heavy metal Hg pollution source based on GeoDetector. Pol J Environ Stud 31(1)

Weissmannová HD, Pavlovský J (2017) Indices of soil contamination by heavy metals–methodology of calculation for pollution assessment (minireview). Environ Monit Assess 189(12):616

Winkel LH, Johnson CA, Lenz M, Grundl T, Leupin OX, Amini M, Charlet L (2012) Environmental selenium research: from microscopic processes to global understanding

Yahaya AI (2011) Nickel in soils: a review of its distribution and impacts. Sci Res Essays 6(33):6774–6777

Yang XE, He ZL, Mahmood Q (2007) Assessing potential dietary toxicity of heavy metals in selected vegetables and food crops. J Zhejiang Univ Sci B 8(1):1–13

Yaylalı-Abanuz G (2011) Heavy metal contamination of surface soil around Gebze industrial area, Turkey. Microchem J 99(1):82–92

Yedjou CG, Patlolla AK, Sutton DJ (2012) Heavy metals toxicity and the environment Paul B Tchounwou. Published in Final Edited Form as EXS 101:133–164

Yolcubal I, Brusseau ML, Artiola JF, Wierenga P, Wilson LG (2004) Environmental physical properties and processes. In: Environmental monitoring and characterization. Academic Press, pp 207–239

Zavoda J, Cutright T, Szpak J, Fallon E (2001) Uptake, selectivity, and inhibition of hydroponic treatment of contaminants. J Environ Eng 127(6):502–508

Zhao P, Capella-Gutiérrez S, Shi Y, Zhao X, Chen G, Gabaldón T, Ma XF (2014) Transcriptomic analysis of a psammophyte food crop, and rice (Agriophyllumsquarrosum) and identification of candidate genes essential for sand dune adaptation. BMC Genomics 15(1):1–14

Zhou ZS, Guo K, Elbaz AA, Yang ZM (2009) Salicylic acid alleviates mercury toxicity by preventing oxidative stress in the roots of Medicago sativa. Environ Exp Bot 65(1):27–34

Chapter 13
Soil Nutrient Content Estimation Using Hyperspectral Remote Sensing

M. Sivasakthi, S. Sathiyamurthi, and S. Praveen Kumar

Keywords Soil nutrient content · Hyperspectral remote sensing · PLSR · ANN · Machine learning

13.1 Introduction

In the last decades, agriculture has transitioned from the machinery era to the information era, and the term "precision agriculture" (PA) has become increasingly prevalent. This term refers to the small-scale, information-based optimization of inputs for increased profit and environmental safety. Precision agriculture includes geographic information systems (GIS), Global Positioning Systems (GPS), and remote sensing (RS), which is widely employed in many discipline. Almost every facet of agricultural production has been incorporated with precision agriculture. As data and computation-intensive technology, PA relies heavily on extremely efficient and trustworthy ways for collecting and processing site-specific field data. RS utilises electromagnetic energy to estimate the attributes of distant objects/material and has the benefits of extensiveness, non-invasiveness, speed, and adaptability.

The soil is most essential elements for agricultural production and can have a significant impact on quality and yields of crop. Farmers have used traditional soil knowledge for generations to make judgments regarding crop management strategies. Typically, soil physical and chemical variables data are gathered through grid

M. Sivasakthi · S. Sathiyamurthi (✉) · S. Praveen Kumar
Department of Soil Science and Agricultural Chemistry, Faculty of Agriculture, Annamalai University, Annamalai Nagar, Chidambaram 608002, India
e-mail: sathya.soil@gmail.com

S. Sathiyamurthi
Horticultural Research Station,, Tamil Nadu Agricultural University, Thadiyankudisai, Perum, Thadiyankudisai, Perumbarai 624212, India

P. P. Adhikary et al. (eds.), *Soil, Water Pollution and Mitigation Strategies*, Environmental Science and Engineering, https://doi.org/10.1007/978-3-031-63296-9_10

sampling. According to Samuelson and co-workers (2002), a soil map with a scale of 1:31,680 costs an average of 4.2 USD ha^{-1}. The typical cost of a 1:3000 scale map may be 47 USD ha^{-1}, a price farmers would not pay. There are few observations at sampling sites and extrapolating soil qualities from a small number of locations causes inaccuracy. With the growth of precision agriculture, which can necessitate high spatial resolution of in-field soil properties, the limitations of grid sampling have grown more obvious, and agricultural scientists and engineers have turned to remote sensing to characterise in-field soil qualities.

Understanding the dynamics of agricultural ecosystems necessitates a spatial assessment of soil attributes. Site-specific data can provide vital information about healthy soils nutrient supply potential, minimising nutrient loss and sediments and assessing the transfer of components like carbon between soil and air. The laboratory estimation of soil determining the fertility status and determining the quantity of compost to be applied. Such laboratory analysis are expensive, intricate, and time-consuming for soil macronutrients analyses, and require highly labour to collect samples for both field and laboratory testing. Conventional plant nutrition diagnosis and fertilizer recommendations are predicated mostly on field sampling and laboratory chemical analysis. Based on these restrictions, it is reasonable to conclude that the technologies listed above are inadequate to meet the increasing expectations of precision agriculture.

In recent decades, HRS has been utilised to discover and evaluate soil nutritional characteristics via the examination of spectral absorption features in order to circumvent these significant challenges. This technique has been accounted for in a number of soil macronutrient estimation studies (Morellos et al. 2016). Many scientists have used HRS in specific wavelength regions to distinguish soil type (Lacerda et al. 2016), soil hydrocarbon pollution (Resendez-Hernández et al. 2019), soil salinity (Mitran et al. 2015), and soil nitrogen prediction (Wang et al. 2017). It gives real-time findings and sample monitoring, even in remote places where in situ sensors are present (Gomez et al. 2016).

Utilizing HRS spectral information datain an efficient manner is a difficult task. Derivative spectroscopy is an effective method for analysing hyperspectral spectral data (Kumar and Ghosh 2017). The techniques to improve hyperspectral data estimation accuracy include use derivative algorithms, removing overlapping spectral features, separating closely related absorption characteristics, improving reflectance spectra contrast, and eliminating background signals. The calculation of derivative spectroscopy incorporates the change in spectral reflectance, which characterises the form of the spectral curve and sensitive to changes in the shape rather than magnitude of spectra. The sensitivity of derivatives to fluctuations in illumination intensity induced by changes in sun angle, cloud cover, or topography should be quite low (Tsai and Philpot 1998).

13.2 Hyperspectral Remote Sensing

Imaging and non-imaging spectroscopy are both terms used to describe hyperspectral remote sensing. Recent advancements in RS and GIS have paved the way for the creation of hyperspectral sensors, a relatively new technology that is now utilised by researchers and scientists for the recognition and identification of materials, crop species, soil analysis, terrestrial vegetation, and artificial materials and background applications. Since more than a century ago, HRS has been utilised to analyse various objects' chemical and biological compositions. The hyperspectral sensor provides an alternative, non-destructive method for analysing the physical and chemical properties of a substance. Spectroscopy may detect unique absorption properties resulting from specific bonding in a solid, liquid, or gaseous substance. As a result of recent technological advancements, image spectroscopy has begun focused on nearly all Earth applications. Since the mid-1980s, geologists have utilised HRS extensively for the mapping of mineral deposits. Typically, hyperspectral sensors catch light between 350 and 2500 nm. It encompasses the visible, NIR, and SWIR frequency bands. The acquisition of hyper-spectral data encompasses tens or hundreds of small spectral bands against multispectral data with less spectral bands.

13.3 Basics of Hyperspectral Sensors

Hyperspectral sensors, also termed as imaging spectrometers, are the successor to multispectral imaging radiometers in spectral imaging. Current examples of satellites in space include LANDSAT, SPOT, IKONOS, and WorldView. Multiple pictures are collected across multiple wavelength bands during spectral imaging. The hyperspectral images model is presented in Fig. 13.1. Hyperspectral images consist of a cube of data in three dimensions which provides data for all measured wavelengths. The X, Y and Z axes indicates the size and wavelength of the spectrum. The value of each pixels called as spectral reflectance which indicates the quantity of optical energy observed by the sensor from objects in the earth. The significance of hyperspectral sensing lies in its ability to collect information on the spatial and spectral properties of the Earth and can be used to interpret a variety of properties of the observed items.

13.4 Type of HRS Platforms

Usually, HRS imagers are typically mounted on lab or field based (stationary or handheld) and aerial or space-based systems. The ground-based systems are employed for measurements requiring exceptionally high spectral resolution, such as spectral signature or bidirectional reflection distribution function studies (BRDF). Airborne

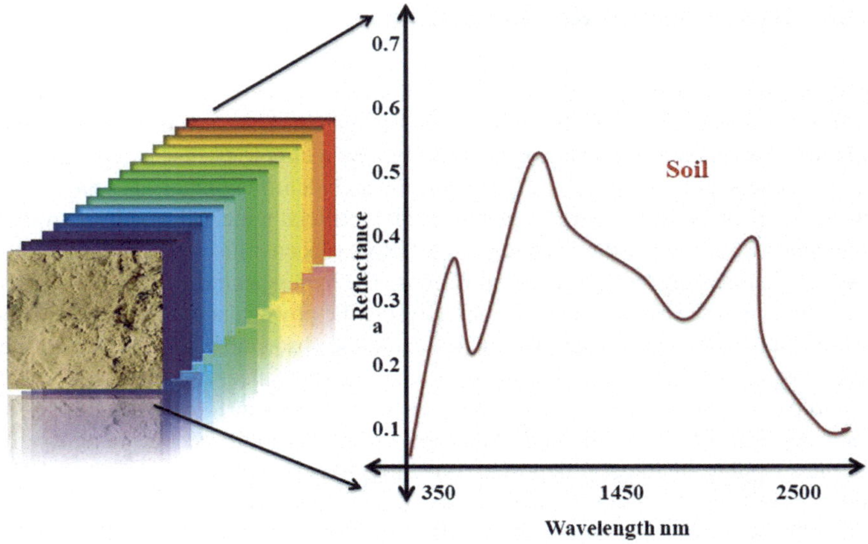

Fig. 13.1 The hyperspectral images model

HRS has a nearly 30-year history, beginning with NASA's Airborne Imaging Spectrometer (AIS) in 1982, which was the first airborne hyperspectral platform to fly (Richter 2005). Recently many airborne hyperspectral imagers, including AVIRIS, CASI, Hyper-CAM, and Hymap, have been developed (Varshney and Arora 2004; Vorovencii 2009). The aerial imagers offer higher spatial and spectral resolution than their space-borne equivalents, their swath widths and spatial coverage are more restricted. Satellite images are able to image any portions of the globe without restriction. However, because of their orbits, satellites can only fly over a region once every few days. In contrast, airborne missions can be organised arbitrarily and encompass numerous passes over a single location, while being more expensive.

Several hyperspectral earth observation satellite sensors, including PROBA-CHRIS, EO-1 HYPERION, and HJ-1A, are currently in orbit. PROBA-CHRIS that operates in the VNIR area and has a spectral resolution ranging from 1.25 to 11 nm (Barnsley et al. 2004). Several hyperspectral satellites, including MSMSIS in South Africa and Belgium, PRISMA in Italy, EnMAP in Germany, HISUI in Japan, HyspIRI in the United States, and Resourcesat-3 in India, are now in development.

HYPERION is able to acquire images in the VNIR and SWIR spectral regions. It has 10 nm spectral resolution (STR) and a 30 m spatial resolution (SPR) (Liao and Jarecke 2000). HJ-1A has a SPR of 100 m and a STR of 5 nm on average. **PRISMA** is designed for research on vegetation, environmental conservation, and global warming and is expected for release in 2012. Its orbital altitude will be 700 km, and its lifespan is anticipated to be five years. The imager will function with a STR of 12 nm in the visible, near-infrared, and short-wave infrared domains. The image mass is less than 90 kg. It has over 210 bands that are connected (Galeazzi et al. 2008).

ENMAP is a German hyperspectral sensor (Kaufmann et al. 2008). ENMAP and PRISMA function in a substantially same spectral area (420–2450 nm). The average STR for VNIR and SWIR is 6.5 nm and 10 nm, respectively; with a SPR of 30 m. The aim of ENMAP is to offer high-resolution images for research in agricultural, forests, coastal regions, and waterways (Kaufmann et al. 2008). NASA designed the 700-km-high **EO-1 HyspIRI** earth observation satellite to monitor ecology, flora, volcanic eruptions and wildfires. **HyspIRI** will monitor 380–2500 nm (VSWIR) using 212 bands at 10 nm interval and also has thermal infrared (TIR) imager. The resolution of both devices at the nadir is 60 m (Chien et al. 2009).

13.5 Advantage of Hyper Spectral Remote Sensing

HRS contains continuous spectral bands precise data over the full spectrum compared with MRSexists only in discrete spectrum patches. Certain chemical compounds exhibit unique absorption characteristics at a particular wavelength. Use HRS for soil investigations, soil categorization, soil nutrient content estimation, soil resource mapping, soil moisture evaluation and soil deterioration mapping. The availability of such a huge pool of spectral data affords the option to predict several soil properties from a single reflectance signature.

13.6 Spectral Reflectance

Remote sensing is dependent on the concept of reflecting or emitted radiation from distinct objects. Light radiation from the sun is redirected or absorbed differently by objects with distinct surface properties. These reflection qualities of a material are dependent on the substance biophysical and biochemical states, surface irregularity, and geometric conditions (e.g., angle of incidence). The most essential surface properties of the material are its colour, structure, and surface roughness. By analysing their patterns of spectral reflectance, or signatures, these modifications make it possible to distinguish between various surface morphology and materials. These characteristics can be represented by so-called spectrum reflectance curves as a function of the wavelength. The diagram below illustrates the spectral signatures of soil, water, and vegetation.

13.7 Mechanism of Soil Reflectance

Soil reflectance (SR) is the aggregate reaction of several soil components (chromophores) to EMR. Physical and chemical chromophores are separated based on the nature of soil spectrum (Ben-Dor 2011). Physical chromophores have an effect across the entire spectrum, whereas chemical chromophores only absorb incident energy at specific wavelengths (Hill et al. 2010). Variations in absorption characteristics may reveal the presence or absence of chemical chromophores, and variations in reflectance spectra account for of physical chromophores. The soil chemical chromophores include water, organic material, clays, and Fe2O3. Likewise, particle size and structure are considered as physical chromophores (Ben-Dor 2011).

The amount of soil moisture and spectral reflectance has an inverse relation. As gravimetric soil moisture content increases, spectral reflectance reduces (Bowers and Hanks 1965; Whiting et al. 2004). The O–H bond in water is the most powerful Near infrared absorption. Liquid water absorption occurs at 3106, 6079, and 2903 nm, which correspond to symmetric O–H stretching, H–O–H bending, and asymmetric O–H stretching, accordingly. The O–H stretching overtones occur about 1400 nm, and the pairing of H–O–H bends and O–H stretching occurs near 1900 nm (Clark 1999). In the range of 2200–2300 nm, metal–OH bending and O–H stretching occur simultaneously (Clark et al. 1990). Lobell and co-workers (2002) stated that soil moisture has a substantial impact on soil reflectance and have an effect on other chromophores. The SR of mineral soils appears to be significantly impacted by soil organic matter and its decomposed products (Aber et al. 1990). If OM is removed from the soil, significant spectrum reflectance increases can be observed (Bowers and Hanks 1965). The absorbance properties correlated with OM at the VNIR wavelength are frequently faint and could be difficult to perceive with the naked eye (Stenberg et al. 2010). In contrast, the VIS area typically reveals a large and noticeable difference in the global reflectance of OM-containing items (Baumgardner et al. 1985). Likewise, clay minerals have a substantial impact on the VNIR component of spectrum reflectance (Ben-Dor et al. 1999). To understand the effect of clay minerals on the spectral reflectance of soil, it is possible to link together all the OH-related absorption properties. Due to the ET of Fe^{2+} and Fe^{3+} present in the soil as Fe_2O_3 or contaminants, iron concentration plays a crucial impact in establishing the shape of reflectance spectra and SR decreases with increasing particle size. Bowers and Hanks (1965) demonstrated that the soil spectral reflectance increases exponentially with decreasing particle size, and that the increase is rapid below 0.4 mm diameter. As a result, the surface of the soil becomes smoother as particle size decreases, they established a relationship between soil reflectance and surface roughness. As the soil surface roughness rises, a significant percentage of the radiant energy becomes retained in the inter-aggregate gaps, resulting in a reduction in reflectance (Sarathjith et al. 2014).

13.8 Estimation of Soil Nutrient Content Using HRS and Statistical Methods

Liu et al. (2008) estimated SOC in northeast China using HSR and MLSR. The wavelength is 350–2500 nm using an "ASD FieldSpec Portable Spectrometer" and spectral derivative analysis is the pre-processing technique. The results demonstrated that SR of black-soil between 400 and 1100 nm to organic matter concentration is more pronounced than between 1100 and 2500 nm. In particular, the principal response range of SOC is between 620 and 810 nm, with a maximum spectral response at 710 nm with a R^2 of 0.93 and RMSE of 0.18%. Hively et al. (2011) evaluated the geographical variation of soil and measured soil parameters using an aerial imaging spectrometer (400–2450 nm). The SOC, texture, and 15 agronomically significant components of soil were analysed. The PLSR derived reflectance spectra was used to estimate concentrations of carbon (0.65), aluminium (0.76), iron (0.75), and silt content (0.75), were predicted with $R^2 > 0.50$.

Anne and co-workers (2014) estimated soil characteristics using HS image in west-central peninsula, Florida. Spectral models utilising Hyperion data predicted particle organic matter, silt + clay, labile carbon, and labile nitrogen adequately (RPD > 1.4). R^2 scores ranged from 0.21 to 0.47 for the best individual bands, from 0.28 to 0.81 for two-band indices, and from 0.53 to 0.96 for PLSR on the Hyperion image data.

Chen et al. (2015) estimated soil dispersion variables, including pH, sodium, kaolinite, illite, and montmorillonite clay concentration, using HRS. Using MLR and PLSR are used to correlate between soil dispersion and its primary factors were established. Between 350 and 2500 nm, the raw spectra were analysed. The continuity-removal SR was computed. According to the result of multiple LR models with the highest R^2 (0.686) was the best at predicting soil dispersion. The CR reflectance was the strongest predictor of Na^+ (0.725), pH (0.852), kaolinite (0.909), illite (0.923), and montmorillonite (0.923). Mzinyane et al. (2016) measured soil nitrogen concentration from a handheld spectroradiometer using Backward stepwise regression. Three to eight centimetres thickness of soil were sampled and analysed for nitrogen. There were substantial connections between spectral indices and soil nitrogen as determined. The prediction capacity of the models based on continuum removed spectra had higher adjusted R^2 of 0.85.

Malavath and Mani (2017) determined some soil characteristics in a semiarid area of Tamil Nadu using hyperspectral remote sensing (350–1050 nm) and MLSR technique. For silt, clay, field capacity, wilting point, AWC, pH, and EC, and $CaCO_3$, R^2 values was between 0.609 and 0.826. The R^2 values for exchangeable soil characteristics and CEC, and chemical composition, including SiO_2 and Fe_2O_3, ranged from 0.759 to 0.906. Al_2O_3 ($R^2 = 0.313$) provided the best fit, followed by organic carbon ($R^2 = 0.220$). The R^2 values for Fe and Mn were 0.49 and 0.49, respectively. Peon et al. (2017) evaluated soil organic carbon content of north-west of Spain using five hyperspectral and multispectral sensors. Validation models derived using PLSR were effective for both hyperspectral sensors and the multispectral sensor MODIS

($R^2 = 0.75$–0.89), indicating that hyperspectral capacity is not required for accurate prediction of topsoil organic C.

Qiu et al. (2017) assessed the K^+ concentration using ASDTM spectrometer and a HySpexTM imager. At 999, 1076, 1206, 1232, 1763, and 1970 nm, the spectral absorption features of carnallite are highly repeatable and distinct from the absorption properties of the halite and anhydrite. With a correlation coefficient of 0.60, the absorption intensity of carnallite at around 1200 nm is positively correlated with its potassium content. Yu et al. (2017) predicted soil parameters using HRS in Shenzha County, Qinghai-Tibet Plateau. A total of 67 sample were used to gather hyperspectral data. The collected soil samples at each site and soil characteristics such as OC, TN, TK, and TP were analysed. The RMSE values for soil OC, TN, TP, and TK were 68.9, 46.3, 31.4, and 45.5%, respectively.

Mustafa and Moursy (2020) evaluated soil samples of Agra, Uttar Pradesh, India for soil pH, EC, and $CaCO_3$ parameters via PLSR and soil hyperspectral signature. The spectral region of 350–2500 nm was used by an analytical spectroradiometer to capture soil hyperspectral signatures. The outcomes demonstrated that the pH and $CaCO_3$ parameters were highly predictable, with R^2 values of 0.69 and 0.83.

Thaler et al. (2019) constructed and evaluated the link between analysed SOC and SR and hyperspectral measured SOC. All samples spectral reflectance was determined using an ASD Labspec Spectrometer. The SOC index predicts SOC concentrations with an RMSE of 1.5, which is similar to estimates based on the SWIR/NIR ratio (RMSE = 1.3) and exceeds estimates based on NIR and red wavelengths (RMSE = 2.8).

Bangelesa et al. (2020) analysed the OC content of hilly soil in eastern Lesotho, southern Africa using ASD Field Spec-4 optical sensor. Important wavelengths that predict SOC content were determined within the visible spectrum in 109 soil samples (400–700 nm). The results suggest that PLSR models may overfit the calibration dataset. All RF models showed superior predictions (RPD > 2.5).

Patel et al. (2020a, b) assessed the essential nutrients in soil using HRS. The methodology has been evaluated using soil sample were collected in the region surrounding Roorkee, India. Using spectroradiometer data, the employed methodology investigates the spectrum reflectance. The best spectrum absorption characteristics have been identified in the ranges of 989.3 nm for pure NPK compost and 2195.1 nm for pure soil. On the basis of a LR model between the actual and predicted abundances, the precision of the estimates has been evaluated. At the 989.3 nm spectral region, R^2 values for compositions of silty clay soil and NPK compost were found to be 0.892, 0.897, and 0.906 for loamy soil and NPK compost, respectively. Similarly, the R^2 values found at spectral regions of 2195 nm for silty clay soil and NPK compost are 0.932, 0.926 for loamy soil and NPK compost compositions, and 0.933 for sandy soil and NPK compost.

Vibhute et al. (2020) analyzed and predicted the soil N content Phulambri Tehsil soils of India using NIR reflectance spectra (350–2500 nm) obtained from the "Field Spec-4 Spectroradiometer" in conjunction with the PLSR technique. Before and after pre-treatment, the coefficient of determination (R^2) values was 0.68 and 0.94, and the RMSEP was 4.34 and 1.56, respectively. The wavelength discovered as sensitive to nitrogen concentration were 480, 511, 653, 997, 1472, 1795, 2210, and 2296 nm.

Zhang et al. (2021) estimate soil physical and geochemical parameters Using hyperspectral image, Sentinel-1and 2, and LiDAR in south-central Wisconsin, USA. For developing correlations between soil and remote sensing data, PLSR model is applied. The topographical features and Sentinel-1 and 2 were used in the predictions; silt and sand content, Fe, Ti, Mn, and Zr were predicted fairly well with $R^2 > 0.50$ or CCC > 0.50. However, silt, sand, TC, Mg, Al, Ca, Ti, Mn, and Zr were not able to be predicted using hyperspectral data alone.

Ma et al. (2022) evaluated soil nutrients in the black soil using HSR. Measurements were taken of SOM, TN, TP, TK, and pH. To estimate soil nutrients, a MLSR was used. The results revealed that of soil total nitrogen content prediction accuracy was about 78.16% and a relative error of 21.84%. The results show that computer vision HRS has better estimation accuracy and optimal reflection.

13.9 Estimation of Soil Nutrient Content Using HRS and Machine Learning Methods

Lin et al. (2014) assessed TP and Olsen-P concentrations using an optimised index in NIR and redwavelength ranges of 350–2500 nm. The optima response was observed at red spectral area (630–690 nm) and the NIR spectral range (690–1550 nm). The best suited wavelength for TP is 695 nm ($r = 0.76$) and Olsen-P is 693 nm ($r = 0.74$). In the NIR, the optimal wavelengths for TP and Olsen-P were 1440 to 1453 nm and 1079 to 1081 nm, respectively. Vohland et al. (2014) assessed a variety of both biological and chemical soil parameters using VIS–NIR diffuse reflectance and DRIFT (VIS–NIR: 400–2500 nm; MIR: 4000–800 cm). Excellent results were achieved for OC, AN, microbial biomass-C (Cmic), and pH values; the RPD for extractable carbon from hot water (Chwe) was 2.60. The accuracy achieved with VIS–NIR data was significantly lower than that obtained with MIR spectra; the best results were retrieved for pH and Cmic, nearly quantitatively, as shown by RPD values ranging between 2.0 and 2.5.

Jia and co-workers (2017) used HSR to analyse soil total nitrogen and classify type of soil of Shangyu City, China. HSR system with a near-infrared wavelength (874–1734 nm), SVM and PLSR were used to classify and modelling techniques. The findings suggested that HSR might be employed for soil type categorization and TN determination. Results indicated paddy soil $R^2 = 0.83$, RPD = 2.5; red soil $R^2 = 0.80$, RPD = 2.3; seashore salty soil $R^2 = 0.83$, RPD = 2.5; and general model $R^2 = 0.76$, RPD = 2.1.

Adak et al. (2018) estimated the quality indices of soil using hyperspectral proximal remote sensing. Soil spectral reflectance using MLSR, PLSR, SVR, RF, MARS, and the R programming language Among the models, the RF could predict 59, 48, and 54% of the observed sand, silt, and clay content, respectively, whereas the SVR model explains 66% of the variance in the MWD. The RF model could explain 81%

of the observed variation in total OC and EC, and the SVR model could predict 59% of the reported variation in soil pH. The RF model could explain 61, 80, and 78%, respectively, of the observed variation in the available NPK levels. Song et al. (2018) evaluated the TN, AP, and AK content in the Pearl River Delta, Zengcheng, China. The Chinese Environmental 1A satellite's HRS images (115 bands of 450–950) and SLR, SVM, RF, and BPNN were used to develop the prediction models. The BPNN model demonstrated superior predictive accuracy for soil total nitrogen ($R^2 = 44.24$), available phosphorus ($R^2 = 42.91$), and available potassium ($R^2 = 48.53\%$) than the SLR, SVM, and RF algorithms.

Gao et al. (2019) predicated of forage in the Tibetan Plateau on HRS using a machine learning algorithm. The FD-IBs + SVM model is the optimal model with R^2 and RMSE values of 0.67 and 0.0472%, respectively. Sestak et al. (2019) assessed the capability of HSR for prediction of soil pH, TC, and TN content in arable stagnosols using liner and non-liner calibration methods. PLSR and ANN prediction models were used to predict soil pH, TC, and TN content. PLSR validation: TC: $R^2 = 0.85$; TN%: $R^2 = 0.76$; soil pH: $R^2 = 0.69$; ANN validation: TC%: $R^2 = 0.88$; TN%: $R^2 = 0.86$; soil pH: $R^2 = 0.74$. ANN models captured the intricate relationship between chosen soil characteristics and soil reflectance spectra more accurately than PLSR models. Mahajan et al. (2020) evaluated the qualities of the salt-affected soils using HSR (350–2500 nm) with PCR, PLSR, MARS, RF, and SVR. The absorption properties of the wavelength 494, 673, 800, 1415, 1748, 1915, 2207, and 2385 nm were remarkable. Predictions of soil parameters such as pH, EC, BD, Av.N, ex.Mg, Av.Zn, and B ranged from good to excellent RPD ranged from 1.48 to 2.06. SVR, PLSR, and PCR were found to be more resilient than MARS and RF among the predicted models.

Patel et al. (2020a, b) used a deep learning network to predict the abundance of urea fertiliser utilising spectroradiometer data and hyperspectral data at Roorkee, Uttarakhand, India. The optimal spectrum absorption characteristics for urea have been determined to be 1899.2 nm and 2195.10 nm, respectively. At 1899.2 nm, R^2value for urea-and-silt–clay-soil-mixed samples was determined to be 0.945, whereas the R^2 for urea-and-loamy-soil-mixed samples was 0.954. Similarly, R^2 was calculated to be 0.953 for urea applied mixed silty clay soil and 0.944 for urea applied loamy soil at 2195.1 nm.

Safanelli et al. (2020) investigated farmland topsoil properties across Europe using VIS–NIR and SWIR (0.4–2.5 m) data. Both SYSIs used machine learning to calibrate prediction models of clay, sand, SOC, $CaCO_3$, CEC, and pH obtained in water by means of the gradient boosting regression technique. A SYSI calculated from the complete set of Landsat images can be used to predict the clay and $CaCO_3$ contents of topsoil with moderate performance (testing R^2, RMSE, and RPIQ of 0.44, 9.59, 1.77, and 0.36, 13.99, 1.55, respectively).

Wang et al. (2020) estimated leaf biochemical parameters using an RFR modelin China's eastern Beibu Gulf. A portable ASD FieldSpec 4 spectroradiometer was used to measure the leaf reflectance spectra (350–2500 nm). The models used wavelet power spectra with the incorporation of species data and soil parameters raised mean R^2 values by 0.34–23.73%, mean RPD values by 0.24–8.16%, and mean RPD values by 0.90–11.95%. Main contributions to the RFR models for estimating leaf SPAD-502 and N were made by the species and soil total carbon components.

Bao et al. (2021) investigated SOC prediction using HSR image. The results revealed, the precision of the model was improved significantly with a determination $R^2$of 0.76, RMSE of 3.16 g kg^{-1} and RPIQ of 2.28. In addition, the use of auxiliary variables enhanced the precision of the SOC models, resulting in R^2 values of 0.79, 0.54, 0.76, and 0.81 for phaeozems, chernozems, and arenosols, respectively. Guo et al. (2021) investigated hyperspectral (Headwall-Hyperspec, 380–1700 nm) and time-series multispectral remote sensing images (Sentinel 2 and Landsat 8) to map SOC stock in southeast of Iowa using PLSR and ELM models. The results revealed that (1) Hyperspectral images have been used effectively to predict the SOC stock, SOC, and SBD using PLSR and ELM, with ELM (RPIQ = 2.03, 1.97, and 1.64) exceeding PLSR (RPIQ = 1.83, 1.97, and 1.53). Multispectral remote sensing images predicted by Sentinel-2 (RPIQ = 1.45, 1.25, and 1.26). Peng et al. (2019) determined soil TN, TP and TK contents using hyperspectral data. Various statistical models, like PLSR, PCC, LASSO, GBDT, linear MLR, RR, nonlinear SVM, and GABP are used to generate characteristics variables. The results revealed the GBDT-GABP was observed as the most precise estimation technique of soil TN (R^2 of 0.69), TP (R^2 = 0.73) and the LASSO-GABP proved to be optimal for soil TK estimations (R^2 of 0.82). Moreover, the highly precise LASSO-GABP-estimated soil TK (R^2 = 0.79) indicates the viability of the LASSO-GABP method for retrieving soil TK content at the regional scale.

13.10 Conclusion

This review explains the potentiality of hyperspectral remote sensing for assessing soil nutrient content. This technique implies low cost, less time, and a high spatial extent, and has covered the capacity of assessing soil nutrient content. This technique will be improved with an advanced model algorithm like multiple regression, PCA, PLS, machine learning algorithms like support vector machine, random forest, artificial neural network, and other deep learning methods. The average model R^2 value to the targeted values lies between – and - - - - - - -. Based on the targeted value, soil quality parameters can be used to delineate the deficiency and sufficiency of the study area's one or more targeted soil nutrients. Further, these studies proved the relationship between the soil macro- and micronutrients. Finally, this review aimed to highlight the potential of hyperspectral remote sensing to assess the quantity of nutrients in the soil.

References

Aber JD, Wessman CA, Peterson DL, Melillo JM, Fownes JH (1990) Remote sensing of litter and soil organic matter decomposition in forest ecosystems. In: Hobbs RJ, Mooney H (eds) Remote sensing of biosphere functioning. SpringerVerlag, New York, pp 87–103

Adak S, Bandyopadhyay KK, Sahoo RN, Purakayastha TJ, Shrivastava M, Mridha N (2018) Assessment of soil health parameters using proximal hyperspectral remote sensing. J Agric Phys 18(1):88–98. ISSN 0973–032X

Anne NJP, Abd-Elrahman AH, Lewis DB, Hewitt NA (2014) Modeling soil parameters using hyperspectral image reflectance in subtropical coastal wetlands. Int J Appl Earth Obs Geoinf 33:47–56. https://doi.org/10.1016/j.jag.2014.04.007

Bangelesa F, Elhadi A, Jasper K, Inos D, Marubini R, Thabiso MM (2020) Predicting soil organic carbon content using hyperspectral remote sensing in a degraded mountain landscape in lesotho. Appl Environ Soil Sci 1–11. https://doi.org/10.1155/2020/2158573

Bao Y, Ustin S, Meng X, Zhang X, Guan H, Qi B, Liu H (2021) A regional-scale hyperspectral prediction model of soil organic carbon considering geomorphic features. Geoderma 403:115263. https://doi.org/10.1016/j.geoderma.2021.115263

Barnsley MJ, Settle JJ, Cutter MA, Lobb DR, Teston F (2004) The PROBA/CHRIS mission: a low-cost smallsat for hyperspectral multiangle observations of the earth surface and atmosphere. IEEE Trans Geosci Remote Sens 42(7)

Baumgardner MF, Silva LF, Biehl LL, Stoner ER (1985) Reflectance properties of soils. Adv Agron 38:1–44

Ben-Dor E, Irons JR, Epema GF (1999) Soil reflectance. In: Rencz NA (ed) Remote sensing for the earth sciences: manual of remote sensing. Wiley, New York, pp 111–188

Ben-Dor E (2011) Characterization of soil properties using reflectance spectroscopy. In: Thenkabail PS, Lyon JG, Huete A (eds) Hyperspectral remote sensing of vegetation. CRC Press, pp 513–558

Bowers S, Hanks R (1965) Reflection of radiant energy from soils. Soil Sci 100:130–138

Chen J, Chen J, Wang Q, Zhang Y, Ding H, Huang Z (2015) Retrieval of soil dispersion using hyperspectral remote sensing. J Indian Soc Remote Sens https://doi.org/10.1007/s12524-015-0530-9

Chien S, Silverman D, Davies D (2009) Mand onboard science processing concepts for the HyspIRI mission. IEEE Intell Syst 24(6):12–19

Clark RN, King TVV, Klejwa M, Swayze GA, Vergo N (1990) High spectral resolution reflectance spectroscopy of minerals. J Geophys Res 95:12653–12680

Clark RN (1999) Spectroscopy of rocks and minerals, and principles of spectroscopy. In: Rencz AN (ed) Remote sensing for the earth sciences: manual of remote sensing. American Society for Photogrammetry and Remote Sensing, pp 3–58

Galeazzi C, Sacchetti A, Cisbani A, Babini G (2008) The Prisma Program. In: IEEE geoscience and remote sensing symposium, IGARSS, vol 4, pp 105–108

Gao J, Meng B, Liang T, Feng Q, Ge J, Yin J, Wu C, Cui X, Hou M, Liu J, Xie H (2019) Modeling alpine grassland forage phosphorus based on hyperspectral remote sensing and a multi-factor machine learning algorithm in the east of Tibetan Plateau, China. ISPRS J Photogrammetry Remote Sens 147:104–117. https://doi.org/10.1016/j.isprsjpr

Gómez D, Lagüela S, Díaz-Vilariño L, Picon I, Arias P (2016) Monitoring of remote and real-time sensors for precision agriculture using unmanned aerial vehicles. Sensors 16(4):379. https://doi.org/10.3390/s16040379

Guo L, Sun X, Fu P, Shi F, Dang L, Chen Y, Linderman M, Zhang G, Zhang Y, Jiang Q, Zhang Q, Zeng C (2021) Mapping soil organic carbon stock by hyper spectral and time-series multispectral remote sensing images in low-relief agricultural areas. *Geoderma* 398:115118. https://doi.org/10.1016/j.geoderma.2021.115118

Hill J, Udelhoven T, Vohland M, Stevens A (2010) The use of laboratory spectroscopy and optical remote sensing for estimating soil properties. In: Oerke E-C et al (eds) Precision crop protection—the challenge and use of heterogeneity. Springer, Dordrecht, pp 67–86

Hively WD, McCarty GW, Reeves JB, Lang MW, Oesterling RA, Delwiche SR (2011) Use of airborne hyperspectral imagery to map soil properties in tilled agricultural fields. Appl Environ Soil Sci https://doi.org/10.1155/2011/358193

Jia S, Li H, Wang Y, Tong R, Li Q (2017) Hyperspectral imaging analysis for the classification of soil types and the determination of soil total nitrogen. Sensors 17(10):2252. https://doi.org/10.3390/s17102252

Kaufmann H, Segl K, Guanter L, Hofer S, Foerster K-P, Stuffler T, Mueller A, Richter R, Bach H, Hostert P, Chlebek C (2008) Environmental mapping and analysis program (EnMAP)—recent advances and status. IEEE Geosci Remote Sens Symp 4:109–112

Kumar V, Ghosh JK (2017) Objects detection in hyperspectral images using spectral derivative. J Indian Soc Remote Sens 45(4):603–610. https://doi.org/10.1007/s12524-016-0627-9

Lacerda MP, Dematte JA, Sato MV, Fongaro CT, Gallo BC, Souza AB (2016) Tropical texture determination by proximal sensing using a regional spectral library and its relationship with soil classification. Remote Sens 8(9):701

Liao L, Jarecke P (2000) Radiometric performance characterization of the hyperion imaging spectrometer instrument. In: Proceedings optical science and technology symposium, earth observing systems V, SPIE 1435

Lin C, Ma R, Zhu Q, Li Q (2014) Using hyper-spectral indices to detect soil phosphorus concentration for various land use patterns. Environ Monit Assess 187:4130. https://doi.org/10.1007/s10661-014-4130-x

Liu H, Zhang Y, Zhang B (2008) Novel hyperspectral reflectance models for estimating black-soil organic matter in Northeast China. Environ Monit Assess 154(1–4):147–154. https://doi.org/10.1007/s10661-008-0385-4

Lobell DB, Asner GP (2002) Moisture effects on soil reflectance. Soil Sci Soc Am J 66:722–727

Ma L, Lia A, Yu H, Chen G (2022) Hyperspectral remote sensing estimation of soil nutrients in the black soil region based on computer vision model. Sci Asia 48:287–293. https://doi.org/10.2306/scienceasia1513-1874.2022.035

Mahajan G, Das B, Gaikwad B, Desai A, Morajkar S, Patel K, Kulkarni R, Murgaonkar D (2020) Using hyperspectral remote sensing to monitor the properties of salt-affected soils. https://doi.org/10.2254/au.158212912.27052084

Malavath R, Mani S (2017) Detection of some soil properties using hyperspectral remote sensing of semi-arid region of Tamil Nadu. Asian J Soil Sci 12(1):191–202. https://doi.org/10.15740/HAS/AJSS/12.1/191-202

Mitran T, Ravisankar T, Fyzee MA, Suresh JR, Sujatha G, Sreenivas K (2015) Retrieval of soil physicochemical properties towards assessing salt-affected soils using hyperspectral data. Geocarto Int 30(6):701–721. https://doi.org/10.1080/10106049.2014.985745

Morellos A, Pantazi XE, Moshou D, Alexandridis T, Whetton R, Tziotzios G, Wiebensohn J, Bill R, Mouazen AM (2016) Machine learning based prediction of soil total nitrogen, organic carbon and moisturecontent by using VIS-NIR spectroscopy. Biosyst Eng 152(104–116). https://doi.org/10.1016/j.biosystemseng.2016.04.018

Mustafa A, Moursy A (2020) Using a multivariate regression model and hyperspectral reflectance data to predict soil parameters of Agra, India. Int J Geogr Geol Environ 2(1):4–9. https://doi.org/10.22271/27067483.2020.v2.i1a.12

Mzinyane T, Aardt JV, Ahmed F, Gebreslasie M (2016) Predicting soil nitrogen content using narrow-band indices from Eucalyptus grandis canopies as proxy. S Afr J Plant Soil 33(1):23–31. https://doi.org/10.1080/02571862.2015.1057771

Patel AK, Ghosh JK, Pande S, Sayyad SU (2020a) Deep-learning-based approach for estimation of fractional abundance of nitrogen in soil from hyperspectral data. J Sel Top Appl Earth Obs Remote Sens 13. https://doi.org/10.1109/JSTARS.2020.3039844

Patel AK, Ghosh JK, Sayyad SU (2020b) Fractional abundances study of macronutrients in soil using hyperspectral remote sensing. Geocarto Int 1(20)

Peng Y, Li Z, Yueming H, Guangxing H, Lu W, Zhenhua L (2019) Prediction of soil nutrient contents using visible and near-infrared reflectance spectroscopy. ISPRS Int J Geo Inf 8(10):437. https://doi.org/10.3390/ijgi8100437

Peon AD, Fernandez S, Recondo CA, Calleja JF (2017) Evaluation of the spectral characteristics of five hyperspectral and multispectral sensors for soil organic carbon estimation in burned areas. Int J Wildland Fire 26(3):230–239. https://doi.org/10.1071/WF16122

Qiu J, Zhang C, Xu Q, Yao J (2017) Mapping of carnallite along with semi-quantitative estimation of potassium content of drill cores using hyperspectral imagery. Remote Sens Lett 8(9):859–868. https://doi.org/10.1080/2150704X.2017.1333651

Resendez-Hernandez LA, Prudencio-Csapek D, Lozano-Garcıa DF (2019) Hyperspectral analysis of soilpolluted with four types of hydrocarbons. Geocarto Int 34(9):925–942. https://doi.org/10.1080/10106049.2018.1451921

Richter R (2005) Hyperspectral sensors for military applications. In: Emerging EO phenomenology", meeting proceedings RTO-MP-SET094, Keynote 1. RTO, Neuilly-sur-Seine, France. Available from http://www.rto.nato.int/abstracts.asp, pp. KN1-1–KN1-4

Safanelli JL, Demattê JAM, Oliveira RB, Rizzo R (2020) Mapping topsoil properties across the European farmland using Sentinel-2 and laboratory spectral data in the vis–NIR–SWIR range (400–2500 nm). Remote Sens Environ 236:111143. https://doi.org/10.1016/j.rse.2019.111143

Samuelson JR, Stelford M, Rooney DJ (2002) The importance and value of soil information. ASABE Paper No. 021093. St Joseph, Mich: ASABE

Sarathjith MC, Das BS, Vasava HV, Mohanty B, Anand SS, Wani SP, Sahrawat KL (2014) Diffuse reflectance spectroscopic approach for the characterization of soil aggregate size distribution. Soil Sci Soc Am J 78:369–376

Sestak I, Boltek LM, Mesic M, Zgorelec Z, Percin A (2019) Hyperspectral sensing of soil pH, total carbon and total nitrogen content based on linear and non-linear calibration methods. J Cent Eur Agric 20(1):504–523. https://doi.org/10.5513/JCEA01/20.1.2158

Song YQ, Zhao X, Su HY, Li B, Hu YM, Cui XS (2018) Predicting spatial variations in soil nutrients with hyperspectral remote sensing at regional scale. Sensors 18(9):3086. https://doi.org/10.3390/s18093086

Stenberg B, Viscarra Rossel RA, Mouazen AM, Wetterlind J (2010) Visible and near infrared spectroscopy in soil science. Adv Soil Sci 107:163–215

Thaler EA, Larsen IJ, Yu Q (2019) A new index for remote sensing of soil organic carbon based solely on visible wavelengths. Soil Sci Soc Am J 83:1443–1450. https://doi.org/10.2136/sssaj2018.09.0318

Tsai F, Philpot W (1998) Derivative analysis of hyperspectral data. Remote Sens Environ 66(1):41–51. https://doi.org/10.1016/S0034-4257(98)00032-7

Varshney PK, Arora MK (2004) Advanced image processing techniques for remotely sensed hyperspectral data

Vibhute AD, Kale KV, Gaikwad SV, Dhumal RK (2020) Estimation of soil nitrogen in agricultural regions by VNIR reflectance spectroscopy. SN Appl Sci 2:15–23. https://doi.org/10.1007/s42452-020-03322-9

Vohland M, Ludwig M, Thiele-Bruhn S, Ludwig B (2014) Determination of soil properties with visible to near- and mid-infrared spectroscopy: effects of spectral variable selection. Geoderma 223–225:88–96

Vorovencii I (2009) The hyperspectral sensors used in satellite and aerial remote sensing. Bull Transilvania Univ Braşov 2(51)

Wang L, Li H, Zheng Z, Du Y, Fu X (2017) Prediction of soil nitrogen using machine learning algorithms based on remote sensing variables. J Integ Agric 16(2):394–403. https://doi.org/10.1016/S2095-3119(16)61435-4

Wang J, Xu Y, Wu G (2020) The integration of species information and soil properties for hyperspectral estimation of leaf biochemical parameters in mangrove forest. Ecol Ind 115:106467. https://doi.org/10.1016/j.ecolind.2020.106467

Whiting ML, Li L, Ustin SL (2004) Predicting water content using Gaussian model on soil spectra. Remote Sens Environ 89:535–552

Yu H, Kong B, Wang G, Du R, Qie G (2017) Prediction of soil properties using a hyperspectral remote sensing method. Arch Agron Soil Sci 1476–3567 ISSN: 0365–0340 (Print). https://doi.org/10.1080/03650340.2017.1359416

Zhang Y, Hartemink AE, Huang J, Townsend PA (2021) Synergistic use of hyperspectral imagery, Sentinel-1 and <scp>LiDAR</scp> improves mapping of soil physical and geochemical properties at the farm-scale. Eur J Soil Sci 72(4). https://doi.org/10.1111/ejss.13086

Chapter 14
Impact of Major Nutrients Fertilizer Application on Soil Pollution and Management Measures

P. Ramamoorthy, S. Sathiyamurthi, A. Pavithra, M. Sivasakthi, and S. Praveen Kumar

Keywords Soil pollution · Fertilizer · Nutrient cycle · Environment · Mineralization · Mitigation

14.1 Introduction

Soils are the universal sink bears the utmost burden of the environment pollution. Soil pollution is the serious global concern defined as the contamination of soil with uncharacteristic concentration of the toxic and hazardous substances in the soil that harbors many health hazards. According to the FAO, the term soil pollution pretains to the degradation of the soil caused by the concentration of the pollutants on the surface becomes so high that can cause detrimental effect on the health of the plant, animal and human. Soil pollution is the component of the land degradation typically caused by the improper disposal of the waste, presence of xenobiotic chemicals, application of agricultural chemicals, industrial waste disposal that can alter the natural properties of soil. Intensification of urbanization and industrialization caused considerable

P. Ramamoorthy · A. Pavithra
Department of Soil Science and Agricultural Chemistry, Palar Agricultural College, Affiliated to Tamil Nadu Agricultural College, Vellore, Tamilnadu 635805, India

S. Sathiyamurthi (✉) · M. Sivasakthi · S. Praveen Kumar
Department of Soil Science and Agricultural Chemistry, Faculty of Agriculture, Annamalai University, Annamalai Nagar, Chidambaram, Tamilnadu 608002, India
e-mail: sathya.soil@gmail.com

S. Sathiyamurthi
Horticultural Research Station, Tamil Nadu Agricultural University, Thadiyankudisai, Perumparai, Tamil Nadu 624212, India

P. P. Adhikary et al. (eds.), *Soil, Water Pollution and Mitigation Strategies*, Environmental Science and Engineering, https://doi.org/10.1007/978-3-031-63296-9_11

amount of waste generation cause major soil pollution. The most soil contaminants include fertilizers, pesticides, other agrochemicals, petroleum hydrocarbons, heavy metals. Soil pollution poses a worrisome threat to the agricultural food production, food safety, food sustainability and plant, animal and human health. There is alarming increase in soil contaminated sites around globally due to wide growth in industrialization and other activities. Of that agricultural soils also contaminated with extensive range of compounds both from point source (direct source of contamination) such as application of xenobiotic chemicals such as mineral fertilizers, pesticides and non point sources (diffuse pollution) such as flooding, atmospheric deposition.

Major nutrients (Nitrogen, Phosphorus and Potassium) are the indispensible element for the sustenance of life on Earth and for agricultural production. Simultaneously the large amount of fertilizers used in the agricultural sector causing severe damage to human health and ecosystem. The nutrients accumulation in the soil can cause various consequences such as eutrophication of water bodies, reduce the quality of surface water by leaching, runoff and infiltration processes, create soil compaction besides that presence of high amount of phosphorus in the soil system creates the mobilization of heavy metals and increase their bioavailability in soil, plant and water continuum (Kleiman et al. 2017; Rickson 2014).

Fertilizer is any inorganic or organic substance, whether it is from a natural source or a manufactured one, that is supply to soil in order to offer one or more nutrients necessary for plant growth is known as fertilizer. India began using fertilisers in 1906, but there was little increase until 1965. A pivotal moment in the evolution of rice and wheat HYVs that respond to fertiliser. After the mid-60s, both the use of fertilisers and the production of food grains quickly increased. In order to ensure self sufficiency in the production of food grains, fertiliser was essential. Only fertilisers account for the 55% increase in food production. Total nutrient consumption (NPK)—25.58 MT (2018) in India. Inorganic fertilizers are nitrate, ammonium, phosphate and potassium salts are the main sources for agriculture production systems. But fertilizer production system we are used high amount of natural radionuclide and heavy metals as a potential sources. It contains a large amount of the heavy metals like Hg, Cd, As, Pb, Cu and Ni (Sonmez et al. 2007) that will cause major soil contamination as well as environment also. Plants absorb nutrients through soil at the time heavy metals also go through the food chain. So fertilizer are leads to water, soil, air and human health are also polluted (Savci 2012).

14.2 Importance of Mineral Fertilizer Usage

NPK is an essential nutrients for plants, playing an important role in their growth and development. In soil, nitrogen in the form of nitrate and ammonical, which results in vegetative growth and all the metabolic activities were increased. In phosphorus is mainly found in the form of phosphate, which is the result of the reaction between calcium and phosphoric acid. Phosphate is essential for plants, as it is used to build cell walls, as well as in energy transfer and photosynthesis. Phosphorus is also important for microbial processes, as it is necessary for microorganisms to produce enzymes and proteins. In soils potassium availability in various forms that will be enhanced the nutrient uptake, nutrient transformation, enzymes actions, water transport and photosynthesis activity were enhanced.

Nutrient cycle in soil is a process of transfer of nutrients from one form to another, from the soil environment to the organism and vice versa. In the soil, nutrient cycle is influenced by several factors, such as soil moisture, pH, temperature, biological activity and mineralization. The availability of nutrient in soil is determined by the amount of nutrient in the soil and its solubility (Kleiman et al. 2017). The solubility of nutrient is influenced by soil pH, which affects the form and solubility of the major nutrients in the soil. Soils with netural pH are more likely to have more soluble in most of the plant nutrients. It interacts with other nutrients in soil in several ways. For example, phosphorus availability is often limited by the presence of iron and aluminum compounds, which can bind to phosphate and reduce its availability to plants and also interacts with calcium, magnesium, and potassium, which can increase or decrease its availability, depending on their concentrations in the soil (Meissner et al. 2008). Healthy soil is the significant component to the food security and maintaining the sustainable future. Soil are the finite resource helps mainly in sustaining the food production, to mitigate and adapt climate change, acts as filtering agents for water, and improve the resilience to floods and droughts (Bartlett and Dedekorkut-Howes 2022; Rickson 2014). Soil pollution is the major concern refers to the degradation of the soil caused by the concentration of the pollutants on the surface becomes so high that can cause detrimental effect on the health of the plant, animal and human. **Scenario of fertilizer consumption**.

The annual average chemical fertilizer usage Netherlands were 665.3 kg NPK/ha, 624.8 kg NPK/ha in Egypt, 372.3 kg NPK/ka in Japan, 301.5 kg NPK/ha in china, 287.5 kg NPK/ha in Brittan, 205.4 kg NPK/ha in Germany, 180.1 kg NPK/ha in France; 160.8 kg NPK/ha in the USA, 126.4 kg NPK/ha in Italy, 121.4 kg NPK/ha in India, 115.4 kg NPK/ha in Greece and 106.9 kg NPK/ha in Indonesia respectively (Savci 2012). A least fertilizer consumption country was Turkey is lesser than other developed and many developing countries. Chemical fertilizer used per hectare in Turkey (N+P+K) are determined as 100.4 (Atılgan et al. 2007) (Tables 14.1 and 14.2).

For the next 10–20 years, more fertilizer will be used to acquire in more products. Excessive use of high amount of mineral and chemical fertilizer will be creating major environmental issues like heavy metal toxicity, green house gas emission, poor water

Table 14.1 Fertilizer usage status in India

• India ranks second –Nutrient consumption • Ideal nutrient ratio of NPK—4:2:1 • But, India has nutrient ratio—5.9:2.8:1 • Tamil Nadu nutrient ratio—3.9:2.2:1	Zone wise: fertilizer use percent • East zone—14% • North zone—28% • South zone—26% • West zone—31%

Toth et al. (2014)

Table 14.2 Fertilizer related pollution to soil

Nitrogen	Phosphours	Potassium
1. Removing the nitrogen through field drains 2. Gaseous loss as nitrogen oxides after Denitrification 3. Volatilization as ammonia	1. P—weekly taken up by crops 2. Remaining—effectively converted to water insoluble form 3. Leaching—negligible P adsorbed on soil particles may be carried into surface waters	1. K—taken up by crops to variable extents 2. Remainder—adsorbed by cation exchange complex of soil 3. Very little loss by—leaching

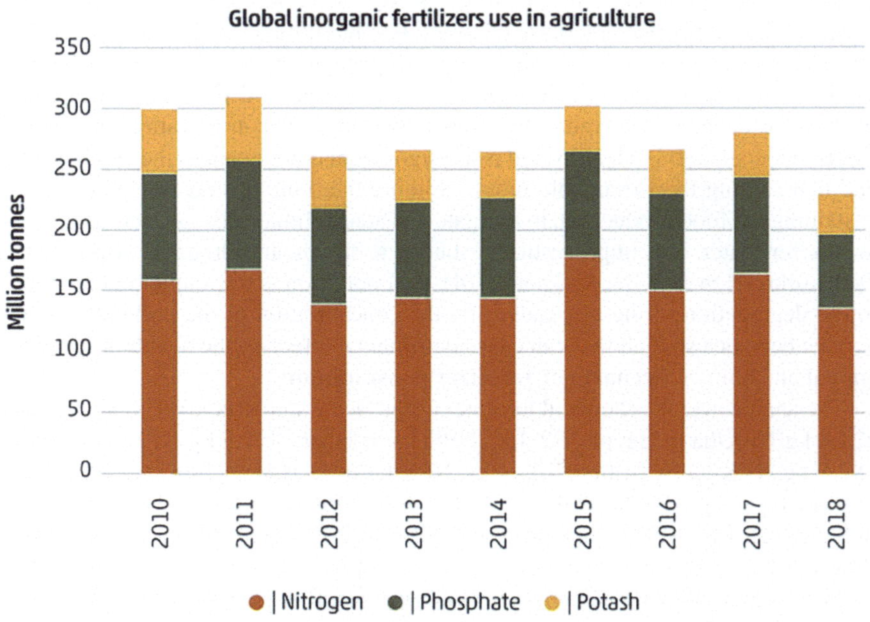

Fig. 14.1 Global inorganic fertilizer use in agriculture

quality, eutrophication, soil compaction, etc., (Sodango et al. 2018; Ngatia et al. 2021) (Fig. 14.1).

14.3 Chemical Fertilizers and Their Impact on Soil Health

The most common agricultural problems such as low soil organic carbon, minimum fertilizer use efficiencies, low imbalance of fertilizer addition and removal. The sustenance of biodiversity and securing sustainable agriculture the management of soil health is necessary. The maintenance and improvement of soil fertility is important for sustaining the global food agricultural production but modern agricultural practices mainly depend on the application of various inputs such as chemical fertilizers, pesticides, hybrid seeds, herbicides, genetically modified products. These inputs favored the improvement in agricultural production but aids in deterioration of soil quality and decline in the soil productivity which is major concern (Dar and Bhat 2020). The chemical fertilizers are the amalgam that encompassing the major nutrient concentration which is essential for the plant growth. Chemical fertilizer has a crucial role in improving the soil fertility, crop productivity and played essential role in feeding global population. But their negative impact on the environment cannot be unnoticed. There is increasing usage of fertilizers in recent decade which continue to bring unavoidable environmental contamination that possess serious risk to animal and human life. Knowledge on efficient fertilizer use is lacking among the farmer community which leads to over application of the fertilizers leading to runoff of the nutrients into water system and ecosystem (Bhatti et al. 2017). Fertilizer consumption varies for different countries globally whereas Netherlands ranked first which use 665.6 kg ha^{-1} of fertilizer and Turkey ranked least using only 100.4 kg ha^{-1} (Pahalvi et al. 2021). Also, chemical fertilizer contains source of heavy metals, radionuclides which is also accumulated in soil system when application. It is evident that heavy metals such as cadmium (Cd), arsenic (As), mercury (Hg), nickel (Ni), lead (Pb), copper (Cu) contamination in agricultural soils have increased in recent decade (Fei et al. 2022; Yong et al. 2015; Kumar and Maiti 2015). This accumulation in soil can make their entry into food chain and possess serious threat to sustenance of life. Also prolonged usage of chemical fertilizers leads to hardening of the soil, decrease soil organic matter, alters soil pH resulting in loss of soil productivity. However, most of the applied proportion of the fertilizer will ran off and leached due to rainfall and other factors leads to environmental pollution and lower the effect of fertilizers (Singh et al. 2020). These chemical fertilizers also produce GHG which leads to the global warming. It is known that 24% of the GHG is produced from agricultural sectors mainly accountable by nitrogenous fertilizers. The most commonly used fertilizers belong to the category of major nutrients such as nitrogen, phosphorus, potassium that are used for increasing the crop productivity.

Soil pollution by the overuse of the fertilizers in some regions and countries lead to various environmental issues such as saturation of nutrients in the soil exchange complex and loss via leaching in groundwater that leads to contamination of drinking water and eutrophication of fresh water bodies. This excessive use of fertilizers can be due to several factors such as

- Lack of updated and accurate fertilizer recommendations for many crops, especially tree crops.
- Lack of awareness and training on the usage of the product.
- Low price contrasted with perceived lack of downside to excessive use, as opposed to decreasing yields.

To address dwindling supplies of vital nutrients like P, to satisfy growing global food demands, and to reduce environmental dangers brought on by excessive fertilizer usage (such as eutrophication), it is critical to increase fertilizer use efficiency (Rahman and Zhang 2018). The FAO developed a Code of Conduct for Sustainable Use and Management of Fertilizers in response to the need for a more sustainable use of fertilizers (Boliko 2019). In terms of nutrient imbalances and soil pollution, the code provides a tool for putting into practice the FAO's 2017 Voluntary Guidelines for Sustainable Soil Management. The code encourages methods for enhancing soil health, such as nutrient recycling and agrono and land management. The European Union is creating a new tool to assist farmers in managing the usage of nutrients on their farms as part of its recommendations for the Common Agricultural Policy for 2021–2027. The Farm Sustainability Tool for Nutrients promotes the digitization of the agricultural industry while assisting all farmers in the European Union in using fertilizers sustainably (European Commission, Agriculture and Rural Development 2020).

14.4 Nitrogen Usage and Soil Pollution

Mineral nitrogen is important for plant growth and other physiological processes. Leaf growth and other reproductive process will often respond to an maximum amount of soil nitrogen availability. But overuse of nitrogenous fertilizers will leads to nitrogen losses from the soil through leaching into ground water through surface runoff and erosion, volatilization, denitrification will cause soil pollution. Depending on the both form of the nitrogen and present soil conditions, nitrogen is transformed in to nitrous oxide (N_2O) by microbial activity which is one of the prominent green house gas responsible for the depletion of the ozone layer. Nitrogenous fertilizer contribute soil pollution by undergoing various transformations in th soil and atmosphere such as soil acidification, acid rainfall and eutrophication (FAO and IFA 1995; Sinha et al. 2017), that are hazardous to both the environment and human health can arise as a result of reactions between nitrogen-containing organisms in the atmosphere (Bittman et al. 2014). In addition to these, there are a number of transformations that occur when nitrogen is added to soil.

1. Nitrification: Nitrification is a chemical process in which nitrogen is converted from ammonium (NH_4^+) to nitrite (NO_2^-) and then to nitrate (NO_3^-). This process is carried out by soil bacteria and is an important part of the nitrogen cycle. Nitrification helps make nitrogen available to plants, and can also lead to increased leaching of nitrates into nearby water systems.
2. Denitrification: Denitrification is a process in which nitrates are converted back into nitrogen gas (N_2) and released into the atmosphere. This process is carried out by soil bacteria, and prevents nitrates from leaching into water systems.

3. Assimilation: Assimilation is the step by which plants take up nitrogen from the soil and convert it into proteins, amino acids and other nitrogen-containing compounds. This process is important for plant growth and development, and is essential for soil fertility.
4. Mineralization: Mineralization is a process in which organic nitrogen is converted into inorganic forms, such as ammonium, nitrate and nitrite. This process is carried out by soil microbes and helps make nitrogen available to plants.
5. Immobilization: Immobilization is the process by which nitrogen is taken up by soil microbes and converted into organic forms, such as proteins and amino acids. This process helps to retain nitrogen in the soil and prevent it from leaching into water systems (Fig. 14.2).

The industrial N fixation process, directly contributes to the nitrogen in terrestrial ecosystems, is where the majority of the nitrogen is primarily derived from. High nitrogen deposition in landscapes dominated by agriculture is a result of this rise in atmospheric nitrogen levels. Reactive nitrogen is mostly sourced from three main sources: nitrogen mineral fertilizers, nitrogen based food, and airborne nitrogen emissions. Reactive nitrogen (38–75%) is primarily found in food products worldwide. Nitrogen fertilizers supply around 11–32% of the reactive nitrogen, while ammonia and nitrogen oxide airborne emissions, as well as the following deposition from the atmosphere, account for about 11–36%. Lack of oxygen in the air, acidification of

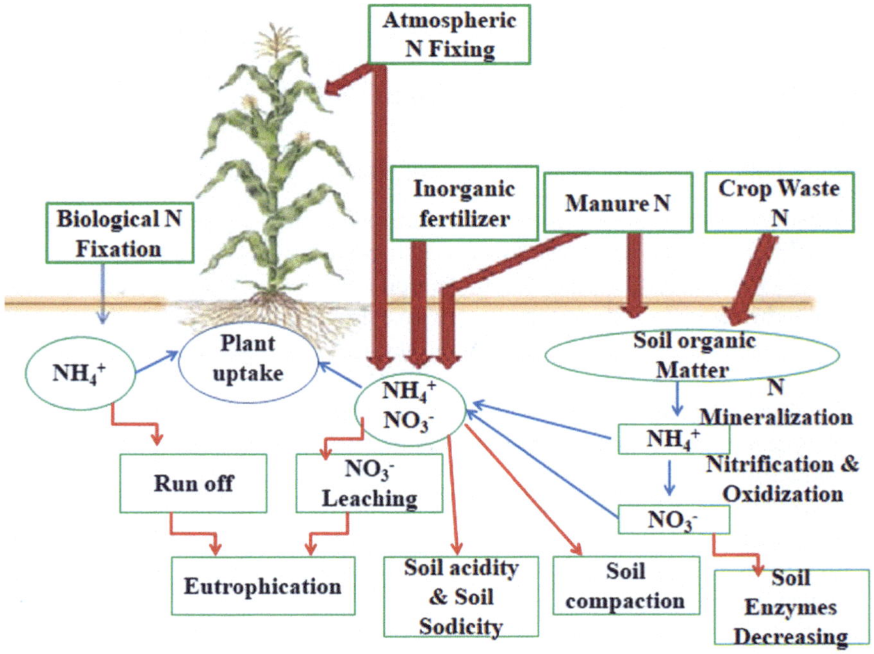

Fig. 14.2 Schematic representation of soil Nitrogen transformation contributes soil pollution

lakes and rivers, interruption of the forestry process, and environmental deterioration are all consequences of an excess of reactive nitrogen.

Loss of nitrogen, a heterotrophic process carried on by facultatively anaerobic organisms using a variety of carbon substrates as electron donors is the nitrogen cycle's key TA phase. It is critical to comprehend denitrification and the kinetics of nitrogen change it involves. Many nitrogen fluxes that produce and consume NO_3 through nitrification and leaching, respectively, have an impact on nitrate, the major nitrogen species for nitrogen loss. Reducing NO_3 to a nonreactive state is necessary for it to be environmentally beneficial (N_2). The three primary biological routes for NO_3 reduction are

- Assimilatory NO_3 reduction into biomass
- Dissimilatory NO_3 reduction to N_2 and
- Dissimilatory NO_3 reduction to NH_4^+.

Soil pollution caused by fertilizer nitrogen occurs when nitrogen from fertilizer accumulates in the soil. This can occur when too much fertilizer is used or when fertilizer is not properly applied. Excess nitrogen affects the soil microbiome and can lead to reduced soil fertility, increased runoff and leaching, and algal blooms. It can also damage local flora and fauna as well as contaminate groundwater and surface water. To prevent soil pollution due to fertilizer nitrogen, farmers should use the correct amount and type of fertilizer, spread it evenly, and use cover crops and crop rotation to minimize nitrogen runoff and leaching.

14.5 Impact of Nitrogen Pollution in Soil

Nitrogen pollution in soil can have a range of effects, including:

1. Altering soil's microbial composition, including increasing the abundance of nitrifying bacteria, which can cause a decrease in soil pH
2. Increasing the level of nitrate in the soil, which can lead to leaching into groundwater and surface water, leading to eutrophication?
3. Decreasing the availability of essential nutrients, including phosphorus, potassium, calcium, and magnesium
4. Increasing the risk of soil erosion and nutrient runoff
5. Reducing the species diversity of soil flora, leading to a decrease in soil health and fertility
6. Increasing the risk of plant diseases, such as root rot and various fungal infections
7. Decreasing crop yields
8. Increasing air pollution, as the excess nitrogen is converted to nitrous oxide, a potent greenhouse gas.

14.6 Soil Pollution by Phosphoric Fertilizers

Phosphorus is an essential nutrient for plants, but excessive amounts can cause serious environmental problems. In soils, phosphorus exists in inorganic and organic forms, and can be found in the form of phosphates, such as orthophosphates, metaphosphates, and polyphosphates. Due to crop absorption and erosion, phosphorus in soil is reduced, and its replenishment through rock weathering and atmospheric deposition is sluggish. Hence, the application of phosphorus in both inorganic and organic forms has become a crucial procedure in crop production (Liu et al. 2014; Chen et al. 2014). Phosphorus from sedimentary ores, which makes up 75–80% of phosphate rock (PR), a non-renewable resource, is essential for modern agriculture. Variable concentrations of radionuclides and trace elements can be found in phosphate rock, particularly in sedimentary ores (Gupta et al. 2014). Trace elements and radionuclides are mobilised during the mining of phosphate, and the dust that is produced is a source of fluoride and radon emissions (Azzi et al. 2017; Reta et al. 2019). When phosphate fertilizer is applied, this results in the addition of hazardous compounds like arsenic, cadmium, chromium, lead, mercury, fluorine, and radionuclides like uranium, radium, and thorium to the soil. These substances may build up in plants and enter the food chain (Tiessen et al. 2011). Frequent fertilizer applications can cause a considerable buildup of potentially harmful substances in the soil (Hamza and Anderson 2005). High concentrations of copper, zinc, and cadmium were found in soils that had previously received phosphorus fertilizer applications (Mihelcic et al. 2011). Overuse of phosphorus fertilizers may result in runoff or erosion that removes it from agriculture (Smith et al. 2012). A large amount of phosphorus is lost by erosion or runoff from agricultural areas, which causes eutrophication of nearby water bodies (Withers et al. 2019). As erosion rates vary greatly even within a single field, it is particularly difficult to quantify phosphorus losses in eroding agricultural soils (Wang et al. 2020). Also, few countries regularly do extensive inventory of their soil erosion (Liu et al. 2014).

Phosphorus pollution can occur from the excessive use of fertilizers and animal manure, as well as from runoff from urban and industrial areas. Inorganic phosphorus pollution is primarily caused by runoff from agricultural land and urban runoff. Fertilizer and manure are the primary sources of phosphorus in the environment. When these materials are applied to soil, the phosphorus can easily leach into surface and ground water sources, causing eutrophication of lakes and rivers. Urban runoff can also introduce phosphorus into the environment, and can come from both point sources (e.g. wastewater treatment plants) and non-point sources (e.g. stormwater runoff). Organic phosphorus pollution is caused by the decomposition of organic matter, such as plant material and animal manure (Glendell et al. 2014; Schoumans et al. 2014; Ngatia et al. 2021). In soil, organic phosphorus is slowly converted to inorganic phosphorus, which can then be leached into surface and ground water sources. When phosphorus is added to soil from fertilizer, sewage, and other sources, it can accumulate to levels that can lead to environmental problems (Carpenter et al. 1998; Kleinman et al. 2011). High levels of phosphorus can lead to excessive growth of aquatic plants, which can lead to reduced water quality, decreased oxygen levels, and other negative effects (Ouyang et al. 2012).

The chemistry of phosphorus pollution in soil is largely driven by the reactions and transformations of the various forms of phosphorus found in soils. In the presence of oxygen, orthophosphate (PO_4) can be oxidized to form more soluble and mobile forms of phosphorus such as dihydrogen phosphate (H_2PO_4) and monohydrogen phosphate (HPO_4). These forms of phosphorus are less mobile and can remain in soils for longer periods of time. In addition to the oxidation and reduction reactions of phosphorus, the mobility of phosphorus in soils can also be affected by adsorption and desorption processes. Adsorption is the binding of phosphorus to soil particles, which can reduce the amount of available phosphorus for uptake by plants and can lead to phosphorus accumulation in soils. Desorption is the reverse process, where phosphorus is released from soil particles, which can increase the mobility of phosphorus in soils and can lead to increased transport and accumulation in aquatic systems (Abrams and Jarrell 1995; Heathwaite et al. 2005; Carpenter et al. 1998).

Reducing phosphorus pollution in soil requires an understanding of the sources and pathways of phosphorus in the environment, as well as the chemical and physical processes that affect phosphorus movement and transformation. Management practices such as using phosphorus-efficient fertilizers, reducing runoff, and maintaining riparian buffer zones can all help to reduce phosphorus pollution in soils (Zhang et al. 2013) (Fig. 14.3).

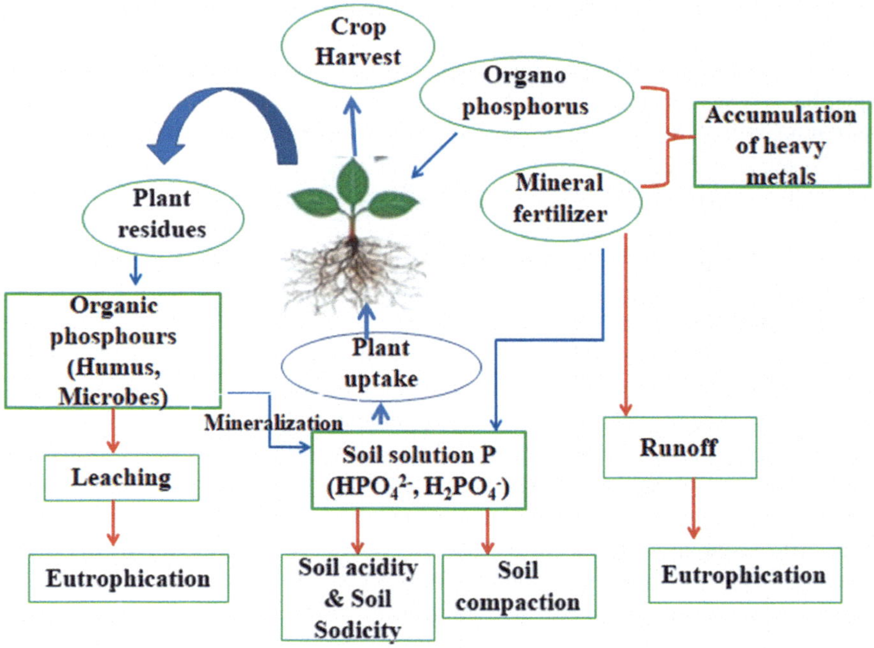

Fig. 14.3 Schematic representation of soil phosphorus transformation contributes soil pollution

14.7 Ways to Mitigate the Phosphorus Pollution

1. **Coated fertilizer**: It is known as the group of fertilizer consisting of pellets of urea coated in low soluble chemicals such as sulfur, polymers, other products or combination. Those fertilizers mitigate a few negative impact of urea fertilization, inclusive of fertilizer burn. The coatings release the urea both whilst penetrated with the water, as with sulfur, and when broken down, as with polymers.

   Examples: Sulfur coated urea, Polymer coated urea and Neem coated urea
2. **Slow release fertilizer**: It is also known as a controlled release fertilizer (CRF). A fertilizer that contains a plant nutrient in a form that delays its availability for uptake and use by plants after application or prolongs its availability to plants for a much longer period of time than the quick release reference nutrient such as ammonium nitrate or urea, ammonium phosphate or potassium chloride. Such a delay in initial availability or continued availability may occur due to various mechanisms. These include controlled solubility of the material in water through semi-permeable coatings, occlusion, proteinaceous materials or other chemical forms, slow hydrolysis of low molecular weight water-soluble compounds, or other unknown means.

   Examples: Isobutylidenediurea and Urea formaldehyde
3. **Soil test based fertilizer application**: Soil test and crop based fertilizer application will be decreased the higher fertilizer application that will help to reduce the leaching losses, volatilization losses and eutrophication of water bodies. At the same time yield the produce will be increased and soil chemical health is also maintained properly.
4. **Use of phosphorus-free fertilizers**: Phosphorus-free fertilizers are a type of fertilizer that contain no phosphorus. They are used to reduce the amount of phosphorus in the soil and prevent it from leaching into nearby water sources.
5. **Crop rotation**: Crop rotation is a put into practice of growing diverse crops in different fields each season. This helps to reduce the amount of phosphorus in the soil as well as reduce the risk of nutrient loss due to leaching.
6. **Cover crops**: Cover crops are planted between cash crops to help reduce soil erosion and increase soil fertility. They also help reduce the amount of phosphorus that leaches into nearby water sources.
7. **Conservation tillage**: Conservation tillage is a practice of minimal tillage where the soil is disturbed as little as possible. This helps reduce soil erosion and the amount of nitrogen and phosphorus that can leach into nearby water sources.
8. **Nutrient-capturing buffers**: Nutrient-capturing buffers are strips of vegetation, such as grasses or shrubs, planted around fields to help reduce the amount of nutrients that leaches into nearby water sources.
9. **Use organic manures**: Use of compost and other organic manures will release nutrient slowly at the same time that will improve the soil physical properties, chemical properties and increased microbial community of the soil ecosystem. Those microbes can involving the nutrient fixation, transformation

and encourage to uptake of nutrients at the same time to neutralize the heavy metal toxicity.

10. **Use green manuring**: Usage of green leaf and green manure crops will improve the soil nutrient status in naturally, so use of chemical fertilizer rates will be decreased. Green manure crops like Sunhemp and Daincha will fix atmospheric nitrogen to.
11. **Biochar**: Biochar is a produced by pyrolysis of natural biomass. It is a char having more number of functional groups that will capture and hold the nutrients are very strongly and then release the nutrient slowly to the crops based on their needs. Now a days biochar blended fertilizers are also available in market itself that will reduced the nitrogen losses and nitrogen emission are also.
12. **Nanotechnology**: Nanofertilizers on controlled release improve the soil productivity, soil health and yield parameters. Effective use of nanopesticides use of nanosensors, controlled release sensors that greatly contribute to the precision farming that reduce the environmental havoc. Nanomaterials in the agriculture greatly reduce the soil pollutant hazards and improve the soil condition (Shang et al. 2019).

14.8 Conclusion

High fertilizer inputs raise the reactive nitrogen level and phosphorus content in the air, soil and water. Reactive unused nitrogen and phosphorus acts as pollutant to the environment and harms the natural resources and bioaccumulation into food chain. The use of new age controlled fertilizers, slow release fertilizers, controlled release fertilizers, nano coated fertilizers are used to slow down the nutrient leaching has been practiced by farming community globally which added advantage to sustain the fertilizer availability in soil for longer time period and reduce the soil pollution. To the date, the greater efforts taken to reduce the production and consumption of fertilizers that moves towards more sustainable agricultural operations. Also greater investments needed in bringing appropriate knowledge on fertilizer overuse among the farming communities, treatment and remediation of contaminated sites, use and management of toxic fertilizer products, knowledge on mitigation strategies promoting more circular, sustainable and create the less polluting economy that shift toward the cleaner soil condition of the country as well.

References

Abrams M, Jarrell W (1995) Soil phosphorus as a potential nonpoint source for elevated stream phosphorus levels. Wiley Online Library

Atılgan A, Coşkan A, Saltuk B, Erkan M (2007) Antalya yöresindeki seralarda kimyasal ve organik gübre kullanım düzeyleri ve olası çevre etkileri. Ekoloji 15(62):37–47

Azzi V, Kanso A, Kazpard V, Kobeissi A, Lartiges B, El Samrani A (2017) Lactuca sativa growth in compacted and non-compacted semi-arid alkaline soil under phosphate fertilizer treatment and cadmium contamination. Soil Tillage Res 165:1–10

Bartlett JA, Dedekorkut-Howes A (2022) Adaptation strategies for climate change impacts on water quality: a systematic review of the literature. J Water Clim Change

Bhatti AA, Haq S, Bhat RA (2017) Actinomycetes benefaction role in soil and plant health. Microb Pathog 111:458–467

Bittman S, Dedina M, Howard C, Oenema O, Sutton M (2014) Options for ammonia mitigation: guidance from the UNECE task force on reactive Nitrogen. NERC/Centre for Ecology & Hydrology

Boliko MC (2019) FAO and the situation of food security and nutrition in the world. J Nutr Sci Vitaminol 65(Supplement):S4–S8

Carpenter SR, Caraco NF, Correll DL, Howarth RW, Sharpley AN, Smith VH (1998) Nonpoint pollution of surface waters with phosphorus and nitrogen. Ecol Appl 8(3):559–568

Chen X, Cui Z, Fan M, Vitousek P, Zhao M, Ma W, Wang Z, Zhang W, Yan X, Yang J (2014) Producing more grain with lower environmental costs. Nature 514(7523):486–489

Dar SA, Bhat RA (2020) Aquatic pollution stress and role of biofilms as environment cleanup technology. Fresh Water Pollut Dyn Rem 293–318

European Commission, Agriculture and Rural Development (2020) Farm sustainability tool for nutrients. Available at: https://ec.europa.eu/info/food-farming-fisheries/sustainability_en

Fei X, Lou Z, Xiao R, Ren Z, Lv X (2022) Source analysis and source-oriented risk assessment of heavy metal pollution in agricultural soils of different cultivated land qualities. J Clean Prod 341:130942

Food and Agriculture Organization (FAO) (2017) Voluntary guidelines for sustainable soil management. Rome: FAO. Available at: https://www.fao.org/policy-support/tools-and-publications/resources-details/en/c/1027927/

Glendell M, Granger SJ, Bol R, Brazier RE (2014) Quantifying the spatial variability of soil physical and chemical properties in relation to mitigation of diffuse water pollution. Geoderma 214:25–41

Gupta D, Chatterjee S, Datta S, Veer V, Walther C (2014) Role of phosphate fertilizers in heavy metal uptake and detoxification of toxic metals. Chemosphere 108:134–144

Hamza M, Anderson WK (2005) Soil compaction in cropping systems: a review of the nature, causes and possible solutions. Soil Tillage Res 82(2):121–145

Heathwaite AL, Quinn P, Hewett C (2005) Modelling and managing critical source areas of diffuse pollution from agricultural land using flow connectivity simulation. J Hydrol 304(1–4):446–461

Kleiman NJ, Lavrinienko A, Kivisaari K, Boratynski Z, Dauer L, Mappes T, Mousseau T (2017) Radiation cataract in chernobyl voles. Invest Ophthalmol vis Sci 58(8):2037–2037

Kleinman PJ, Sharpley AN, McDowell RW, Flaten DN, Buda AR, Tao L, Bergstrom L, Zhu Q (2011) Managing agricultural phosphorus for water quality protection: principles for progress. Plant Soil 349:169–182

Kumar A, Maiti SK (2015) Assessment of potentially toxic heavy metal contamination in agricultural fields, sediment, and water from an abandoned chromite-asbestos mine waste of Roro hill, Chaibasa, India. Environ Earth Sci 74:2617–2633

Liu J, Chen L, Cui H, Zhang J, Zhang L, Su C-Y (2014) Applications of metal–organic frameworks in heterogeneous supramolecular catalysis. Chem Soc Rev 43(16):6011–6061

Meissner R, Leinweber P, Rupp H, Shenker M, Litaor M, Robinson S, Schlichting A, Koehn J (2008) Mitigation of diffuse phosphorus pollution during rewetting of fen peat soils: a trans-European case study. Water Air Soil Pollut 188:111–126

Mihelcic JR, Fry LM, Shaw R (2011) Global potential of phosphorus recovery from human urine and feces. Chemosphere 84(6):832–839

Ngatia LW, Moriasi D, Grace III JM, Fu R, Gardner CS, Taylor RW (2021) Land use change affects soil organic carbon: an indicator of soil health. Environ Health

Ouyang W, Huang H, Hao F, Shan Y, Guo B (2012) Evaluating spatial interaction of soil property with non-point source pollution at watershed scale: the phosphorus indicator in Northeast China. Sci Total Environ 432:412–421
Pahalvi HN, Rafiya L, Rashid S, Nisar B, Kamili AN (2021) Chemical fertilizers and their impact on soil health. Springer Nat 1 Switz 2:1–20
Rahman KA, Zhang D (2018) Effects of fertilizer broadcasting on the excessive use of inorganic fertilizers and environmental sustainability. Sustainability 10(3):759
Reta GL, Dong X, Su B, Hu X, Bo H, Wan H, Liu J, Li Y, Peng T, Ma H (2019) The influence of large scale phosphate mining on the water quality of the Huangbaihe River basin in China: dominant pollutants and spatial distributions. Mine Water Environ 38(2):366–377
Rickson R (2014) Can control of soil erosion mitigate water pollution by sediments? Sci Total Environ 468:1187–1197
Savci S (2012) Investigation of effect of chemical fertilizers on environment. Apcbee Proc 1:287–292
Schoumans O, Chardon W, Bechmann M, Gascuel-Odoux C, Hofman G, Kronvang B, Rubæk G, Ulén B, Dorioz J-M (2014) Mitigation options to reduce phosphorus losses from the agricultural sector and improve surface water quality: a review. Sci Total Environ 468:1255–1266
Shang Y, Hasan MK, Ahammed GJ, Li M, Yin H, Zhou J (2019) Applications of nanotechnology in plant growth and crop protection: a review. Molecules 24(14):2558
Singh TB, Ali A, Prasad M, Yadav A, Shrivastav P, Goyal D, Dantu PK (2020) Role of organic fertilizers in improving soil fertility. Contam Agric Sources Impacts Manage 61–77
Sinha E, Michalak A, Balaji V (2017) Eutrophication will increase during the 21st century as a result of precipitation changes. *Science* 357(6349):405–408
Smith LM, Haukos DA, McMurry ST (2012) High plains playas
Sodango TH, Li X, Sha J, Bao Z (2018) Review of the spatial distribution, source and extent of heavy metal pollution of soil in China: impacts and mitigation approaches. J Health Pollut 8(17):53–70
Sonmez I, Kaplan M, Sonmez S (2007) Investigation of seasonal changes in nitrate contents of soils and irrigation waters in greenhouses located in Antalya-Demn region. Asian J Chem 19(7):5639
Tiessen H, Ballester MV, Salcedo I (2011) Phosphorus and global change. Phosphorus Action Biol Process Soil Phosphorus Cycl 459–471
Tóth G, Guicharnaud R-A, Tóth B, Hermann T (2014) Phosphorus levels in croplands of the European Union with implications for P fertilizer use. Eur J Agron 55:42–52
Wang L, Liu Y, Lu H, Huang Z (2020) Recycling of phosphorus-containing plastic based on the dual effects of switchable hydrophilicity solvents. Chemosphere 259:127402
Withers PJ, Vadas PA, Uusitalo R, Forber KJ, Hart M, Foy RH, Delgado A, Dougherty W, Lilja H, Burkitt LL (2019) A global perspective on integrated strategies to manage soil phosphorus status for eutrophication control without limiting land productivity. J Environ Qual 48(5):1234–1246
Yong L, Huifeng W, Xiaoting L, Jinchang L (2015) Heavy metal contamination of agricultural soils in Taiyuan, China. Pedosphere 25(6):901–909
Zhang P, Liu Y, Pan Y, Yu Z (2013) Land use pattern optimization based on CLUE-S and SWAT models for agricultural non-point source pollution control. Math Comput Model 58(3–4):588–595

Chapter 15
Diverse Agro-Climatic Zone Impacts the Soil's Nutrient Content and the Fertility Condition of Odisha's Agricultural Diversity

Jiban Kumar Behera, Pabitra Mishra, Anway Kumar Jena, Manojit Bhattacharya, and Bhaskar Behera

Keywords Agriculture · Coastal · Health status · Nutrients · Odisha · Soil

15.1 Introduction

Nutrient imbalance is a major global threat to soil health, which can have disastrous effects on the environment, society, and the economy. Despite 195 million metric tons of fertilizer being used annually, 768 million people were still affected by hunger in the year 2020 (Organization 2021). Alternative methods, such as organic farming, should be explored to ensure sustainable food production for future generations. It is necessary to maintain and improve the nutrient content of soils, plants, animals, and humans via sustainable soil management (SSM) (Raj et al. 2019). Nutrient-rich cultivars of staple crops are a technological advancement for reducing hunger when they are interlinking cultivate or twirling with other species like legumes or native plant. To preserve soil fertility, fertilizer extraction, and processing should not harm the ecosystem. The loss of mineral reserves brought on by mining for fertilizer also underlines the need for an efficient, safe, and sustainable use of soil nutrients (Duchemin 2022). Safe and nutrient-rich food produce capability of soil is

J. K. Behera · P. Mishra · A. K. Jena · M. Bhattacharya
Department of Zoology, Fakir Mohan University, VyasaVihar, Balasore, Odisha 756020, India

B. Behera (✉)
Department of Biosciences and Biotechnology, Fakir Mohan University, VyasaVihar, Balasore, Odisha 756020, India
e-mail: drbhaskarbehera@gmail.com

P. P. Adhikary et al. (eds.), *Soil, Water Pollution and Mitigation Strategies*,
Environmental Science and Engineering, https://doi.org/10.1007/978-3-031-63296-9_12

essential for both humans and nature. Therefore, 95% of the nutrients in a person's food come from soils, which naturally give crops what they need to grow (Anwar et al. 2018). However, certain soils can offer nutrients while others cannot, leading to nutritional inequality. The soils in certain locations are unsuitable for farming due to the way they were created, while the soils in other places have lost some fertility due to a lack of nutrients. Poor management of nutrients can lead to soil, air, and water contamination, as well as a negative effect on biodiversity (Wall et al. 2015). These two starkly different situations of nutritional imbalance cause food insecurity and are neither environmentally, economically, nor socially sustainable, leading to global warming and greenhouse gas emissions (Garnett 2014). Furthermore, there are few sources via which the nutrients are deposited in the soil; these are as follows: (a) Weathering of rocks forming soil and minerals, (b) Application of inorganic fertilizers, (c) Precipitation, (d) Decomposition of organic matters, (e) Biological nitrogen fixation.

When the soil removes the nutrients, its fertility decreases, and this process is gradually occurred by many factors, including leaching, soil erosion, growing crops, and excessive rainfall etc. Odisha has covered 15.57 million hectares geographical are with tropical belt in the eastern part of India, extending from 17,049′–22,034′ North latitude and 81,027′–87,029' East longitude (Sahoo 2012; Dash and Sahoo 2020). This state has four geographical regions: Central river basins, the Coastal plain covering 480 km from East-South, the Northern Plateau, and the Eastern ghats (Pany et al. 2016). Further, its climate is tropical, so it contains high temperatures, high humidity, medium rainfall, and short and mild winter, also divided into four seasons (Mary et al. 2006). Moreover, about 1500 mm annual rainfalls with 26.2 °C annual temperature are present in Odisha (Meena et al. 2022). About 76% of people are engaged in agricultural activities and have 87, 46,000 ha of the cropped area (Hota and Behera 2016). Based on soil, weather and other climatic characteristics, agro-climatic zone of Odisha has been subdivided into 10 parts (Fig. 15.1a) (Singh et al. 2020). Further its geographical land division classified into 3 parts, i.e. 40% of up-lands, 33.6% of medium, and 25.6% of low land, respectively (PATRA). However, various soil ranges have been divided into 8 broad soil groups of 4 orders, 10 suborders, and 18 groups (Bhattacharyya et al. 2021). In 8 broad soil groups, some are red, black, and brown forest soil in the eastern part, red-loamy and black soil in the central region, fertile, alluvial, deltaic soil in the coastal plains, and low fertility and medium texture area of northern Plateau region has red and yellow soil are present (Fig. 15.1b) (Kumar et al. 2020). Further, that region has different bio-inductor activity like highly acidic to slightly alkaline and light sandy to stiff clay (Gulati et al. 2021).

Soil health is important for long-term farming, as it can help farmers make smart choices about how to take care of their soil. Biomarkers and macromolecular probes have come a long way in the last few years, making it possible to quickly and accurately measure the number of microorganisms in the soil. About two times more carbon is present in soil than the atmosphere, but two-to-three times more than all living things combined, making it the greatest carbon reservoir on Earth's surface. It acts as a home and food source for all living things, filters water, recycles waste, and may be used as a building material. This paper discusses soil nutrition and how it aids

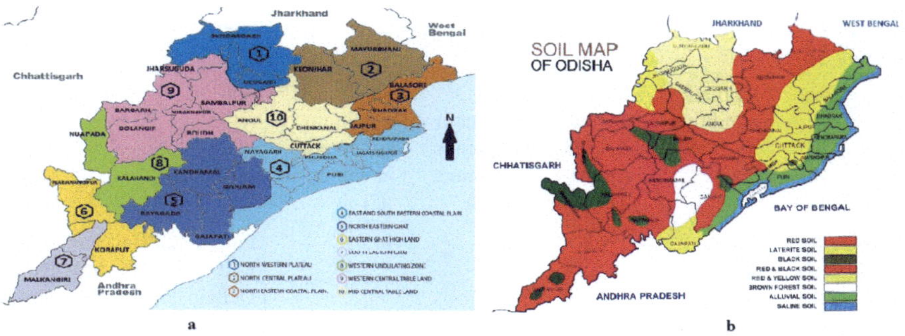

Fig. 15.1 (**a**) The map represent as 10 agro-climatic zones of Odisha. https://agri.odisha.gov.in/media-gallery/photo-gallery/odisha-map-various-category. (**b**) The geographical map represent as 8 broad soil groups, some are red, black, and brown forest soil in the eastern part, red-loamy and black soil in the central region, fertile, alluvial, deltaic soil in the coastal plains, and low fertility and medium texture area of northern Plateau region has red and yellow soil are present. https://agri.odisha.gov.in/media-gallery/photo-gallery/odisha-map-various-category

in the structural stability of plants as well as the retention and release of water and nutrients in the soil, the current crop of soil health vaccines, the biological, chemical, and physical characteristics of soil, and the nutrition of Odisha's soil, its buildings, and their implications on health. It also provides an overview of soil functioning and the state of soil conservation in Odisha. This chapter will discuss soil nutrition and health status in Odisha's fourth part and 10 agro-climatic zones (Table 15.1).

We are risking the health of other species and the environment by making too much food, which can lead to environmental damage, poverty, hunger, migration, illness, and climate change. Is it possible to grow more food organically without harming the environment? This chapter answers these questions. Things that happen on scales ranging from molecules to the whole planet affect soil fertility and nutrition. Although human interference in these systems may exacerbate existing global issues, it also has the potential to make them more manageable. From the perspectives of food production, food security, and the consequences of improper and excessive fertilizer usage on the environment and climate change, this chapter illuminates our comprehension of soil health processes. Finally, it discusses the most pressing problems and necessary actions for restoring nutritional balance to our food production systems. Further it emphasizes the importance of a holistic approach to soil health management, considering local and global impacts. By implementing sustainable practices and reducing our reliance on harmful fertilizers, we can improve soil fertility and nutrition while mitigating the negative effects of agriculture on the environment and climate.

Table 15.1 Odisha diverse districts and weather fall under one of 10 agro-climatic zones, each with vast soil types

Agro-climatic zone	Climatic condition	Mean annual rainfall (in mm)	Mean maximum temperature in summer (in °C)	Mean minimum temperature in winter (in °C)	Soil types	Districts
North eastern coastal plain	Moist sub-humid	1568	36.0	14.8	Lateritic, Red, Deltaic alluvial, Coastal alluvial and saline	Balasore, Bhadrak, Parts of Jaipur, and hatidihi block of Keonjhar
East and south eastern coastal plain	Hot and humid	1577	39.0	11.5	Lateritic, Red and Mixed Red and Black, Alluvial and saline	Kendrapara, Khurda, Jagatsinghpur, Part of Cuttack, Puri, Nayagarh and Part of Ganjam
Western central table land	Hot and moist sub-humid	1614	40.0	12.4	Red and Yellow, Red and Black, Black, Brown Forest, Lateritic	Bargarh, Bolangir, Boudh, Sonepur, Parts of Sambalpur and Jharsuguda
Mid central table land	Hot and moist sub-humid	1421	38.7	14.0	Lateritic, Red and Mixed Red and Black, Alluvial	Angul, Dhenkanal, Parts of Cuttack and Jajpur
North central plateau	Hot and Moist Sub-humid	1534	36.6	11.1	Lateritic, Red and Yellow, Mixed Red and Black	Mayurbhanj, Major parts of Keonjhar (except Anandapur and Ghasipura)

(continued)

Table 15.1 (continued)

Agro-climatic zone	Climatic condition	Mean annual rainfall (in mm)	Mean maximum temperature in summer (in °C)	Mean minimum temperature in winter (in °C)	Soil types	Districts
North western plateau	Hot and moist sub-humid	1600	38.0	15.0	Red, Brown forest, Red and Yellow, Mixed Red and Black	Sundargarh, Parts of Deogarh, Sambalpur and Jharsuguda
South eastern ghat	Warm and humid	1710	34.1	13.2	Lateritic, Red, Black	Malkangiri and Part of Keonjhar
North eastern ghat	Warm and humid	1597	37.0	10.4	Brown Forest, Lateritic, Alluvial, Mixed Red and Black	Phulbani, Rayagada, Gajapati, Part of Ganjam and small patches of Koraput
Easter ghat high land	Hot and moist sub-humid	1522	34.1	7.5	Red and Yellow, Red and Black, Red	Major parts of Koraput, Nabarangpur
Western undulating zone	Warm and humid	1352	37.8	11.9	Mixed Red and Black, Red and Black	Kalahandi and Nuapada

15.2 Essential Nutrients in Soils

Soil fertility is a soil's ability to help plants grow by giving them the nutrients they need and having good chemical, physical, and biological properties (Thies and Rillig 2012). Values measured and indicated by biological, physical, and chemical makersmay be used to categorize a number of soil qualities as sufficient or insufficient (Table 15.2). Organic matter, water, air, and mineral granules are the three basic types of organic matter. Water and air provide plants with the three crucial nutrients: carbon, hydrogen, and oxygen. The two kinds of nutrients that are present in soil are macronutrients and micronutrients (Fig. 15.2) (Jones et al. 2013; Bhatla et al. 2018). It is important to strike a balance between the two, as too few can cause the plant to lose its color and develop slowly. There is a nutrient shortfall when a growing plant needs more of an important nutrient to meet its needs, and when a plant gets more of a nutrient than it needs, the extra nutrient turns poisonous and hurts the plant's growth or quality. Soil nutrition is important because it tells us how the soil works as a whole and what happens when there are too many or too few of many important substances (Lynch and Brown 2001).

While the availability of certain nutrients, like nitrogen (N), seems limitless, other nutrients, including phosphorus (P), potassium (K), and zinc (Zn), have limited worldwide supplies (Pahalvi et al. 2021). One of the most important things soils do is store and transport nutrients, which help plants, grow food and fiber. Soils are important for making food and feed that people and animals need to eat because they increase the amount and quality of food and feed that can be made (Horrigan et al. 2002). Because farming has become more intensive in many places over the last few decades, the amount of organic matter in agricultural soils has decreased. Therefore, the regulatory functions that check the soil, air, and water quality have been hurt. So, rich soils need to be protected and brought back to life where they have been lost so that more nutritious biomass can be grown and negative environmental effects can be avoided. The term "fertility" is used in the field of soil science to describe the soil's capacity to foster and sustain plant development via the regulation of nutrient inputs. Nutrient sorption, availability or scarcity of air and water in soil mapping through three physical and chemical processes: (1) storing nutrients in the organic matter of the soil, (2) recycling nutrients from their organic forms into forms that plants can use, and (3) physical–chemical processes. Soil fertility and productivity are affected by the mineral matrix of soil, plants, and microbes, which work together to maintain and make nutrients available. Unfortunately, our soils are under much stress because agriculture has become more intensive because of better farming technology and rising food demand for a growing population. That has led to a loss of nutrients, physical damage, and decreased biodiversity. Moreover, they will not be able to meet the needs of future generations as well as they could. Also, this has caused micronutrient shortages in soils all across the globe, which harm both animal and human health (Brevik et al. 2020). More than 2 billion individuals throughout the globe are affected by micronutrient deficiencies. That means micronutrient deficiencies are a significant source of illness (Black 2003). Nearly one-third of the 800 million people without access to food that satisfies their basic energy needs worldwide live in India, where most people eat cereals with low nutritional value (Jensen 2002). South Asia has a high prevalence of micronutrient deficiency since cereals are the main staple meal and are often rotated with other grains. These concerns are critical to health and safety and must be addressed promptly (Nandagopal).

Some soils have toxic substances like cadmium and arsenic, but many soils worldwide are missing minerals important for animal and human health (Havugimana et al. 2017). As a result, surface water and groundwater can worsen, and plants can be harmed when these substances enter the food chain through contaminated soil. Additionally, it is crucial to note that only about 50% of soils used to grow cereals have adequate plant-available Zn levels, demonstrating the urgent need to increase Zn and other micronutrient concentrations in foods made from cereal (Cakmak 2009). According to mathematical models, adding Zn to cereal-based meals can save the

Table 15.2 Several soil properties may be classified as adequate or deficient based on the values measured and expressed by biological, physical, and chemical markers

Parameter	Value range	Classification	Citric limit	Unit of measurement	Indicator	Qualitative soil health characteristics
	< 4.4	Extreme to ultra-acidic				
	4.5–5.5	Strong to very strong acidic				
	5.6–6.5	Slight to moderate acidic				
	6.6–7.3	Neutral				
	7.4–8.4	Slight to moderately alkaline				
	> 8.4	Strong to very strong alkaline				
Electrical conductivity (dS/m)	< 0.8	Normal	< 1 (Normal)	dS m^{-1}		Soil water potential and Salinity
	0.8–1.6	Citric to salt sensitive crops				
	1.6–2.5	Citric to salt tolerant crops				
	> 2.5	Injurious to crop				
Organic carbon (%)	< 0.5	Low	0.5%	% or g kg^{-1}	Carbon cycling	Color and smell
	0.5–0.75	Medium				
	> 0.75	High				
Available phosphorous (mg/kg)	< 5	Low	22.5 kg ha^{-1}	mg kg^{-1} soil or kg ha^{-1}	Phosphorous cycling	Fertility
	5–10	Medium	55.5–56 kg ha^{-1}			
	> 10	High	56 kg ha^{-1}			

(continued)

Table 15.2 (continued)

Parameter	Value range	Classification	Citric limit	Unit of measurement	Indicator	Qualitative soil health characteristics
Exchangeable potassium (mg/kg)	< 50	Low	108 kg ha^{-1}	mg kg^{-1} soil or kg ha^{-1}	Potassium cycle	Fertility
	50–100	Medium	108–280 kg ha^{-1}			
	> 100	High	280 kg ha^{-1}			
Available sulfur (mg/kg)	< 10	Deficient	10 ppm	kg ha^{-1}	Sulfur cycle	Fertility
	> 10	Sufficient				
Available boron (mg/kg)	< 0.58	Deficient	0.5 ppm	mg kg^{-1}	Boron cycle	Fertility
	>= 0.58	Sufficient				
Available zinc (mg/kg)	< 0.75	Deficient	0.6 ppm	mg kg^{-1}	Zinc cycle	Fertility
	>= 0.75	Sufficient				

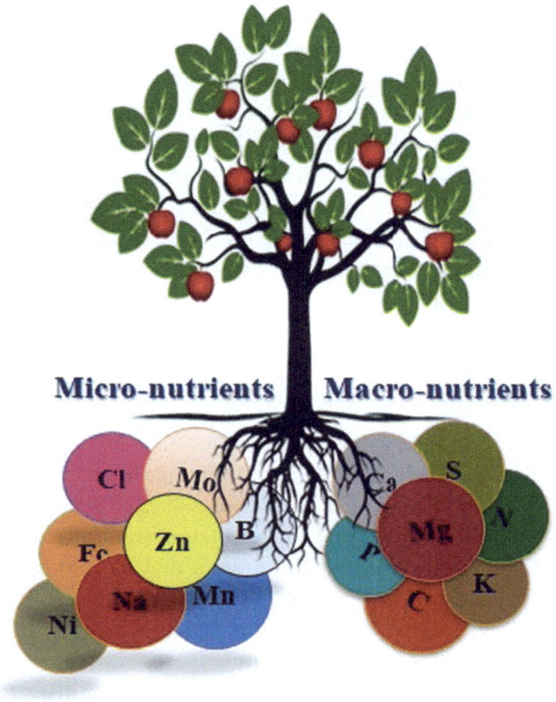

Fig. 15.2 Diagram presented as particulate matters of two kinds of nutrients that are present in soil are macronutrients and micronutrients

lives of approximately 50,000 children each year. Zinc is an essential micronutrient that plays a crucial role in children's immune systems, growth, and development. However, many children in low-income countries suffer from zinc deficiency due to poor diets and limited access to nutrient-rich foods (Akhtar et al. 2013). So, fixing nutrient deficiencies in the soil can greatly affect people's health and death rates, especially in developing countries where malnutrition is common. Using sustainable methods to manage soil, such as crop rotation, intercropping, and organic fertilization, can help improve the health of the soil and make it easier for crops to get the nutrients they need.

15.3 Role of Agro Climatic Zones in Various Soil of Odisha

Although the green revolution has boosted the agricultural field and provided a certain level of food security to most of the population, India, one of the top producers of all agricultural commodities, still needs to be higher. Therefore, another boosting up is required in that field. The Agro-climatic Zone is a key tool for boosting production from the available resources and prevailing climatic conditions. Thus, it can be defined as an extension of the climatic classification, keeping the point of the perfect to agriculture. The agro-climatic zone aims to maximize agricultural output, raise farm revenue, create more rural jobs, wisely manage existing water resources, and reduce regional status-related inequities in agriculture.

India's landmass is split into 15 distinct agro-climatic zones, each of which is further subdivided into 72 distinct subzones. Zone-VII, which includes the states of Maharashtra, Uttar Pradesh, Orissa, and West Bengal, is characterized as the country's "eastern plateau and hilly area." This region includes the Chotanagpur plateau, the Rajmahal hills, the Chhattisgarh plains, and the Dandakaranya. The Agro-climatic zone of Odisha is further subdivided into 10 sub-zones (Hota et al. 2018a, b) and these are as follows: North Eastern Coastal Plain, North Western and North Central Plateau, North Eastern Ghat, East and South Eastern Coastal Plain, Eastern Ghat High Land, Western Undulating Zone, Western Central Table Land, South Eastern Ghat and Mid-Central Table Land (Table 15.1).

I. **North Western Plateau**

Sundargarh and adjoining parts of Deogarh, Sambalpur, and Jharsuguda are all included in this hot, wet, and sub-humid climatic zone. Around 1600 millimeters (m) of precipitation (rainfall) falls annually on average. It gets as hot as 38 °C in the summer and as cold as 15 °C in the winter on average every year. There is a wide spectrum of soil colors from red to brown to red to yellow to black in this area (Hota et al. 2018a, b).

II. **North Central Plateau**

This area includes most of Keonjhar (except for the Anadapur and Ghasipura blocks) and Mayurbhanj. It receives an average of 1534 millimeters of precipitation each year and shares a climate with the neighboring North Western Plateau region. In the hottest months, temperatures may reach as high as 36.6 °C, while the coldest months can drop as low as 11.1 °C. The soil quality is the same as in the northwest, and the lateritic type may also be found in this area (Hota et al. 2018a, b).

III. **North Eastern Coastal Plain**

It includes the districts of Balasore, Bhadrak, a portion of Jajpur, and the Hatidihi block of Keonjhar, where the climate is characterized as a wet sub-humid. It is observed that there is an average annual rainfall of 1568 mm, a maximum summer temperature of 36 °C, and the lowest winter temperature of 14.8 °C. Several distinct kinds of soil can be seen, including red, lateritic, deltaic, coastal alluvial, and saline (Priyadarshini et al. 2017).

IV. **East and South Eastern Coastal Plain**

Kendrapara, Jagatsinghpur, Khurda, and parts of Cuttack, Nayagarh, Puri, and Ganjam are among the districts that get an average of 1577 mm of precipitation each year. Therefore, this place might become hot and sticky later on. The coldest temperature ever recorded was just 11.5 °C, while the warmest was 39 °C in the middle of winter. Soils ranging from saline to lateritic to alluvial, red to reddish-black are present (Haplaquepts and Ustochrepts 2016).

V. **North Eastern Ghat**

This area encompasses parts of the districts of Phulbani, Rayagada, Gajapati, Ganjam, and Koraput. Summer highs average about 37 °C, while winter lows average around 10.4 °C. The average annual rainfall is predicted to be about 1597 mm, making this location hot and humid with a sub-humid climate. Several varieties of soil, including red, mixed red, and black, may be found in this area, in addition to a speck of brown forest dirt (Singh et al. 2013).

VI. **Eastern Ghat High Land**

A warm and humid environment prevails due to the average annual rainfall of 1522 mm and the presence of red, mixed red and black, and yellow soil types. The provinces of Koraput and Nabarangpur are both heavily represented in this region. It has reached a high of 34.1 °C in the summer and a low of 7.5 °C in the winter.

VII. **South Eastern Ghat**

This area includes the warm and humid climate zones of Malkangiri and some of Keonjhar, with a mean annual rainfall of 1710 mm, highest summer temperatures of 34.1 °C, and lowest winter temperatures of 13.2 °C. Soil colors in these areas range from red to lateritic to black.

VIII. **Western Undulating Zone**

The average annual rainfall for this zone is 1614 mm, including the districts of Klahandi and Nuapada. Here, you may find red, mixed red and black, and black soil, and the climate is often hot and humid, and swampy. The average annual high is 37.8 °C, while the average annual low is 11.9 °C.

IX. **Western Central Table Land**

Cities such as Bargarh, Bolangir, Boudh, Sonepur, and parts of Sambalpur and Jharsuguda are located in this region, which is hot and humid due to its sub-humid climate and high annual rainfall of 1614 mm. The average summer temperature is above 40 °C, while the average winter temperature is little over 12 °C. Red and black, red and yellow, brown, forest, and lateritic are only few of the many hues and textures of soil.

X. **Mid Central Table Land**

This region includes Angul, Dhenkanal, and portions of Cuttack and Jajpur. The mean annual rainfall is 1421 mm, and the weather is hot, humid, and sub-humid. Here, the summertime high is measured at 38.7 °C, and the wintertime low is approximately 14 °C. Some soil types present here include alluvial, red, lateritic, mixed red, and black (Singh et al. 2011c).

15.4 Verities of Soil and Their Management Practices

I. **Red soil**

The state of Odisha has 7.14 million hectares of red soil spread over the districts of Nawrangpur, Koraput, Keonjhar, Rayagada, Malkanagiri, Ganjam, Dhenkanal, Bolangir, Kalahandi, Mayurbhanj, and Nuapada (Table 15.3) (Rath et al. 2011).

These soils are typically heavier in texture than those in the remaining districts and have angular or sub-angular blockiness. They range in acidity from very acidic to somewhat acidic and have a low to medium level of organic matter and a limited ability to retain water. As a result, both nitrogen and phosphorus levels are low, and the soils lack essential micronutrients like boron and molybdenum. To improve soil acidity, 1–2 tons per hectare of paper mill waste should be spread, and winter vegetable and groundnut crops should be sprayed with 10–15 kg of borax per hectare. In addition, for the sulfur needs of oil seed crops, 200 kg of phosphogypsum should be spread. Mango trees, jackfruit trees, guava trees, papaya trees, and sapota trees, as well as rice, finger millet, minor millets, potato, and brinjal trees, all thrive in these climates.

II. **Yellow and mixed red soil**

The 5.5 million hectares of red and yellow soil in Sambalpur, Bargarh, Deogarh, and Sundargarh are home to these soils, which are often shallow and coarse-textured (Table 15.3) (Parida et al. 2016). Most highland soils are somewhat acidic, while lowland soils are moderately acidic. Lowland soils in the Hirakud command area are zinc poor, whereas upland light-textured soils are boron deficient. The highland soils are ideal for growing potatoes, pointed guard, rice, tomatoes, finger millet, brinjal, and sugarcane. At the same time, paddy may be grown in the lowlands and on the appropriate soils after pulse crops have been harvested. Mango, guava, and banana trees flourish on these soils.

III. **Black soil**

Black soils are found in various part of Odisha such as Angul, Sambalpur, Puri, Boudh, Ganjam, Sonepur, Malkangiri, Bolangiri, Kalahandi, and Nuapada (Table 15.3) (Mishra et al. 2021). Titaniferous magnetite, humins, bituminous minerals, and so on all contribute to the dark hue of the soil. The clay percentage in certain soils is higher than 30%, making them denser. Alkaline conditions prevail, with free calcium carbonate nodules in the soil profile. Drought causes soil moisture stress to develop rapidly, making it challenging to manage these soils. Water retention can be improved by using green manure and applying large quantities of fertilizer. These soils can cultivate various crops, including rice, jowar, bajra, maize, Bengal gram, safflower, and mustard.

IV. **Laterite soil**

Lateritic soils cover a total area of 0.7 million hectares in Sambalpur, Puri, Mayurbhanja, Khurda, Keonjhar, Nayagarh, Dhenkanal and Cuttack (Table 15.3) (Das 2012). As a result of the presence of fungible aluminum and manganese, these soils are low in organic matter and fertility. Honeycomb-shaped degraded laterites may be discovered in Khurda and Cuttack, hard clay lurks in the subsoil, and crusting plagues the upland literite. Excellent soil management techniques such as liming, applying balanced fertilizer using paper mill waste (1–2 t/ha), spreading organic manure, and growing green manure crops can increase crop yields. These techniques can also improve soil health, increase nutrient availability, and reduce soil erosion, leading to

Table 15.3 Various kinds of macro- and micronutrients exist in soil, depending on its type and color. As a result, there are benefits and drawbacks to the distribution and vegetation patterns created by these soils

Type of soil	Colour	Soil contain	Areacover	District	Vegetation	Limitation
Red soil	Red	High iron and aluminum, low to medium level of organic matter, nitrogen and low phosphorus	7.14 million hectares	Koraput, Rayagada, Nawrangpur, Malkanagiri, Keonjhar, Ganjam, Kalahandi, Nuapada, Bolangir, Dhenkanal, and Mayurbhanj	Fruits like mango, jackfruit, guava, papaya, sapota trees, rice, finger millet, minor millets, tube vegetables	Rainfed farming
Yellow and mixed red soil	Upland red and yellowish red, down land yellow grayish and olive	Organic matter, iron oxide, and humic acid	5.5 million hectares	Sambalpur, Bargarh, Deogarh, and Sundargarh	Rice, finger millet, sugarcane, potatoes, brinjal, tomatoes, pointed guard, Mango trees, guava trees, and banana trees	Deficient in nitrogen and lime
Black soil	Black	Titaniferous magnetite, humins, bituminous minerals, rich calcium carbonate, low–moderate N&K	96 million hectares	Puri, Ganjam, Malkangiri, Kalahandi, Nuapada, Bolangir, Sonepur, Boudh, Sambalpur, Bargarh, and Angul	Wide variety of crops including rice, jowar, bajra, maize, bengal gram, safflower, mustard, and more	Low fertility
Laterite soil	Reddish to yellow	90–100% of aluminium, iron, titanium, and manganese oxides, lower content of potassium, phosphorus, nitrogen, lime, and magnesia	0.7 million hectares	Puri, Khurda, Nayagarh, Cuttack, Dhenkanal, Keonjhar, Mayurbhanja, and Sambalpur	Groundnuts, pulses, Mango, jackfruit, banana, guava, and sapota trees	Generally coarse in texture and cannot retain moisture

(continued)

Table 15.3 (continued)

Type of soil	Colour	Soil contain	Areacover	District	Vegetation	Limitation
Deltaic soil	Orange/ reddish mottles amidst grey patches	Oxides of Fe and Al, organic matter	0.67 million hectares	Balasore, Bhadrak, Jajpur, Kendrapara, Jagatsinghpur, Cuttack, Puri, Gajapati, and Ganjam	Rice in the kharif season and groundnuts, mustard greens, potatoes, and other vegetables in the rabi season. Green gram, tandeblack gram, and black gram	More prone
Alluvial and coastal saline soil	Light grey to ash grey	Low—medium phosphorus levels, high in nitrogen and potassium. sulfate, boron, molybdenum, and chloride, sodium sulphate, magnesium	0.254 million hectares	Balasore, BhadrakJagatsinghpur, Kendrapara, Puri, Khurda, and Ganjam	Rice like Sola, pateni, and cuttackchandi, lunisharee, SR-26B and Mohan. Best for all green vegetables	Less fertile and poor in organic matter content
Brown forest soil	Brown	Medium to high organic matter and nitrogen, moderate potash and phosphorus content, and high micronutrients	0.17 million hectares	Phulbani, Kandhamal, Rayagada, as well as portions of Ganjam and Nayagarh,	Tapioca, ginger, and turmeric. fruit crops such as citrus, jackfruit, mango, guava, and other horticultural crops are established successfully	Vegetation-rich swards where light levels permit
Black and mixed red soil	Black and red	Calcium, iron and low zinc	0.16 million hectares	Sambalpur, Bargarh, Sonepur, and Bolangir	Rice, cotton and maize	Low water holding capacity

sustainable and profitable agriculture. Therefore, farmers should adopt these practices to enhance productivity and protect the environment. These acidic soils would be ideal for growing groundnuts and pulses, while phosphatic fertilizers, both water-soluble and insoluble, can increase crop output. Legume harvests may be enhanced by applying boron to the soil or molybdenum to the seeds. Mango, jackfruit, banana, guava, and sapota trees thrive in this climate and soil type.

V. **Deltaic soil**

Alluvial deltaic soils may be found in Ganjam, Khurada, Puri, Kendrapara, Jagatsinghpur, Bhadrak, Balasore, Cuttack, and Gajapati, along the banks of rivers such the Rushikullya, Mahanadi, Subarnarekha, Brahamani, and Baitarani (Table 15.3) (Jin et al. 2018). They span an area of around 0.67 million hectares and are composed of different types of soil, from coarse sand to clay. The granular or platy structure is possible, but clay is more difficult to work with pH ranges from acidic to neutral, and coarse-grained soils are low in nitrogen, phosphorus, potassium, and sodium. They are ideal for growing rice in the kharif season and groundnuts, mustard greens, potatoes, and other vegetables in the rabi season. Green, tandeblack, and black gram thrived in the soggy leftovers.

VI. **Alluvial and coastal saline soil**

Districts around Odisha's coast, including Ganjam, Khurada, Puri, Kendrapara, Jagatsinghpur, Bhadrak, and Balasore, have these soils over an area of roughly 0.254,000 square kilometers (Table 15.3). Even alluvial soils have an EC-4 ds/m salt concentration, making them unsuitable for agriculture (Joshi et al. 2016). These soils have a columnar structure and a lump of clay-to-clay loam texture and have a pH range of 6.0–8.0 and a summer conductivity of 10–40 dS/m. They have low to medium phosphorus levels and are high in nitrogen and potassium. The salty soils are adequate in terms of sulfate, boron, molybdenum, and chloride. The most prevalent causes of widespread starvation include plasmolysis of germination-stage seeds and roots, seedling mortality, low potassium, calcium, and magnesium intake due to excessive sodium toxicity, and injury from hydrogen sulfide. The kharif season has consistent and widespread rainfall, making rice the most common crop. Native saline-resistant rice varieties such as Cuttack Chandi, Pateni, and Sola outperform high-yielding salt-tolerant cultivars such as Mohan, Lunisharee, and SR-26B. Salt embankments, draining away soluble salts, erecting sluices across streams, installing a suitable drainage system, and raising and closing salty belts are all potential preventative measures that reduce the problems connected with salinity. One cultural method of protecting crops from salt is to plant at the outside edges of alternating ridges of irrigated furrows while leaving the central furrow fallow. In coastal saline soils where exchangeable sodium is moderate but acidic reactions are low, lime may replace some Na^+ or H^+ with Ca^{++}, allowing reclamation.

VII. **Brown forest soil**

About 0.17 million hectares of these soils, found in the districts of Phulbani, Kandhamal, and Rayagada, as well as portions of Ganjam and Nayagarh, are associated with forested areas (Table 15.3) (Sarap 2013). They have medium to high levels of organic matter and nitrogen, moderate potash and phosphorus content, and high concentrations of micronutrients. *Shifting cultivation* is a common practice that degrades the land. The best soils for growing tapioca, ginger, and turmeric require proper moisture conservation. Nitrogen fertilizer effectively raises the Nile in marginal lands, and established fruit crops such as citrus, jackfruit, mango, guava, and other horticultural crops are established successfully.

VIII. **Black and mixed red soil**

These soils are a mix of red and black soil, with small patches of black soil scattered through the red soil. Sambalpur, Bargarh, Sonepur, and Bolangir are four western districts where the soil covers around 0.16 million hectares of land (Table 15.3) (Schmidt and Heckenberger 2009). Black soils with a lot of calcium and red soils with much iron make a chain-like structure in the ground. The soils are moderately fertile and deep. The zinc content of the lowland soils used for growing rice is low. Orissa has among the world's most climate-, topography-, and parent-material-influenced soils, making its average agricultural yields among the lowest. Accelerating agricultural yields is possible if particular soil concerns are identified and effectively managed.

15.5 Soil Health

Soil health is nurturing plant and animal life while protecting and enhancing the ecosystem. It is determined by its tilth (structure and texture) and its ability to drain internally, hold nutrients, and retain water. To be considered healthy, the soil needs sufficient nutrient reserves, an appropriate elemental concentration and balance, and a lack of pests and diseases (Singh et al. 2011b). Soil quality is evaluated quantitatively by identifying and measuring soil characteristics; soil health is evaluated subjectively by calculating a composite soil health index and monitoring several biological markers. Ecosystem services are the benefits humans get from ecosystems, such as cleaner air and water, a more pleasant environment, and a more moderate temperature. Regarding maintaining soil health and ecosystem services, soil organisms are crucial drivers of the components and qualities (Table 15.2). Abiotic and biotic components of the soil ecosystem interact with one another to ensure the ecosystem's health. Abiotic components consist of inorganic materials such as nutrients, minerals, water, gases, and air. Soil organisms are important in reducing carbon sequestration, GHG emissions, nitrogen cycling, organic matter decomposition, and soil restoration. Ecosystem services that rely on soil organisms include water storage, the breakdown of harmful substances, and controlling pests and diseases (Singh et al. 2011a).

Soil microorganisms, called soil microflora, are crucial to developing a healthy soil ecosystem. Due to their large surface area, fungi are the most successful soil organisms. Soil microorganisms can be put into three broad groups: ecosystem engineers, chemical engineers, and biological regulators (Giri and Varma 2020). Ecosystem engineers design and build porous aggregates that can withstand the elements and serve as breeding grounds for new species. Chemical engineers break down dead organisms and recycle them into useful nutrients. Biological regulators control the spatial and temporal dynamics of their respective food webs. Microorganisms have a crucial role as major decomposers of organic matter and drivers of the most important biogeochemical cycles in soil (Giri and Varma 2020).

15.5.1 Role of Soil Fertility on Human and Crop Nutrition

The health of humans is profoundly impacted by soil, both directly and indirectly. Recent studies have revealed that soils impact human health in various ways, including exposing humans to contaminants and viruses, influencing the availability and quality of food, and influencing the process through which nutrients are transferred from the Earth to the human body (Burras et al. 2013). Malnutrition is a major factor in the relationship between soils and human health, and micronutrient deficiency is a root cause of many other health problems. Essential for growth and development are micronutrients, such asvitamin A, C and B-25 complex, Zinc, Calcium and Iodine (WHO 2022). Zinc deficiency may negatively affect the neurological, digestive, immunological, skin, reproductive, and skeletal systems (Jurowski et al. 2014). It was projected that half the world's population suffered from iron insufficiency and that a comparable proportion of the population was zinc deficient in 2000 (Welch and Graham 2005; Alloway 2008). People who do not get enough micronutrients can have a weaker immune system, making them more likely to get infections and diseases. A balanced diet and supplements can help prevent or treat these issues (Knez and Graham 2012).

Over 2 billion people worldwide have "hidden hunger," caused by a lack of micronutrients from soils and crops that do not have enough nutrients (Organization 2011). That causes serious health problems that cannot be seen. Malnutrition is made worse by farming practices that are not sustainable and put crop production ahead of food quality (Mayer and Wang 2018). The most important factors affecting crop micronutrient concentration are the soil's amount of organic matter, temperature, pH, and topography (Bermudez et al. 2020). An abundance of calcareous soils, Drought, restricted availability, a lack of fertilizers, low SOM content, and a lack of nutrients in the soil all contribute to the complex issue of food insecurity that plagues many regions of the globe, including the Middle East and West Asia. In addition, the ongoing conflicts and political unrest in these areas, which have disrupted food production and distribution systems, worsen this problem. As a result, many people in these areas need help accessing sufficient and nutritious food daily.

15.5.2 Impact of Soil Nutrition-Depleted and Overloaded on Environmental Pollution and Climate Change

About 38% of the planet's livable surface is made up of agricultural soils, and 70% of the freshwater on the planet is used for irrigation (Foley et al. 2011). The overuse of fertilizers has polluted the air, water, and soil, a major source of greenhouse gases that cause global warming (Erisman et al. 2008). After leaving the plant-soil system, pollution makes it difficult and costly to recover lost soil nutrients. Such fertilizers have provided advantages and are vital for human society, but they also pose a danger to human and animal health (Fig. 15.3). The stability of the Earth is being threatened by the widespread environmental disruptions brought on by intensive, unsustainable agriculture practices (Sutton et al. 2011).

Unsustainable agricultural methods have affected the nitrogen and phosphorus biogeochemical cycles, leading to nitrogen levels reaching beyond the threshold and phosphorus levels reaching halfway above the threshold (Fig. 15.4) (Schütz et al. 2018). Low efficiency in their usage has been linked to both nitrogen (N) and phosphorus (P) input increases, leading to major environmental contamination and human health concerns (Lory and Scharf 2003; Greer and Pittelkow 2018). The issue of fertilizer overuse is complicated by inequalities in its usage, with only 50 nations making significant use of the available synthetic fertilizer. Farmers in wealthy nations have built a substantial reserve of residual phosphorus in agricultural soils over the last decades (Bouwman et al. 2017).

Nitrogen fertilizers are a major source of greenhouse gas N_2O, which has a 298 times larger global warming potential than CO_2 (Secretariat et al. 2021). The amount

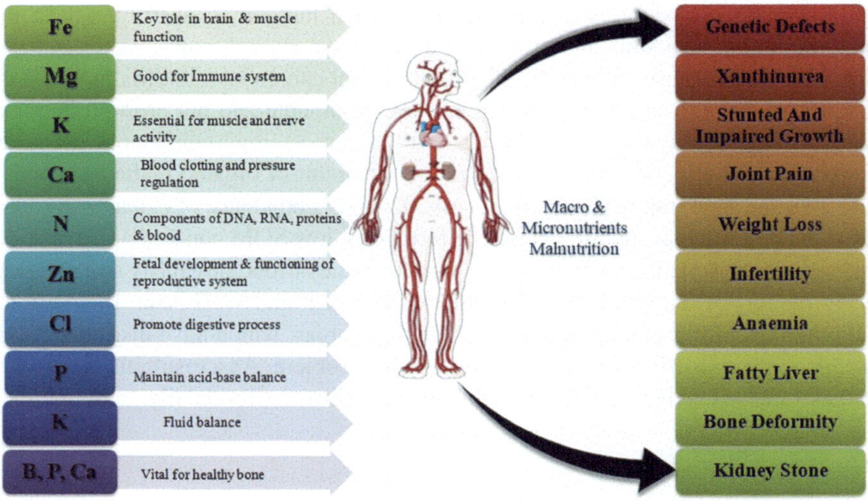

Fig. 15.3 The schematic diagram represent as mechanisms of micro and macronutrients in human body

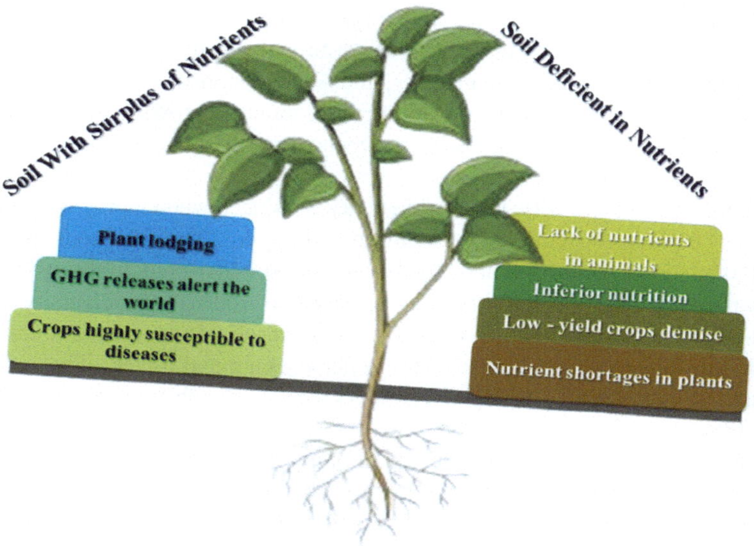

Fig. 15.4 Diagram presented as soil with nutrition-depleted and overloaded effect on agriculture

of N_2O released into the atmosphere annually by humans has reached 7.3 Tg yr^{-1} (1 Tg = 10^{12} g). Almost half of all anthropogenic agricultural emissions come from farming, with nitrogen (N) additions to croplands being the primary contributor (Tian et al. 2020). N_2O emissions may also be caused by the improper or excessive application of organic fertilizers, the deposition of urine and dung by grazing animals, the decay of organic crop residues, and the breakdown of soil organic matter.

15.6 Many Issues with Soil Minerals and How They Affect Agricultural Systems' Ability to Provide Healthy Food

(I) **Degrading Resource Base**: The post-green revolution era has resulted in food contamination with hazardous chemicals, groundwater pollution, soil quality degradation, and harm to agriculturally beneficial microorganisms (Manjaiah et al. 2019). That has been caused by intensified agriculture, indiscriminate irrigation water usage, and careless fertilizer use. In addition, both surface and groundwater are becoming unsafe for consumption due to the high concentration of pesticide residues.

(II) **Climate Change**: Rising greenhouse gas levels have caused global warming, with the temperature expected to rise by 1.8–4.0 °C by the end of this century (Ding et al. 2007). That has a direct and indirect impact on agriculture, with

India bearing a disproportionately large burden due to its reliance on agriculture, dwindling natural resources, an alarming rise in the number of people and animals, shifting land-use patterns, and socioeconomic factors. For example, for every 1 °C increase in temperature throughout the growing season, research suggests the likelihood of losses in the wheat output of 4–5 million metric tons (Ding et al. 2007).

(III) **Minimal Biodiversity**: Genetic diversity is reduced when more contemporary ones replace conventional agricultural and regional animal breeds. It is even though these novel types and breeds are better suited to contemporary intensive agriculture. The increased farming density also erodes the biodiversity of flora and fauna in agricultural ecosystems, such as rice–wheat monoculture in the Indo-Gangeticplains. Large-scale pesticide usage and a lack of recycling of agricultural wastes also negatively affect the soil microbiota in the area (Altieri 2004).

(IV) **Integrated farming systems come in many different forms**: Odisha farmers often turn to integrated farming methods to augment their demands for food, fuel, fiber, and finances (Pandey et al. 2018). However, there are no examples of integrated agricultural systems that are scientifically developed, commercially viable, and socially acceptable, leading to subsistence-based efforts that often need to be more reputable.

(V) **Agricultural Resource Recycling is not very common**: Organic waste from farms, factories, and cities goes to waste due to a lack of knowledge about recycling. However, reusing agricultural wastes can be an organic source to maintain soil health and improve physiochemical soil characteristics, creating a better environment for crop development.

(VI) **Gaps in technological adoption**: People's involvement in the ideation and development of technologies is essential to make new ones and improving the ones we already have. Our research had never been about the farm family, so we had a bad impression of the problems they tried to solve. One reason there is little sharing of agricultural technologies is that farmers, service providers, and technological and financial institutions need to talk to each other better.

15.7 Future Perspective

From the beginning of civilization, soils have been crucial to people's ability to grow food and important resources. However, with the increasing global population and demand for food, soil degradation has become a major concern that threatens food security and the sustainability of agriculture. Urbanization and the growing separation of people from the land have made forests even more vital. To meet these needs, managers must focus strongly on monitoring and evaluating to provide adaptive management and a steady flow of ecosystem services. One of our global goals must be to teach everyone how important forests and soils are to keep strong ecosystems and have good soils. We are all affected by urban, managed, or wild woods. Long-term agricultural production requires management strategies that come from different fields. That leads to better use of resources, decomposition and nutrient cycling, nitrogen fixation, water storage, protection against pollution and land degradation, and nitrogen fixation. Too many pesticides can lead to soil compaction and other changes that hurt soil biodiversity. It has been shown that switching to an organic or sustainable system increases SOM buildup and microbial activity. Sustainable management practices can bring back the organic carbon lost during intensive agriculture. We can learn more about biodiversity and find new microorganisms with unrealized potential in bioremediation, soil restoration, and medicine if we develop methods that do not rely on growing bacteria in the laboratory.

15.8 Conclusion

The rapid depletion of soils and biodiversity throughout the globe threatens people's health everywhere. This chapter compared a "reductionist" way of using public health and soil biodiversity in agricultural systems with a "holistic" way of doing things. The holistic approach focuses on keeping soil ecosystems and their variety alive so that the soil can do its many jobs better. Conventional farming methods, which have been shown to hurt the soil, the organisms that live in it, and the many ecosystem services that the soil and its variety provide, are still used because they are cost-effective. However, soil organisms are now known to be an important part of life that needs to be protected and grown. Farmers struggle to meet consumer demand for cheap food and societal expectations for a clean environment. Studies suggest that sustainable practices that boost soil biodiversity seldom benefit farmers financially. Integrated public health and food security policies at the national and international levels should include incentives for sustainable practices that increase soil biodiversity to maximize agricultural productivity and human health. However, a strategy to tackle climate change in the United States has resulted in societal inequalities due to incentives like carbon banks for farmers. Regenerative and sustainable agriculture needs channels outside public institutions, government policies, and incentives to convey their varying financial demands to the investing community. Incentives, legislation, and investors may be needed to make agriculture economically feasible and sustainable. However, we must avoid paternalism if we want farmers to have autonomy over their land. To reduce financial risk, farmers should work with many groups in their area to coordinate activities and take baby steps toward their goals.

Further evidence of the advantages of sustainable and regenerative agriculture may encourage more people to become involved. The most important details in this text are that scientists must properly disseminate their results to the general public and that soil scientists and human health experts need to be more typically trained in effective large-scale public outreach. To bridge these gaps, we need create multidisciplinary teams comprised of experts from several fields like chemistry, agronomy, communications, biology, toxicology, social science, medicine, sociology and public health. Additionally, a change of perspective on how we address problems related to human health is necessary for the kind of cooperation needed.

Declaration of Conflicts of Interest Authors declare no conflicts of interest.

References

Akhtar S, Ismail T et al (2013) Micronutrient deficiencies in South Asia–current status and strategies. Trends Food Sci Technol 31(1):55–62

Alloway BJ (2008) Zinc in soils and crop nutrition. published by IZA and IFA. Brussels, Belgium and Paris, France, p 139

Altieri MA (2004) Linking ecologists and traditional farmers in the search for sustainable agriculture. Front Ecol Environ 2(1):35–42

Anwar C, Annisa S et al (2018) The relationship between soil transmitted helminthes (STH) infection and nutritional status in students of state elementary school number (SDN) 200 Palembang Indonesia. Bioscientia Med J Biomed Transl Res 2(2):42–53

Bermudez A, Calderon-Ramirez S et al (2020) A first glance to the quality assessment of dental photostimulable phosphor plates with deep learning. In: 2020 international joint conference on neural networks (IJCNN). IEEE

Bhatla SC, Lal MA et al (2018) Plant mineral nutrition. Plant physiology, development and metabolism, pp 37–81

Bhattacharyya T, Wani SP et al (2021) Scaling-up agro-technologies using agro-eco sub-regions in the target states. Scaling-up Solutions Farmers Technol Partnerships Convergence 55–119

Black R (2003) Micronutrient deficiency: an underlying cause of morbidity and mortality. SciELO Public Health 81:79–79

Bouwman A, Beusen A et al (2017) Lessons from temporal and spatial patterns in global use of N and P fertilizer on cropland. Sci Rep 7(1):1–11

Brevik EC, Slaughter L et al (2020) Soil and human health: current status and future needs. Air Soil Water Res 13:1178622120934441

Burras C, Nyasimi M et al (2013) Soils, human health, and wealth: a complicated relationship. Soils and human health. CRC Press, Boca Raton, pp 215–226

Cakmak I (2009) Enrichment of fertilizers with zinc: an excellent investment for humanity and crop production in India. J Trace Elem Med Biol 23(4):281–289

Das S (2012) Status paper on rice in Orissa. Rice Knowledge Management Portal, Directorate of Rice Research, Hyderabad. http://www.rkmp. co. in/sites/default/files/ris/rice-statewise/Status%20Paper%20on%20Rice% 20in%20Orissa.pdf

Dash B, Sahoo BK (2020) Impact of development projects on grazing resources of Odisha

Ding Y, Ren G et al (2007) Detection, causes and projection of climate change over China: an overview of recent progress. Adv Atmos Sci 24:954–971

Duchemin B (2022) The sustainability of phytomass-derived materials: thermodynamical aspects, life cycle analysis and research perspectives. Green Chem 24(7):2653–2679

Erisman JW, Sutton MA et al (2008) How a century of ammonia synthesis changed the world. Nat Geosci 1(10):636–639

Foley JA, Ramankutty N et al (2011) Solutions for a cultivated planet. Nature 478(7369):337–342

Garnett T (2014) Three perspectives on sustainable food security: efficiency, demand restraint, food system transformation. What role for life cycle assessment? J Clean Prod 73:10–18

Giri B, Varma A (2020) Soil health. Springer

Greer KD, Pittelkow CM (2018) Linking nitrogen losses with crop productivity in maize agroecosystems. Front Sustain Food Syst 2:29

Gulati A, Roy R et al (2021) Revitalizing Indian agriculture and boosting farmer incomes. Springer Nature
Haplaquepts F, Ustochrepts H (2016) Soil fertility status of some villages under east and south eastern coastal plain agroclimatic zone of Odisha. J Indian Soc Coast Agric Res 34(1):63–67
Havugimana E, Bhople BS et al (2017) Soil pollution–major sources and types of soil pollutants. Environ Sci Eng 11:53–86
Horrigan L, Lawrence RS et al (2002) How sustainable agriculture can address the environmental and human health harms of industrial agriculture. Environ Health Perspect 110(5):445–456
Hota P, Behera B (2016) Opencast coal mining and sustainable local livelihoods in Odisha, India. Miner Econ 29:1–13
Hota A, Biswal S et al (2018a) Seroprevalence of PPR among sheep and goats of different agroclimatic zones of Odisha. Int J Livest Res 8(4):296–302
Hota A, Biswal S et al (2018b) Seroprevalence of Capripoxvirus infection in sheep and goats among different agro-climatic zones of Odisha, India. Vet World 11(1):66
Jensen HH (2002) Food insecurity and the food stamp program. Am J Agr Econ 84(5):1215–1228
Jin L, Whitehead PG et al (2018) Simulating climate change and socio-economic change impacts on flows and water quality in the Mahanadi River system, India. Sci Total Environ 637:907–917
Jones DL, Cross P et al (2013) Nutrient stripping: the global disparity between food security and soil nutrient stocks. J Appl Ecol 50(4):851–862
Joshi P, Roy D et al (2016) Technologies for maize, wheat, rice and pulses in marginal districts of Bihar and Odisha. Technological and institutional innovations for marginalized smallholders in agricultural development, pp 323–367
Jurowski K, Szewczyk B et al (2014) Biological consequences of zinc deficiency in the pathomechanisms of selected diseases. J Biol Inorg Chem 19:1069–1079
Knez M, Graham RD (2012) The impact of micronutrient deficiencies in the impact of micronutrient deficiencies in agricultural soils and crops on the nutritional health of humans. Essentials of medical geology: revised edition. Springer, pp 517–533
Kumar KA, Lalitha M et al (2020) Major soil types and classification. The soils of India, pp 81–108
Lory J, Scharf P (2003) Yield goal versus delta yield for predicting fertilizer nitrogen need in corn. Agron J 95(4):994–999
Lynch JP, Brown KM (2001) Topsoil foraging–an architectural adaptation of plants to low phosphorus availability. Plant Soil 237:225–237
Manjaiah KM, Mukhopadhyay R et al (2019) Clay minerals and zeolites for environmentally sustainable agriculture. Modified clay and zeolite nanocomposite materials. Elsevier, pp 309–329
Mary S, Nair N et al (2006) Analysis of genetic diversity among Saccharum spontaneum L. from four geographical regions of India, using molecular markers. Genet Resour Crop Evol 53:1221–1231
Mayer A-K, Wang H (2018) Regulations concerning pesticides and fertilizers. Handbook of agri-food law in China, Germany, European Union: food security, food safety, sustainable use of resources in agriculture, pp 277–346
Meena MK, Nayak L et al (2022) Integrated nutrient management in rainfed maize (Zea mays L.) under Eastern Ghat High Land Zone of Odisha. Int J Plant Soil Sci 102–108
Mishra MC, Senapati S et al (2021) Geotechnical characteristics of soils and rocks of India (Odisha). CRC Press, pp 511–527
Nandagopal R. Diversification: the key to better livelihood opportunities
Organization WH (2011) Vitamin and mineral nutrition information system (VMNIS). Anaemia database by country. Available online http://www.who.int/vmnis/database/anaemia/countries/en/#E. Accessed 27 Feb 2017
Organization WH (2021) The state of food security and nutrition in the world 2021: transforming food systems for food security, improved nutrition and affordable healthy diets for all. Food & Agriculture Org
Pahalvi HN, Rafiya L et al (2021) Chemical fertilizers and their impact on soil health. Microbiota and biofertilizers. Ecofriendly Tools Reclam Degraded Soil Environs 2:1–20

Pandey P, Sharma H et al (2018) Integrated farming system: only way to increase farmer's Income in a sustainable manner. J Pharmacognosy Phytochem 7(2S):210–214

Pany SS, Biswas SK et al (2016) Antigenic evidence of bluetongue virus from small ruminant population of two different geographical regions of Odisha, India. Vet World 9(3):304

Parida U, Bastia T et al (2016) Variation on the physical parameters of rice husk depending on the texture of quality of different types of soil on Odisha. Asian J Water Environ Pollut 13(3):103–105

PATRA SP. A case study on pani panchayat of lift-irrigation point of khurda district. Orissa University of Agriculture and Technology

Priyadarshini P, Saren S et al (2017) Soil fertility status of some villages under North-Eastern Coastal Plain Agro climatic Zone of Odisha. J Indian Soc Coast Agr Res 35(2):42–47

Raj A, Jhariya MK et al (2019) Soil for sustainable environment and ecosystems management. Sustainable agriculture, forest and environmental management, pp 189–221

Rath S, Mishra NC et al (2011) Upland rice in Orissa. Upland rice in India, p 284

Sahoo KC (2012) Antibiotic resistance and environmental factors: focusing on the situation in Odisha, India. Karolinska Institutet (Sweden)

Sarap K (2013) Forests and livelihoods in Orissa. Forests people and power. Routledge, pp 283–323

Schmidt MJ, Heckenberger MJ (2009) Amerindian anthrosols: Amazonian dark earth formation in the upper Xingu. Amazonian dark earths: Wim Sombroek's vision, pp 163–191

Schütz L, Gattinger A et al (2018) Improving crop yield and nutrient use efficiency via biofertilization—a global meta-analysis. Front Plant Sci 8:2204

Secretariat I, Gullino M et al (2021) Scientific review of the impact of climate change on plant pests. FAO on behalf of the IPPC Secretariat

Singh A, Parmar N et al (2011a) Bioaugmentation, biostimulation, and biocontrol in soil biology. Springer

Singh BP, Cowie AL et al (2011b) Soil health and climate change. Springer

Singh R, Mishra S et al (2011c) Mineral profile of feeds, fodders and animals in mid-central table land zone of Orissa. Anim Nutr Feed Technol 11(2):177–184

Singh S, Joshi RK et al (2013) Identification of elite genotypes of turmeric through agroclimatic zone based evaluation of important drug yielding traits. Ind Crops Prod 43:165–171

Singh G, Panda RK et al (2020) Regional scale trend and variability of rainfall pattern over agro-climatic zones in the mid-Mahanadi river basin of eastern India. J Hydro Environ Res 29:5–19

Sutton MA, Oenema O et al (2011) Too much of a good thing. Nature 472(7342):159–161

Thies JE, Rillig MC (2012) Characteristics of biochar: biological properties. Biochar for environmental management. Routledge, pp 117–138

Tian H, Xu R et al (2020) A comprehensive quantification of global nitrous oxide sources and sinks. Nature 586(7828):248–256

Wall DH, Nielsen UN et al (2015) Soil biodiversity and human health. Nature 528(7580):69–76

Welch RM, Graham RD (2005) Agriculture: the real nexus for enhancing bioavailable micronutrients in food crops. J Trace Elem Med Biol 18(4):299–307

WHO (2022) Micronutrients (webpage accessed on the 15th of march 2022). https://www.who.int/health-topics/micronutrients#tab=tab_1

Chapter 16
Exploring the Recent Trends, Progresses, and Challenges in the Application of Artificial Intelligence in Water Quality Assessment and Monitoring in Nigeria: A Systematic Review

Michael E. Omeka

Keywords Artificial intelligence · Drinking water quality · Machine learning · Nigeria · Pollution

Abbreviations

SAR	Sodium absorption ratio
RMSE	Root mean square error
MLP-ANN	Multi-layer perceptron neural network
TP	Total phosphorus
NN-CS	Neural network trained by Cuckoo search algorithm
TS	Total solids
TDS	Total dissolved solids
ANFIS	Adaptive Neuro-fuzzy inference system
BOD	Biochemical oxygen demand
TN	Total nitrogen
TA	Total alkalinity
TH	Total hardness
COD	Chemical oxygen demand
DO	Dissolved oxygen
TSS	Total suspended solids

M. E. Omeka (✉)
Department of Geology, University of Calabar, P.M.B. 1115, Calabar, Cross River State, Nigeria
e-mail: omeka.ekuru@unical.edu.ng

P. P. Adhikary et al. (eds.), *Soil, Water Pollution and Mitigation Strategies*,
Environmental Science and Engineering, https://doi.org/10.1007/978-3-031-63296-9_13

BPNN	Back propagation neural network
R^2	Coefficient of determination
PI	Permanganate index
SVM	Support vector machine
W-BPNN	Wavelet and back propagation neural network
W-ANFIS	Wavelet and adaptive Neuro-fuzzy inference system

16.1 Introduction

Water occupies about three-quarters of the total earth's crust. Among the total water mass, about 68.7 and 30.3% are trapped in icecaps or glaciers and seawater respectively. Only 1% is considered fresh (Al Aani et al. 2019). In recent times globally, there has been an upward decline in the quality of water resources. This has been reflective of the upsurge in anthropogenic activities and global climate variability (Ayejoto et al. 2023; Agbasi et al. 2023). The quality of water is also driven by natural hydrological, hydrogeologic, and geogenic processes which tend to modify the chemistry of the water (Omeka et al. 2022). The rate of modification (or pollution) of the water quality will depend largely on the prevalent pollution precursors peculiar to the area, either regionally or locally. Moreover, the processes and sources of water quality deterioration are usually complex and ubiquitous (Omeka2023). For instance, natural processes variations in aquifer characteristics, changes in stream dynamics as a response to pollutant influx, variation in rainfall intensity, water–rock interaction, chemical speciation, etc. anthropogenic processes may be sourced from both points (discharges from industrial and mine effluents), and non-point sources (e.g., run-off from agricultural lands, infiltration of leachates from municipal dumps into groundwater, run-off from open defecation, etc.) (Xiao et al. 2021; Okamkpa et al. 2022). As a result of this complexity, appraising the water quality is usually challenging and rigorous in terms of data collection, laboratory analysis, and data analysis. In recent decades, water quality assessment and monitoring have been carried out through the use of conventional and non-conventional approaches. While the conventional approaches rely on the use of local and international benchmark standards for quality checks, the non-conventional approach involves the use of multiple numerical indices, parametric and non-parametric statistical models, health risk assessment models, and spatiotemporal models (using GIS). The drawback however with the use of multiple numerical models is that the computation of sub-indices for model output is usually time-consuming and requires expert knowledge. As a consequence, this will result in errors in result output and by extension influence the decision-making process which may have a huge impact on the economy (Kouadri and Samir 2021).

Among the black African countries, Nigeria is made up of the highest human population; with a population density of over one-hundred and ninety-nine million (Ighalo et al. 2020). However, a staggering number of about 66.3 million do not have access to portable and safe drinking water (Ighalo et al. 2020; Okamkpa et al. 2022).

Additionally, due to the high population density, and increasing socio-economic activities a lot of stress is being put on the available water sources. Furthermore, seasonal changes (in temperature and rainfall intensity) due to climate changes are also known to have a significant shift in the quality of water resources (Ayejoto et al. 2023; Agbasi et al. 2023). In this context, the increasing anthropogenic activities in the region (e.g. hydrocarbon exploration, industrialization, land use, and population increase) will indirectly result in increased climate change variability. This in turn will influence the frequency of aquifer recharge/discharge rates (Omeka and Egbueri 2022; Ricolf et al. 2020). Therefore, for adequate and sustainable water management and protection, efforts and attention should be shifted toward establishing a relationship between the different water quality precursors peculiar to the region. This will entail the design of novel and cost-effective approaches for robust and holistic water quality appraisal and monitoring. This can be accomplished through the use of artificial intelligence (AI).

Globally, there has been some progress from the conventional water quality assessment and monitoring techniques towards artificial intelligence (AI) and Internet of things (IoT) technologies (Geetha and Gouthami 2016). AI was first introduced in the early 1950s and has since gone through significant modifications and gained applications in various fields in data simulation and forecasting of events outcomes (Al Aani et al. 2019). The use of AI in water quality assessment and monitoring started in the early 1990s and has since then gained the attention of many researchers and environmental stakeholders (Hasan et al. 2011). As a result of the complexity, ubiquity, and nonlinearity usually associated with water quality datasets, a challenge may arise in the prediction of water quality through the use of conventional approaches. This is attributed to increasing sources of anthropogenic pollution sources and varying climatic changes. Therefore, developing a model that can efficiently predict and monitor water quality remains crucial, especially in the present Anthropocene age where they seem to be no sight of solution to the impending global pollution crisis.

As part of her industrial revolution plan, Nigeria projects to increase its industrial output from 4 to 10% of GDP by 2020 (NIRP 2014). This will mean that more pressure will be put on the environment and its valued components (such as water, air, and soil). The big question remains how the country will maintain a balance between its industrial revolution projections and the protection of valued environmental components. It is feared that with the continuous increase in population and impeding industrial revolution, Nigeria will be put under greater strain in terms of climatic variations. Additionally, the quantity and quality of water resources will experience a further decline and continuous shifts (Ayejoto et al. 2023). It becomes imperative therefore for more conscious action plans to be put toward the assessment and prediction of the quality of water resources for informed decision-making and amelioration of contaminated sites through the use of cutting-edge approaches. As a response, in recent times, some studies have tilted towards the use of AI in water quality prediction in Nigeria, with some recorded progress. On a global scale, Ighalo et al. (2020) have carried out a comprehensive review of the application of different AI models in the assessment and monitoring of surface water in the last decade (2011–2020). Their study found that two AI models- Artificial Neural Networks (ANN) and

Adaptive Neuro-Fuzzy Inference System (ANFIS) were the most applied model for water quality assessment and monitoring within the period. The study also showed that globally, Southeast Asia and Iran are among the countries with the most studies on the use of these models in water quality prediction.

Studies in Nigeria on the application of AI in water quality prediction are very few in the literature. Among these studies, most have focused on the use of standalone machine learning (ML) algorithms (Egbueri 2022; Egbueri and Agbasi 2022; Omeka 2023) while some have integrated ML and AHP in water quality prediction and prioritization (Ifediegwu 2022; Omeka et al.2023). For instance, Egbueri (2022) has integrated a multilayer perceptron artificial neural network (MLP-ANN) and multiple linear regression (MLR) with corrosion and scaling potential indices for the prediction of industrial water quality in southeastern Nigeria. This study concluded that the incorporation of ML with indexical models can provide robustness in industrial water quality assessment and management. In another study in the region, MLR and MLP-ANN were incorporated to predict groundwater quality using two numerical models- overall index of pollution (OIP) and water quality index (WQI) as predicted or output variables. The MLP-ANN showed higher efficiency of prediction as lower modeling errors were observed in their validation metrics (Egbueri and Agbasi 2022). In an agricultural-mine district in southeastern Nigeria, a composite irrigation suitability zonation map was generated through the integration of a GIS-based analytical hierarchy process (AHP) and ML algorithms (Omeka et al. 2023). The model successfully identified Mg, HCO_3, Cl, and SO_4 as the major contributors to the deterioration of the irrigation water quality in the region. In a similar study, simple kriging (SK) and inverse distance weighting (IDW) GIS-based spatiotemporal models have been used in combination with supervised ML algorithms for a composite assessment of the irrigation water quality for crop yield in a mining region in southeastern Nigeria (Omeka 2023). This study revealed that soil is more susceptible to sodicity through the application of surface water for irrigation due to the higher pH of the surface water compared to groundwater.

Indeed, some efforts have been put toward the use of AI in water quality monitoring in Nigeria, especially, in the southeastern parts. It is however recommended for more research be channeled toward this area by embracing the application of hybridized (ensemble) ML algorithms. Due to the nonlinearity and randomness of water quality data, and the impending climate variability, the need for the use of improved ML models is imperative. The relationship between water quality and its controlling factors is known to vary over time (due to variations in anthropogenic factors and seasonal fluctuations), hence, the use of a predictive model that functions under the framework of historical data becomes inefficient over time (Khan et al. 2021). The use of hybridization in other parts of the world has shown very promising results. For instance, in Pakistan, the forward-rolling artificial intelligence algorithm has been used to generate a model that will show resiliency to variability in climate change in a hydrologic setting (Khan et al. 2021). Two ML algorithms—the recursive feature elimination and support vector machine learning (RFESVM) have been combined to monitor the variability in rainfall in Malaysia (Pour et al. 2020). In Penang, Malaysia, the support vector machines (SVM), backpropagation neural network (BP-NN), and radial basis function neural network (RB-NN) have been incorporated to monitor the quality of surface water in constructed wetlands (Mohammadpour et al. 2016). Nayak (2020) combined and compared the use of multiple linear regression (MLP) with correlation models in assessing the trend in the surface water quality of the Brahmani River. The MLR models were found to show better results in water quality trends. In the semi-enclosed Tolo Harbour water bay in Hon Kong, random forest machine (RF) and ANN machine learning models were combined to assess the spatial variations in eutrophication in the water. The ensemble model identified suspended solids, dissolved nutrients, and organic pollutants as significant contributors to water quality (Deng et al. 2022). In Iran, Sun et al. (2021), have combined two hybrid ML models—the multivariate adaptive regression spline (MARS) and intrinsic time-scale decomposition (ITD) with the multi-step supervised-based machine learning approach (MSMLEA) for the prediction of the water quality of the Tajan River in Iran. From the results, the hybrid ML models showed higher performance accuracy in water quality prediction.

In the last decades, Nigeria has recorded severe cases of waterborne diseases including diarrhea, cholera, and other gastrointestinal illnesses resulting from the ingestion of contaminated water (Martins et al. 2016). According to the WHO/UNICEF (2010) joint report on monitoring programs for water supply and sanitation, Nigeria is ranked third; just behind China and India, as the countries without access to safe drinking water. It is estimated that about 130,000 children in Nigeria die annually due to waterborne diseases (WHO/UNICEF 2010). It is feared that this number would rather increase by the year 2040 if drastic measures are not taken (WHO/UNICEF 2010). According to a report on cancer patterns in Nigeria, reported by the Nigeria National System of Cancer Registries (2016), pollution contributes to about 80% of cancer risks in the country. Of this, 30% is attributed to the ingestion of contaminated water (Morounke et al. 2017). This seems to agree with the 1996 Harvard report on cancer prevention, that among all the cancer-causing risk factors, cancer risks from the ingestion of dissolved heavy metals in drinking water (e.g. Ni, Cd, As, and Pb) accounts for about 2% of cancer risks globally. Hence, there is a need for a continuous review of the novel and cutting-edge approaches toward adequate and holistic water quality monitoring and prediction. In developed countries of the world, data on water quality and seasonal changes is well documented compared to developing countries such as Nigeria. This can be linked to a range of factors including lack of funding from the government and inaccessibility of laboratory facilities or lack thereof. Therefore, in this study, a systematic literature review (SLR) on the application of machine learning models in water quality monitoring and assessment was carried out for the last two decades (2003–2023). Considering the importance of clean water for the sustainable development and protection of human health, it is imperative to embrace the trend in machine learning, which is a new research area and has shown great promise in water quality prediction. Additionally, an in-depth assessment of progresses from the present well-observed trends to more hypothetical and advanced machine learning model simulation and comparative analysis has also been provided. The first section takes into account the prevalent factors influencing the quality of water in Nigeria as a function of contaminant sources and an insight into possible emerging contaminants. The second section takes into account the past and expected future trends in water quality monitoring and management techniques in southeastern Nigeria as well as their implication for prospects. Additionally, insight has been provided into the application of various ML models in water quality monitoring, with a focus on the frequency of input variables, and the accuracy of validation metrics used to present the results. Recommendations for future water quality research prospects and studies have also been discussed.

16.2 Study Area

16.2.1 Location

Nigeria is bounded to the north by Niger, towards the east by Chad and Cameroon, towards the south by the Atlantic Ocean's Gulf of Guinea, and to the west by Benin (Encyclopædia Britannica 2023) (Fig. 16.1).

Fig. 16.1 Location Map of Nigeria showing the different geographic locations. *Source* Encyclopædia Britannica (2023)

16.2.2 Relief

Nigeria's terrain is characterized by lowlands in the north and south, which are punctuated by plateaus and mountains in the center of the nation. The Sokoto Plains are located in the country's northwest region, whereas the Borno Plains are located in the country's northeastern corner and extend to the Lake Chad basin. Porous, geologically young sedimentary strata underpin the Lake Chad basin and coastal regions, along with the Niger River delta and western sections of the Sokoto region in the extreme northwest (Encyclopædia Britannica 2023; Ayejoto et al. 2023). The plateaus' distinctive features are high plains with vast, shallow valleys interspersed with several hills or solitary summits known as inselbergs; the base rocks are crystalline, with sandstones appearing in river basins. Some degraded features, including the Udi-Nsukka escarpment, rise suddenly above the plains at least 1000 feet in height (300 m) (Okamkpa et al. 2022).

16.2.3 Climate

The climate of Nigeria is tropical, with varying rainy and dry seasons depending on location. The southeast is hot and humid most of the year, but dry in the southwestern and further inland. The west and north have a savanna climate with distinct dry and wet seasons, whereas the extreme north has a steppe environment with limited precipitation.

The frequency of the rainy season generally reduces from south to north. The rainy season runs from March to November in the south, except for mid-May to September in the extreme north. The rains stop completely in the south during August, resulting in a brief dry season known as the "August break." Precipitation is greater in the south, particularly the southeastern, which experiences over 120 in. (3000 mm) of rain every year, compared to around 70 in. (1800 mm) in the southwest. Temperature and humidity stay largely consistent in the south throughout the year, but seasons fluctuate significantly in the north; and during the northern dry season, the average temperature range also increases.

16.3 Methodology

16.3.1 Systematic Literature Review

The research method used in the present study was the systematic literature review. To carry out the research, the following research questions were raised: (1) what are the major sources of water contamination in Nigeria? (2) what are the most frequently used machine learning models in water quality analysis in the last two decades? (3) what are the most accurate machine learning models based on their model predict metrics (such as coefficient of determination [R^2]) (4) what are the most frequently monitored water quality parameters (5) what are the implications of the use of ML models in water quality monitoring and their implication for future research prospects? The systematic literature review was comprised of published articles from Web of Science and Scopus databases. The search terms used for the search in the databases included the *sources of contamination in the aquatic environment in Nigeria and the application of machine learning in water quality prediction.*

In the Scopus database, each term was entered in the search directory including the abstract, title, and keywords. The filters used were the year of publication (within the date range of 2003–2023), within the thematic area of environmental sciences, access type (open and none-open access articles), and the document type (article or review). In the Web of Science database, the search filter was based on Abstract, title, author keywords, and study within the date range of 2003–2023. Two thematic categories- machine learning in water quality analysis and environmental science.

The articles were selected based on the following criteria: (1) articles that give a detailed explanation of the major sources of water pollution (2) articles that explains the use of hybridized or standalone machine learning algorithms in water quality monitoring and (3) articles that explain the peculiarities of the different validation metrics (e.g. coefficient of determination, standard error of estimate, the sum of square errors and residual error). The exclusion was made on certain criteria such as (1) articles that appear outside the scope of the study (2) articles that do not explicitly address the use of machine learning algorithms in water quality monitoring and (3) articles concerning other valued environmental components outside of water.

16.3.2 Creating a Bibliographic Map Using the VOSviewer Software

The VOSviewerwas used in creating a bibliographic map for keywords in machine learning and water quality between the period of 2003–2023. Bibliographic data from Scopus and Web of Science databases were extracted with a filter in the area of environmental sciences. A total of 729 and 2234 articles were respectively returned for the search terms *for sources of contamination in the aquatic environment in Nigeria and the application of machine learning in water quality prediction* in the Scopus database. While for the Web of Science database, 521 and 1672 articles respectively were returned. The articles were then exported from the database in CSV format to the VOSviewer. The bibliographic map was then created in the VOSviewer environment using keywords with a minimum of five occurrences selected (van Ech and Waltman 2020).

16.4 Results and Discussion

16.4.1 Results from Databases

The variation in the number of articles using the two search items (*sources of contamination in aquatic environment in Nigeria and application of machine learning in water quality prediction*) in the two databases showed an upward increase in articles on the term *application of machine learning in water quality prediction* compared to the former. The Scopus database showed a higher number of publications compared to the Web of Science database. This is because the Scopus database has a greater number of indexed journals compared to the Web of Science database.

The number of articles returned with the use of the search term *sources of contamination in the aquatic environment in Nigeria and application of machine learning in water quality prediction* were 729 and 2234 respectively in the Scopus database. While for the Web of Science database, 521 and 1672 articles respectively were returned from the search (Table 16.1).

The number of articles considered for review was 35 and 20 for the terms *sources of contamination in the aquatic environment in Nigeria and application of machine learning in water quality prediction* in the Scopus database. While 25 and 18 were selected for the Web of Science database (Table 16.1). It is worth mentioning that due to a large number of returned articles for the Scopus database, the titles and abstracts of articles read were for the first 600.

Table 16.1 Number of articles analyzed and considered for the SLR

Search terms	Returned articles	Databases	Selected articles in the search
Sources of contamination in the aquatic environment in Nigeria	729	Scopus	35
Application of machine learning in water quality prediction	2234	Scopus	20
Sources of contamination in the aquatic environment in Nigeria	521	Web of Science	25
Application of machine learning in water quality prediction	1672	Web of Science	18
The sum of articles considered from abstracts and titles			98
The sum of articles considered after full-text analysis			40

For inclusion and exclusion criteria, only articles that did not appear in the two databases were included for the review; as some articles appeared in both the Scopus and Web of Science databases. Therefore, a total of 98 articles were selected for reading of titles and abstracts while a total of 40 articles were selected for full-text analysis while others were discarded (Table 16.1). The discarded articles did not meet the established criteria.

Based on the search term "*application of machine learning in water quality prediction*", out of the 257 publications that were considered, China occurred in 61, the United States in 57, India in 32, Malaysia in 15, south Korea in 14, Australia in 13, the United Kingdom and Canada in 10, Vietnam in 9, Iran in 8, Brazil and Iraq occurred in 7, Poland and Taiwan occurred in 6, Russian Federation, France, Germany, Italy, and Saudi Arabia occurred in 5; Morocco, Sweden, Bangladesh, Netherlands, and Hong Kong occurred in 4; Singapore, Spain, United Arabs Emirates, Norway, Palestine, Pakistan, and Greece occurred in 3; Croatia, Egypt, Pietro Florentine, Serbia, Belgium, Algeria, Nigeria, South Africa, Turkey, Columbia occurred in 2; Dakota, Philippines, Tunisia, Austria, Uganda, Romania, Chile, Denmark, Czech Republic, Luxembourg, Switzerland, Finland, Georgia, Peru, Sri Lanka, Portugal, State of Libya, Israel, and Mexico occurred in 1 had one publication each (Table 16.2). It can therefore be deduced frequency in the application of ML in this countries may be due availability of experts in the use of neural networks in engineering and environmental studies. Although this may be speculative, the author hope that this observation will spur the interest of researchers in the field for more discuss and enlightenment.

16.4.2 Bibliographic Mapping Using VOSviewer

The bibliographic map was created using the VOSviewer software. Co-occurrence was chosen as the type of analysis, while All keywords were chosen as the unit of analysis, with fractional counting selected as the preferred counting method. For the *application of machine learning in the water quality prediction* search file, the software returned 377 keywords with 186 meeting the threshold and occurring at least six times across all analyzed articles. For each of the 186 keywords, the total strength of the co-occurrence links with other keywords was calculated. The keyword with the greatest link strength was then selected for generating the bibliographic map. Links between keywords are an indication of the relationship between them (the link strength) represented by a numerical value. Implying that the higher the link strength, the stronger the association where the two terms occur together (van Eck and Waltman 2020). Terms were then grouped into clusters of a different colour; with each colour representing the strength of the association.

For the *application of machine learning in water quality prediction* terms, analysis was made based on both the countries of study and the type of machine learning algorithm frequently used. The bibliographic map based on the overlay visualization (with the year of publication) and density visualization is shown in Fig. 16.2 based on the type of machine learning model. The graphical representation of the

Table 16.2 Number of articles per country occurring from the systematic literature review

Country	No. of Publication
China	61
United States	57
India	32
Malaysia	15
South Korea	14
Australia	13
United Kingdom	10
Canada	10
Vietnam	9
Iran	8
Brazil, Iraq	7
Poland, Taiwan, Poland	6
Russian Federation, France, Germany, Italy, Saudi Arabia	5
Morocco, Sweden, Bangladesh, Netherlands, Hong Kong	4
Singapore, Spain, United Arabs Emirates, Norway, Palestine, Pakistan, Greece	3
Croatia, Egypt, Pietro Florentine, Serbia, Belgium, Algeria, Nigeria, South Africa, Turkey, Columbia	2
Dakota, Philippines, Tunisia, Austria, Uganda, Romania, Chile, Denmark, Czech Republic, Luxembourg, Switzerland, Finland, Georgia, Peru, Sri Lanka, Portugal, State of Libya, Israel, and Mexico	1

variation in the number of publications and the occurrence of the ML algorithms in water quality prediction is shown in Fig. 16.3. *Artificial neural networks* had the highest occurrence and strongest link strength in all analyzed publications followed by *regression analysis and random forest*, while *Bayes theorem* and the *convolutional neural network* had the lowest occurrence and weakest link strength (Figs. 16.2 and 16.3). From the bibliographic map, high density and link strengths seem to occur mostly with *Artificial neural networks, support vector machines, regression analysis, decision trees, and random forests*. However, *fuzzy systems, long short-term memory, convolutional neural network, hybrid, and ensemble* ML models had the weakest links and strength, with the number of publications on the subject seeming to be increasing since 2021. This result implies that the use of hybrid ML models in water quality prediction has not been well explored globally; a majority of the prediction has been based on the use of artificial neural networks.

Based on the countries of study, China, the United States, and India had the highest number of publications and link strength concerning the term *application of machine learning in water quality prediction* (Fig. 16.4). The total number of publications involving the review developed by country was 61, 57, and 15 respectively for China, the United States, and India (Table 16.2; Fig. 16.4). This implies that countries like

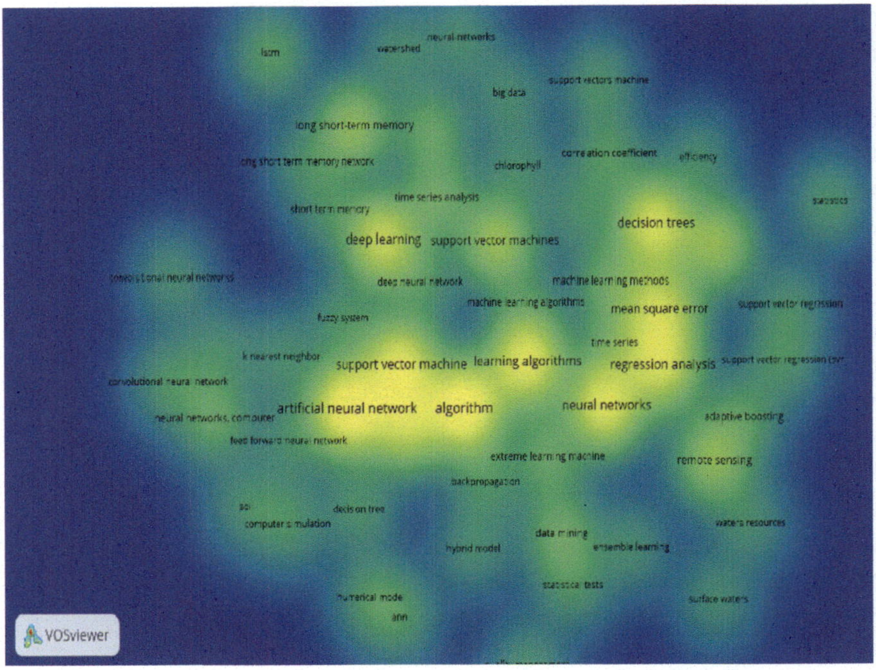

Fig. 16.2 Bibliographic coupling map showing results for frequently used machine learning in water quality prediction

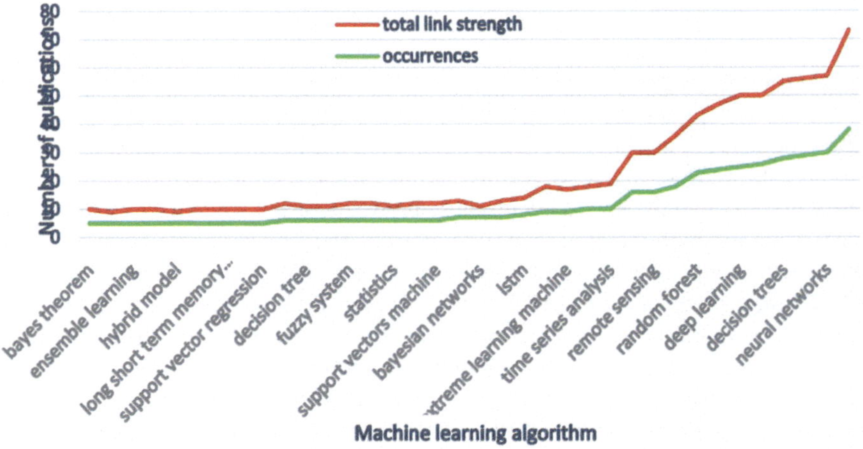

Fig. 16.3 Graphical representation of the frequently used machine learning model

Croatia, Egypt, Pietro Fiorentini, Serbia, Belgium, Algeria, Nigeria, South Africa, Turkey, Columbia, Dakota, Philipines, Tunisia, Austria, Uganda, Romania, Chile, Denmark, Czech Republic, Luxembourg, Switzerland, Finland, Georgia, Peru, Sri Lanka, Portugal, State of Libya, Isareal and Mexico are grossly lacking behind in the use of ML in water quality monitoring and assessment.

For the *sources of contamination in the aquatic environment in Nigeria* search term, analysis was made based on the relationship between anthropogenic and natural/geogenic sources of contamination and the number of publications within the last two decades (2003–2023). Based on the relationship between pollutant-type, heavy metals, Lead, Cadmium, Copper, Chromium, and Zinc was reported in more publications and had the greatest link strength among others. Meanwhile, Antibiotic agent, aromatic compounds, bitumen, Cl, combustion, dumpsite, EC, mineralization, municipal solid wastes, organic carbon, phosphates, rainwater, saline intrusion, urbanization, and suspended particulate matter showed the lowest link strength and was reported in fewer publications (Fig. 16.5; Table 16.3). Through the analysis

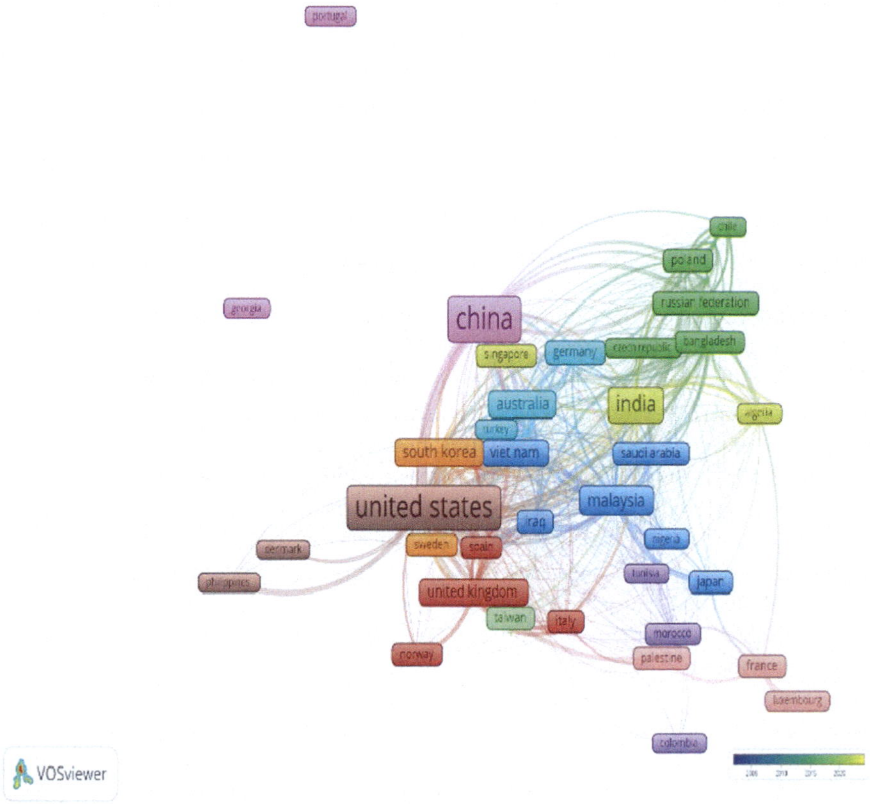

Fig. 16.4 Bibliographic coupling map of countries with the highest publication in the application machine learning in water quality prediction

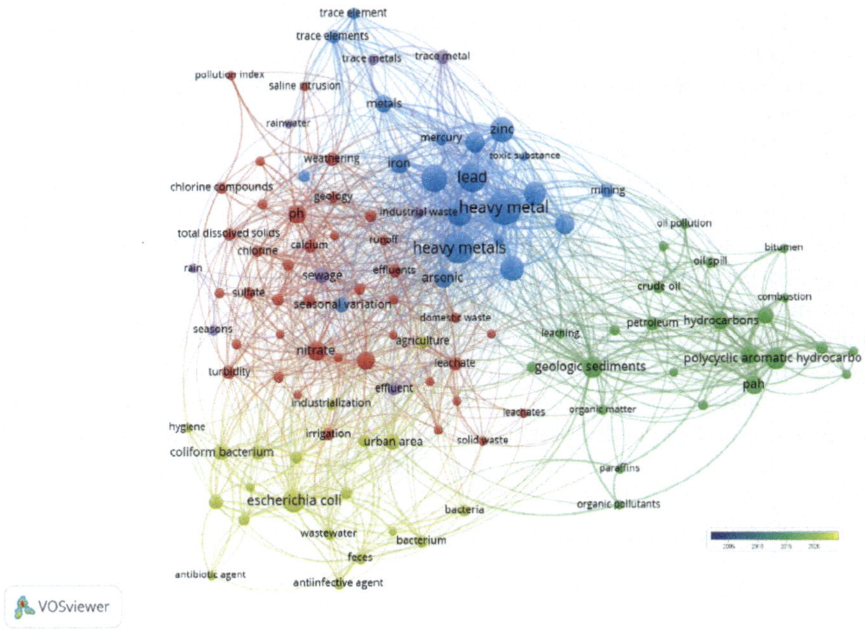

Fig. 16.5 Bibliographic coupling map for the sources of aquatic contamination in Nigeria

of extracted literature, for the period of 2003–2023, a relationship was established between anthropgenic and natural sources of contamination among all the reported articles water quality contamination in Nigeria. From the results it was observed that most studies attributed the occurrence of contamination in water to mining followed by lithogenic and anthropogenic processes (Fig. 16.6).

These results are in agreement with reports by Omeka and Egbueri (2022), that due to an upsurge in anthropogenic activities in the country such as mining, industrialization, and socio-economic activities, high concentration of heavy metal(loids) and potentially toxic elements have been reported in drinking water in the country. As reported by Obasi and Akudinobi (2020), there is an upward increase in the concentration of Cd, Pb, Cr, Zn, and cobalt in waters from the solid-mineral-rich southern Benue trough in southeastern Nigeria. This has been attributed to the open-cast mining method that is prevalent among most mining operations in the country (Okolo et al. 2018; Omeka and Igwe 2023). In southeastern Nigeria, Ajala et al. (2022), provided a critical review of potential sources of heavy metals in water and aquatic fish as well as the human health risks from their consumption. High concentration of Cd, As, Cu, and Zn was reported in both the drinking water and the shell of fishes. In Nsukka, southeastern Nigeria, they have been a reported high concentration of As, Cd, Hg, and Cr in water; attributed to anthropogenic influxes and run-off from domestic solid and sewage wastes (Nnaji et al. 2023).

Table 16.3 Number of publications based on sources of aquatic contamination in Nigeria

Pollutant	Occurrences per publication	Total link strength
Heavy metal	67	62
Heavy metals	59	52
Lead	48	47
Cadmium	40	39
Copper	36	36
Chromium	35	35
Zinc	34	34
Escherichia coli	30	26
Geologic sediments	25	24
Polycyclic aromatic hydrocarbons	25	21
Manganese, Nickel	23	23
Arsenic, Iron	22	22
Nitrate	21	20
PAHs, pH	20	20
Electrical conductivity	19	16
Hydrocarbons	18	17
Coliform bacterium, sewage	14	13
Bacterium contamination, seasonal variation	13	13
Microbial contamination	12	10
Effluents, leachates, mercury, mining, petroleum, trace elements, trace metals, turbidity, weathering	11	11
Agriculture, Calcium, total dissolved solids	10	10
Aluminum, crude oil, feces, irrigation, sulfate, wastewater	9	9
Anti-infective agents, chlorine, fecal coliform, industrial waste, land use, NO3, Na, population density, wastewater	8	6
Chlorine compounds, DO, geology, industries, Mg, oil spill, K	7	7
Co, F, landfill, leachates, leaching, organic matter, runoff, sulfur compounds, hardness, domestic wastes	6	6
Antibiotic agents, aromatic compounds, bitumen, Cl, combustion, dumpsite, EC, mineralization, municipal solid wastes, organic carbon, phosphates, rainwater, saline intrusion, urbanization, suspended particulate matter	5	4

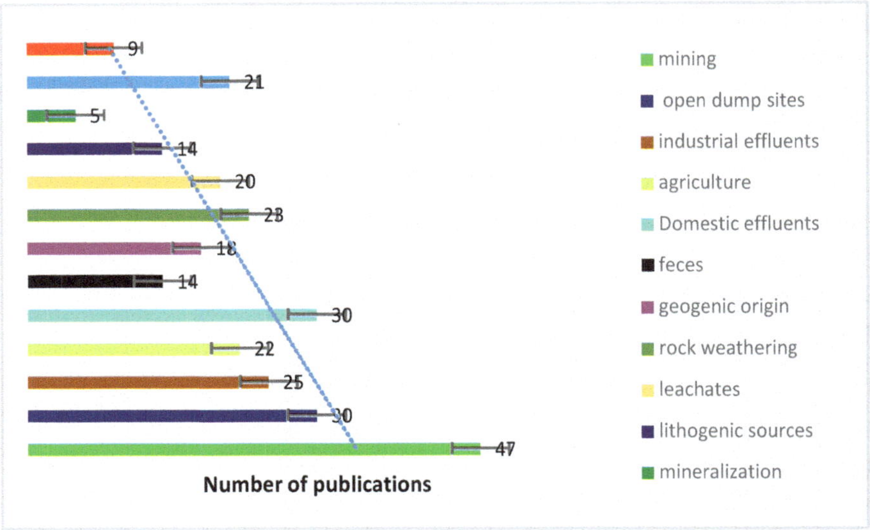

Fig. 16.6 Relationship of anthropogenic and natural sources of contamination in aquatic systems and number of publications between 2003 and 2023

Many studies in southwestern Nigeria have attributed the concentration of heavy metals in drinking water sources to industrialization. According to an extensive literature review carried out by Balogun et al. (2022) in the region, the long-standing sources of groundwater pollution cases have been attributed majorly to industrial effluents. In some parts of the Ibadan metropolis, southwest Nigeria, the concentration of heavy metals in shallow had-dug wells was observed to occur in the order of Zn > Fe > Pb > Cd > Mn; attributed to poor sanitation and industrialization (Ganiyu et al. 2021a). In the same region, the water quality of shallow aquifers was investigated for their metal content and bacterial load. The water was reportedly polluted due to pathogens and heavy metals; with the heavy metals occurring in the order of Cd > Pb > Zn > Fe > Mn, attributed to varying anthropogenic activities in the area(Ganiyu et al. 2021b). In the densely populated parts of the southwestern region, such as Lagos, water pollution has been associated with poor waste management practices, lowering of the water table (due to over-abstraction), and influx from industrial effluents (Ogundiran and Afolabi 2008). Some studies have also attributed the presence of heavy metals to both geogenic processes (such as weathering and leaching of subsurface geology) and anthropogenic sources (from industrial effluents). A study on the water quality assessment of Asa River in Ilorin, southwestern Nigeria revealed a high concentration of chromite ($FeCr_3O_4$) and pyrite (FeS) in river sediments, attributed to weathering of subsurface geology. Heavy metals such as Zn, Fe, Cr, and Mn were also reportedly found in elevated concentration dueto the influx of industrial effluents in sediments (Adekola and Elleta 2007). Although some studies in the region have reported the presence of heavy metals in water due to vehicular emissions and related repair products, domestic sewage, and effluents

(Tijani et al. 2004), most have been mostly due to industrial and agricultural effluents (Olaojo et al. 2018; Emenike et al. 2020).

In the northern and central parts of Nigeria, very few studies have reported water pollution due to industrial effluents Most reports on heavy metal concentration in water have been attributed to mining activities. In the Anka gold mining area of northwestern Nigeria, they have been reported cases of Pb poisoning found in the hairs and nails of children within the vicinity of the mine site; attributed to the ongoing mining activities within the region (Adebwumi 2020). Although a study by Lar and Gusikit (2015) reported the concentration of heavy metals (such as Zn, V, Pb, Cu, Co, Be Cr, As, Cd, Sb, and Se) in drinking water wells due to volcanic eruption in the Panyam volcanic province, majority of the reports on metal concentration in the water have been attributed to mining and poor disposal of its effluents. Pb, Ni, and Hg poisoning have been reported in analyzed surface and groundwater samples from the Bagega gold mine province in Zamfara state, Northwestern Nigeria. The high elevated concentration of carcinogenic elements was attributed to artisanal gold mining (Nuhu et al. 2014).

In southeastern Nigeria, the sources of water pollution are highly ubiquitous-ranging from a wide range of anthropogenic sources (e.g., mining, poor waste disposal activities, poor hygiene, industries, population increase, land use, etc.) and geogenic sources (precipitation, chemical weathering, weathering, and rock-water interaction) (Omeka et al. 2023; Aghamelu et al. 2022). Although most of the reports on contamination have been from both anthropogenic and geogenic activities (Nnorom et al. 2019; Edet et al. 2003), the majority have been from mining and poor waste disposal activities. Reports from mining have been majorly from the solid mineral-rich zones of the lower Benue trough; with high concentrations of ore minerals occurring in association with Pb–Zn (Omeka and Igwe 2023; Adamu et al. 2015). This has been mainly reported from Ebonyi, Enugu, and Cross River states. A review carried out by Umeoguaju et al. (2022) for two decades (2000–2020) revealed that the concentration of heavy metals in most water sources in southeastern Nigeria is attributed majorly to anthropogenic influences from mining and oil exploration. This is in agreement with a study by Opuene and Agbozu (2008) on the heavy metal assessment in fish from Taylor creek, southern Nigeria.

16.4.3 Application of Neural Networks in Water Quality Monitoring: Implications for Prospects

This section addresses the efficacy of artificial neural networks (ANNs) in water quality assessment, the most frequently used ANN models, the input or predictor variables, and the model accuracy based on the coefficient of determination (R^2). To achieve this, a literature search was conducted on the application of neural networks in water quality modeling and prediction. The search was conducted on papers from the Scopus and Web of Science databases between 2011 and 2023, using the keywords "water quality and artificial intelligence" with emphasis on both surface and groundwater. The results from the search were compiled and presented in Table 16.4, and plots were generated based on the results in Table 16.4.

As observed earlier from the bibliographic coupling map in Fig. 16.2, artificial neural networks appear to be the most frequently used machine learning model in water quality monitoring and assessment globally, followed only by support vector machines and decision trees. The wide usage of ANN in water quality studies has been attributed to its efficiency and versatility in predictions even in systems with poor computational strength (Egbueri et al. 2023; Omeka et al. 2023). According to Ghavidel and Montazeri (2014), the unique ability of the ANN model to accurately match a broad range of nonlinear variables makes it a widely accepted ML model in most water quality studies. The ANN architecture is designed to mimic the human neural system; with the unique ability to quickly learn and send signals about a range of linear and nonlinear datasets through its interwoven parts known as "neurons" (Ozel et al. 2020). This makes it stand out among other water quality modeling techniques. Water quality data are usually nonstationary, random, nonlinear, and unpredictable (Beven 2016). This will mean that the relationship between water quality and its controlling factors will tend to vary over time (as a result of varying anthropogenic and seasonal fluctuations), hence, the application of predictive models that works on historical data will become ineffective over time (Khan et al. 2021), hence more focus have been put on ANN due to its unique ability to process complex datasets.

16.4.3.1 Model Utilization

It can be observed from Fig. 16.7 that the multilayer perceptron artificial neural network (MLP-NN) and back-propagated neural network (BP-NN) are the most frequently used neural network algorithm in water quality monitoring and prediction within the period of 2011–2023.This was closely followed by the adaptive neuro-fuzzy inference system (ANFIS), and other neural networks such as the Wavelet and BPNN (WNN) hybrid model, Neural Network trained by Cuckoo Search (NNCS), Neural Network trained by Genetic Algorithm (NN-GA), Elman Neural Network (ENN) and Wavelet-Adaptive Neural Fuzzy Interference System (WANFIS) in decreasing order. The high frequency of the MLP-NN and BP-NN in water quality monitoring and prediction is in line with a study by Rajaee et al. (2020). The wide acceptance of these algorithms has also been attributed to their ability to attain high modeling accuracy with fewer input parameters compared to other white-box models (Yetilmezsoy et al. 2011; Ighalo et al. 2020; Omeka 2023).

16.4.3.2 Model Validation Accuracy

In this assessment, only studies that used the coefficient of determination (R^2) as a validation metric were considered. The exclusion was made for the root mean

Table 16.4 Summarized table showing the application of different machine learning algorithms in water quality prediction

Year	Location	ANN model	Input parameters	Highest R^2 (during the testing stage)	References
2023	1. Surface and groundwater, Ojoto suburb, southeastern Nigeria	MLP-ANN	TH, T, pH, TDS EC, Cl, Ca, SO_4, Pb, HCO_3, Zn Fe	0.878	Egbueri et al. (2023)
	2. Groundwater samples from Osisioma, southeastern Nigeria	MLP-ANN	Cu, Pb, Fe As, Cr, Benzene, Ethylbenzene m-Xylene, Toluene, and o-Xylene	0.896	Akakuru et al. (2023b, a)
	3. Groundwater samples from Egbema, southeastern Nigeria	MLP-ANN	Fe, Zn, Ni, Cd, Cu and Pb	0.966	Akakuru et al. (2023b)
	4. Surface and groundwater from Okurumutet-Iyamitet mine province, southeastern Nigeria	MLP-ANN	HCO_3^-, SO_4^{2-}, NO_3^-, Cl^-, Mg^{2+}, Ca^{2+}, K^+, and Na^+	0.861	Omeka et al. (2023)
2022	1. Bouregreg watershed, Morocco	BPNN	pH and EC	0.87	Bilali and Taleb (2020)
	2. Groundwater from Agartala municipality, India	BPNN	Na^+, Mg^{2+}, Ca^{2+}, EC, HCO_3^-, and B	0.990	Mallik et al. (2022)
	3. Groundwater from El Kharga Oasis, Western Desert of Egypt	ANFIS	EC, pH, T∘, TDS, Na^+, Mg^{2+}, K^+, Ca^{2+}, Cl^-, CO_3^{2-}, SO_4^{2-}, $NO_3^-$and HCO_3^-,	0.997	Ibrahim et al. (2023)
2021	(1) Groundwater wells from Illizi county, Algeria	BP-NN	pH, TH, EC, TDS, Ca, Mg, Na, K, HCO_3, Cl, SO_4, and NO_3	0.8957	Kouadri et al. (2021)
	(2) groundwater from El Merk is an oil field in the SOUTHEAST of Algerian	MLP-ANN	TH, NO_3, and NO_2	0.9967	Kouadri and Samir (2021a, b)
2020	1. The Elbe River, Germany	Wavelet and BPNN (W-BPNN)	Flow, pH, Fe, and DO	0.780	Li et al. (2020)

(continued)

Table 16.4 (continued)

Year	Location	ANN model	Input parameters	Highest R^2 (during the testing stage)	References
	2. Groundwater from Gaza strip Palestine	MLP-NN	Abstraction average rate (AVR), relative humidity (RH), depth from the surface to well screen (DSWS), aquifer thickness (AT), recharge rate (RR), Initial chloride concentration (ICC), and groundwater level (GWL)	0.9770	Kassem (2020)
	3. Yipin River, China	MLP-NN	pH, TP, Temperature, EC, PI, NH_3-N, COD and TN	0.717	Zhu and Heddam (2020)
2019	Water samples from the Shivganga River basin, India	BP-NN	EC, pH, TH, TDS, Mg, Ca, K, Na, HCO_3,PO_4 NO_3, Cl, and SO_4	0.932	Kadam et al. (2019)
	Lakes, Tezpur University, India	MLP-ANN	BOD and TSS	0.783	Ahamad et al. (2019)
2018	1. Gorganrood Basin, Iran	ANFIS	SAR	0.99	Azad et al. (2018)
	2. Gorganrood Basin, Iran	ANFIS	EC	0.99	Azad et al. (2018)
2017	1.Hooghly river, India	NN-CS	Chlorides, turbidity, pH, TA, and residual chlorine TH	–	Chatterjee et al. (2017)
	2. Hooghly river, India	NN-GA	Chlorides, pH, residual chlorine TA TH,and turbidity	–	Chatterjee et al. (2017)
	3. Saint John River, Canada	BPNN	TSS	0.976	El Din and Zhang (2017)
	4. Langat River and Klang River, Malaysia	BPNN	BOD, DO, COD, Ph, NH3-N, and TSS	0.7267	Hameed et al. (2017)
2016	1.Aji-Chay River, Iran	W-ANN	Salinity (EC)	0.9960	Barzegar et al. (2016)
	2. Aji-Chay River, Iran	W-ANFIS	Salinity (EC)	0.9958	Barzegar et al. (2016)

(continued)

Table 16.4 (continued)

Year	Location	ANN model	Input parameters	Highest R^2 (during the testing stage)	References
2015	1. Hilo Bay, Hawaii, USA	MLP	Salinity, DO, and temperature	0.80	Alizadeh and Kavianpour (2015)
	2. Hilo Bay, Hawaii, USA	W-ANN	Temperature, DO, and salinity	0.967	Alizadeh and Kavianpour (2015)
	3. Dahan River, Taiwan	BPNN	NH_3–N	0.979	Chang et al. (2015)
	4. Nadong River, South Korea	W-ANN	Water level		Seo et al. (2015)
2014	Karoon River, Iran	MLP-NN	DO	0.85	Emamgholizadeh et al. (2014)
	Karoon River, Iran	MLP-NN	COD	0.74	Emamgholizadeh et al. (2014)
2013	Johor River, Malaysia	MLP-NN	EC, TDS, and turbidity	0.799	Najah et al. (2013)
	Jishan Lake, China	W-ANN	DO, Temperature, and pH	–	Xu and Liu (2013)
	Jishan Lake, China	ENN	DO, Temperature, and pH		Xu and Liu (2013)
2012	Kinta River, Malaysia	MLP-NN	36 parameters	0.765	Gazzaz et al. (2012)
2011	Nile River, Egypt	MLP-NN	33 parameters	–	Khalil et al. (2011)

square error (RMSE). The RMSE was excluded because of the difficulty of directly comparing the units to different parameters. This implies that its accuracy is dependent on the size (value) of the parameter being considered. This would mean that a parameter with a large numerical value will show a high RMSE value; resulting in inaccuracy (Ighalo et al. 2020). Unlike the RMSE, the R^2 is a measure of the extent of variation of a particular dataset, therefore giving better accuracy of parameters in a particular dataset (Adeniyi et al. 2020).

An observation of Table 16.4 shows that the adaptive neuro-fuzzy inference system (ANFIS), Wavelet-Adaptive Neural Fuzzy Interference System (W-ANFIS), and the Wavelet and BPNN (W-NN) hybrid models are the most accurate neural networks for both surface and groundwater monitoring and prediction. These findings are in agreement with the observation of Ighalo et al. (2020) on the frequently used neural networks for surface water prediction between 2011 and 2020. This study has revealed that although the MLP-NN and BP-NN seem to be the most popularly used algorithms in water quality prediction, however, they have low modeling accuracy compared to other hybrid models. The high accuracy of the W-ANFIS and W-NN

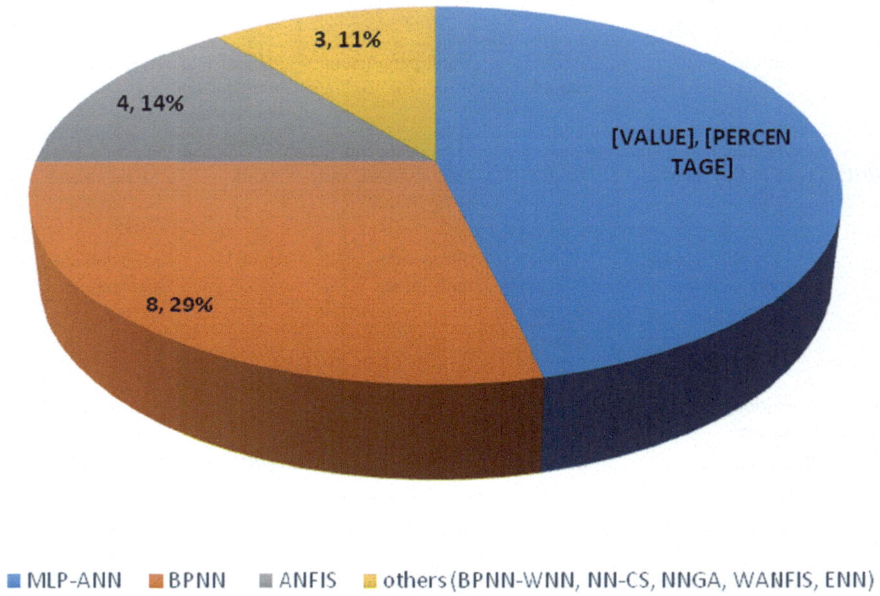

Fig. 16.7 Frequency in the application of ANN algorithms in water quality prediction

hybrid models has been attributed to the in-cooperation of the wavelet decomposition of covariates (predictor variables) into detail and approximate components (Barzegar et al. 2016). On the other hand, the high accuracy of the ANFIS is because its architecture integrates the learning ability of both the Fuzzy Inference System and ANN (Emamgholizadeh et al. 2014).

16.4.3.3 Predicted Water Quality Parameters

In this section, the most frequently analyzed water quality parameters for both surface and groundwater have been discussed. An observation of Table 16.4 shows that temperature, dissolved oxygen (DO), pH, Electrical conductivity (EC; as a function of salinity), and total dissolved solids (TDS) are the most frequently investigated parameters in water quality prediction. A possible reason for their high prevalence is due to significance in the determination of the overall water quality in both surface and groundwater sources. Ighalo et al. (2020) have attributed the prevalence of DO, TDS, and pH to the high availability and low cost of measuring equipment. Moreover, these parameters are readily measured insitu before laboratory analysis because of their high susceptibility to surface environmental changes (Igwe and Omeka 2021).

16.5 Conclusion

With the aid of meta-analysis from Scopus and Web of Science databases and bibliometric coupling for data extraction and visualization, the present study has explored the recent trends to more hypothetical and advanced model simulation in the application of machine learning (ML) models in water quality monitoring and prediction in Nigeria for the last two decades (2003–2023). Emphasis has been made on the type of ML frequently used, the accuracy of the model in water quality prediction, and the frequently monitored water quality parameters. It was revealed that the use of hybridized ML models in water quality prediction has not been well explored globally; a majority of the prediction has been based on the use of artificial neural networks (ANN). The adaptive neuro-fuzzy inference system (ANFIS), and Wavelet-Adaptive Neural Fuzzy Interference System (W-ANFIS) hybrid models were observed to be the most accurate ANN algorithms, in prediction. Based on analysis of the literature, temperature, dissolved oxygen (DO), pH, conductivity (EC) and total dissolved solids (TDS) appear to be the most frequently predicted parameters. China and the United States had the highest number of publications concerning the application of ML in water quality prediction; while Nigeria appears among the countries grossly lacking behind. Based on a literature analysis of the sources of contaminants in aquatic systems in Nigeria, heavy metals (Pb, Cd, Cu, Cr, and Zinc) were reported in more publications; linked majorly to mining and industrial effluents. This study anticipates that with the high complexity of water quality data as a result of the impending variations in anthropogenic activities and seasonal changes (due to global climatic variations), future water quality monitoring and prediction studies in Nigeria should take advantage of the rapidly evolving field of machine learning; with more emphasis placed on the hybridized machine learning algorithms.

Although this study only considered the Scopus and Web of Science databases in its analysis, it is recommended that future studies should explore other databases such as Science Direct, PubMed, Cochrane Library, Lens, and Dimensions. Other bibliometric analytic techniques such as the Hierarchical Density-Based Spatial Clustering of Applications with Noise (HDBSCAN) should be explored as well.

References

Adamu CI, Nganje TN, Edet A (2015) Heavy metal contamination and health risk assessment associated with abandoned barite mines in Cross River State, southeastern Nigeria. Environ Nanotechnol Monit Manage 3:10–21

Adekola FA, Eletta OAA (2007) A study of heavy metal pollution of Asa River, Ilorin. Nigeria; trace metal monitoring and geochemistry. Environ Monit Assess 125:157–163

Adeniyi AG, Ighalo JO, Marques G (2020) Utilisation of machine learning algorithms for the prediction of syngas composition from biomass bio-oil steam reforming. Int J Sustain Energy. https://doi.org/10.1080/14786451.2020.1803862

Adewumi AJ (2020) Contamination, sources and risk assessments of metals in media from Anka artisanal gold mining area, Northwest Nigeria. Sci Total Environ 718:137235

Agbasi JC, Egbueri JC, Ayejoto DA, Unigwe CO, Omeka ME, Nwazelibe VE, Ighalo JO, Pande CB, Fakoya AA (2023) The impact of seasonal changes on the trends of physicochemical, heavy metal and microbial loads in water resources of southeastern Nigeria: a critical review. In: Egbueri JC, Ighalo JO, Pande CB (eds) Climate change impacts on Nigeria. Springer, Cham. https://doi.org/10.1007/978-3-031-21007-5_25

Aghamelu OP, Omeka ME, Unigwe CO (2022) Modeling the vulnerability of groundwater to pollution in a fractured shale aquifer in SE Nigeria using information entropy theory, geospatial, and statistical modeling approaches. Model Earth Syst Environ 1–22

Ahamad KU, Raj P, Barbhuiya NH, Deep A (2019) Surface water

Ajala OA, Oke MR, Ajibade TF, Ajibade FO, Adelodun B, Ighalo JO, Silva LF et al (2022) Concentrations, bioaccumulation, and health risk assessments of heavy metals in fishes from Nigeria's freshwater: a general overview. Environ Sci Pollut Res 29(55):82660–82680

Akakuru OC, Adakwa CB, Ikoro DO, Eyankware MO, Opara AI, Njoku AO, Usman A (2023) Application of artificial neural network and multi-linear regression techniques in groundwater quality and health risk assessment around Egbema, Southeastern Nigeria. Environ Earth Sci 82(3):77

Akakuru OC, Njoku UB, Obinna-Akakuru AU, Akudinobi BE, Obasi PN, Aigbadon GO, Onyeanwuna UB (2023) Non-carcinogenic health risk assessment and predicting of pollution indexing of groundwater around Osisioma, Nigeria, using artificial neural networks and multi-linear modeling principles. Stoch Environ Res Risk Assess 1–31

Al Aani S, Bonny T, Hasan SW, Hilal N (2019) Can machine language and artificial intelligence revolutionize process automation for water treatment and desalination? Desalination 458:84–96

Alizadeh MJ, Kavianpour MR (2015) Development of wavelet-ANN models to predict water quality parameters in Hilo Bay Pacific Ocean. Mar Pollut Bull 98:171–178

Ayejoto DA, Egbueri JC, Agbasi JC, Omeka ME, Unigwe CO, Nwazelibe VE, Ighalo JO, Pande CB (2023) Influence of seasonal changes on the quality of water resources in Southwestern Nigeria: a review. In: Egbueri JC, Ighalo JO, Pande CB (eds) Climate change impacts on Nigeria. Springer, Cham. https://doi.org/10.1007/978-3-031-21007-5_22

Azad A, Karami H, Farzin S, Saeedian A, Kashi H, Sayyahi F (2018) Prediction of water quality parameters using ANFIS optimized by intelligence algorithms (case study: Gorganrood River). KSCE J Civ Eng 22:2206–2213

Balogun II, Sojobi AO, Galkaye E (2017) Public water supply in Lagos State, Nigeria: review of importance and challenges, status and concerns and pragmatic solutions. Cogent Eng 4(1):1329776

Barzegar R, Adamowski J, Moghaddam AA (2016) Application of wavelet-artificial intelligence hybrid models for water quality prediction: a case study in Aji-Chay River, Iran. Stoch Environ Res Risk Assess 30:1797–1819

Beven K (2016) Advice to a young hydrologist. Hydrol Process 30(20):3578–3582. https://doi.org/10.1002/hyp.10879

Chang F-J, Tsai Y-H, Chen P-A, Coynel A, Vachaud G (2015) Modeling water quality in an urban river using hydrological factors– data-driven approaches. J Environ Manage 151:87–96

Chatterjee S, Sarkar S, Dey N, Ashour AS, Sen S, Hassanien AE (2017) Application of cuckoo search in water quality prediction using artificial neural network. Int J Comput Intell Stud 6:229–244

Deng T, Duan HF, Keramat A (2022) Spatiotemporal characterization and forecasting of coastal water quality in the semi-enclosed Tolo Harbour based on machine learning and EKC analysis. Eng Appl Comput Fluid Mech 16(1):694–712. https://doi.org/10.1080/19942060.2022.2035257

Edet AE, Merkel BJ, Offiong OE (2003) Trace element hydrochemical assessment of the Calabar Coastal Plain Aquifer, southeastern Nigeria using statistical methods. Environ Geol 44:137–149

Egbueri JC (2022) Incorporation of information entropy theory, artificial neural network, and soft computing models in the development of integrated industrial water quality index. Environ Monit Assess 194(10):693

Egbueri JC, Agbasi JC (2022) Data-driven soft computing modeling of groundwater quality parameters in southeast Nigeria: comparing the performances of different algorithms. Environ Sci Pollut Res 29(25):38346–38373

Egbueri JC, Igwe O, Omeka ME, Agbasi JC (2023) Development of MLR and variedly optimized ANN models for forecasting the detachability and liquefaction potential index of erodible soils. Geosyst Geoenvironment 2(1):100104

El Bilali A, Taleb A (2020) Prediction of irrigation water quality parameters using machine learning models in a semi-arid environment. J Saudi Soc Agric Sci. https://doi.org/10.1016/j.jssas.2020.08.001

El Din ES, Zhang Y (2017) Improving the accuracy of extracting surface water quality levels (Swqls) using remote sensing and artificial neural network: a case study in the saint John river, Canada. International archives of the photogrammetry. Rem Sensing Spat Inf Sci XLII-4/W4:245–249

Emamgholizadeh S, Kashi H, Marofpoor I, Zalaghi E (2014) Prediction of water quality parameters of Karoon River (Iran) by artificial intelligence-based models. Int J Environ Sci Technol 11:645–656

Emenike PC, Tenebe IT, Neris JB, Omole DO, Afolayan O, Okeke CU, Emenike IK (2020) An integrated assessment of land-use change impact, seasonal variation of pollution indices, and human health risk of selected toxic elements in sediments of River Atuwara, Nigeria. Environ Pollut 265:114795

Encyclopædia Britannica, Inc. (2023)

Ganiyu SA, Mabunmi AA, Olurin OT, Adeyemi AA, Jegede OA, Okeh A (2021a) Assessment of microbial and heavy metal contamination in shallow hand-dug wells bordering Ona River, Southwest Nigeria. Environ Monit Assess 193:1–22

Ganiyu SA, Oyadeyi AT, Adeyemi AA (2021b) Assessment of heavy metals contamination and associated risks in shallow groundwater sources from three different residential areas within the Ibadan metropolis, southwest Nigeria. Appl Water Sci 11(5):81

Gazzaz NM, Yusof MK, Aris AZ, Juahir H, Ramli MF (2012) Artificial neural network modeling of the water quality index for Kinta River (Malaysia) using water quality variables as predictors. Mar Pollut Bull 64:2409–2420

Geetha S, Gouthami S (2016) Internet of things enabled real-time water quality monitoring system Smart. Water 2:1. https://doi.org/10.1186/s40713-017-0005-y

Ghavidel SZZ, Montazeri M (2014) Application of different data-driven methods for the prediction of total dissolved solids in the Zarinehroud basin. Stochastic Environ Res Risk Ass 28(8):2101–2118. https://doi.org/10.1007/s00477-014-0899-y

Hameed M, Sharqi SS, Yaseen ZM, Afan HA, Hussain A, Elshafe A (2017) Application of artificial intelligence (AI) techniques in water quality index prediction: a case study in the tropical region, Malaysia. Neural Comput Appl 28:893–905

Hasan R, Raghav A, Mahmood S, Hasan MA (2011) Artificial intelligence in river quality assessment. In: 2011 international conference on information management, innovation management and industrial engineering, 2011. IEEE, pp 491–495

Ibrahim H, Yaseen ZM, Scholz M, Ali M, Gad M, Elsayed S, Khadr M, Hussein H, Ibrahim HH, Eid MH et al (2023) Evaluation and prediction of groundwater quality for irrigation using an integrated water quality indices, machine learning models and GIS approaches: a representative case study. Water 15:694. https://doi.org/10.3390/w15040694

Ifediegwu SI (2022) Assessment of groundwater potential zones using GIS and AHP techniques: a case study of the Lafa district, Nasarawa State, Nigeria. Appl Water Sci 12:10. https://doi.org/10.1007/s13201-021-01556-5

Ighalo JO, Adeniyi AG, Marques G (2020) Artificial intelligence for surface water quality monitoring and assessment a systematic literature analysis. Model Earth Syst Environ. https://doi.org/10.1007/s40808-020-01041-z

Igwe O, Omeka ME (2021) Hydrogeochemical and pollution assessment of water resources within a mining area, SE Nigeria, using an integrated approach. Int J Energy Water Resour 1–22

Kadam AK, Wagh VM, Muley AA, Umrikar BN, Sankhua RN (2019) Prediction of water quality index using artificial neural network and multiple linear regression modeling approaches in Shivganga River basin. Model Earth Syst Environ, India. https://doi.org/10.1007/s40808-019-00581-3

Kalamdhad A, Singh J, Dhamodharan K (eds) (2019) Surface water quality modeling by regression analysis and artificial neural network. Advances in waste management. Springer, Singapore. https://doi.org/10.1007/978-981-13-0215-2_15

Khalil B, Ouarda T, St-Hilaire A (2011) Estimation of water quality characteristics at ungauged sites using artificial neural networks and canonical correlation analysis. J Hydrol 405:277–287

Khan N, Shahid S, Ismail TB, Behlil F (2021) Prediction of heat waves over Pakistan using support vector machine algorithm in the context of climate change. Stochastic Environ Res Risk Assess. 1–19. https://doi.org/10.1007/s00477-020-01963-1

Kouadri S, Elbeltagi A, Islam ARM, Kateb S (2021) Performance of machine learning methods in predicting water quality index based on the irregular data set: application on Illizi region (Algerian southeast). Appl Water Sci 11(12):1–20

Kouadri S, Samir K (2021) Hydro-chemical study with geospatial analysis of groundwater quality Illizi region, South-East of Algeria. Iran J Chem Chemical Eng (IJCCE) 40(4):1315–1333. https://doi.org/10.30492/ijcce.2020.39800

Lar UA, Gusikit RB (2015) Environmental and health impact of potentially harmful elements distribution in the Panyam (Sura) volcanic province, Jos Plateau, Central Nigeria. Environ Earth Sci 74:1699–1710

Mallik S, Chakraborty A, Mishra U, Paul N (2022) Prediction of irrigation water suitability using geospatial computing approach: a case study of Agartala city, India. Environ Sci Pollut Res 1–16

Martins P, Forstinus NO, Ikechukwu NE (2016) Water and waterborne diseases: a review

Mohammadpour R, Shaharuddin S, Zakaria N, Ghani A, Vakili M, Chan N (2016) Prediction of water quality index in free surface constructed wetlands. Environ Earth Sci 75:1–12.https://doi.org/10.1007/s12665-015-4905-6

Morounke SG, Ayorinde JB, Benedict AO, Adedayo FF, Adewale FO, Oluwadamilare I, Sokunle SS, Benjamin A (2017) Epidemiology and incidence of common cancers in Nigeria. J Cancer Biol Res 5(3):1105

Najah A, El-Shafe A, Karim OA, El-Shafe AH (2013) Application of artificial neural networks for water quality prediction. Neural Comput Appl 22:187–201

Nayak S (2020) Assessment of water quality of Brahmani river using correlation and regression analysis

Nigerian National System of Cancer Registries (2013) Federal ministry of health of Nigeria, Federal Secretariat Complex Shehu Shagari Way, Garki, Abuja P.M.B. 083, GarkiAbuja. ISBN: 978-978-955-030-2

Nnaji CC, Chibueze CV, Mama CN, Alum OL, Afangideh CB (2023) Impact of anthropogenic and environmental conditions on surface run-off quality: a case study of Nsukka, Eastern Nigeria. Int J Environ Sci Technol 1–12

Nnorom IC, Ewuzie U, Eze SO (2019) Multivariate statistical approach and water quality assessment of natural springs and other drinking water sources in Southeastern Nigeria. Heliyon 5(1):e01123

Nuhu AA, Sallau MS, Majiya MH (2014) Heavy metal pollution: the environmental impact of artisanal gold mining on Bagega village of Zamfara state, Nigeria. Res J Pharm Biol Chem Sci 5(6):306–313

Obasi PN, Akudinobi BB (2020) Potential health risk and levels of heavy metals in water resources of lead–zinc mining communities of Abakaliki, southeast Nigeria. Appl Water Sci 10:184. https://doi.org/10.1007/s13201-020-01233-z

Ogundiran OO, Afolabi TA (2008) Assessment of the physicochemical parameters and heavy metals toxicity of leachates from the municipal solid waste open dumpsite. Int J Environ Sci Technol 5:243–250

Okamkpa JR, Omeka ME, Igwe O, Iyiokwu MU (2022) An integrated geochemical and spatiotemporal assessment of groundwater resources within an industrial suburb, Southeastern Nigeria. Int J Energy Water Resour 1–20. https://doi.org/10.1007/s42108-022-00183-3

Okolo CC, Oyedotun TDT, Akamigbo FOR (2018) Open cast mining: a threat to water quality in the rural community of Enyigba in south-eastern Nigeria. Appl Water Sci 8:1–11

Olaojo AA, Oladunjoye MA, Sanuade OA (2018) Geoelectrical assessment of polluted zone by sewage effluent in University of Ibadan campus southwestern Nigeria. Environ Monit Assess 190:1–12

Omeka ME (2022). Evaluation and prediction of irrigation water quality of an agricultural district, SE Nigeria: an integrated heuristic GIS-based and machine learning approach. Environ Sci Pollut Res. https://doi.org/10.1007/s11356-022-25119-6

Omeka ME (2023) Evaluation and prediction of irrigation water quality of an agricultural district, SE Nigeria: an integrated heuristic GIS-based and machine learning approach. Environ Sci Pollut Res. https://doi.org/10.1007/s11356-022-25119-6

Omeka ME, Egbueri JC (2022) Hydrogeochemical assessment and health-related risks due to toxic element ingestion and dermal contact within the Nnewi-Awka urban areas, Nigeria. Environ Geochem Health 1–29. https://doi.org/10.1007/s10653-022-01332-7

Omeka ME, Egbueri JC, Unigwe CO (2022) Investigating the hydrogeochemistry, corrosivity, and scaling tendencies of groundwater in an agrarian area (Nigeria) using graphical, indexical, and statistical modeling. Arab J Geosci 15(13):1–24. https://doi.org/10.1007/s12517-022-10514-7

Omeka ME, Igwe O (2023) Heavy metals concentration in soils and crop plants within the vicinity of abandoned mine sites in Nigeria: an integrated indexical and chemometric approach. Int J Environ Anal Chem 103(16):4111-4129

Omeka ME, Igwe O, Onwuka OS, Nwodo OM, Ugar SI, Undiandeye PA, Anyanwu IE (2023) Efficacy of GIS-based AHP and data-driven intelligent machine learning algorithms for irrigation water quality prediction in an agricultural mine district within the Lower Benue Trough, Nigeria. Environ Sci Pollut Res pp 1–30

Opuene K, Agbozu IE (2008) Relationships between heavy metals in shrimp (Macro brachium felicinum) and metal levels in the water column and sediments of Taylor Creek

Ozel HU, Gemici BT, Gemici E, Ozel HB, Cetin M, Sevik H (2020) Application of artificial neural networks to predict the heavy metal contamination in the Bartin River. Environ Sci Pollut Res. https://doi.org/10.1007/s11356-020-10156-w

Pour SH, Wahab AKA, Shahid S (2020) Physicalempirical models for prediction of seasonal rainfall extremes of Peninsular Malaysia. Atmos Res 233:104720. https://doi.org/10.1016/j.atmosres.2019.104720

Rajaee T, Khani S, Ravansalar M (2020) Artificial intelligence-based single and hybrid models for prediction of water quality in rivers: a review. ChemometrIntell Lab Syst 200:103978

Ricolfi L, Barbieri M, Muteto PV, Nigro A, Sappa G, Vitale S (2020) Potential toxic elements in groundwater and their health risk assessment in drinking water of Limpopo National Park, Gaza Province, Southern Mozambique. Environ Geochem Health 1–13

Seo Y, Kim S, Kisi O, Singh VP (2015) Daily water level forecasting using wavelet decomposition and artificial intelligence techniques. J Hydrol 520:224–243

Sun K, Rajabtabar M, Samadi S, Rezaie-Balf M, Ghaemi A, Band S, Mosavi A (2021) An integrated machine learning, noise suppression, and population-based algorithm to improve total dissolved solids prediction. Eng Appl Comput Fluid Mech 15(1):251–271. https://doi.org/10.1080/19942060.2020.1861987

Tijani MN, Jinno K, Hiroshiro Y (2004) Environmental impact of heavy metals distribution in water and sediments of Ogunpa River, Ibadan area, southwestern Nigeria. J Min Geol 40(1):73–83

Umeoguaju FU, Akaninwor JO, Essien EB, Amadi BA (2022) Heavy metal profile of surface and groundwater samples from the Niger Delta region of Nigeria: a systematic review and meta-analysis. Environ Monit Assess 194:1–37

van Ech NJ, Waltman L (2020) VOSviewer Manual. Universiteit Leiden, The Centre for Science and Technology Studies, Meaningful metrics. https://www.vosviewer.com/. Accessed Feb 2023

World Health Organisation/United Nations Children Fund. (WHO/UNICEF) (2010) Progress on household drinking water, sanitation and hygiene 2000-2022: Special focus on gender

Xiao Y, Liu K, Hao Q, Li J, Zhang Y, Cui W, Pei Q (2021) Hydrogeochemical features and genesis of confined groundwater and health perspectives for sustainable development in urban Hengshui, North China Plain. J Chem

Xu L, Liu S (2013) Study of short-term water quality prediction model based on wavelet neural network. Math Comput Model 58:807–813

Yetilmezsoy K, Ozkaya B, Cakmakci M (2011) Artificial intelligence-based prediction models for environmental engineering. Neural Netw World 21:193

Zhu S, Heddam S (2020) Prediction of dissolved oxygen in urban rivers at the Three Gorges Reservoir, China: extreme learning machines (ELM) versus artificial neural network (ANN). Water Qual Res J 55:106–118

Chapter 17
Justifiable Intermediations and Approaches to Alleviate Water Crunch in the Agro-climatic Zones of Chhattisgarh State in India—An Appraisal

Anadi Gayen and V. S. Joji

Keywords Dry farming · Ecology · Fishbone · Integrated approach · Sustainable management · SWOT

17.1 Introduction

Water and land resources are depleting at rapid rate. To address the pertinent issues and challenges like water resources management along with land resources, climatic aberration, conservation of wetlands, waste water management and water pollution are of strong concern mainly in the rainfed agricultural sectors in Chhattisgarh. Water management in a judicious manner is of utmost challenge the organisations dealing with groundwater perspectives. At present erratic rainfall followed by drought has become a prominent issue to be addressed for to ensure the sustainable supply of water with potable quality.

Water conservation through the mode of rain water harvesting is becoming widely accepted option as far as the technical feasibility and economical viabilities are

A. Gayen (✉)
Rajiv Gandhi National Ground Water Training and Research Institute, Central Ground Water Board, Ministry of Jal Shakti, Government of India, Naya Raipur, Chhattisgarh, India
e-mail: anadigayen1968@gmail.com

V. S. Joji
Department of Marine Geology & Geophysics, School of Marine Sciences, Cochin University of Science and Technology, Lake Side Campus, Fine Arts Avenue, Kochi, Kerala 682016, India

P. P. Adhikary et al. (eds.), *Soil, Water Pollution and Mitigation Strategies*, Environmental Science and Engineering, https://doi.org/10.1007/978-3-031-63296-9_14

concerned. The water conservation may fulfil the desired solution to optimise efficiency, equity and economy of the rainfed agriculture. The marginal and small-scale farmers holding up to 2 ha land are in despair and losing faith on cultivation on account of industrialization and urbanisation. The site-specific interventions may welcome in consideration with topological differences, terrain condition, soil texture, moisture retention capacity and more importantly hydrogeological condition of the region. Hence, identification of spatio-temporal variation of water problems, an appreciable cropping regime and improved water management strategies can be developed to achieve optimum agricultural growth. Groundwater Potential delineation in semi-arid basin was attempted by Castillo et al. (2022) using the tools of Remote Sensing, GIS, and AHP Approaches. Jhariya et al. (2021) **attempted to delineate** groundwater potential zone Raipur city, Chhattisgarh by using GIS-based multi-influencing factor (MIF), multi-criteria decision analysis (MCDA) and electrical resistivity survey techniques and found that high potential zones reported from west and north-eastern portions and low to medium in the central and eastern portions of the Raipur city.

17.1.1 Justification of the Work

The Chhattisgarh state needs to pay more attention towards agricultural growth and socio-economic development. Around 80% of the population are belonging to under-privileged section and are agricultural dependent. The 44% geographical area of the state is covered by forest and other natural vegetation. Sustainable and rational planning, utilization, conservation and scientific management of water and land resources are required to uplift the socio-economic status of the people. In this context, the study with emphasis on sustainable interventions and socially adoptable strategies to combat the water in the state deserves utmost relevance.

17.1.2 Physiographic Configuration and Climate

Chhattisgarh is situated between North Latitude 17°47′ to 24°06′ and East Longitude 80°14′ to 84°24′ (refer to Fig. 17.1). Geographically, it can be categorized into three distinct regions—the Bastar plateau in the south, the Chhattisgarh Plain in the center, and the Northern hilly region in the north (refer to Fig. 17.2).

The southern part of the state includes Bastar, Kondagaon, Narayanpur, Kanker, Bijapur, Sukma, and Dantewada districts, constituting the Bastar plateau region. This area is characterized by dense evergreen forests and hilly terrain, primarily in the Indravati river basin. Landforms consist of high-level plateaus, structural hills, valleys, pediments, and pediplains, with altitudes ranging from 400 to 600 m above mean sea level (amsl). The central parts, including the plains of the Indravati River and along the Shabri River in the southeast, have altitudes varying from 250 to 300 m amsl.

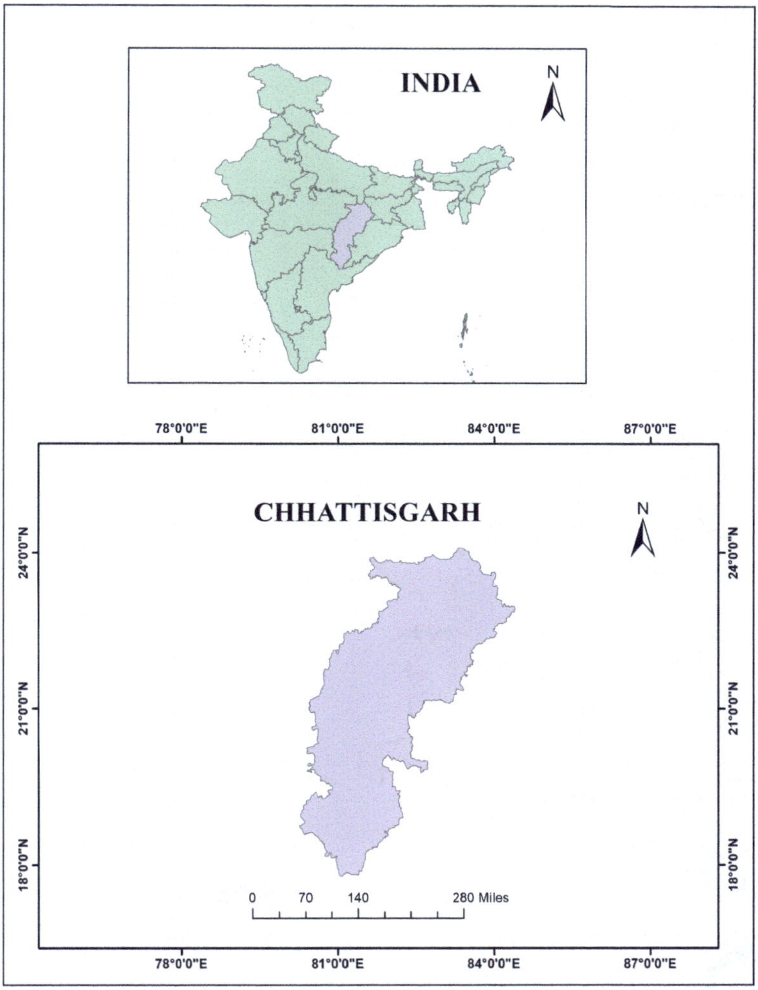

Fig. 17.1 Location map of Chhattisgarh State

Moving to the central part of the state, the Chhattisgarh Plain region encompasses districts like Bilaspur, Mungeli, Janjgir-Champa, Mahasamund, Dhamtari, Raipur, Balodabazar, Gariyaband, Durg, Balod, Bemetara, Rajnandgaon, and Kawardha. This area features structural plains on Proterozoic rocks and Pediplain, with remnants of isolated hills and ridges among the floodplains of various Mahanadi River tributaries. The terrain is gently undulating and flat, with altitudes ranging from 750 m amsl in the northwest to 284 m amsl in the southeast.

The Northern hilly region covers the north-central part of the state and spans Sarguja, Balrampur, Surajpur, Koriya, Korba, Bilaspur, Jashpur, and Raigarh districts.

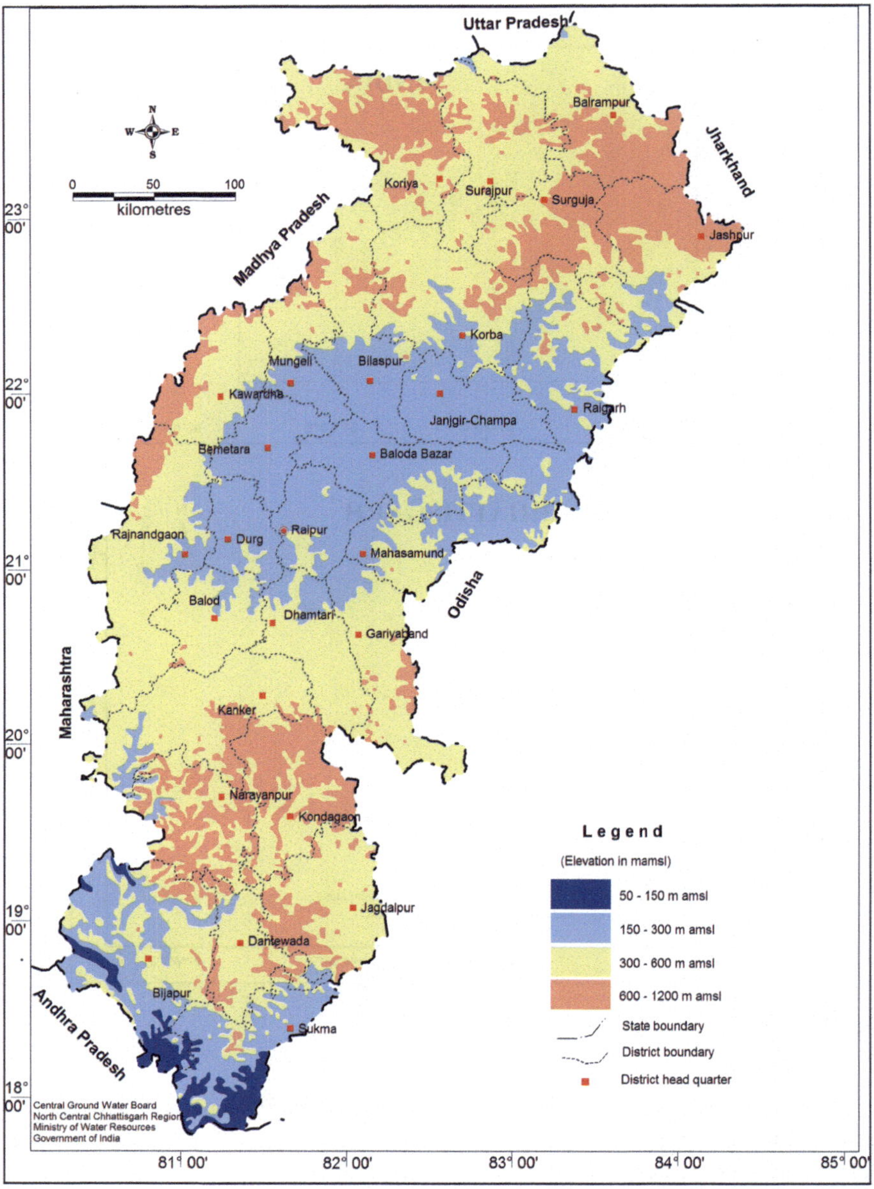

Fig. 17.2 Physiographic characteristics of Chhattisgarh (Anon 2020)

This region is part of the Maikal and Hazaribagh hill ranges in central India, characterized by structural plains of Gondwana rocks, pediment, pediplains, structural and denudational plateaus, structural and denudational hills, and valleys. It supports north-flowing tributaries of the Son River and south-flowing rivers like Hasdeo and others joining the Mahanadi River. The Narmada, a west-flowing river in central India, originates from Amarkantak in the central part of this physiographic unit. The highest point in the state is at 1197 m amsl in the Tulisi Dongri range in Dantewada district, while the lowest point is at 50 m amsl in Konta, Dantewada district.

Various water sources, including dug wells for shallow phreatic aquifer systems and tube wells and bore wells for deeper semi-confined aquifer systems, are utilized for water withdrawal. The agro-climatic zones of the area are depicted in Fig. 17.3.

17.1.3 River Network

The Mahanadi River, along with its tributaries Seonath, Hasdeo, Mand, and Arpa, traverses through parts of Raipur, Durg, Rajnandgaon, Bilaspur, Raigarh, and Surguja districts. The Indravati River, functioning as a tributary to the Godavari River, acts as the drainage system for Kanker, Bastar, and Dantewada districts. Most of these rivers maintain a continuous flow throughout the year. In a broader perspective, the drainage patterns in the region display dendritic, parallel, angular, and radial configurations, as indicated by Anon (2020). The Son River, a tributary of the Ganges River, drains areas in Sarguja and Koriya districts. For a visual representation, please refer to the river basin map in Fig. 17.4 within the study area.

17.1.4 Climatic Condition

The region features a sub-tropical monsoon climate marked by three distinct seasons: summer, monsoon, and winter. The southwest monsoon typically begins in June and persists until mid-September, followed by the winter season from October to February. The summer season spans from March to mid-June. Rainfall constitutes the primary source of groundwater recharge, with the area receiving the majority (85%) of its rainfall during the southwest monsoon season. Winter rainfall is relatively scant, accounting for only a small fraction of the total. The state encounters a sub-tropical climate characterized by intense summers and milder winters. Summer prevails from March to mid-June, with May being the hottest month. The mean daily maximum temperature in May can soar up to 46 °C.

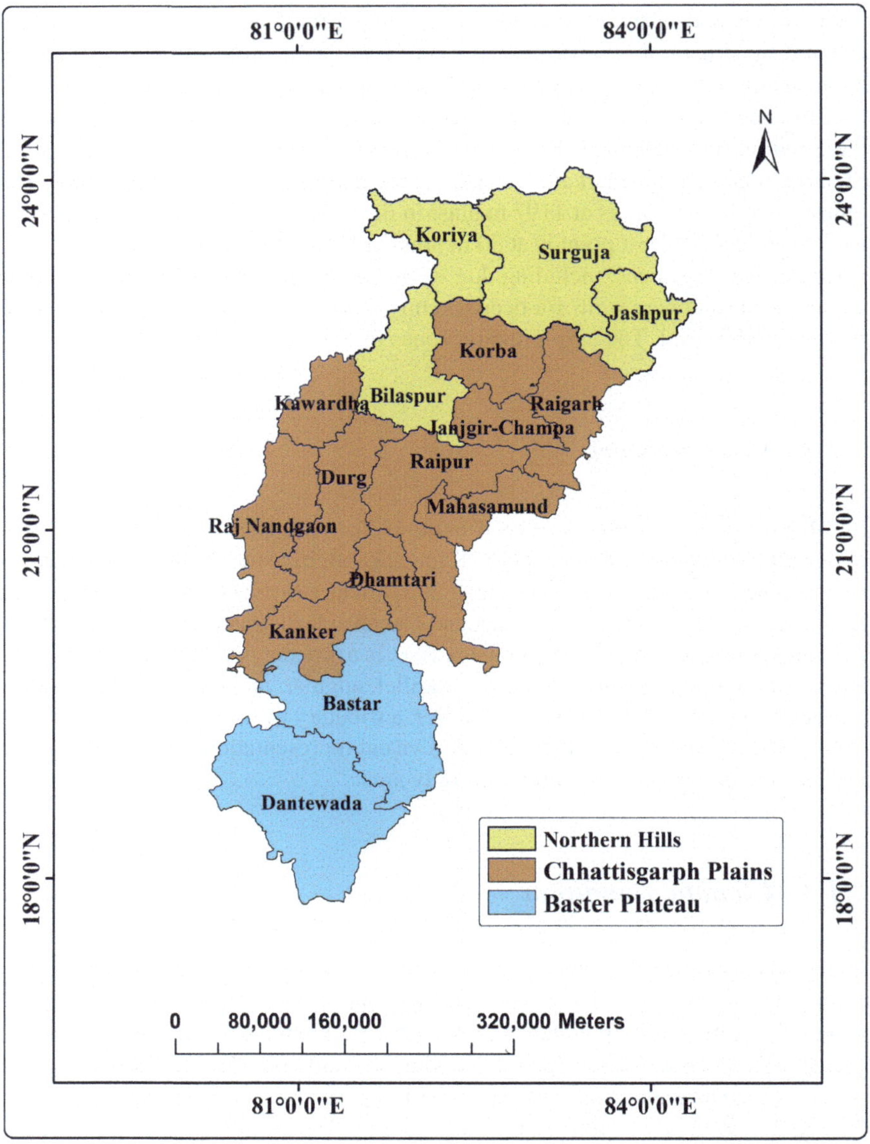

Fig. 17.3 Agro-climatic zones of Chhattisgarh

17.1.5 Soil and Land Use/Land Cover

There are 8 types of soil reported in the study area namely Red gravelly soil, Red loamy soil, Red sandy soil, shallow black soil, lateritic soil, red and yellow soil, deep black soil and medium black soil (Anon 2020) and the soil map is shown in

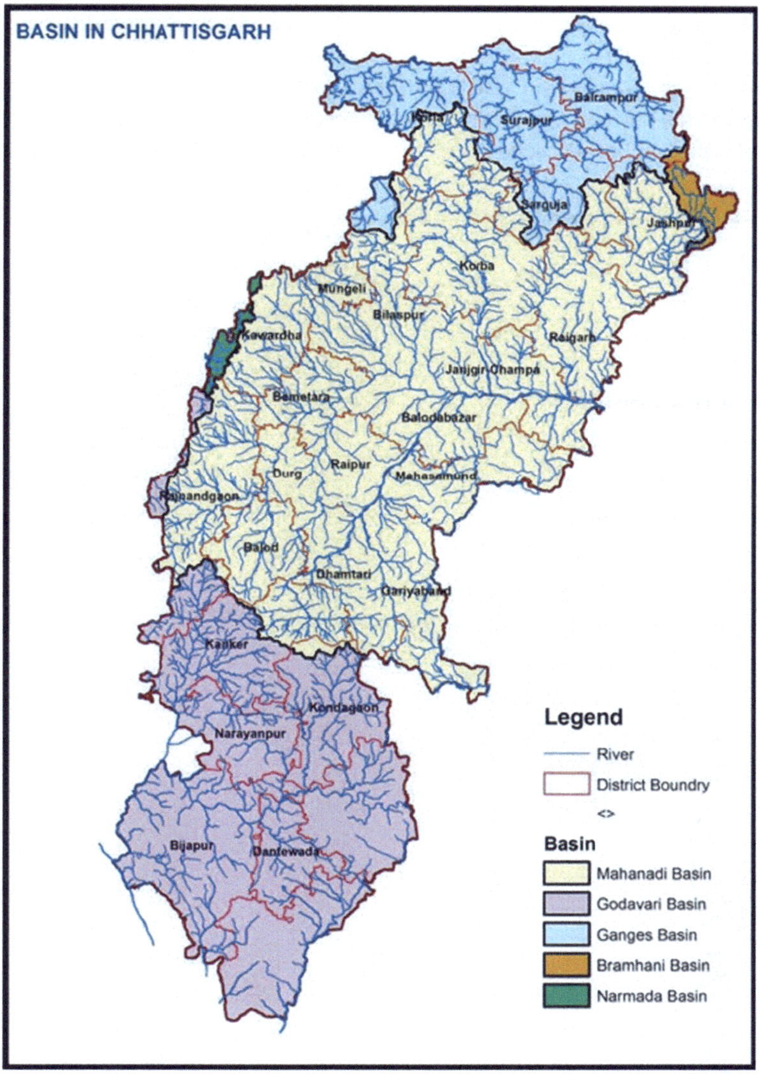

Fig. 17.4 River basin map of Chhattisgarh (http://hydrologyproject.cgwrd.in)

Fig. 17.5. The soils in the upstream areas of the drainage with limited thickness and these areas witnessed more riverine erosive activities by first, second and third order Hortonian streams. The changes in the properties of soil have been reported in the mature stages of the river basins in Chattisgarh as eroded materials from the upstream areas undergone redeposition. Anon (2020) reported increase in the concentration of calcium and magnesium, depth increase, increase of water holding capacity, colour change from red to dark brown, variation of soil texture from sandy loam to clayey, and sticky to very sticky. The types of soil in the area and crop suitability have been compiled (Table 17.1).

The Land use Land cover (LU-LC) indicates the physical and biological cover over the earth's surface together with water bodies, natural and cultivable vegetation, soil types and infrastructural facilities made by human beings (Areendran et al.

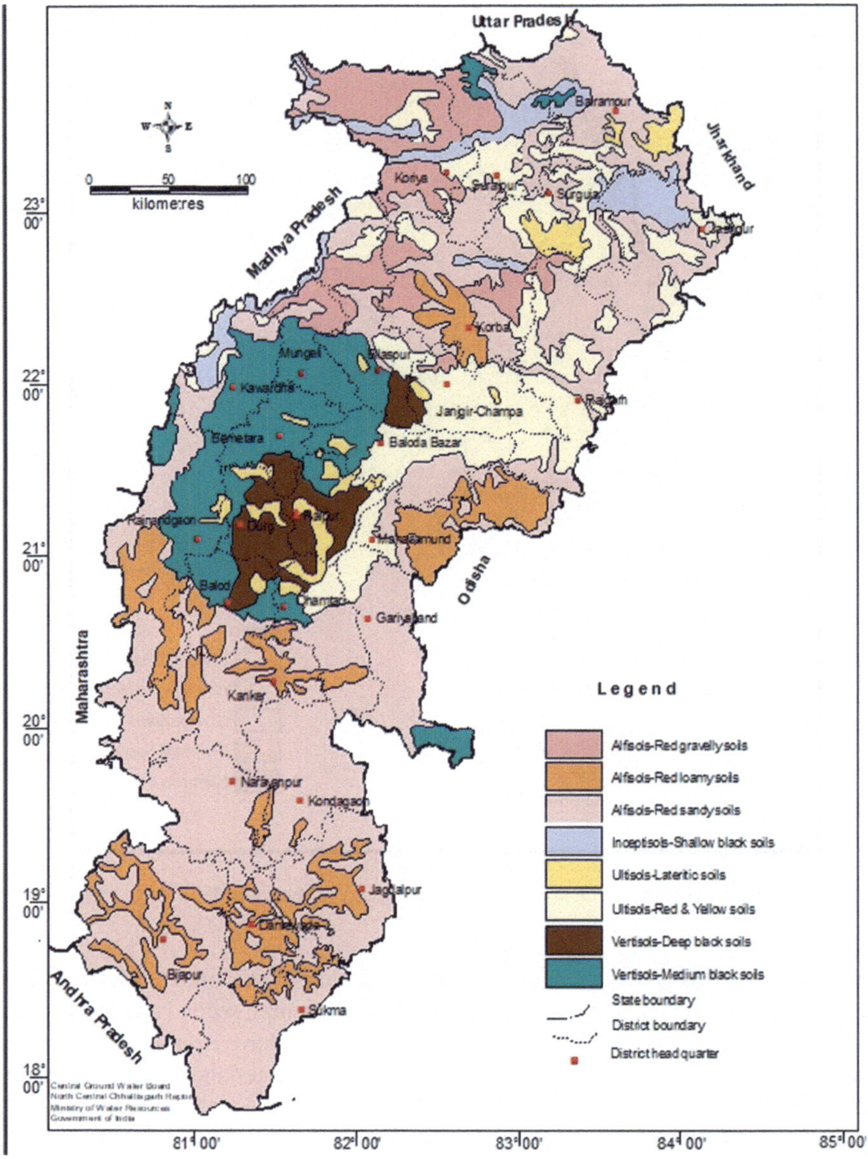

Fig. 17.5 Soil map of Chhattisgarh (Anon 2020)

Table 17.1 Soil type wise suitable crops

Type of soil	Parent rock	Distribution (districts/ tehsils)	Suitable crops
Red-yellow soil (Matasi)	Gondwana, Chhattisgarh Supergroup	Surguja, Koriya, Jashpur, Raigarh, Korba, Bilaspur Kawardha, Durg, Raipur, Dhamtari and and Mahasamund districts	Paddy
Red-sandy soil	Archaean granite	Bastar, Dantewada, Kanker, Durg, Rajnandgaon and Dhamtari districts	Kodo-Kutki, Jawar, Maize, Potato Coarse grains etc
Red-domat soil	Archaean granite	Dantewara and Konta tehsils	Paddy
Laterite soil	Mixed	Bagicha, Samri, Sitapur, Ambikapur, Kawardha, Chhui-Khaddan, Saja, Bemetera and Jagdalpur tehsils	Potato, Jawar, Kuddo-Kutti, Oilseeds, Pulses etc.
Black soil	Mixed	Mungeli, Ariya, Raipur, Rajim, Mahasamund, Kurud and Kawardha tehsils	Paddy, Wheat, Cotton, Gram, Sugarcane and Rabi crops

2011). Areendran et al. (2011) identified dense forest, open forest, open scrub forest, agricultural areas, water bodies and built up areas as LU-LC categories in the northern districts of Chhattisgarh. Khan and Jhariya (2016) identified LU-LC categories in parts of Chattisgarh (Raipur Urban Agglomeration) by visual interpretation of satellite images and classified them as settlement, road net works, cultivable area, water bodies, open land area, vegetation and industry. As per the Directorate of Economics and Statistics, Govt. of Chhattisgarh, 46% of the State (6,352,413 ha) is covered by natural vegetation (forests). The district of Narayanpur tops in forest cover 85.14% (638801Ha) and Bemetara district with lowest forest cover area (0.015%, 40 ha) of the state. Ahmad et al. (2019) analyzed health of forest of the study area in the change in climatic conditions by using the tools of Remote Sensing and Arc GIS and the land use-land cover map of the study area is depicted as Fig. 17.6.

17.2 Materials and Methods

Chhattisgarh state water policy report, IGKV reports on water issues in the state and Central Ground Water Board (CGWB) reports have been consulted and compiled the information for this paper. Fishbone (Cause-Effect) analysis (Ishikawa 1968) has been carried out to diagnose the causative factors for irrigation water challenges.

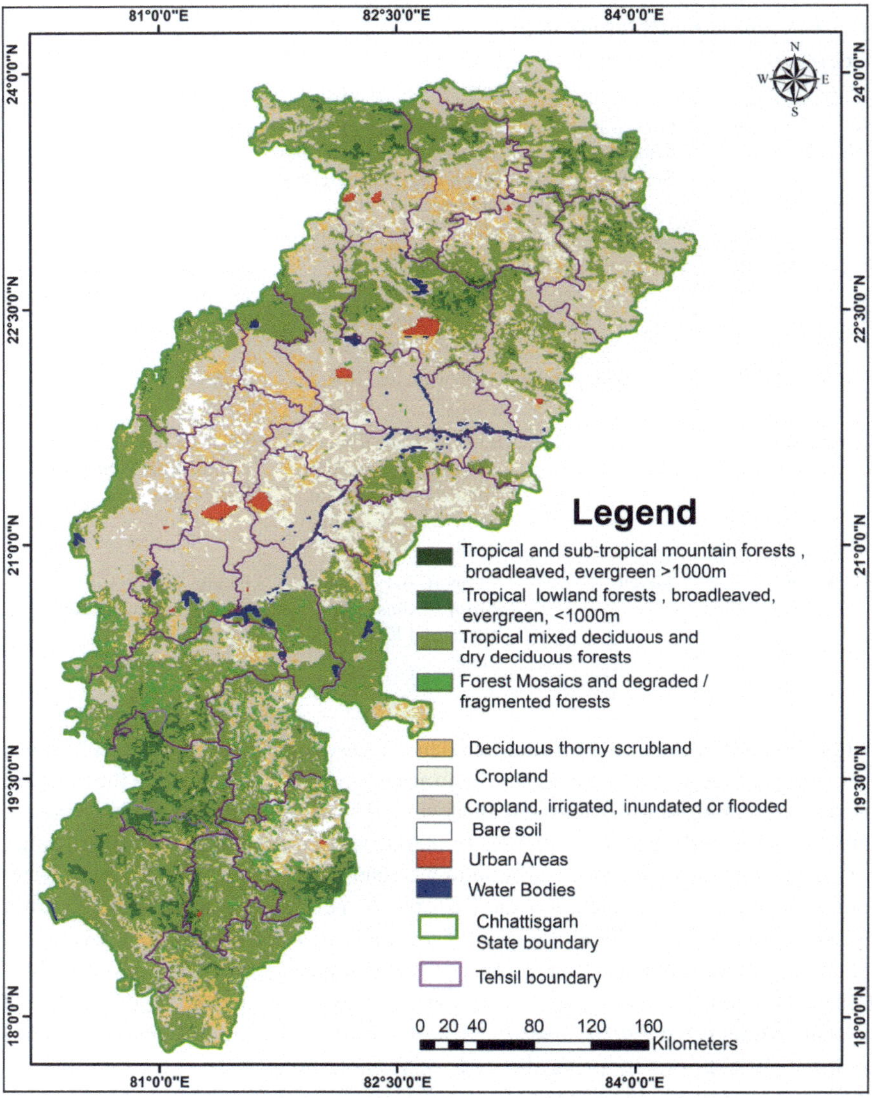

Fig. 17.6 Land use-land cover map of Chhattisgarh (Ahmad et al. 2019)

The SWOT analysis helps in understanding the gravity of the internal and external responsible factors for irrigation and other water problems in Chhattisgarh.

17.2.1 Hydrogeological Framework

The occurrence and movement of groundwater in the state are intricately linked to its geological composition. The state's subsurface is characterized by a diverse range of rock types spanning from Azoic to Quaternary periods. Approximately 58% of the state is dominated by crystalline rocks, while the Chhattisgarh Group of rocks covers around 27%. The semi-consolidated Gondwana Supergroup of rocks occupies 13% of the area, and the remaining 2% comprises Deccan trap, Lameta, Laterite, and River Alluvium.

The Quaternary-age unconsolidated formation consists of alluvium, clay, silt, and laterite, forming a thin and extensive unconfined aquifer in isolated patches along major river courses. The thickness can reach up to 30 m below ground level along rivers such as Mahanadi, Arpa, Hasdeo, Seonath, Kharun, Mand, and Kelo. From a hydrogeological perspective, the rock types in the state can be broadly categorized into consolidated formations, semi-consolidated formations, and unconsolidated formations.

Chemical analysis indicates that the groundwater in the region tends to be neutral to low alkaline. However, there are instances of high arsenic levels observed in the Ambagarh Chowki block of Rajnandgaon district. Elevated fluoride content (> 1.5 mg/l) has been reported in Raighar, Surgpur, Korba, Bilaspur, Durg, Mahasamund, and Jashpur districts (Anon 2020). A study conducted by Khageshwar Singh Patel et al. in 2017 analyzed arsenic and fluoride in groundwater in Rajnandgaon District, Chhattisgarh. The highest fluoride concentration, 3.2 mg/l, was recorded in Hanumangarh village, Premnagar block of Surgpur district. Refer to Fig. 17.7 for the hydrogeological map of Chhattisgarh.

17.2.2 Long Term Groundwater Level Trend (2009–2018)

The long-term water level trend (2009–2018) of the phreatic aquifer was analyzed for both pre and post-monsoon periods. During the pre-monsoon period, the majority of Chhattisgarh exhibited a water level trend ranging from − 10 to + 10 cm/yr, considered safe. However, many areas in Jashpur, Surguja, Korba, Janjgir-Champa, Kawardha, and Rajnandgaon displayed a significant decline of more than 20 cm/yr, raising concerns. The post-monsoon decadal water level trend map of the phreatic aquifer paints an even more alarming picture. According to Anon (2021), extensive regions in Surguja, Koriya, Jashpur, Kawardha, and Rajnandgaon experienced a notable decline in water level of more than 20 cm/yr over the last decade (refer to Fig. 17.8).

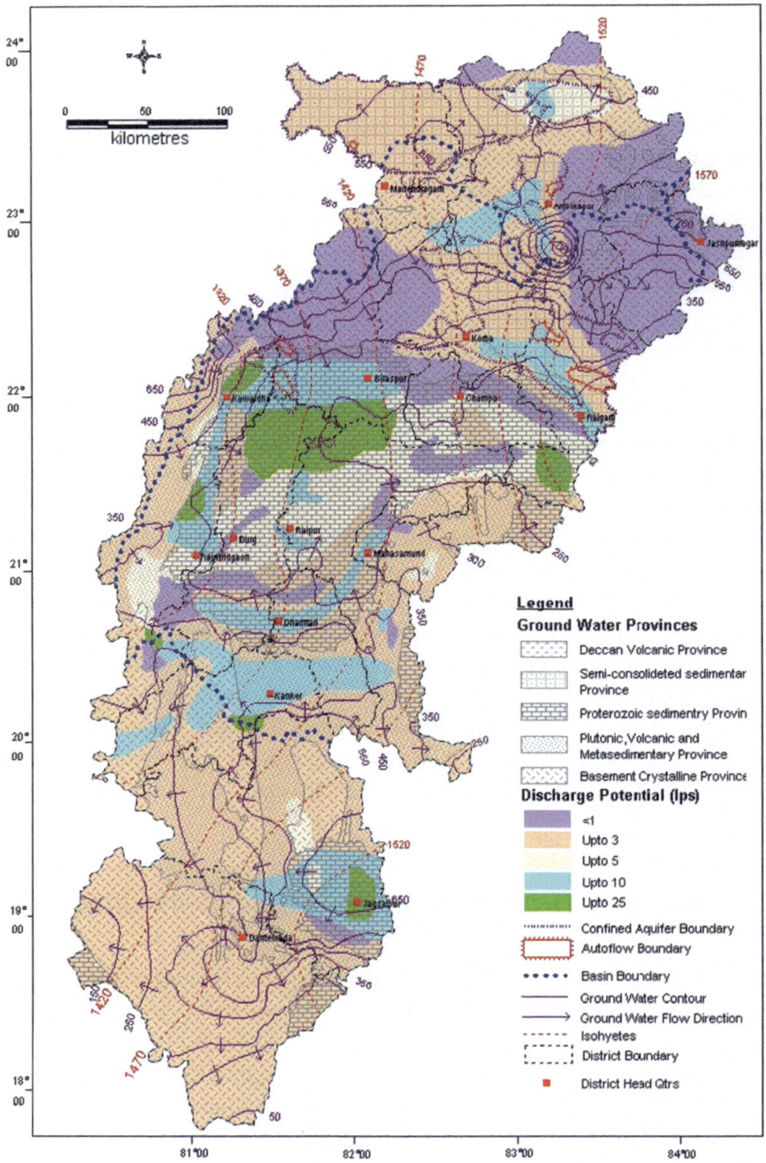

Fig. 17.7 Hydrogeological map of Chhattisgarh (Anon 2020)

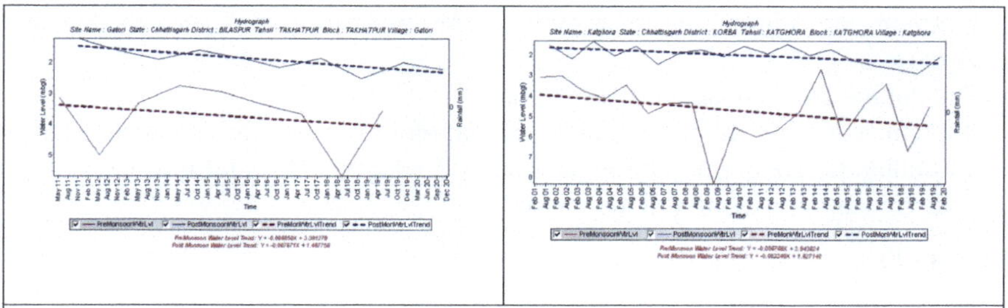

Well hydrographs of Gatori in Bilaspur district (Northern Hill Agro-climatic Zone) & Katghora in Korba district (Central Chhattisgarh Plain Agro-climatic Zone)

Fig. 17.8 Long term trend of groundwater regime in Chhattisgarh (Anon 2021)

17.2.3 SWOT Analysis of Chhattisgarh Water Problems

The SWOT Analysis (Bell and Rochford 2016; Dess et al. 2012; Madsen 2016; Minsky and Aron 2021 has been carried out to identify S, W, O and T components—[Strength (S), Weakness (W), Opportunities (O) and Threat (T)] related to the water related issues and management aspects. The SWOT components have been selected on the basis of parameters controlling and affecting ground water potential in the different agro-climatic zones of the study area. The internal and external responsible factors for irrigation and water problems in the area are also taken into consideration under SWOT.

Strength (S)

- The 85% rainfall receives during monsoon
- Extensive potential unconfined aquifer is available along major River courses.
- Rivers are perennial in nature
- Soils topology facilitates various crops.

Weakness (W)

- Approximately 58% of the State is covered by crystalline and metamorphic rocks and around 27% is covered by potential groundwater bearing Chhattisgarh Group of rocks
- Problems associated with cultivation in *bunded* and unbunded fields, in different soil types, in each of the three ecological zones
- Diverse topograpghy and terrain condition across the state
- Deterioration in the quality of groundwater and water table depletion
- 46% of the total area in the State is covered by forests
- Net sown area for Chhattisgarh is only 33.87% (4,671,469 ha)
- Climatic aberration involves sub-tropical climate with extreme summer and moderate winter.

Opportunities (O)

- Pre-monsoon depth to the water table (DTW) of major part of shows safe water level trend
- The double cropped area is 1019386 ha
- Implementation of micro- and minor irrigation projects would be encouraged in the undulating terrain of tribal areas along with adoption of dry farming technology hold greater applicability in the region.

Threat (T)

- Post-monsoon DTW of major part of the state shows falling trend
- Harmful mind set for the application of big dam water
- Insufficient irrigation and agriculture data base
- Faulty selection of agricultural crops
- Around 37% of the net sown area has irrigation facilities.

17.2.4 Fishbone Analysis for Chhattisgarh Irrigation Water Challenges

Cause and consequent effect of Chhattisgarh water problems have been analysed by fishbone analysis (Ishikawa 1968) and the details are compiled (Table 17.2). Fishbone analysis diagram or Ishikawa diagram is tool to consolidate reasons of a particular problem or effect by explicitly demonstrating in detail and proposing causative relations. Ishikawa diagram can provide the possible causal factors like the bones of a fish.

17.3 Discussion and Results

The water availability and crop yield in different agro-climatic zones, soil erosion details and related issues are discussed in brief.

17.3.1 Inadequate Governance

The irrigation department was in-charge of construction, operation and maintenance of all irrigation works built at public cost, from the head works down to pipe outlets. The government had a bureaucratic approach and limited capacity to intervene at the field level, resulting in several distributional inefficiencies and poor repair and maintenance because of lack of funds and tariff collection (Fig. 17.9).

In order to break out of this vicious cycle, the study recommends a three pronged strategy: tariff rationalisation to cover at least the entire O&M cost; private sector participation and beneficiary participation through water users associations (WUAS).

Table 17.2 Fishbone analysis for Chhattisgarh irrigation water challenges

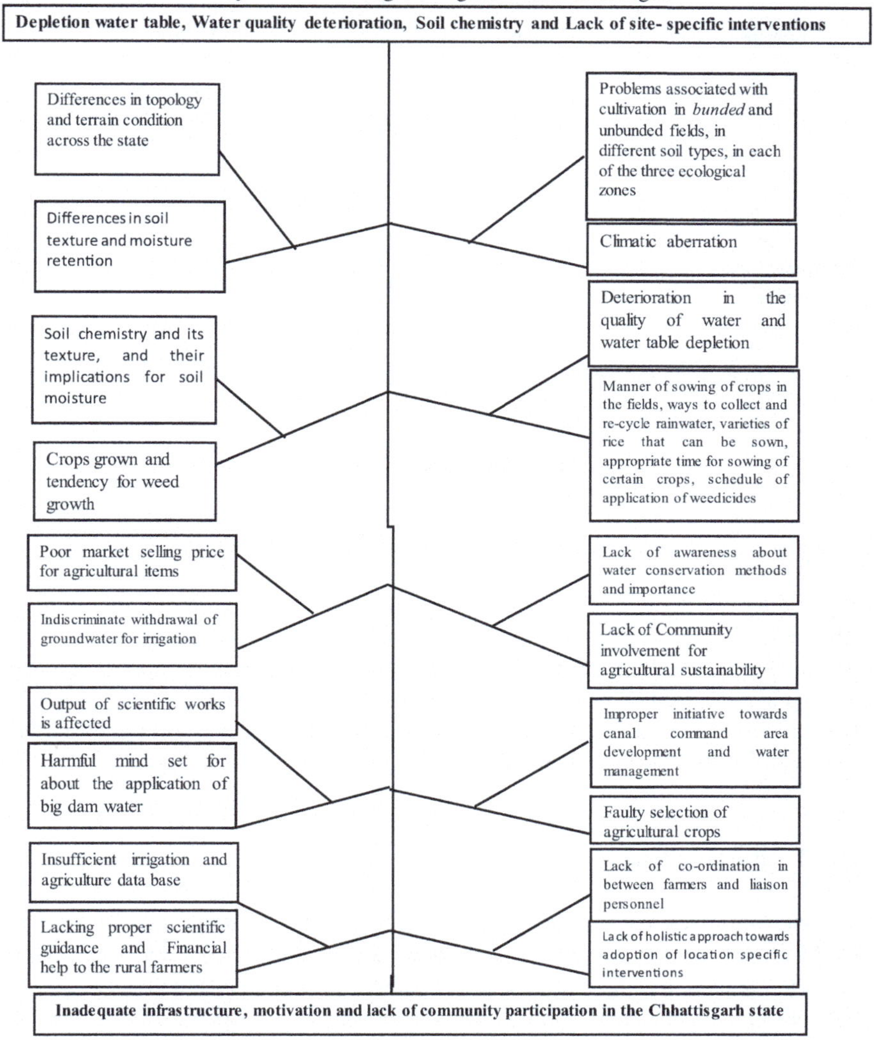

It speaks most approvingly of the attempt to increase the collection of user charges through beneficiary management by the Government interventions to be adopted in different agro-climatic zones

17.3.1.1 Chhattisgarh Plains

This zone consists of the five sub agro-climatic zones include Bhata soils, *Unbunded* black soils, *Bunded* rice fields (rainfed), Utera and Rice bunds. Due to low capacity

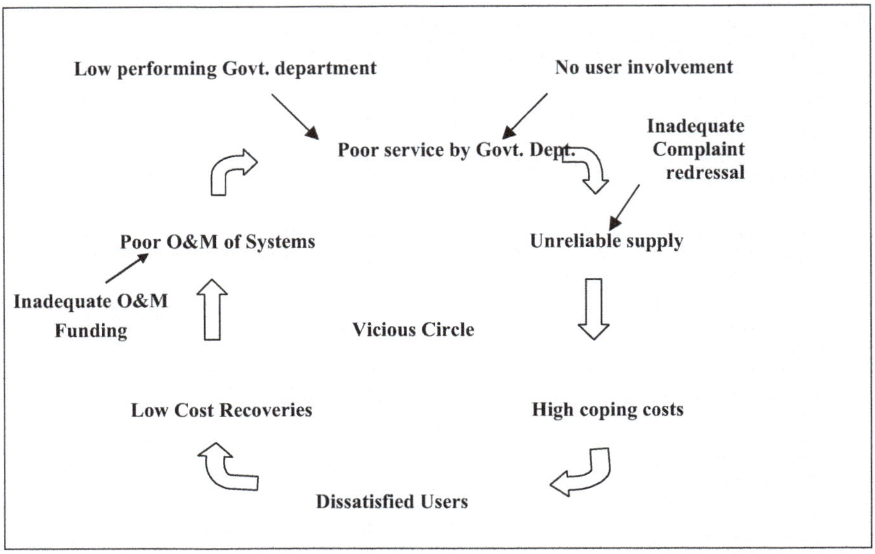

Fig. 17.9 Role of governance in water Crisis (Anon 2019)

of Bhata soils it is kept as fallow land. As the demand is increasing this type of land can be converted into rice fields and other crops by *bunding*. *Unbunded* black soils are characterised by poor drainage and less fertility. Drainage facilities may be enhanced with water harvesting in the farm ponds with broad bed and furrow (BBF) system with fishing provision may be effective.

17.3.1.2 Bastar Plateau

This zone comprises five sub-agro-climatic zones: Marhan, Tikra, Mal, Gabhar, and Mixed Farming. In the Mixed Farming zone, an integrated approach is essential, tailored to the land holding and financial capacity. In Gabhar, productivity is limited by waterlogging conditions. Mal exhibits a mono-cropping pattern with low rice productivity due to issues like pests, diseases, weeds, and water stress. The Tikra zone faces challenges with low soil fertility and acidity, while the Marhan zone experiences low crop productivity due to unfavorable soil physical conditions.

Given these agro-climatic conditions in the Bastar Plateau, there's a necessity to identify more remunerative alternative crops to diversify agriculture. Introducing different crop combinations is crucial for optimizing yields compared to the current poor yield. Identifying crops and varieties suitable for acidic soils with drought tolerance is essential since soil amendments may not be economically viable. Farm pond technology is recommended to collect and recycle runoff water, mitigating water stress conditions in rice and enabling the cultivation of a second crop in the rabi season. Efficient irrigation cum fish pond technology is necessary, along with studies to reduce water losses and enhance the efficient utilization of stored water.

As per the recommendations of IGKV, Raipur, rice varieties like Kranti, Mahamaya, and Madhuri are suggested for shallow water, while Mahsuri is recommended for medium water submergence. Rice cum fish culture technology is recommended for regions with medium and deep submergence. Additionally, there's a need to develop crops and mixed farming system varieties that align with the preferences and requirements of the tribal population.

17.3.1.3 Northern Hills Zone

Soil erosion is the major problem in the Northern Hilly zone. To overcome the problem of soil erosion and provide sustainable source of income, agro-forestry may be helpful strategy. This intervention facilitates the improvement of the physical structure of soil. Intensive efforts are required to transfer this technology on the fields of farmers and to identify suitable tree species and crops varieties in a location specific manner.

17.3.2 Emphasis on Micro- and Minor Irrigation

In 1972, the Irrigation Commission had strongly advocated for importance of minor irrigation in the eastern Madhya Pradesh (now Chhattisgarh).The Commission explained the water demand in the following manner—"Wherever we went in the State, there was a demand for minor irrigation schemes, arising possibly because of the delay in the execution of major and medium projects. There was a demand for the construction of small diversion works across streams to irrigate up to 40 ha, and for action to maintain and repair tanks, which exist in very large numbers, but which, through lack of maintenance and repair, are going out of use. We would recommend that the repair and maintenance of tanks should be given very high priority particularly in areas, where almost the entire irrigation is through wells and small tank. In Bastar district also, where there is little scope for big project and where a small population is thinly spread over a large area, minor irrigation worse, particularly, small tanks, have the greatest utility."

In undulating terrain of Chhattisgarh, micro- and major irrigation options are appreciable propositions. Large storage dams are very useful for irrigation, but the creation of such type huge structure and full-fledged commissioning includes numerable problems like time consuming, loss of cultivable lands and millions of people residing over that area will become homeless. Moreover, canal system irrigates in a very slow pace and make harm to the surrounding environment as well. The management of groundwater resources, development and recharge are found as challenging in the state, which supposed to be implemented in a holistic manner. Hence, the dam structure is not advisable in the state like Chhattisgarh. In this circumstance, various measures have been suggested (See Recommendations and way ahead).

17.3.3 Integrated Approach Towards Water Conservation and Irrigation

The arrangements for water harvesting sites and conservation are to implements with due care with an annual rate of silt deposition component in any groundwater basin. It is to be careful that the longevity of any water conservation reservoir depends on silt deposition management—the more the silt deposition, the lesser the life of a conservation structure. The design of water conservation project would consider utmost care to minimise the silt load through curbing of large scale deforestation, soil moisture conservation, and restricted soil erosion. Mega scale recharge to be emphasised in the sub-surface zone of aeration (vadose zone) to combat the water crisis. Vadose zone modelling would be the appreciable proposition before stepping into water conservation in the water table depleted areas. The area of irrigation can be increased through accelerated use of water harvesting, water storage, watershed management and reversal of ecological degradation.

17.3.4 Scope of Managed Aquifer Recharge

The surface run-off erodes the fertile soil and destroys soil topology. Hence, soil–water conservation is of pre-requisite in Chhattisgarh. Initiative is required to make running water to walk, walking water to stop and stopped water to percolate to recharge groundwater resources. The structures like check dam, trenches and contour bunds are the recommended managed aquifer recharge structures in the state of Chhattisgarh.

17.3.5 Recommendations and Way Ahead

- Big dams of huge capacity in Chhattisgarh may not be constructed for irrigation,
- For marginal farmers, rainfall during monsoon is the only option for their cultivation. But, due to lack of awareness the rainwater is escaping as surface run-off, which needs to be arrested for utilization of farmers in their agricultural fields. Conjunctive use of surface water and ground water may be the alternative option for this. Further, small scale water development may be promoted through decentralization of water resources development programme,

- Less water developed areas can be demarcated along with identification of arid and semi-arid regions with an aim to allocation of the water in the need based areas of the state,
- To enhance the soil moisture content, forest areas should be protected,
- Large scale adoption of Dry Farming Technology would be useful,
- The renovation, maintenance and management of the existing water conservation structures are to be prioritised,
- Frequent checking of erosion from canal banks and siltation aspects for canals and existing dams are important,
- Measures need to be undertaken for canal command area development with emphasis on groundwater management perspectives,
- Unirrigated farm lands are to be recognised to increase the agricultural yield with state supported rural development programme,
- Recommendation of soil type specific crops is advised to achieve desired crop yield,
- Precision agriculture is advised to be promoted throughout the state,
- Micro level water management technology needs to be promoted in a community participation mode to uplift the rural economy and increase the employment generation in manifold,
- Participation of progressive farmers and NGOs are to be ensured in the water harvesting process and related management issues,
- The special projects are to be introduces for the uplands and fallow s in the midland areas with increased water supply in consideration of hydrogeological perspectives,
- State water policy with subsidy provision may be there to ensure water for irrigation till food security and livelihood of poor rural people improved. Investment from state government for irrigation projects may facilitate the agricultural growth,
- Panchayat Raj Institutions (PRI) and other associated state governments organisations may come forward to make awareness to the rural people to develop their ownership issues and accountability for water conservation structures,
- To reduce drought vulnerability, water resources development may be planned in integration with land us/ land cover aspects along with agricultural development and maximum retail price (MSP) for the agricultural market, and
- To have expected outcome and sustainability, the implementing authority and the grass root level performers would come into a consensus about the desired policy and approaches about the irrigation perspectives.

17.4 Conclusions

The study aims to assess the physiographic characteristics, soil topology and climatological impact on water scenario of Chhattisgarh in conjunction with hydrogeological condition, site specific water problems and suitable sustainable interventions. The SWOT analysis is useful in depicting the intrinsic and extrinsic factors responsible for water crisis in the region. Fishbone analysis helps in understanding the cause-effect analysis of water challenges in the state.

Dry farming technology would be beneficial to ensure more crop yield during lean periods with proper maintenance of dams, canals, and other water conservation structures. This subsistence farming technology would help to the desired growth of water consuming crops like rice and sugarcane in the regions of water deficit, without irrigation facilities and deficit in soil moisture. Proper scientific governance of water resources from the side of Government and enhancement of micro- and minor irrigation facilities would be beneficial to the solution of the problem.

Crop diversification in relation to soil texture and climatic conditions, site specific conjunctive use of surface water and groundwater resources, water harvesting & conservation, and utilization of managed aquifer recharge facilities are important to uphold the environmental sustainability, economic viability and ecological upliftment. Proper monitoring of groundwater regime, replenishment of water resources and long term trend of water table along with proper understanding of hydrogeological framework would pave the way to formulate the sustainable water management strategies in different terrains of Chhattisgarh. Community participation and sensitisation to the small land holder farmers about the modern agricultural technologies would be useful to address the water issues at grass root level.

Acknowledgements The authors are thankful to the Chairman of CGWB, Member (RGI) and Regional Director, RGI, Naya Raipur for according necessary permission to publish the paper in Journal.

References

Ahmad F, Goparaju L, Uddin MM (2019) Analysis of forest health and socioeconomic dimension in climate change scenario and its future impacts: remote Sensing and GIS approach. Spat Inf Res 27(4):385–397. https://doi.org/10.1007/s41324-019-00245-5

Anon (2019) Government of Chhattisgarh. Chhattisgarh Water Management Bill, p 14

Anon (2020) Ground water year book of Chhattisgarh 2019–20. Central Ground Water Board, NCCR, Raipur

Anon (2021) Ground water year book of Chhattisgarh 2020–21. Central Ground Water Board, NCCR, Raipur

Areendran G, Raj K, Mazumdar S, Munsi M, Govil H, Sen PK (2011) Geospatial modeling to assess elephant habitat suitability and corridors in northern Chhattisgarh, India. Trop Ecol 52(3):275–283

Bell GG, Rochford L (2016) Rediscovering SWOT's integrative nature: a new understanding of an old framework. Int J Manag Educ 14(3):310–326. https://doi.org/10.1016/j.ijme.2016.06.003

Castillo JLU, Cruz DAM, Leal JAR, Vargas JT, Tapia SAR, Celestino AEM (2022) Delineation of groundwater potential zones (GWPZs) in a semi-arid basin through remote sensing, GIS, and AHP approaches. Water 14:2138. https://doi.org/10.3390/w14132138

Dess GG, Lumpkin GT, Eisner AB, McNamara G (2012) The limitations of SWOT analysis. Strategic management: text and cases, 6th edn. McGraw-Hill/Irwin, New York, p 82. ISBN 9780078029318. OCLC 740281685

Ishikawa K (1968) Guide to quality control. JUSE, Tokyo

Jhariya DC, Mondal KC, Kumar T, Indhulekha K, Khan R, Singh VK (2021) Assessment of groundwater potential zone using GIS-based multi-influencing factor (MIF), multi-criteria decision analysis (MCDA) and electrical resistivity survey techniques in Raipur city, Chhattisgarh, India. Aqua Water Infrastruct Ecosyst Soc 70(3):375–400. https://doi.org/10.2166/aqua.2021.129

Khan R, Jhariya DC (2016) Land use land cover change detection using remote sensing and geographic information system in Raipur Municipal Corporation Area, Chhattisgarh. SSARSC Int J Geo Sci Geo Inf 3(1):1–5

Madsen DQ (2016) SWOT analysis: a management fashion perspective. Int J Bus Res 16(1):39–56

Minsky L, Aron D (2021) Are you doing the SWOT analysis backwards? Harvard Business Review

Patel KS, Sahu BL, Dahariya NS, Bhatia A, Patel RK, Matini L, Sracek O, Bhattacharya P (2017) Groundwater arsenic and fluoride in Rajnandgaon District, Chhattisgarh, northeastern India. Appl Water Sci 7:1817–1826

Chapter 18
Arsenic Contaminants of Groundwater and Its Mitigation

Khalil Ahmad, Habib Ur Rahman Shah, Khizar Qureshi, Hammad Majeed, Ifzan Arshad, Tabinda, Muhammad Ashfaq, Muhammad Zubair, and Tehreema Iftikhar

Keywords Arsenic toxicity · Arsenic mitigation · COFs · MOFs

18.1 Introduction

Arsenic, denoted by the chemical symbol As, atomic number 33, is a metalloid with a silver-gray appearance that exhibits properties of both metals and nonmetals. At room temperature, it appears as a brittle and crystalline solid. In the earth crust it is found naturally in different minerals forms like realgar, orpiment and arsenopyrite. Boiling and melting points of arsenic are in the range of 817 °C and 613 °C respectively and it classified as semiconductor which means its conductance is better than insulators and lower than conductors due to which its electrical conductivity is highly dependent on temperature (Antman 2001; Tian et al. 2011). Despite of its

K. Ahmad (✉) · K. Qureshi · H. Majeed · I. Arshad · Tabinda
Department of Chemistry, University of Management and Technology (UMT), Sialkot Campus, Sialkot, Punjab, Pakistan
e-mail: khalilnoorpur@gmail.com

H. U. R. Shah · M. Ashfaq
Institute of Chemistry, Baghdad Ul Jadeed Campus, Islamia University of Bahawalpur, Bahawalpur, Punjab, Pakistan

M. Zubair
University of Agriculture Faisalabad, Faisalabad, Punjab, Pakistan

T. Iftikhar
Department of Botany, GC University Lahore, Lahore 54000, Pakistan

K. Ahmad
Department of Chemistry, Emerson University Multan, Multan, Pakistan

P. P. Adhikary et al. (eds.), *Soil, Water Pollution and Mitigation Strategies*, Environmental Science and Engineering, https://doi.org/10.1007/978-3-031-63296-9_15

toxicity, arsenic commonly used in many applications like agriculture, industries, lead-acid batteries, preservation of wood, metallurgical operations, and in semiconductor devices. It is also utilized as a pesticide to control pests and weeds and in semiconductor industry serves as doping agent for high performance microchips as well as in glass industry to produce high refractive index glass. Finally, certain chemicals made up of arsenic are also used as medicines traditionally and are still applied now a days to treat various types of diseases (Francesconi and Kuehnelt 2004; Cao et al. 2007; Wang et al. 2013). Despite the crucial role of arsenic and its compounds in different fields, exposure to them can results in harmful impacts on ecosystem. It is regarded as toxic heavy metal because it causes serious health issues including respiratory problems, skin lesions, kidney and liver disorders as well causes cancer (Buchanan 1962; Naranmandura et al. 2011; Hosseini et al. 2013). For this reason, it is essential to apply this toxic metal and its related compounds cautiously and control its practice for the prevention of toxic effects on ecosystem and human beings. Furthermore, continuous research is being conducted for identification and manufacturing of new more suitable compounds which would be more effective for diverse purposes and will be ecofriendly.

18.2 Arsenic Causes and Accumulation

No doubt arsenic is found naturally but human activities are also major contributor of arsenic production in the ecosystem including water bodies. Rocks and minerals containing arsenic when dissolves in water bodies then arsenic enters in water environment and man-made sources which mainly contribute the percolation of arsenic in ecosystem are industries, mining, smelting, acid lead batteries, semiconductor devices, insecticides, pesticides, fungicides, and glass industry which leads to contamination of arsenic in water sources (Styblo and Thomas 2001; Karadjova et al. 2008; Mudhoo et al. 2011). In addition, treated wood products can leach arsenic, further contributing to water contamination. Groundwater in some regions of the world can also naturally contain high levels of arsenic. Once in water bodies, arsenic can dissolve and pass in the food chain via marine animals and fishes. It is also reserved in plants, consumed by animals, and afterward ingested by humans. Furthermore, industrial processes, food additives, and animal feed can also release arsenic into the environment. This arsenic has a tendency to accumulate in the parts of bodies in animals and human beings like tissues especially in seafood tissues cereals and rice plants. In seas and rivers mainly arsenic percolates due to sewerage of polluted water from industries and pesticides (Francesconi and Kuehnelt 2004). Overall many sources contribute for disposal of arsenic in water bodies and environment and some of these are given as (Fig. 18.1 also showed arsenic sources).

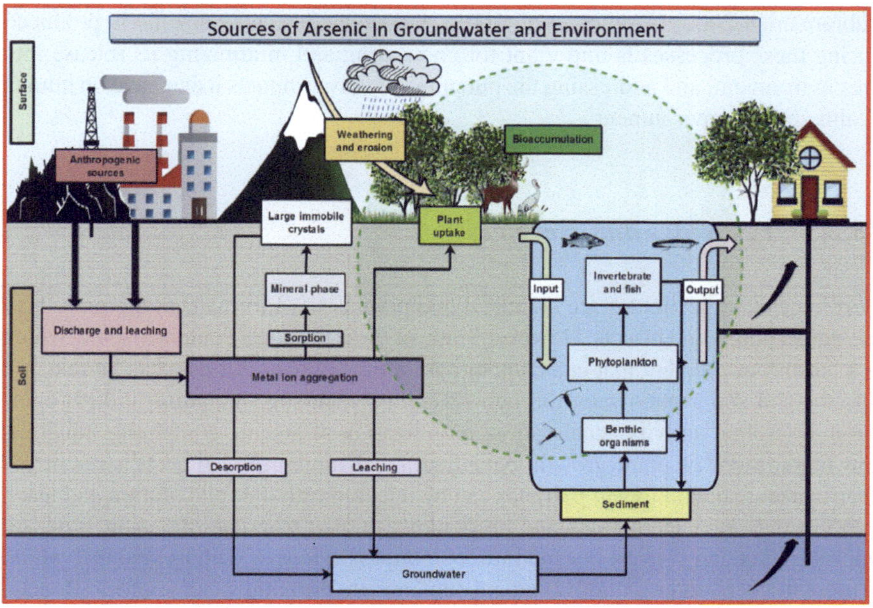

Fig. 18.1 Sources of arsenic in groundwater and environment

18.2.1 Contaminated groundwater

Millions of people rely on groundwater as their primary source of drinking water globally. However, in certain regions, naturally occurring arsenic may exist in groundwater at significant levels. The phenomenon where arsenic is released and becomes mobile in certain areas due to geological and hydrological conditions. These conditions are found in areas with volcanic rocks, high levels of organic matter, and reducing groundwater conditions. This leads to a prevalence of arsenic contamination in such areas, which can have negative impacts on human health (Jiang et al. 2013; Gillispie et al. 2015; Smith et al. 2018a, b). Understanding these conditions and their relationship to arsenic mobility is important for addressing and mitigating the problem of arsenic contamination.

18.2.2 Mining and Smelting

Mining and smelting operations for different metals like silver, iron, gold and copper can produce arsenic as a consequence due to the presence of arsenic-containing minerals. Smelting can release arsenic as a gas or solid waste, while mining operations can release arsenic through the use of explosives. Arsenic presence in minerals can make it an unwanted byproduct of mining and smelting operations (Madhavan and

Subramanian 2000; Garelick et al. 2008). Understanding how arsenic is produced during these processes is important for controlling and minimizing its release into the environment, and addressing the potential negative impacts it can have on human health and the environment.

18.2.3 Pesticides and Herbicides

Herbicides and pesticides are chemical compounds used to manage the growth of unwanted plants and insects. However, some of these products contain arsenic, which is a hazardous element that can contaminate water sources. Arsenic can be released into the soil and groundwater through several mechanisms, including soil pH alteration, microbial activity acceleration, mobilization of arsenic-containing minerals, and impairment of plant growth (Smedley and Kinniburgh 2012). When soil pH drops, arsenic bound to soil particles can be released into the soil solution and leach into groundwater. Herbicides and pesticides can also stimulate the growth of soil microbes that can convert arsenic into its more toxic forms, such as arsenite, which is more likely to enter groundwater and contaminate drinking water sources (Bencko and Yan Li Foong 2017). Additionally, some herbicides and pesticides can dissolve and mobilize arsenic-containing minerals in the soil, increasing the concentration of arsenic in the soil solution and promoting its transport to groundwater. Herbicides and pesticides that inhibit plant growth can also reduce the uptake of arsenic by plants, leading to higher concentrations of arsenic in the soil over time (Meharg and Hartley-Whitaker 2002). Overall, the use of herbicides and pesticides that contain arsenic can have negative consequences for the environment and human health due to their potential to contaminate water sources.

18.2.4 Wood Preservatives

Arsenic-based compounds are used to preserve and protect wood from insect infestations and degradation. When these woods are useless are thrown in environment where bacterial decomposition takes place and these arsenic compounds are entered into water bodies through rain and into plants body through routes by dissolving in water and ultimately becomes part of human body and animals when they consumed this water, plants, vegetables and fruits (Larsen et al. 1992; Hingston et al. 2001).

18.2.5 Coal Combustion

The burning of fossil fuels like coal can release arsenic into the atmosphere, which can later settle on soils and water bodies during rains and directly enters into humans and animals through respiration and causes respiratory problems and skin problems (Wang et al. 2018, 2021).

18.2.6 Industrial Processes

Arsenic is a widely used industrial chemical in processes like ceramics, glass, and semiconductor production, but it can cause elevated levels of contamination in nearby water sources, especially groundwater, due to leaching and runoff. Such arsenic contamination can contaminate the soil, surface water, and groundwater, which can cause various adverse health impacts (Piver 1983; Ratnaike 2003). Thus, the management and treatment of arsenic-contaminated water sources are essential to ensure the safety of the environment and nearby communities. Adequate measures should be taken to prevent further contamination, and effective treatment technologies must be used to remediate the contaminated water sources.

18.2.7 Contaminated Water

Arsenic can contaminate groundwater through both natural processes and human activities such as mining and farming. Arsenic is a toxic element that can be harmful to human health and can enter the body through drinking contaminated water (Palma-Lara et al. 2020). Therefore, it is important to monitor and prevent arsenic contamination in groundwater sources.

18.2.8 Natural Deposits

Geological processes are the primary sources for the arsenic metal in ecosystem like soil and water bodies. Regions which have maximum concentration of minerals like sulfides and arsenide's mostly have elevated concentrations of arsenic in the soil and water. For this reason, these areas which are present near these minerals rich in arsenic compounds are mostly posed to higher concentrations of arsenic in soil and water bodies (Welch et al. 1988; Saunders et al. 2005). So, it is concluded that these geological deposits also contribute the pollution of arsenic in water bodies and ecosystem.

18.2.9 *Tobacco and Narcotics*

Arsenic can also be found in substantial amounts in tobacco and narcotics, such as opium. The presence of arsenic in tobacco and opium can be due to various factors, including processing and smoking equipment (Smith et al. 2018a, b). Tobacco and opium can be processed using chemicals containing arsenic, which can result in the occurrence of this toxic element in the ultimate artefact. The instruments which are used for opium and tobacco are mostly made up of lead and arsenic metal which are mixed with smoke causes toxic effects to consumers as well as to those persons which are affected by are exposed to smoke (Anwar et al. 2022). So, it is very necessary for monitoring and controlling of opium, tobacco usage and their instruments as well as their smoke to ensure safe consumption if necessary.

18.3 Types of Arsenic

Different types of arsenic are existed in nature having different compounds and each compound have characteristics and variable degrees of harmfulness. The typical arsenic containing chemicals includes:

18.3.1 *Arsenate*

Arsenate have oxidation state of (+5) that is naturally occurring in the environment due to weathering of minerals, biological processes, and volcanic activity. This compound can be found in contaminated water sources, particularly groundwater, and its prolonged ingestion can be harmful to both humans and animals (Vahter and Concha 2001; Hughes 2002).

18.3.2 *Arsenite*

Arsenite is an arsenic compound that occurs naturally due to volcanic activity and human activities, including mining and smelting. It exists in the lowest oxidation state of +3 and is highly toxic. Exposure to high levels of arsenite can lead to several health issues, such as skin lesions, cancer, and cardiovascular diseases (Abedin and Meharg 2002).

Fig. 18.2 Some organic arsenic compounds used for various applications

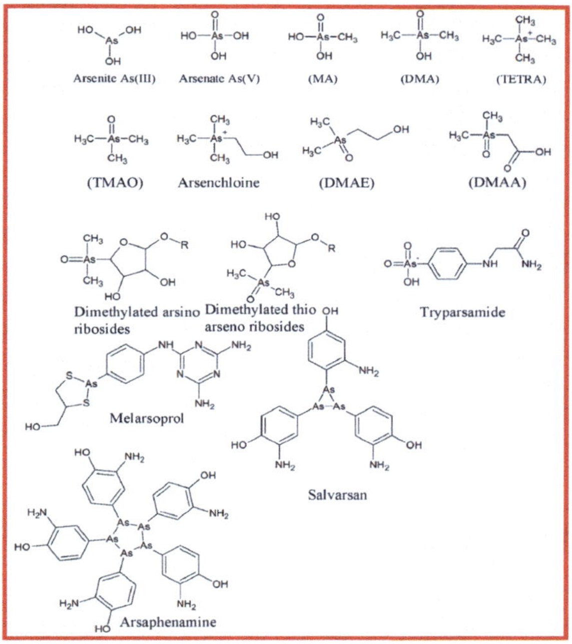

18.3.3 Organic Arsenic Compounds

Organic arsenic compounds are characterized by the presence of carbon-arsenic bonds (Vahter 1994), (Some examples of organic arsenic compounds used are given in Fig. 18.2). These compounds are frequently used in pesticides, herbicides, fungicides and in different conventional medicines used for the treatments of diseases. These compounds frequently detected in many foods especially food derived and composed of marine edible animals like shellfish, fish and other species which are utilized as food in different parts of World (Sakurai 2002). These organic compounds containing arsenic are generally considered to be not as much of toxic than inorganic compounds containing arsenic counterparts. Anyhow, long contact and continuously usage of these organic compounds containing arsenic to elevated amounts over an extended period can also causes many lethal and adverse effects on humans as well as marine lives (Edmonds et al. 1993).

18.3.4 Inorganic Compounds Containing Arsenic

Inorganic arsenic compounds refer to chemicals that do not contain a carbon-arsenic bond. They can exist in different forms, such as arsenite, arsenate, and arsenic trioxide, and are naturally present in the environment. These compounds are applied in many industries like ammunition, paints, fertilizers, insecticides, lab analysis and synthesis of different chemicals, cosmetics and semiconductors (Irgolic et al. 1977). Long usage and ingesting high levels of inorganic arsenic compounds can lead to various health issues, including cancer, cardiovascular disease, and skin lesions, due to their high toxicity (Khairul et al. 2017).

18.3.5 Arsenic Trioxide

Arsenic trioxide is also an inorganic compound of arsenic which is frequently utilized for production of several goods like semiconductors, glass, paints, varnishes and pigments. If ingested or inhaled in high concentrations, it can be highly poisonous and result many health issues (Antman 2001; Fang and Zhang 2020).

18.4 Toxicity of Arsenic

Arsenic is regarded as highly toxic heavy metal and almost it affects every organ of body in humans. Most common toxic effects of arsenic are severe pain in abdominal area, vomiting, nausea, loose motion, black clumpy stool, diabetes, and heart problems. Long exposure and usage of arsenic leads to multiple organ failure, kidneys damages, respiratory problems, premature birth in pregnancy, mental retardation, liver problems, and skin diseases (Rhoads and Sanders 1985; Jomova et al. 2011). Arsenic toxicity also causes cancer of skin, liver, bladder, lungs and intestine. There are several routes and mechanism by which arsenic causes side effects in humans and others living organism and among these the most common mechanism is that arsenic disrupt the cellular metabolization and interfere with repairing of DNA mechanisms, which leads to produce mutagenesis and abnormal dysfunction of chromosomes (Fig. 18.3) (Ahmat et al. 2021a; Rhoads and Sanders 1985; Akter et al. 2008). These effects of arsenic mostly reported due to consumption of polluted water, food and marine foods with arsenic for long time or inhalation of arsenic compounds for long periods of time as well exposure of humans or animals on arsenic contaminated places.

In addition to its harmful effects on human health, arsenic can also cause a number of harmful effects on the yields, growth and production of crops, fields and plants. When concentration of arsenic is high then plants growth reduced badly which ultimately reduce height and biomass of plants, crops results in premature growth of shoots, flowers, fruits and leaves of plants (Requejo and Tena 2006). Due to arsenic toxicity yellowing of leaves takes place which causes death of leaves and due to death of leaves chlorophyll production disrupted and this chlorophyll disruption causes abnormality in photosynthesis, functioning of cell membranes and also cause

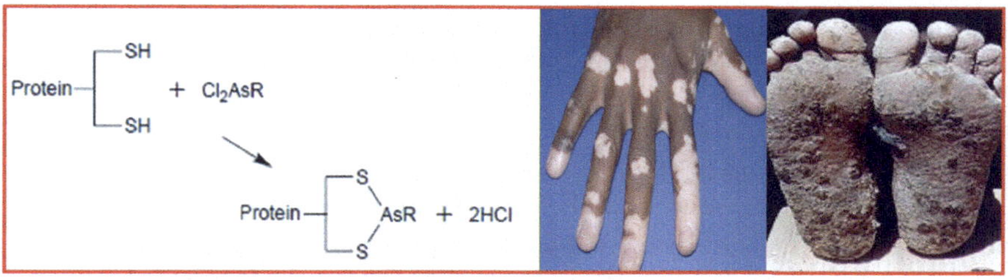

Fig. 18.3 Mechanism of arsenic toxicity causing skin lesions' and skin cancer

interference in uptake of nutrients required for growth. For this reason, arsenic toxicity can expressively cause reduction in plant growth and crop yields (Finnegan and Chen 2012).

Aquatic animals living in water bodies are also suffered with arsenic compounds by using water contaminated with arsenic and food which is present in water or sediments where they get their foods if contains arsenic then this arsenic directly enters in their body. After entering into marine lives this arsenic accumulation takes place in their tissues by different mechanism and interactions and cause toxicity. This toxicity greatly depends upon concentration of arsenic pollution in water, size, shape and species of marine lives as well as arsenic compounds present in water bodies (Edmonds and Francesconi 1977; Fallatah et al. 2022; Aziz et al. 2023). Arsenic cause toxicity in marine lives in such a way that cause reduction in their activity level and changes their swimming parameters due to which their behavioral actions changes. Arsenic also cause reduction in their reproductive rate and sperm mobility due to which their survival rate is decreased to a greater extent and their growth rate is reduced and even those which survive they would not be able to grow properly and ultimately affect their developmental stages. Arsenic toxicity also cause mortality, and impair their immune system in aquatic environment (Edmonds and Francesconi 1993). To mitigate the risks of arsenic toxicity in marine lives, it is very important to regulate and monitor the sewerage water disposal in marine, river and ponds water to stop these harmful effects of arsenic species (Table 18.1).

To control and limit exposure to arsenic, it is imperative to avoid polluted air, soil, water and foods products. Further chelation therapy can also be used to limit and control the arsenic from these sources. Proper safety measures should be taken when it is necessary to work in that area or region, or water or soil contaminated with arsenic to minimize exposure to arsenic. In conclusion, arsenic is regarded as extremely lethal element which causes serious health effects to humans, plants, and marine lives, and it is important to take appropriate safety measures to prevent its harms (Kaya et al. 2020).

Table 18.1 Arsenic toxicity in humans and the environment: It is important to note that the effects of arsenic exposure can vary depending on the level and duration of exposure, as well as the individual's age, health status, and other factors

Aspects	Effects of arsenic
Human exposure (acute)	Ingestion of high levels of arsenic can cause nausea, vomiting, abdominal pain, and diarrhea. Severe exposure can lead to shock, coma, and even death
Human exposure (chronic)	Chronic exposure to arsenic can lead to a range of health problems, including skin lesions, cancers of the skin, ling, bladder, and kidney, cardiovascular disease, diabetes, and developmental effects
Environmental (water)	Arsenic contamination in water can have serious environmental and health consequences. It can lead to the death of aquatic life and bioaccumulate in fish and other organisms, which can then be consumed by humans, causing health effects
Environment (soil)	Arsenic contamination in soil can limit the growth and yield of crops, leading to reduced agricultural productivity. It can also contaminate groundwater and surface water, leading to environmental and health impacts
Environment (air)	Arsenic is released into the air through natural processes, such as volcanic activity, as well as human activities, such as mining and smelting. Arsenic exposure through inhalation can lead to respiratory problems, skin lesions, and cancer of lungs, bladder and skin

18.5 Determination of Arsenic

The determination of arsenic in various samples, such as soil, water, and food, is critical to ensure safe human and environmental health. The selection of an appropriate method to identify arsenic in a sample be subject to on, quite a lot of factors, together with the level of accuracy required, available resources and time, and specific analysis requirements. It is advisable to consult an experienced specialist prior showing arsenic analysis from any sample.

To calculate the concentration of arsenic species in water samples, a water sample is collected and subjected to laboratory testing using analytical methods like quantitative analysis, chelation therapy, HPLC, AAS or ICP-MS which gives arsenic concentrations in mg/L or ppm or ppb (Farzana Akter et al. 2005). From these results we can check whether drinking water is safer for drinking purposes or not.

Among these techniques ICP-MS and AAS are highly sensitive and accurate method for detecting and quantifying arsenic in a sample. ICP-MS involves vaporizing and ionizing the sample in an inductively coupled plasma, separating the ions by their mass-to-charge ratio, and detecting and quantifying arsenic based on its specific mass-to-charge ratio. Calibration standards and quality control measures can improve the accuracy and precision of the measurement (Kim et al. 2013).

AAS involves exposing a sample containing arsenic at specific wavelength with a light beam, atomizing sample, and exciting the arsenic atoms. The absorption of light is directly related with the concentration of Arsenic in sample solution and by this absorption of light we can calculate the Arsenic concentration in any given sample. For exact calculation standard solution curve is made and it is calibrated with sample from which exact concentration is measured quantitatively using AAS (Sperling et al. 1991).

In conclusion, accurate and sensitive methods for the determination of arsenic in various samples are essential to ensure safe human and environmental health. ICP-MS and AAS are highly sensitive and accurate methods for quantifying arsenic in samples, with their choice depending on several aspects like the type of sample, required accuracy level, available incomes and time. Regular monitoring and mitigation of As concentrations in samples should be done regularly and it is always advisable to consult an experienced specialist before conducting any analysis.

18.6 Adsorbents for Arsenic Mitigation

In recent years, researchers have explored advanced novel materials for treating heavy metals in aqueous systems (Fig. 18.4) shows arsenic sources, toxicity as well as adsorbents used for mitigation of arsenic from groundwater). Along with conventional materials, advanced materials are also studied to remove As from water bodies (Lata and Samadder 2016). This section will focus on the potential of advanced materials like CNTs, GO, COFs, MOFs and their hybrid composites for remediation of arsenic through the adsorption method.

Additionally, Table 18.2 provides information on various examples of adsorbents other than MOFs for As elimination.

18.6.1 Graphene Oxide (GO)

Graphene oxide (GO) has 2D structural material that incorporates oxygen functional groups such as epoxide, carbonyl, carboxyl, and hydroxyl groups. It is a modified form of graphene, which is a key structural component of several materials, including graphite, charcoal, carbon nanotubes, and fullerenes (Su et al. 2017). GO-based hybrid materials have a wide range of applications for treating water, particularly for removing heavy metal ions, including As (Siddiqui and Chaudhry 2017). Iron-based magnetic GO materials, such as magnetite-GO (M-GO) and magnetite-reduced GO (M-rGO), have been found to be highly effective in eliminating As from contaminated water. These materials have been shown to significantly improve the elimination potentials for heavy metal ions, especially As. The polar oxygen units in GO and rGO can have both positive and negative effects on their effectiveness in As treatment. However, the presence of functional groups on M-GO, such as epoxide, carbonyl, carboxyl, and hydroxyl groups, enhances their removal capability for both As(III) and As(V) (Su et al. 2017).

In literature it is also highlighted that the importance of iron-based modifications in increasing the adsorptivity of GO-based hybrid materials for heavy metal ions, especially As. The real adsorption capacity of As is essentially determined by the

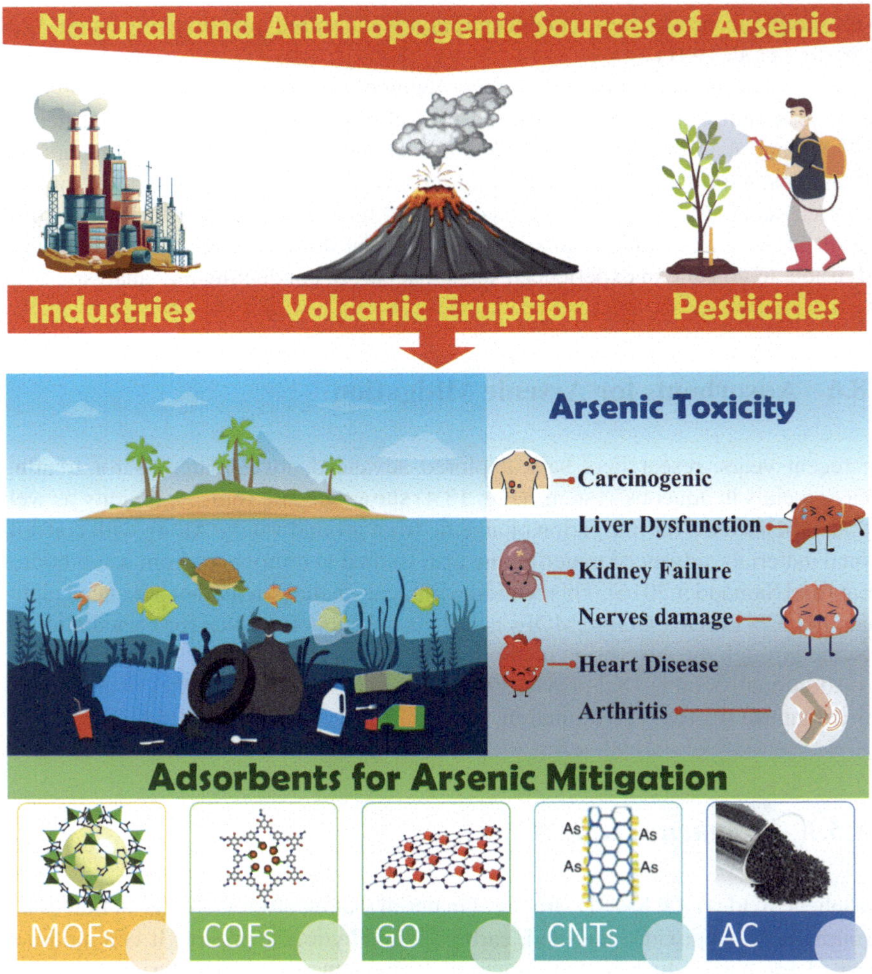

Fig. 18.4 Sources of arsenic in groundwater and environment, toxicity of arsenic and some advances adsorbents used for arsenic mitigation

iron content in a sorbent modified with iron. Due to the presence of more functional groups on M-GO than on M-rGO, M-GO displayed a greater removal capability for As(III) and As(V). As(III) was converted into As(V) before adsorption due to additional challenges in sorptive treatment of As(III) compared to that of As(V). Other magnetically modified GO materials have also demonstrated effective effluent treatment capabilities for As-containing effluent. For instance, GO modified by iron-manganese binary oxide (FeMnOx/$_R$GO) has been investigated for the removal of As from aqueous solutions, and it exhibited excellent reusability for both As. The direct removal of As(III) from rGO-Fe_3O_4-TiO_2 based nanocomposites has tremendous potential (Benjwal et al. 2015; Bibi et al. 2021).

Table 18.2 Some reported adsorbents other than MOFs used for the mitigation of arsenic

Adsorbents	Kinetic parameters	Adsorption capacity (mg/g)	References
Activated alumina	P2O, FM	14.7 = As(III) 39.5 = As(V)	Dutta and Gupta (2022)
Iron oxide	P2O, LM	21.6 = As(III) 37.1 = As(V)	Thirunavukkarasuet al. (2003)
Manganese dioxide	P2O, LM	32.3 = As(III) 50.4 = As(V)	Bajpai and Chaudhuri (1999)
Hydrous oxide	P1O, LM	33.6 = As(III) 28.9 = As(V)	Ghosh and Yuan (1987)
Activated carbon	P2O, FM	2.89 = As(III) 5.29 = As(V)	Arcibar-Orozco et al. (2014)
Zerovalent iron	P2O, LM	14.9 = As(III) 29.5 = As(V)	Lackovic et al. (2000)
Zeolite	P2O, LM	22.4 = As(III) 38.9 = As(V)	Shevade and Ford (2004)

In this table P2O = Pseudo 2nd order, P1O = Pseudo 1st order, FM = Freundlich model, LM = Langmuir model

In conclusion, modified magnetic-based GO materials have exceptional As removal capabilities and can be effective in treating water contaminated with heavy metal ions, particularly arsenic. The use of these materials can provide a practical solution for water purification, especially in areas with limited access to clean water.

18.6.2 Carbon Nanotubes (CNTs)

Carbon nanotubes (CNTs) are known for their high binding energies due to van der Waals interactions, making them ideal candidates for a wide range of environmental applications. Researchers have explored the potential of CNT-based materials for arsenic (As) removal from water systems (AlOmar et al. 2016). The development of a nano-composite of hydrated zirconium oxide-coated CNTs ($ZrO(OH)_2$/CNTs) using a filtration-steam hydrolysis method has demonstrated its potential for As remediation. The Langmuir model was used to determine the maximum adsorption capabilities of this nano-composite, which were found to be 78.2 and 124.6 mgg^{-1} for As(III) and As(V), respectively. $ZrO(OH)_2$/CNTs showed a faster rate of absorption for both As(III) and As(V) ions compared to $ZrO(OH)_2$ nanoparticles due to their uniform dispersion and increased number of adsorption sites. $ZrO(OH)_2$/CNTs exhibited a removal efficiency of over 97% even after six cycles, demonstrating its potential for As remediation (Siddiqui et al. 2019).

Fe-based substances such as goethite and natural iron oxide have also shown promise as adsorbents for the removal of As from effluent. A GO-CNT aerogel was coated with α-FeOOH (iron oxyhydroxide) to create α-FeOOH@GCA using a self-assembly method by in-situ Fe^{2+} reduction method. This material was used to treat three distinct As-containing species, including As(V), dimethylarsinic acid: DMA, and p-arsanilic acid (p-ASA). The maximum removal capabilities of α-FeOOH@GCA for these As species were significantly higher than those by α-FeOOH nanoparticles, showing at 56.4, 24.4, and 102.1 mgg^{-1}, respectively. Ligand exchange was used to explain the differences in adsorption capacities of various As species by α-FeOOH@GCA. This process ultimately led to the creation of inner sphere complexes with multiple As species' structural variations. The adsorption findings also agreed well with the Freundlich and Langmuir isotherm models, which showed dynamic affinities between the sorbent and As species (Pervez et al. 2021).

The research on altering certainly happening iron-based minerals with GO-CNT carriers provides new possibilities for promoting the immobilization of As species from polluted water. The studies on various modified CNT-based materials have shown considerable potential as powerful adsorbents for As remediation. The regeneration of adsorbents is an essential performance measure for assessing their financial worth. The $ZrO(OH)_2$/CNTs material (Liu et al. 2018), and α-FeOOH@GCA both showed high removal efficiencies even after several cycles, validating their ability to remove As from water systems. These findings provide valuable insights into the development of efficient and cost-effective As removal techniques for water treatment applications. Iron-based carbon nanotubes (CNTs) have been identified as potential materials for the removal of arsenic (As) from water systems. Two types of magnetic CNT composites (MI/CNTs and MIO-CNTs) were synthesized and tested for As removal capacity. MI/CNTs were synthesized using a multi-step, one-pot technique that involved the use of as-prepared CNTs and potassium hydroxide (KOH). The resulting MI/CNTs had a surface area of 662 m^2g^{-1}, which was five times greater than that of pure CNTs, and moderate maximum removal capacities of 9.74 mgg^{-1} for As(V) and 8.13 mgg^{-1} for As(III) (Chen et al. 2014). On the other hand, MIO-CNT composite was synthesized using p-CNTs as a template through a catalytic chemical vapor deposition technique. The MIO-CNT composite had a smaller surface area of 209.8 m^2g^{-1} but outperformed MI-CNT composite, with higher removal capacities of 47.41 mgg^{-1} for As(V) and 24.05 mgg^{-1} for As(III) (Chen et al. 2014; Liu et al. 2020a). The oxygen-containing groups of the CNT composite played a vital role in As removal by forming hydroxylated surface complexes with FeOH and AsOH that increased the number of adsorption sites. Other modified CNT-based materials, including CNT/CuO nanocomposites (Singh et al. 2016), TiO_2-coated CNTs (Liu et al. 2014), and Ce-Fe oxide-decorated CNTs (Chen et al. 2013), have also demonstrated promising As removal capabilities in aqueous systems. However, the Langmuir isotherm model estimated moderate maximum removal capacities of As(V) and As(III) for MI/CNTs and MIO-CNTs composites. Thus, further research is required to optimize the As removal capacity of CNT-based materials for efficient As remediation.

In conclusion, CNT-based materials such as $ZrO(OH)_2$/CNTs and Fe-based materials such as α-FeOOH@GCA have shown great potential for As removal from water systems. The unique properties of these materials make them attractive options for environmental remediation, and further research is necessary to explore their capabilities and limitations in real-world scenarios.

18.6.3 MOFs for Arsenic Mitigation

Adsorption is a process in which molecules of a substance are attracted and bound onto the surface of a solid material. Adsorbents are materials that are used to remove contaminants from water by adsorption. MOFs are a class of materials made up of metal ions or clusters linked by organic ligands to form a porous, crystalline structure (Hafsa et al. 2023; Parveen et al. 2023). They have a wide range of potential applications, including adsorption, gas storage, catalysis, and drug delivery. MOFs have several advantages over other adsorbents, such as high surface area, tunable pore size, and the ability to selectively adsorb molecules (Wang et al. 2019; Ahmad et al. 2020, 2021; Khan et al. 2020; Rehman Shah et al. 2021). MOFs can be designed to selectively adsorb specific water pollutants, such as heavy metals, organic molecules, and dyes, and can be regenerated for reuse. They can also be designed to have specific chemical properties that make them highly selective for certain heavy metals. The adsorption process can be influenced by factors such as pH, temperature, and the chemical properties of the MOF (Ahmad et al. 2021, 2023; Nazir et al. 2022). MOFs have shown promise for the removal of heavy metals such as arsenic, lead, cadmium, and mercury from water and other industrial effluents. They have also shown great potential for arsenic removal from water due to their high surface area and tunable pore sizes. The use of MOFs for arsenic removal from water was first reported in 2007 by researchers at the University of California, Berkeley. They used a zirconium-based MOF, UiO-66, functionalized with thiol (–SH) groups to selectively capture arsenic ions from water (Ahmad et al. 2019; Awad et al. 2021; Ayub et al. 2022).

Since then, numerous studies have been conducted to explore the use of various MOFs for arsenic removal, including MIL-101, MIL-53, and MOF-74. One of the advantages of using MOFs for arsenic removal is that they can be easily modified to enhance their selectivity and efficiency. Functionalizing MOFs with specific chemical groups, such as amine (–NH_2) or carboxylic acid (–COOH) groups, can improve their arsenic adsorption capacity and selectivity. MOFs such as MIL-101 (Li et al. 2019), UiO-66 (Wang et al. 2015), and ZIF-8 (Jian et al. 2015) have been studied for arsenic removal and have demonstrated high adsorption capacities and selectivity for arsenic species. ZIF-MOF (zeolitic imidazolate framework metal–organic framework) can remove arsenic from water through a process called adsorption, where the arsenic ions are trapped and bound to the surface of the material (Fig. 18.5 shows some possible mechanism for the mitigation of arsenic from water).

The high surface area and unique pore structure of ZIF MOFs allow for efficient and selective removal of arsenic from water. Additionally, ZIF MOFs can be easily synthesized and modified to enhance their arsenic removal capabilities. Some examples of MOFs for arsenic removal from water include ZIF-8, MOF-74, MIL-53, MIL-101, UiO-66, MOF-808, and HKUST-1 (Haldar et al. 2020). These MOFs

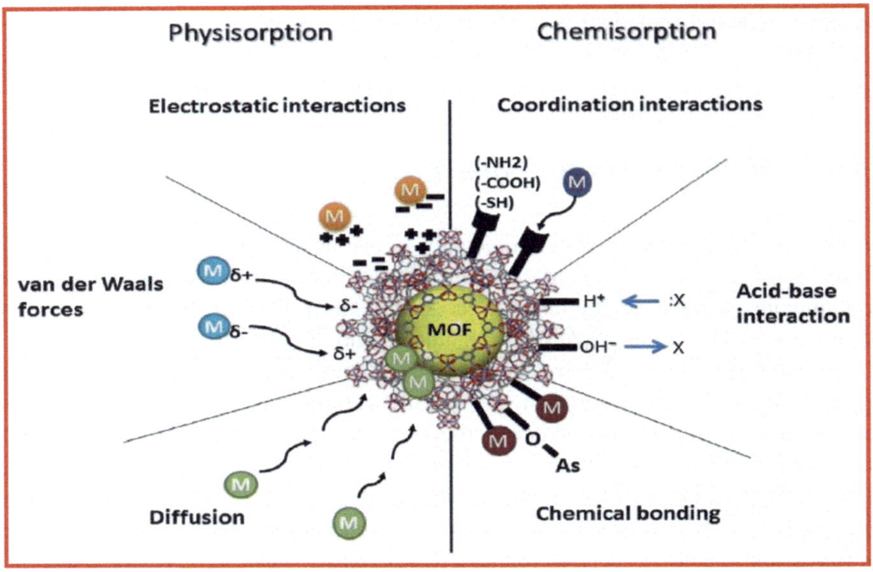

Fig. 18.5 Mechanism for mitigation of arsenic removal from water using MOFs

have been modified with various functional groups to enhance their arsenic removal capacity. HKUST-1 is a copper-based MOF that has been modified with amine or thiol groups to enhance its arsenic removal capacity. Some examples for arsenic mitigation using MOFs are given in Table 18.3.

Table 18.3 Some reported MOFs for arsenic removal:

MOF	Kinetic parameters	Adsorption capacity (mg/g)	References
UiO-66-NH_2	P2O, FM	128.6	Chang et al. (2019)
MIL-101(Cr)	P2O, LM	108.7	Rastkari et al. (2021)
UiO-66-NH_2@Fe_3O_4	P2O, LM	152.7	(Lee 2021)
MIL-101(Cr)@Fe_3O_4	P2O, LM	122.1	Li et al. (2023)
MIL-101(Cr)@GO	P1O, FM	170.9	Folens et al. (2016)
MIL-101(Cr)@SiO_2	P2O, LM	112.4	Abdollahi et al. (2022)
ZIF-8	P2O, LM	65.2	Jian et al. (2015)
MIL-53(Al)	P2O, LM	27.3	Li et al. (2014)
MIL-101(Cr)@GDQs	P2O, LM	175.4	Dervisevic et al. (2019)
ZIF-67	P1O, LM	64.7	Zhao et al. (2020)

In this table P20 = Pseudo 2nd order, P1O = Pseudo 1st order, FM = Freundlich model, LM = Langmuir model, UiO = University of Oslo, MIL = Material Institute The Lavoisier, ZIF = zeolitic imidazolate frameworks

In conclusion, MOFs are promising materials for water pollutant removal due to their high surface area, tunable pore size, and ability to selectively adsorb molecules. Functionalizing MOFs with specific chemical groups can enhance their arsenic removal capacity and selectivity, making them highly effective for arsenic removal from water. They have demonstrated high adsorption capacities and selectivity for arsenic species, and research in this area is ongoing.

18.6.4 COFs for Arsenic Mitigation

Covalent organic frameworks (COFs) are a type of porous material that have been investigated for their potential in removing arsenic from water. COFs are a class of crystalline, porous polymers that are composed of organic building blocks connected by covalent bonds. The resulting structure has a high surface area, which makes it an attractive material for adsorption applications, including arsenic removal (Liu et al. 2020b; Chen et al. 2022).

One of the advantages of COFs is that they can be designed to have specific pore sizes and surface chemistries, which can be tailored to enhance their adsorption properties for a particular contaminant. COFs also have good chemical stability, which makes them suitable for use in water treatment applications. There have been several studies that have investigated the use of COFs for arsenic removal from water. For example, one study published in the journal ACS Applied Materials and Interfaces (Haldar et al. 2020), reported the synthesis of a COF with an imine-linked structure that was found to have high adsorption capacity for both arsenite and arsenate. The researchers attributed this high adsorption capacity to the presence of amino groups on the COF surface, which can form strong chemical bonds with arsenic species. Another study published in the journal Nanoscale reported the synthesis of a COF with a three-dimensional architecture that showed high adsorption capacity for arsenic (Li et al. 2015). The researchers found that the COF was able to remove over 99% of arsenic from a solution with an initial concentration of 50 μg/L. While the use of COFs for arsenic removal is still in the research stage, these studies demonstrate the potential of COFs as an effective and promising material for removing arsenic from water. There have been several COFs that have been reported for arsenic removal from water. Here are some reported COFs for arsenic removal along with the year of publication: 2015: COF-300, COF-320, COF-500, COF-505, and COF-520. 2017: COF-5, COF-102, COF-108, and COF-109. 2019: COF-200 and COF-300. 2020: COF-LZU1, COF-LZU2, and COF-LZU3 (Ayub and Raza 2021; Gendy et al. 2021; Fu et al. 2022; Huang et al. 2022). These COFs have been shown to have high adsorption capacity for arsenic species and have been evaluated under different conditions, including batch and column tests. However, it is important to note that the performance of these COFs may vary depending on the specific water source and treatment conditions. Further research is needed to optimize COF structures and to evaluate their effectiveness for arsenic removal under real-world conditions.

18.6.5 Composites (MOF@COF) for Arsenic Mitigation

MOFs@COFs composites are hybrid materials that consist of both metal–organic frameworks (MOFs) and covalent organic frameworks (COFs). These materials have attracted attention for their potential in various applications, including catalysis, gas storage, and sensing. In the context of arsenic removal, MOFs@COFs composites have also been explored as a potential solution. These composites have the potential to combine the high adsorption capacity of MOFs with the excellent stability and tunability of COFs, leading to improved performance in arsenic removal (Ma et al. 2023).Several studies have reported the synthesis and characterization of MOFs@COFs composites for arsenic removal. For example, a study published in the Journal of Materials Chemistry A in 2019 reported the synthesis of a MOF@COF composite based on MIL-101 and COF-5 for efficient arsenic removal (Solomos et al. 2019). The study showed that the composite had a high adsorption capacity for arsenic, and the presence of the COF-5 layer enhanced the selectivity and stability of the composite. Another study published in the journal Applied Surface Science in 2020 reported the synthesis of a MOF@COF composite based on UiO-66 and COF-5 for arsenic removal (Lv et al. 2019). The study showed that the composite had a high adsorption capacity for both As(III) and As(V), and the COF-5 layer enhanced the stability and reusability of the composite (Gu et al. 2022).

Overall, the use of MOFs@COFs composites for arsenic removal is an active area of research, and further studies are needed to optimize their structures and evaluate their effectiveness under real-world conditions.

18.7 Systems Available on Commercial Scale for Arsenic Mitigation

There are several devices available on the market that can remove arsenic from water. Here are a few examples.

18.7.1 Reverse Osmosis (RO) Systems

RO systems are one of the most effective methods for removing arsenic from water. These systems use a semi-permeable membrane to filter out contaminants, including arsenic.

18.7.2 Activated Alumina Filters

Activated alumina filters are another effective method for removing arsenic from water. These filters use a special type of alumina that binds with arsenic to remove it from the water.

18.7.3 Ion Exchange Systems

Ion exchange systems can also remove arsenic from water. These systems use ion exchange resin to attract and bind with arsenic, removing it from the water.

18.7.4 Adsorption Filters

Adsorption filters are designed to remove a variety of contaminants, including arsenic. These filters use activated carbon or other adsorbent materials to attract and remove contaminants from the water.

18.7.5 Distillation Systems

Distillation systems are another option for removing arsenic from water. These systems boil water to create steam, which is then condensed back into water. Since arsenic does not evaporate, it is left behind in the boiling chamber.

It is important to note that not all devices are equally effective at removing arsenic from water, and some may be better suited for specific types of water sources or levels of arsenic contamination. It is recommended that you consult with a water treatment professional to determine the best option for your specific needs.

18.8 Toxicity Leaching Characteristics Procedures (TCLP)

Toxicity leaching characteristics (TCLP) procedures are used to determine the potential for hazardous waste to leach contaminants into the environment. The procedure involves mixing a sample of the waste with an acidic solution that simulates the conditions of a landfill, and measuring the concentration of various contaminants that may leach out of the waste. Here are the TCLP procedures for arsenic metal: sample preparation: A representative sample of the waste is collected and prepared according to established protocols. The sample is then ground to a specific particle size to ensure consistency. Acid digestion: The sample is mixed with an acidic solution, typically a mixture of acetic acid and hydrochloric acid. The solution is agitated for a period of time to simulate the conditions of a landfill. Filtration: The solution is then filtered to remove any solid particles. Analysis: The concentration of arsenic in the filtered solution is measured using an analytical technique, such as inductively coupled plasma mass spectrometry (ICP-MS). The result is reported in milligrams of arsenic per liter of leachate. Comparison with regulatory limits: The concentration of arsenic in the leachate is compared to regulatory limits established by the relevant authority. If the concentration of arsenic exceeds the regulatory limit, the waste is considered hazardous and may be subject to special handling and disposal requirements. TCLP procedures for arsenic metal are important for ensuring that hazardous waste containing arsenic is properly managed and disposed of to prevent contamination of the environment (Nikolaidis et al. 2003; Ghosh et al. 2004).

Cal-WET is a laboratory test method used to evaluate the performance of adsorbents for arsenic removal. The test is designed to simulate real-world water treatment conditions and provide information on the capacity, selectivity, and kinetics of the adsorbent for arsenic removal. Here are the basic steps of the Cal-WET test for arsenic removal: Prepare the adsorbent: The adsorbent is typically prepared in the form of small particles, such as granules or pellets. The adsorbent is washed and dried to remove any impurities or contaminants. Prepare the test solution: A synthetic test solution containing a known concentration of arsenic is prepared. The solution is usually buffered to a specific pH and may contain other dissolved ions to simulate real-world water conditions. Conduct the batch adsorption test: The adsorbent is added to the test solution and mixed for a specific amount of time, typically several hours. The solution is then filtered to separate the adsorbent from the solution, and the remaining arsenic concentration in the solution is measured. Analyze the results: The amount of arsenic removed by the adsorbent is calculated by subtracting the final arsenic concentration from the initial concentration. The capacity of the adsorbent is determined by dividing the amount of arsenic removed by the mass of the adsorbent. The selectivity of the adsorbent for arsenic is evaluated by comparing its performance with other competing ions in the test solution. The kinetics of the adsorption process can also be analyzed by measuring the rate of arsenic removal over time. The Cal-WET test for arsenic removal provides valuable information on the performance of adsorbents under specific water treatment conditions (Meng et al. 2001; Narasimhan 2005). However, it is important to note that the test is conducted in the laboratory, and the performance of the adsorbent may vary under different water treatment scenarios. Therefore, it is necessary to evaluate the performance of adsorbents under actual field conditions before implementing them in large-scale water treatment applications.

18.9 Challenges for Arsenic Mitigation

However, there are still some challenges that need to be addressed for the widespread use of adsorbents in arsenic removal applications. For example, the production cost of some adsorbents like MOFs, COFs, composites etc. is still relatively high, and there are concerns about the long-term stability and durability of these materials in real-world water treatment applications. Additionally, the scale-up of adsorbents-based systems for industrial applications is still in the early stages of development. Nonetheless, ongoing research and development in the field of adsorbents for arsenic removal is expected to overcome these challenges, and continue to improve the performance and reliability of these materials. As a result, it is likely that these adsorbents especially MOFs, COFs and composites will play an increasingly important role in the treatment of arsenic-contaminated water in the future. In conclusion, while there are still some challenges to be overcome, the use of advance adsorbents for arsenic removal is a promising technology with significant potential for improving access to clean and safe drinking water.

18.10 Conclusion

The article discusses the risks of arsenic contamination in the environment and its potential harm to public health. Inorganic arsenic is known to cause serious health problems such as cancer, skin lesions, cardiovascular diseases, and neurodevelopmental disorders. The Toxicity Characteristic Leaching Procedure (TCLP) is used to measure the concentration of arsenic in waste materials and classify waste as hazardous or non-hazardous. The use of advanced materials such as CNTs, GO, MOFs, COFs, and their hybrids offer a promising solution to arsenic contamination by effectively removing it from water and soil. However, the effectiveness of these materials depends on various factors such as their structure, surface chemistry, and the presence of other contaminants in the water. Further research is necessary to optimize the properties of these materials and to develop practical applications for their use in real-world scenarios. The article emphasizes the need for continued research on arsenic and its removal from the environment to ensure public health and environmental sustainability.

Acknowledgements The authors are very thankful to the Department of Chemistry, University of Management and Technology (UMT), Sialkot Campus, Punjab, Pakistan, for providing necessary facilities to complete this book chapter.

References

Abdollahi B, Salari D, Zarei M (2022) Synthesis and characterization of magnetic $Fe_3O_4@SiO_2$-MIL-53(Fe) metal-organic framework and its application for efficient removal of arsenate from surface and groundwater. J Environ Chem Eng 10(2):107144

Abedin MJ, Meharg AAJ (2002) Relative toxicity of arsenite and arsenate on germination and early seedling growth of rice (*Oryza sativa* L.). Plant Soil 243:57–66

Ahmad K, Naseem HA, Parveen S, Shah H-u-R, Shah SSA, Shaheen S, Ashfaq A, Jamil J, Ahmad MM, Ashfaq M (2019) Synthesis and spectroscopic characterization of medicinal azo derivatives and metal complexes of Indandion. J Mol Struct 1198:126885

Ahmad K, Nazir MA, Qureshi AK, Hussain E, Najam T, Javed MS, Shah SSA, Tufail MK, Hussain S, Khan NAJ (2020) Engineering of Zirconium based metal-organic frameworks (Zr-MOFs) as efficient adsorbents. Mater Sci Eng B 262:114766

Ahmad K, Shah H-u-R, Ashfaq M, Shah SSA, Hussain E, Naseem HA, Parveen S, Ayub A (2021a) Effect of metal atom in zeolitic imidazolate frameworks (ZIF-8 & 67) for removal of Pb^{2+} & Hg^{2+} from water. Food Chem Toxicol 149:112008

Ahmad K, Shah H-U-R, Parveen S, Aziz T, Naseem HA, Ashfaq M, Rauf A (2021b) Metal organic framework (KIUB-MOF-1) as efficient adsorbent for cationic and anionic dyes from brackish water. J Mol Struct 1242:130898

Ahmad Z, Barutçular C, Zia Ur Rehman M, Sabir Tariq RM, Afzal M, Waraich EA, Ahmad A, Iqbal MA, Bukhari MA, Ahmad K, El Sabagh A, Raza A, Nawaz H (2023) Pod shattering in canola reduced by mitigating drought stress through silicon application and molecular approaches—a review. J Plant Nutr 46(1):101–128

Akter MS, Ahmed MK, Akhand MA, Islam MMJ (2008) Acute toxicity of arsenic and mercury to fresh water climbing perch, *Anabas testudineus* (Bloch). World J Zool 3(1):13–18

AlOmar MK, Alsaadi MA, Hayyan M, Akib S, Hashim MAJ (2016) Functionalization of CNTs surface with phosphonuim based deep eutectic solvents for arsenic removal from water. Appl Surf Sci 389:216–226

Antman KHJ (2001) Introduction: the history of arsenic trioxide in cancer therapy. Oncologist 6(S2):1–2

Anwar N, Arshad A, Fatima N, Shaheen S, Bukhari S, Shamsi TJ (2022) Environmental and occupational determinants of myelodysplastic syndrome: a case–control study from Pakistan. Cancer Rep 5(10):e1580

Arcibar-Orozco JA, Josue D-B, Rios-Hurtado JC, Rangel-Mendez JR (2014) Influence of iron content, surface area and charge distribution in the arsenic removal by activated carbons. Chem Eng J 249:201–209

Awad FS, Bakry AM, Ibrahim AA, Lin A, El-Shall MSJ (2021) Thiol-and amine-incorporated UIO-66-NH2 as an efficient adsorbent for the removal of Mercury (II) and phosphate ions from aqueous solutions. Ind Eng Chem Res 60(34):12675–12688

Ayub A, Raza ZAJ (2021) Arsenic removal approaches: a focus on chitosan biosorption to conserve the water sources. Int J Biol Macromol 192:1196–1216

Ayub A, Irfan A, Raza ZA, Abbas M, Muhammad A, Ahmad K, Munwar A (2022) Development of poly(1-vinylimidazole)-chitosan composite sorbent under microwave irradiation for enhanced uptake of Cd(II) ions from aqueous media. Polym Bull 79(2):807–827

Aziz T, Nasim HA, Ahmad K, Shah H-u-R, Parveen S, Ahmad MM, Majeed H, Galal AM, Rauf A, Ashfaq M (2023) Rational synthesis, biological screening of azo derivatives of chloro-phenylcarbonyl diazenyl hydroxy dipyrimidines/thioxotetrahydropyrimidines and their metal complexes. Heliyon 9(1):e12492

Bajpai S, Chaudhuri M (1999) Removal of arsenic from ground water by manganese dioxide & coated sand 125(8):782–784

Bencko V, Yan Li Foong FJ (2017) The history of arsenical pesticides and health risks related to the use of Agent Blue. Ann Agric Environ Med 24(2)

Benjwal P, Kumar M, Chamoli P, Kar KKJ (2015) Enhanced photocatalytic degradation of methylene blue and adsorption of arsenic (iii) by reduced graphene oxide (rGO)–metal oxide (TiO_2/Fe_3O_4) based nanocomposites. RSC Adv 5(89):73249–73260

Bibi S, Ahmad A, Anjum MAR, Haleem A, Siddiq M, Shah SS, Al Kahtani AJ (2021) Photocatalytic degradation of malachite green and methylene blue over reduced graphene oxide (rGO) based metal oxides (rGO-Fe_3O_4/TiO_2) nanocomposite under UV-visible light irradiation. J Environ Chem Eng 9(4):105580

Buchanan WDJ (1962) Toxicity of arsenic compounds

Cao A-M, Monnell JD, Matranga C, Wu J-M, Cao L-L, Gao DJ (2007) Hierarchical nanostructured copper oxide and its application in arsenic removal. J Phys Chem C 111(50):18624–18628

Chang Z-W, Lee Y-J, Lee D-J (2019) Adsorption of hydrogen arsenate and dihydrogen arsenate ions from neutral water by UiO-66-NH2. J Environ Manage 247:263–268

Chen B, Zhu Z, Ma J, Qiu Y, Chen JJ (2013) Surfactant assisted Ce–Fe mixed oxide decorated multiwalled carbon nanotubes and their arsenic adsorption performance. J Mater Chem A 1(37):11355–11367

Chen B, Zhu Z, Ma J, Yang M, Hong J, Hu X, Qiu Y, Chen JJ (2014) One-pot, solid-phase synthesis of magnetic multiwalled carbon nanotube/iron oxide composites and their application in arsenic removal. J Colloid Interface Sci 434:9–17

Chen H, Liu W, Cheng L, Meledina M, Meledin A, Van Deun R, Leus K, Van Der Voort PJ (2022) Amidoxime-functionalized covalent organic framework as simultaneous luminescent sensor and adsorbent for organic arsenic from water. Chem Eng J 429:132162

Dervisevic M, Dervisevic E, Şenel M (2019) Recent progress in nanomaterial-based electrochemical and optical sensors for hypoxanthine and xanthine. A review. Microchim Acta 186(12):749

Dutta N, Gupta A (2022) Development of arsenic removal unit with electrocoagulation and activated alumina sorption: field trial at rural West Bengal, India. J Water Process Eng 49:103013

Edmonds JS, Francesconi KA (1977) Methylated arsenic from marine fauna. Nature 265(5593):436–436

Edmonds J, Francesconi K, Stick RJ (1993) Arsenic compounds from marine organisms. Nat Prod Rep 10(4):421–428

Edmonds JS, Francesconi KA (1993) Arsenic in seafoods: human health aspects and regulations. Mar Pollut Bull 26(12):665–674

Fallatah AM, Shah HUR, Ahmad K, Ashfaq M, Rauf A, Muneer M, Ibrahim MM, El-Bahy ZM, Shahzad A, Babras A (2022) Rational synthesis and characterization of highly water stable MOF@GO composite for efficient removal of mercury (Hg^{2+}) from water. Heliyon 8(10):e10936

Fang Y, Zhang ZJ (2020) Arsenic trioxide as a novel anti-glioma drug: a review. Cell Mol Biol Lett 25(1):1–13

Farzana Akter K, Chen Z, Smith L, Davey D, Naidu R (2005) Speciation of arsenic in ground water samples: a comparative study of CE-UV, HG-AAS and LC-ICP-MS. Talanta 68(2):406–415

Finnegan P, Chen W (2012) Arsenic toxicity: the effects on plant metabolism. Front Physiol 3

Folens K, Leus K, Nicomel NR, Meledina M, Turner S, Van Tendeloo G, Laing GD, Van Der Voort P (2016) Fe3O4@MIL-101—a selective and regenerable adsorbent for the removal of as species from water. Eur J Chem 2016(27):4395–4401

Francesconi KA, Kuehnelt DJ (2004) Determination of arsenic species: a critical review of methods and applications, 2000–2003. Analyst 129(5):373–395

Fu X, Song X, Zheng Q, Liu C, Li K, Luo Q, Chen J, Wang Z, J. J. I. J. o. E. R. Luo and P. Health, (2022) Frontier materials for adsorption of antimony and arsenic in aqueous environments: a review. Int J Environ Res Publ Health 19(17):10824

Garelick H, Jones H, Dybowska A, Valsami-Jones EJ (2008) Arsenic pollution sources international perspectives on arsenic pollution remediation. Rev Environ Contam 197:17–60

Gendy EA, Ifthikar J, Ali J, Oyekunle DT, Elkhlifia Z, Shahib II, Khodair AI, Chen ZJ (2021) Removal of heavy metals by covalent organic frameworks (COFs): a review on its mechanism and adsorption properties. J Environ Chem Eng 9(4):105687

Ghosh MM, Yuan JR (1987) Adsorption of inorganic arsenic and organoarsenicals on hydrous oxides. Environ Prog 6(3):150–157

Ghosh A, Mukiibi M, Ela WJ (2004) TCLP underestimates leaching of arsenic from solid residuals under landfill conditions. Environ Sci Technol 38(17):4677–4682

Gillispie EC, Sowers TD, Duckworth OW, Polizzotto MLJ (2015) Soil pollution due to irrigation with arsenic-contaminated groundwater: current state of science. Curr Pollut Rep 1:1–12

Gu H, Liu X, Wang S, Chen Z, Yang H, Hu B, Shen C, Wang XJ (2022) COF-based composites: extraordinary removal performance for heavy metals and radionuclides from aqueous solutions. Rev Environ Contam Toxicol 260(1):23

Hafsa HUR, Shah K, Ahmad MA, Oku H (2023) Free radical scavenging, antibacterial potentials and spectroscopic characterizations of benzoyl thiourea derivatives and their metal complexes. J Mol Struct 1272:134162

Haldar D, Duarah P, Purkait MKJ (2020) MOFs for the treatment of arsenic, fluoride and iron contaminated drinking water: a review. Chemosphere 251:126388

Hingston J, Collins C, Murphy R, Lester JJ (2001) Leaching of chromated copper arsenate wood preservatives: a review. Environ Pollut 111(1):53–66

Hosseini M-J, Shaki F, Ghazi-Khansari M, Pourahmad JJ (2013) Toxicity of arsenic (III) on isolated liver mitochondria: a new mechanistic approach. Iran J Pharm Res IJPR 12(Suppl):121

Huang W, Xu H, Liu X, Wang L, Li S, Ji L, Qu Z, Yan NJ (2022) Surface protection method for the magnetic core using covalent organic framework shells and its application in As (III) depth removal from acid wastewater. J Environ Sci 115:1–9

Hughes MFJ (2002) Arsenic toxicity and potential mechanisms of action. Toxicol Lett 133(1):1–16

Irgolic K, Woolson E, Stockton R, Newman R, Bottino NR, Zingaro R, Kearney P, Pyles R, Maeda S, McShane WJ (1977) Characterization of arsenic compounds formed by Daphnia magna and Tetraselmis chuii from inorganic arsenate. Environ Health Perspect 19:61–66

Jian M, Liu B, Zhang G, Liu R, Zhang XJ (2015) Adsorptive removal of arsenic from aqueous solution by zeolitic imidazolate framework-8 (ZIF-8) nanoparticles. Colloids Surf A Physicochem Eng Aspects 465:67–76

Jiang J-Q, Ashekuzzaman S, Jiang A, Sharifuzzaman S, Chowdhury SRJ (2013) Arsenic contaminated groundwater and its treatment options in Bangladesh. Int J Environ Res Publ Health 10(1):18–46

Jomova K, Jenisova Z, Feszterova M, Baros S, Liska J, Hudecova D, Rhodes C, Valko MJ (2011) Arsenic: toxicity, oxidative stress and human disease. J Appl Toxicol 31(2):95–107

Karadjova IB, Slaveykova VI, Tsalev DLJ (2008) The biouptake and toxicity of arsenic species on the green microalga *Chlorella salina* in seawater. Aquat Toxicol 87(4):264–271

Kaya C, Ashraf M, Alyemeni MN, Corpas FJ, Ahmad P (2020) Salicylic acid-induced nitric oxide enhances arsenic toxicity tolerance in maize plants by upregulating the ascorbate-glutathione cycle and glyoxalase system. J Hazard Mater 399:123020

Khairul I, Wang QQ, Jiang YH, Wang C, Naranmandura HJ (2017) Metabolism, toxicity and anticancer activities of arsenic compounds. Oncotarget 8(14):23905

Khan NA, Najam T, Shah SSA, Hussain E, Ali H, Hussain S, Shaheen A, Ahmad K, Ashfaq MJ (2020) Development of Mn-PBA on GO sheets for adsorptive removal of ciprofloxacin from water: kinetics, isothermal, thermodynamic and mechanistic studies. Mater Chem Phys 245:122737

Kim J-Y, Kim W-I, Kunhikrishnan A, Kang D-W, Kim D-H, Lee Y-J, Kim Y-J, Kim C-T (2013) Determination of arsenic species in rice grains using HPLC-ICP-MS. Food Sci Biotechnol 22(6):1509–1513

Lackovic JA, NPN, Dobbs GM (2000) Inorganic arsenic removal by zero-valent iron. Environ Eng Sci 17(1):29–39

Larsen EH, Moseholm L, Nielsen MMJ (1992) Atmospheric deposition of trace elements around point sources and human health risk assessment. II: uptake of arsenic and chromium by vegetables grown near a wood preservation factory. Sci Total Environ 126(3):263–275

Lata S, Samadder SR (2016) Removal of arsenic from water using nano adsorbents and challenges: a review. J Environ Manage 166:387–406
Lee D (2021) Experimental and computational approaches for prevention and remediation of gas hydrates in the presence of novel inhibitors, Ulsan National Institute of Science and Technology
Li J, Wu Y-N, Li Z, Zhu M, Li F (2014) Characteristics of arsenate removal from water by metal-organic frameworks (MOFs). Water Sci Technol 70(8):1391–1397
Li R, Zhang L, Wang P (2015) Rational design of nanomaterials for water treatment. Nanoscale 7(41):17167–17194
Li Z, Liu X, Jin W, Hu Q, Zhao YJ (2019) Adsorption behavior of arsenicals on MIL-101 (Fe): the role of arsenic chemical structures. J Colloid Interface Sci 554:692–704
Li X-Y, Ding W-Q, Liu P, Xu L, Fu M-L, Yuan B (2023) Magnetic Fe_3O_4/MIL-101 composite as a robust adsorbent for removal of p-arsanilic acid and roxarsenic in the aqueous solution. Colloids Surf A 662:131014
Liu H, Zuo K, Vecitis CDJ (2014) Titanium dioxide-coated carbon nanotube network filter for rapid and effective arsenic sorption. Environ Sci Technol 48(23):13871–13879
Liu D, Deng S, Maimaiti A, Wang B, Huang J, Wang Y, Yu GJ (2018) As (III) and As (V) adsorption on nanocomposite of hydrated zirconium oxide coated carbon nanotubes. J Colloid Interface Sci 511:277–284
Liu B, Kim K-H, Kumar V, Kim SJ (2020a) A review of functional sorbents for adsorptive removal of arsenic ions in aqueous systems. J Hazard Mater 388:121815
Liu X, Xu H, Wang L, Qu Z, Yan NJ (2020b) Surface nano-traps of Fe0/COFs for arsenic (III) depth removal from wastewater in non-ferrous smelting industry. Chem Eng J 381:122559
Lv S-W, Liu J-M, Wang Z-H, Ma H, Li C-Y, Zhao N, Wang SJ (2019) Recent advances on porous organic frameworks for the adsorptive removal of hazardous materials. J Environ Sci 80:169–185
Ma ML, Wang X, Lu S, Wang Y, Guo, Liang XJ (2023) One-step fabrication of hydrophobic metal-organic framework@ covalent organic framework hybrid as sorbent for high-performance solid-phase extraction of flavonoids. J Chromatogr A 463814
Madhavan N, Subramanian VJ (2000) Sulphide mining as a source of arsenic in the environment. Curr Sci 702–709
Meharg AA, Hartley-Whitaker JJ (2002) Arsenic uptake and metabolism in arsenic resistant and nonresistant plant species. New Phytol 154(1):29–43
Meng X, Korfiatis GP, Jing C, Christodoulatos CJ (2001) Redox transformations of arsenic and iron in water treatment sludge during aging and TCLP extraction. Environ Sci Technol 35(17):3476–3481
Mudhoo A, Sharma SK, Garg VK, Tseng C-HJ (2011) Arsenic: an overview of applications, health, and environmental concerns and removal processes. Crit Rev Environ Sci Technol 41(5):435–519
Naranmandura H, Carew MW, Xu S, Lee J, Leslie EM, Weinfeld M, Le XCJ (2011) Comparative toxicity of arsenic metabolites in human bladder cancer EJ-1 cells. Chem Res Toxicol 24(9):1586–1596
Narasimhan R (2005) Landfill disposal considerations for arsenic treatment residuals. Residuals Biosolids Conference 2005, Water Environment Federation
Nazir MA, Yasar A, Bashir MA, Siyal SH, Najam T, Javed MS, Ahmad K, Hussain S, Anjum S, Hussain E, Shah SSA, Rehman AU (2022) Quality assessment of the noncarbonated-bottled drinking water: comparison of their treatment techniques. Int J Environ Anal Chem 102(19):8195–8206
Nikolaidis NP, Dobbs GM, Lackovic JAJ (2003) Arsenic removal by zero-valent iron: field, laboratory and modeling studies. Water Res 37(6):1417–1425
Palma-Lara I, Martínez-Castillo M, Quintana-Pérez J, Arellano-Mendoza M, Tamay-Cach F, Valenzuela-Limón O, García-Montalvo E, Hernández-Zavala AJ (2020) Arsenic exposure: a public health problem leading to several cancers. Regul Toxicol Pharmacol 110:104539

Parveen S, Naseem HA, Ahmad K, Shah H-U-R, Aziz T, Ashfaq M, Rauf A (2023) Design, synthesis and spectroscopic characterizations of medicinal hydrazide derivatives and metal complexes of malonic ester. Curr Bioact Compd 19(4):31–46
Pervez MN, Fu D, Wang X, Bao Q, Yu T, Naddeo V, Tian H, Cao C, Zhao YJ (2021) A bifunctional α-FeOOH@ GCA nanocomposite for enhanced adsorption of arsenic and photo Fenton-like catalytic conversion of As (III). Environ Technol Innov 22:101437
Piver WTJ (1983) Mobilization of arsenic by natural and industrial processes. Biol Environ Effects Arsenic 6(1)
Rastkari N, Akbari S, Brahmand MB, Takhvar A, Ahmadkhaniha R (2021) Synthesis and characterization of tetraethylene pentamine functionalized MIL-101(Cr) for removal of metals from water. J Environ Health Sci Eng 19(2):1735–1742
Ratnaike RNJ (2003) Acute and chronic arsenic toxicity. Postgrad Med J 79(933):391–396
Rehman Shah HU, Ahmad K, Naseem HA, Parveen S, Ashfaq M, Rauf A, Aziz T (2021) Water stable graphene oxide metal-organic frameworks composite (ZIF-67@GO) for efficient removal of malachite green from water. Food Chem Toxicol 154:112312
Requejo R, Tena MJ (2006) Maize response to acute arsenic toxicity as revealed by proteome analysis of plant shoots. Proteomics 6(S1):S156–S162
Rhoads K, Sanders CJ (1985) Lung clearance, translocation, and acute toxicity of arsenic, beryllium, cadmium, cobalt, lead, selenium, vanadium, and ytterbium oxides following deposition in rat lung. Environ Res 36(2):359–378
Sakurai TJ (2002) Biological effects of organic arsenic compounds in seafood. Appl Organomet Chem 16(8):401–405
Saunders JA, Lee MK, Uddin A, Mohammad S, Wilkin RT, Fayek M, Korte NEJ (2005) Natural arsenic contamination of Holocene alluvial aquifers by linked tectonic, weathering, and microbial processes. Geochem Geophys Geosyst 6(4)
Shevade S, Ford RG (2004) Use of synthetic zeolites for arsenate removal from pollutant water. Water Res 38(14):3197–3204
Siddiqui SI, Chaudhry SAJ (2017) Arsenic removal from water using nanocomposites: a review. Curr Environ Eng 4(2):81–102
Siddiqui SI, Naushad M, Chaudhry SAJ (2019) Promising prospects of nanomaterials for arsenic water remediation: a comprehensive review. Process Safety Environ Prot 126:60–97
Singh DK, Mohan S, Kumar V, Hasan SHJ (2016) Kinetic, isotherm and thermodynamic studies of adsorption behaviour of CNT/CuO nanocomposite for the removal of As (III) and As (V) from water. RSC Adv 6(2):1218–1230
Smedley P, Kinniburgh DG (2012) Arsenic in groundwater and the environment. Essentials of Medical Geology: Revised Edition. Springer, pp 279–310
Smith AH, Marshall G, Roh T, Ferreccio C, Liaw J, Steinmaus CJ (2018) Lung, bladder, and kidney cancer mortality 40 years after arsenic exposure reduction. JNCI J Nat Cancer Inst 110(3):241–249
Smith R, Knight R, Fendorf SJ (2018b) Overpumping leads to California groundwater arsenic threat. Nat Commun 9(1):2089
Solomos MA, Claire FJ, Kempa TJJ (2019) 2D molecular crystal lattices: advances in their synthesis, characterization, and application. J Mater Chem A 7(41):23537–23562
Sperling M, Yin X, Welz B (1991) Differential determination of arsenic(III) and total arsenic using flow injection on-line separation and preconcentration for graphite furnace atomic absorption spectrometry. Spectrochim Acta, Part B 46(14):1789–1801
Styblo M, Thomas DJJ (2001) Selenium modifies the metabolism and toxicity of arsenic in primary rat hepatocytes. Toxicol Appl Pharmacol 172(1):52–61
Su H, Ye Z, Hmidi NJC, Physicochemical SA, Aspects E (2017) High-performance iron oxide–graphene oxide nanocomposite adsorbents for arsenic removal. Colloids Surfaces A Physicochem Eng Aspects 522:161–172
Thirunavukkarasu OS, Viraraghavan T, Subramanian KS (2003) Arsenic removal from drinking water using iron oxide-coated sand. Water Air Soil Pollut 142(1):95–111

Tian Y, Wu M, Lin X, Huang P, Huang YJ (2011) Synthesis of magnetic wheat straw for arsenic adsorption. J Hazard Mater 193:10–16

Vahter MJ (1994) Species differences in the metabolism of arsenic compounds. Appl Organomet Chem 8(3):175–182

Vahter M, Concha GJ (2001) Role of metabolism in arsenic toxicity. Pharmacol Toxicol Minireview 89(1):1–5

Wang T, Zhang L, Wang H, Yang W, Fu Y, Zhou W, Yu W, Xiang K, Su Z, Dai SJ (2013) Controllable synthesis of hierarchical porous Fe_3O_4 particles mediated by poly (diallyldimethylammonium chloride) and their application in arsenic removal. ACS Appl Mater Interfaces 5(23):12449–12459

Wang C, Liu X, Chen JP, Li KJ (2015) Superior removal of arsenic from water with zirconium metal-organic framework UiO-66. Sci Rep 5(1):16613

Wang C, Liu H, Zhang Y, Zou C, Anthony EJJ (2018) Review of arsenic behavior during coal combustion: volatilization, transformation, emission and removal technologies. Prog Energy Combust Sci 68:1–28

Wang C, Luan J, Wu CJ (2019) Metal-organic frameworks for aquatic arsenic removal. Water Res 158:370–382

Wang Y, Yu J, Wang Z, Liu Y, Zhao YJ (2021) A review on arsenic removal from coal combustion: advances, challenges and opportunities. Chem Eng J 414:128785

Welch AH, Lico MS, Hughes JLJ (1988) Arsenic in ground water of the western United States. Groundwater 26(3):333–347

Zhao P, Jian M, Xu R, Zhang Q, Xiang C, Liu R, Zhang X, Liu H (2020) Removal of arsenic(iii) from water by 2D zeolitic imidazolate framework-67 nanosheets. Environ Sci Nano 7(11):3616–3626

Chapter 19
Spatial Variation of Groundwater Nitrate Concentration in Guntur District, Andhra Pradesh, India: Human Health Risk and Mitigation Strategy

Ch. Jyotiprava Dash, Prachi Yadav, Randhir Kumar, Jyotirmayee Lenka, Basit Raza, Prativa Sahu, R. H. Ankhila, and Partha Pratim Adhikary

Keywords Andhra Pradesh · Groundwater · Nitrate pollution · Nitrate pollution index

19.1 Introduction

India's population is growing, the agricultural sector is developing, and industrialization are all contributing to a significant increase in the country's requirement for fresh water (Raju et al. 2011; Ramana and Rao 2017; Balamurgan et al. 2020). As a result, groundwater is becoming a more important source of water supply. Groundwater contamination results from improper and unsustainable methods of groundwater exploration and development, which modify the physical, chemical, and biological characteristics of water (Reddy 2013). One of them is groundwater nitrate contamination. The main causes of nitrate pollution of groundwater are excessive use of chemical fertilizers in agriculture, agricultural runoff, domestic and industrial effluents, municipal sewer leaks, inadequate septic system construction, and wastes from dairy and poultry farming that cause high nitrate concentration in groundwater (WHO 2006; Dash et al. 2010, 2016; Adimalla 2020; Ahada and Suthar 2018; Ramalingam

Ch. J. Dash (✉) · P. Yadav · R. Kumar · J. Lenka · B. Raza
ICAR-Indian Institute of Soil and Water Conservation, Research Centre, Sunabeda, Koraput, Odisha 763003, India
e-mail: jyoti.dash@icar.gov.in

P. Sahu · R. H. Ankhila · P. P. Adhikary
ICAR-Indian Institute of Water Management, Rail Vihar, Chandrasekharpur, Bhubaneswar, Odisha 751023, India

P. P. Adhikary et al. (eds.), *Soil, Water Pollution and Mitigation Strategies*,
Environmental Science and Engineering, https://doi.org/10.1007/978-3-031-63296-9_16

et al. 2022). Many experts have documented evidence of excessive groundwater nitrate levels throughout India (Singh and Sekhon 1976; Dash et al. 2010; Gupta et al. 2011; Ahada and Suthar 2018; Adimalla and Li 2018) where Andhra Pradesh is an important hotspot.

According to some studies, nitrate concentrations in groundwater varied spatially in different parts of Andhra Pradesh. For example, in the Srikakulam district it varies from below detection limit to 450.0 mg l^{-1} (Rao 2006), in Krishna delta from 10.0 to 135.0 mg l^{-1} (Mondal et al. 2008), in Nalgonda district from below detection limit to 879.6 mg l^{-1} (Brindha et al. 2012), in Guntur district from 2.0 to 1150 mg l^{-1} (Central Ground Water Board 2013), in Anantapur district from 0.27 to 212.1 mg l^{-1} (Ganesh et al. 2015), in Kadapa district from 2.5 to 760.1 mg l^{-1} (Suvarna et al. 2020) and in Nellore district, nitrate concentration varies from 35.0 to 75.0 mg l^{-1} (Reddy et al. 2022).

The maximum amount of nitrate that can be present in drinking water is 45 mg l^{-1} (WHO 2017). In Yerraguntala mandal of Andhra Pradesh, it was revealed that roughly 45% of groundwater sample tests were above the allowable limit of nitrate concentration (Suvarna et al. 2020). Similarly, 39% of samples from the Krishna delta region's drinking water indicated excessive nitrate concentrations that were beyond the allowable limits (Mondal et al. 2008). Hence, the presence of nitrate above the allowed levels is dangerous to human health (Chettri and Smith 1995; Canter 1997).

As nitrate is transformed into nitrite in the human body via a chemical reaction with water, nitrate becomes an issue. Methemoglobinemia is a result of methemoglobin formation in red blood cells as a result of nitrite acting as the oxidising agent (Blue baby syndrome). Moreover, drinking high-nitrate concentrated water causes nausea, vomiting, sickness, dizziness, and weakness and may result in gastrointestinal cancer (Sunitha 2013).

Among all the districts of Andhra Pradesh, Guntur district is highly affected by groundwater nitrate pollution (Central Ground Water Board 2013). Because of this, the current study concentrates on the spatial distribution of nitrate content in groundwater over the Guntur area and uses hazard quotient to evaluate any potential health risks associated with consuming water with a high nitrate concentration.

19.2 Materials and Methods

The study area is the Guntur district of Andhra Pradesh, India. Its coordinates are 15°18′ and 16°50′ in the North and 79°10′ and 80°55′ in the East (Fig. 19.1). The district's population density is 432 people per km^2, with a total geographic area of 11,328 km^2 (Census 2011). Physically, the research area is separated into three groups: plains, coastal regions, and undulating hills. Its elevation ranges from 0 to 606 m above mean sea level (Fig. 19.2a). Tropical with hot summers and mild winters best describes the study area's climate. In the summer, the typical daily high and low temperatures are 48.5 °C and 16.8 °C, respectively (Central Ground Water Board, 2013). The district receives roughly 890 mm of rainfall annually, with the south-west and north-east monsoons contributing 59% and 26%, respectively (Sudharani et al. 2021). Black cotton, red loamy, and sandy loam soils are the main types of soil present in the research region. Cropland is the primary land use, followed by both sparse and dense vegetation (Fig. 19.2b). The district's net sown area is 637,671 ha. The main crop, paddy, occupies 38% of the net cultivated area. Paddy is followed by cotton (19.6%), maize (9.2%), and chilies (7.4%). Canals irrigate around 0.32 million ha of land, and tube/bore/dug wells irrigate about 76,828 ha of land (Central Ground Water Board 2013). The area has been endowed with a generous 1.72 lakh ha groundwater resource. In the research area, agriculture and domestic usage of groundwater predominate. The archaean crystalline formation is the main source of water, and there is little primary porosity (Central Ground Water Board 2013). The development of secondary porosity is mostly caused by fracturing and subsequent weathering over long periods of time, which becomes water bearing at hydrogeologically favourable places. The fractured aquifer's transmissivity and storativity values ranged from 21 to 253 m^2 day^{-1} and 0.75×10^{-5} to $3.9.5 \times 10^{-3}$, respectively.

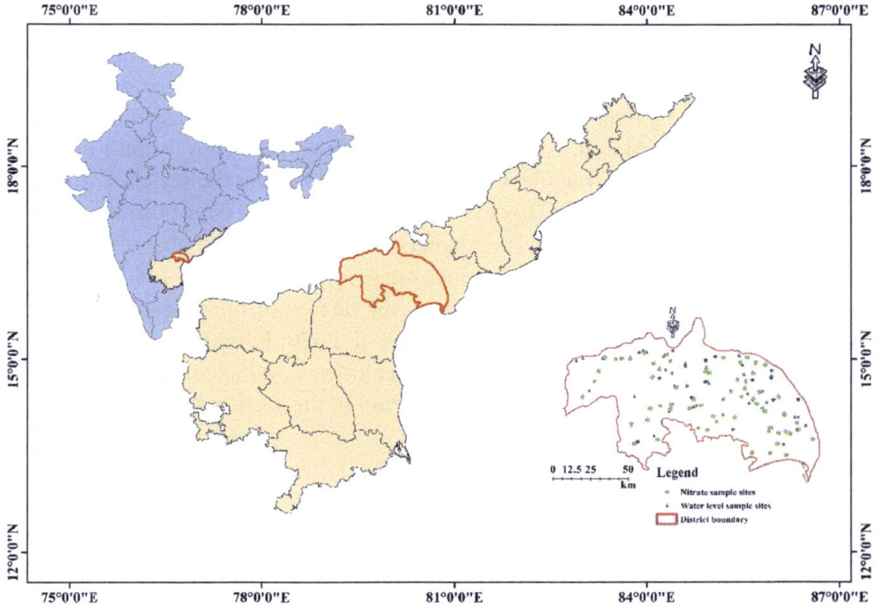

Fig. 19.1 Map of the study area showing nitrate and groundwater level sampling sites

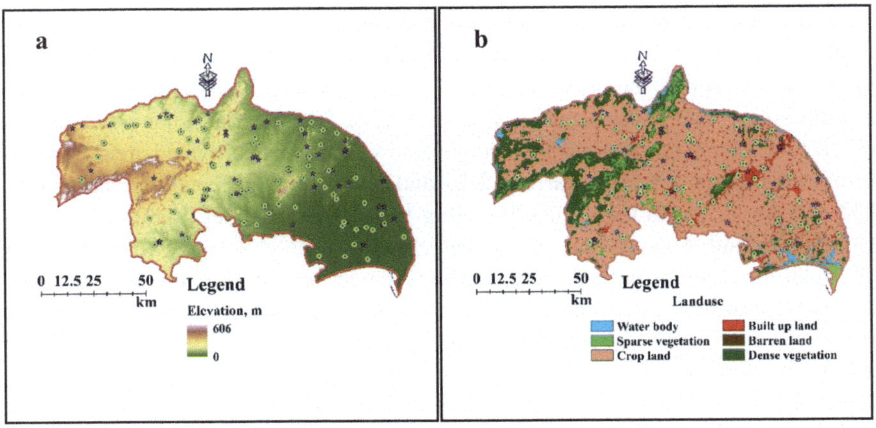

Fig. 19.2 Elevation (**a**) and land use (**b**) map of study area

19.2.1 Data and Methodology

The Central Ground Water Board (CGWB) provided the information for this study for the year 2018 (https://indiawris.gov.in/wris/#/groundwater). There were 96 and 67 data points for the groundwater depth and nitrate concentration, respectively. Both pre-monsoon (month–March, April and May) and post-monsoon (month-October, November and December) water level data were used to assess the seasonal variation of groundwater in the study area. However, groundwater nitrate concentration data are a single-time observation. Figure 19.1 depicts the position of the observation wells within the study region. Based on an investigation of the nitrate pollution index (NPI), as reported by numerous researchers, the degree of nitrate pollution in the groundwater was determined (Panneerselvam et al. 2020; Ramalingam et al. 2022). Similar to this, the effect of nitrate on human health was evaluated by estimating the hazard quotient, which is discussed in more detail in the section below.

(a) **Nitrate Pollution Index (NPI)**

This index was developed to assess the degree of groundwater pollution brought on by nitrate. The nitrate pollution index is calculated using the following equation (Ramalingam et al. 2022).

$$NPI = \frac{C_{ng} - C_{tng}}{C_{tng}}. \tag{19.1}$$

where C_{ng} is the concentration of nitrate in groundwater sample, and C_{tng} is the maximum permissible concentration of nitrate in grounder, and is considered as 20 mg l^{-1}.

(b) **Hazard Quotient due to nitrate** (HQ_{n})

Table 19.1 Human health risk parameter value for different group of people

Parameter	Unit	Male	Female	Children	Infants
Ingestion rate (IR)	l/day	2.0	2.0	0.78	0.30
Exposure duration (ED)	yr	40	40	12	1
Exposure frequency (EF)	days	365	365	365	365
Average body weight (ABW)	kg	78	65	20	10
Average exposure time (AET)	days	14,600	14,600	4380	365

The assessment of health risk in human due to the excess concentration of nitrate in the groundwater wasanalyzed through estimation of hazard quotient ($HQ_{\rm n}$) reported by Suvarna et al. (2020) and Ramalingam et al. (2022).

$$HQ_n = \frac{CDI}{Rfd}. \tag{19.2}$$

where CDI is the chronic daily intake (mg kg^{-1} day^{-1}), and Rfd is the reference dose of nitrate contamination. The value of Rfd is considered as 1.6 mg kg^{-1} day^{-1} (USEPA 2012). The $HQ_{\rm n}$ value greater than 1 is referred as significant risk, whereas less than 1 is considered as safe on human health due to nitrate concentration in the groundwater.

The chronic daily intake is calculated based on different health risk parameters as reported by (Suvarna et al. 2020; Ramalingam et al. 2022)

$$CDI = \frac{C_{ng} \times IR \times ED \times EF}{ABW \times AET}. \tag{19.3}$$

where IR is ingestion rate (l day^{-1}), ED is exposure duration (yr), EF is exposure frequency (day yr^{-1}), ABW is average body weight of human (kg) and AET denotes average exposure time (day).The human health risk parameter value for different group of people such as infants, children, female and male is presented in Table 19.1.

19.2.2 Mapping of Spatial Variation

Arc GIS Software 10.0 was used to map the spatial variation of groundwater depth, nitrate concentration in groundwater, NPI, and hazard quotient owing to nitrate over the study area (Environmental Systems Research Institute 2010). Due to its simplicity and precise prediction, ordinary kriging is one of the most often used geostatistical interpolation techniques in spatial investigation of groundwater depth and water quality indicators (Adhikary et al. 2010; Dash et al. 2010, 2023; Narjary et al. 2017; Shahraki et al. 2021).

Table 19.2 Descriptive statistics of pre-monsoon, post-monsoon ground water level and nitrate concentration in study area

Parameter	Water level (m)		Nitrate concentration (mg l^{-1})
	Pre-monsoon	Post-monsoon	
No. of samples	96	96	67
Minimum	0.2	0.2	1.2
Maximum	39.5	22.4	665.0
Mean	4.8	3.0	113.4
SD	4.2	2.3	98.4
CV (%)	86.7	76.6	86.8
SEM	0.4	0.2	12.0

SEM standard error of mean, *SD* standard deviation, *CV* coefficient of variation

19.3 Results and Discussion

19.3.1 Descriptive Statistics

In order to make a preliminary conclusion on the research area's nitrate concentration and groundwater level, a descriptive statistical analysis was conducted. Pre-monsoon groundwater level ranged from 0.2 to 39.5 m, with a mean of 4.8 m for the Guntur district (Table 19.2). Pre-monsoon groundwater level's coefficient of variation (CV) was found to be 86.7%. Pre-monsoon observations revealed that the south of the district's groundwater depth exceeded 10 m (Fig. 19.3a). During the post-monsoon, the mean groundwater level for the entire study area was 3.0 m. Just a small pocket of groundwater had a level more than 10 m during the post-monsoon, indicating groundwater recharging from the south-west monsoon, which is further supported by the finding that the groundwater level increased by 1.8 m throughout the whole study region (Fig. 19.3b). In the Guntur district, an average of 113.4 mg l^{-1} of nitrate was found in the groundwater. In the study area, the co-efficient of variation for nitrate level in groundwater exceeded 80%.

19.3.2 Spatial Variation of Nitrate Concentration and NPI

Figure 19.4 shows the spatial variation in nitrate content in the groundwater of the Guntur district as well as the groundwater sample's nitrate pollution index. The World Health Organization's recommended acceptable limit of 45 mg l^{-1} was exceeded in around 60% of the sample locations (WHO 2017). In shallow aquifers, 75% of the samples had nitrate concentrations that were over acceptable limits, compared to 32% in deeper aquifers, according to CGWB (2013). Drinkable groundwater is present in the eastern, south-eastern, south, and north-west parts of the district; however, it is

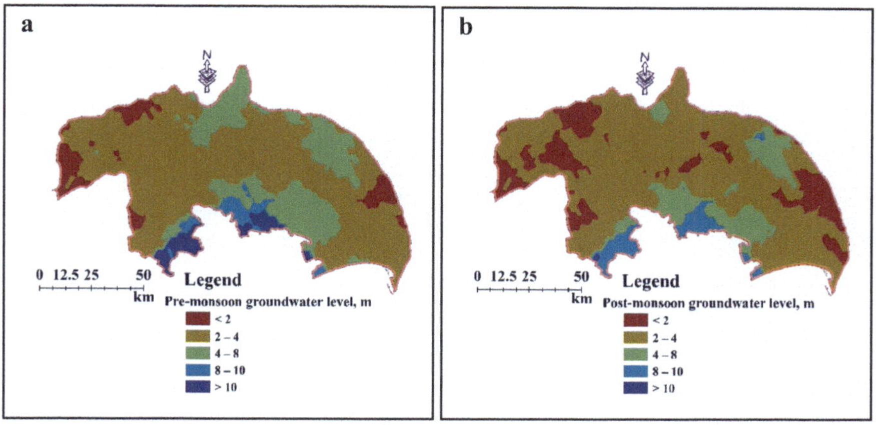

Fig. 19.3 Groundwater level in the study area during pre-monsoon (**a**) and post-monsoon (**b**)

above the acceptable level in the rest of the study area. Only one-third of the district's entire geographical area is appropriate for drinking, according to a spatial analysis of nitrate concentration, while the remaining two-thirds of the region are not good for drinking (Fig. 19.4a).

The NPI classification of groundwater samples for the study area were calculated and presented in Table 19.3. According to the NPI classification of groundwater, the district's 26.9% samples were clean, 10.5% samples had light pollution, 10.4% samples showed moderate pollution, 19.4% samples showed major pollution, and 42.8% samples showed very serious pollution levels. The spatial analysis's findings indicated that the most heavily polluted area was spread over 5701 km^2, followed by considerable pollution (2015 km^2) and light pollution (1932 km^2) (Fig. 19.4b).

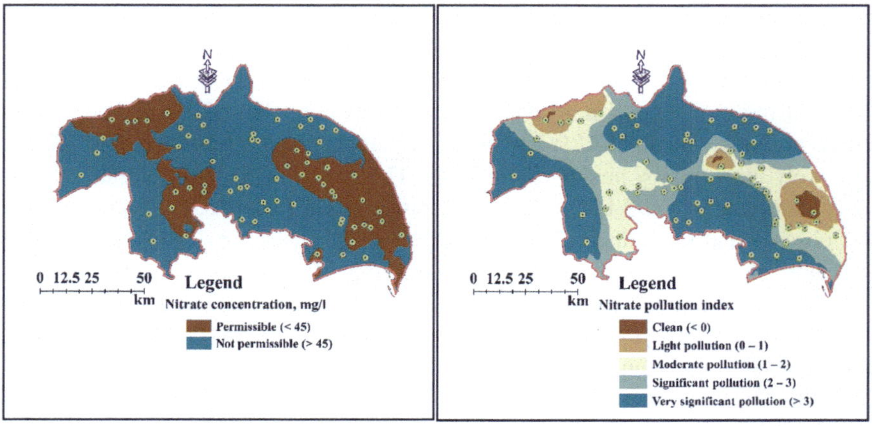

Fig. 19.4 Map showing nitrate concentration in groundwater (**a**) and nitrate pollution index classification (**b**) in the study area

Table 19.3 NPI classification of groundwater sample in the study area

Range	Class of water	No. of samples	Percentage of samples	Area (km^2)	Area (%)
< 0	Clean	18	26.9	817	7.2
0–1	Light pollution	7	10.5	1932	17.1
1–2	Moderate pollution	7	10.4	863	7.6
2–3	Significant pollution	13	19.4	2015	17.8
> 3	Very significant pollution	22	42.8	5701	50.3

In terms of nitrate pollution, only an area of 817 km^2 has safe groundwater. It shows that the concentration and leaching rate of nitrate into groundwater are increased by diffuse sources of nitrate, such as excessive use of synthetic fertilizers for crop production and incorrect management of nitrogen sources in agriculture (Obeidat et al. 2012; Central Ground Water Board 2013; Panneerselvam et al. 2022).

19.3.3 Human Health Risk

According to the United States Environmental Protection Agency (2012), nitrate concentrations above the permissible level of 45 mg l^{-1} can result in major health problems for people, including thyroid disease, colon cancer, blue baby syndrome, and birth defects in newborns. Male, female, children, and infants are the four categories into which the human population has been divided in the current study. The area under the safe or risk zone for each category of humans was assessed based on the nitrate hazard quotient for each group. The study's findings showed that for male, female, children, and infants, respectively, the chronic daily intake (CDI) ranged from 0.03 to 17.05, 0.04 to 20.46, 0.05 to 25.94, and 0.04 to 19.95. It was revealed that 49.3% of the samples can negatively impact on male health, 56.7% can negatively impact on female health, 62.7% can negatively impact on kid health, and 55.7% can negatively impact baby health (Table 19.4). The spatial study showed that groundwater is present in 7675 km^2 of area that poses a risk for male, 3385 km^2 of area that poses a risk for female, 2758 km^2 of area that poses a risk for children, and 3441 km^2 of area that poses a risk for infants (Fig. 19.5; Table 19.4). Around 94% and 100% of the adult and juvenile populations, respectively, are at danger in the Malwa region of Punjab, India, according to Ahada and Suthar's (2018) research. Similar results were found by Suvarna et al. (2020) in Kadapa district of Andhra Pradesh, who found that 35% of samples from a group of male were at danger and 42.5% of samples from groups of babies, children, and females were at risk. In the Salem district of Tamilnadu, 100% of the groundwater samples could have a negative impact on children's and infants' health, while 90%, 57.50%, 55.05 and 57.5% of the

Table 19.4 No. of samples and area under nitrate hazard quotient for male, female, children and infants in the study area

Human	No. of samples		Percentage of samples		Area (km^2)		Area (%)	
	HQ < 1	HQ > 1	HQ < 1	HQ > 1	HQ < 1	HQ > 1	HQ < 1	HQ > 1
Male	34	33	50.7	49.3	3653	7675	32.2	67.8
Female	29	38	43.3	56.7	3385	7943	29.9	70.1
Children	25	42	37.3	62.7	2758	8570	24.3	75.7
Infants	19	38	43.3	56.7	3441	7887	30.4	69.6

samples could have a negative impact on people's health for 10–15, 15–20, 20–60, and more than 60 years, respectively (Ramalingam et al. 2022).

19.3.4 Possible Impact of High Nitrate Concentred Water on Human Health

Figure 19.6 shows the source of nitrate in groundwater, its effects on human health, and potential corrective actions. According to Gupta et al. (2001), both adults and children under the age of six are more likely to develop respiratory tract infections, recurrent diarrhoea, and recurrent stomatitis when nitrate levels are high (Ahada and Suthar 2018). Blue baby syndrome, thyroid disease, and gastrointestinal cancer are just a few of the serious health implications that prolonged exposure to nitrate through drinking water may have on people (Sunitha 2013).

19.4 Conclusions

In Guntur district, Andhra Pradesh, India, the current study was conducted to determine the spatial variation in groundwater depth and nitrate concentration as well as the effects of nitrate on human health across different categories of individuals including male, female, children, and infants. The average groundwater level before the monsoon was 4.8 m, but it was 3.0 m after the monsoon, showing that the south-west monsoon had recharged it. Almost 60% of groundwater samples with nitrate concentrations over the allowable limits of 45 mg l^{-1} were found. According to the NPI classification of groundwater, the district's 10.4% of sample locations have moderate pollution, 19.4% have considerable pollution, and 42.8% have very serious contamination. The spatial analysis of NPI revealed that 5701 km^2 of land were subject to very major pollution, while 2015 km^2 were subject to significant pollution. Comparable risk assessments for human health based on the nitrate hazard quotient showed that 49.3, 56.7, 62.7, and 55.7% of the samples, respectively, can

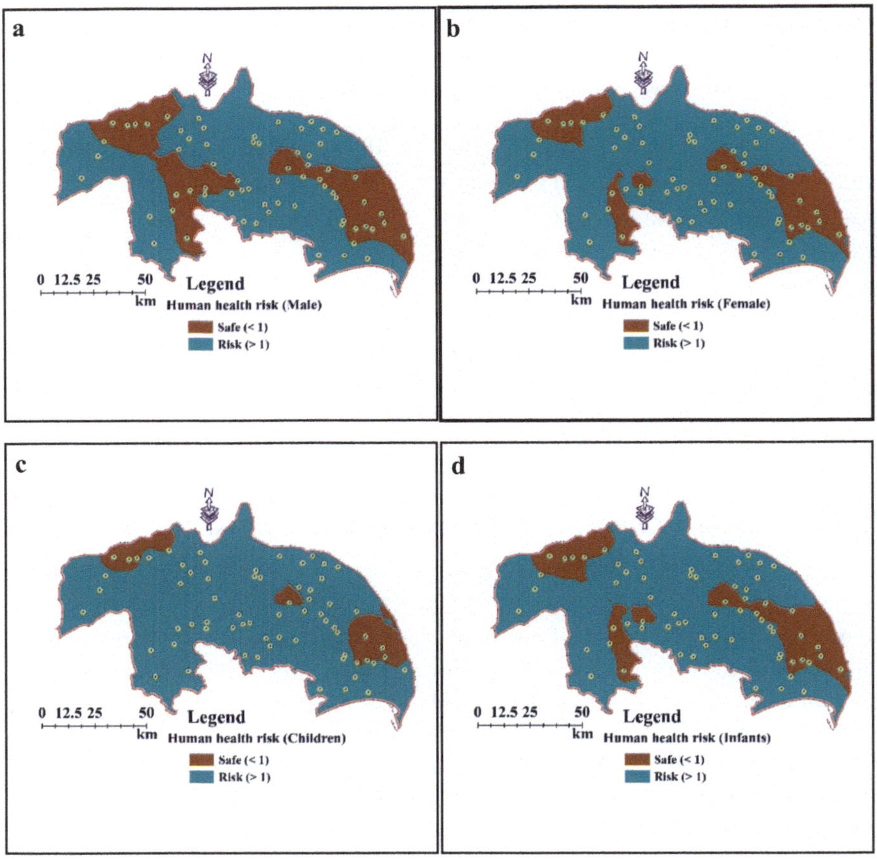

Fig. 19.5 Map showing human health risk (**a** male, **b**: female, **c** children and **d** infants) due to nitrate concentration in groundwater

have an adverse influence on the health of male, female, children, and infants. Due to consumption of nitrate-concentrated water, there is a risk of methemoglobinemia, gastrointestinal cancer, cardiovascular hypertrophy, heart attacks, heart disorders, myocardial infraction, and lipid peroxidation in the retina of the peoples residing in the study area.

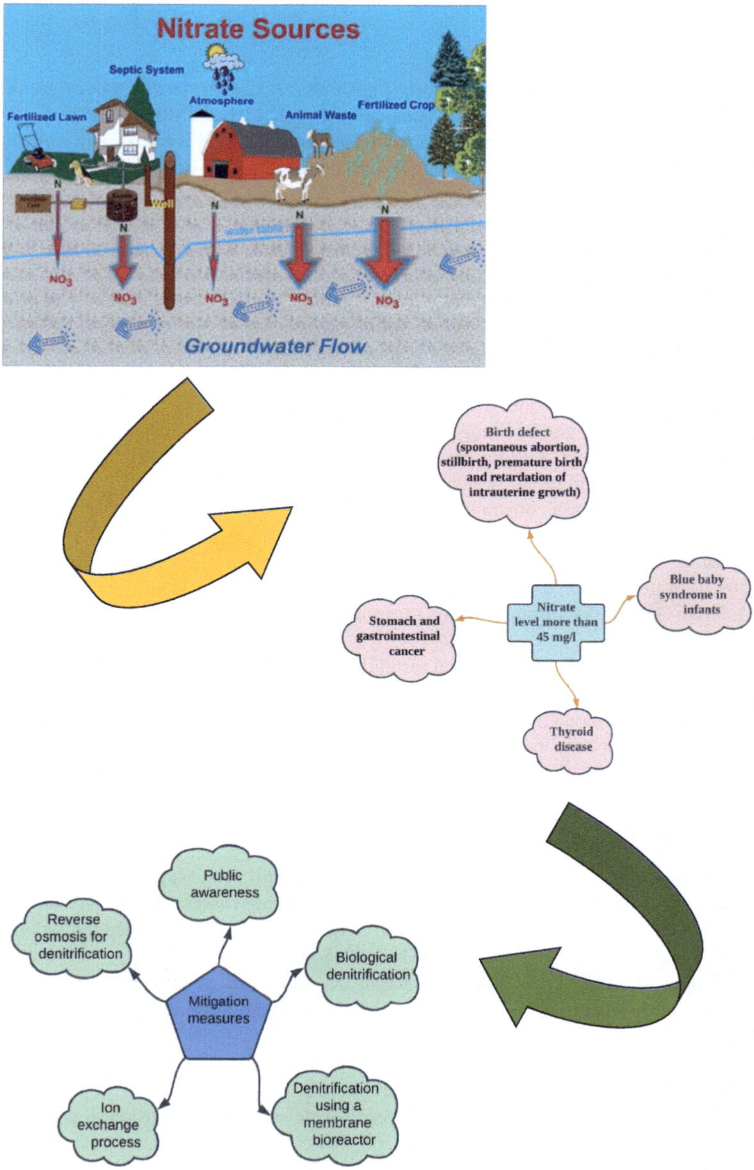

Fig. 19.6 Source, health impact on human, and remedial measures of nitrate pollution in groundwater

References

Adhikary PP, Chandrasekharan H, Chakraborty D, Kamble K (2010) Assessment of groundwater pollution in West Delhi, India using geostatistical approach. Environ Monit Assess 167:599–615

Adimalla N (2020) Spatial distribution, exposure, and potential health risk assessment from nitrate in drinking water from the semi-arid region of South India. Hum Ecol Risk Assess Int J 26(2):310–334

Adimalla N, Li P (2018) Occurrence, health risks, and geochemical mechanisms of fluoride and nitrate in groundwater of the rock-dominant semi-arid region, Telangana State, India. Hum Ecol Risk Assess Int J 25(1–2):81–103

Ahada CPS, Suthar S (2018) Groundwater nitrate contamination and associated human health risk assessment in southern districts of Punjab, India. Environ Sci Pollut Res 25(25):25336–25347

Balamurugana P, Kumar PS, Shankar K, Kumar PJS (2020) Impact of climate and anthropogenic activities on groundwater quality for domestic and irrigation purposes in Attur region, Tamilnadu, India. Desalin Water Treat 208:172–195

Brindha K, Rajesh R, Murugan R, Elango L (2012) Nitrate pollution in groundwater in some rural areas of Nalgonda district, Andhra Pradesh, India. J Environ Sci Eng 54(1):64–70

Canter LW (1997) Nitrate in groundwater. Lewish Publishers, New York

Census (2011) Primary census abstracts, Registrar General of India, Ministry of Home Affairs. Government of India, New Delhi

Central Ground Water Board (2013) Groundwater brochure, Guntur district, Andhra Pradesh, Ministry of Water Resources, Government of India

Chettri M, Smith GD (1995) Nitrate pollution in groundwater in selected district of Nepal. Hydrogeol J 3:71–76

Dash JP, Sarangi A, Singh DK (2010) Spatial variability of groundwater depth and quality parameters in the National Capital Territory of Delhi. Environ Manage 45(3):640–650

Dash JP, Sarangi A, Adhikary PP, Singh DK (2016) Simulation of nitrate leaching under maize–wheat cropping system in a semiarid irrigated area of the Indo-Gangetic Plain, India. J Irrig Drain Eng 142(2):04015053

Dash ChJ, Adhikary PP, Lenka J (2023). Assessment of groundwater salinity risk in coastal belt of Odisha using ordinary Kriging and its management. https://doi.org/10.1016/B978-0-323-99963-2.00017-1

Environmental Systems Research Institute (ESRI) (2010) ArcGIS ArcMap Version 10.0. Redlands

Ganesh KM, Suryanarayana Gand Janardhana C (2015) GIS aided assessment of physico-chemical properties of the groundwater from the villages of Puttaparthi mandal, Anantapur district, Andhra Pradesh (India). Int J Adv Res 3(Apr). ISSN 2320-5407

Gupta SK, Gupta RC, Gupta AB, Seth AK, Bassin JK, Gupta A, Sharma ML (2001) Recurrent diarrhea in children living in areas with high levels of nitrate in drinking water. Arch Environ Health 56(4):369–373

Gupta I, Salunkhe A, Rohra N, Kumar R (2011) Groundwater quality in Maharashtra, India: focus on nitrate pollution. J Environ Sci Eng 53(4):453–462

Mondal NC, Saxena VK, Singh VS (2008) Occurrence of elevated nitrate in groundwaters of Krishna delta, India. Afr J Environ Sci Technol 2(9):265–271

Narjary B, Meena ML, Pathan AI, Kumar N, Kumar S, Kamra SK, Sharma DK (2017) Assessing spatio-temporal variations in groundwater table depth and salinity using geostatistics and its relation with groundwater balance in a salt affected soil. Indian J Soil Conserv 45(3):235–243

Obeidat MM, Awawdeh M, Al-Rub FA, and Al-Ajlouni A (2012) An innovative nitrate pollution index and multivariate statistical investigations of groundwater chemical quality of Umm Rijam Aquifer (B4), North Yarmouk River basin, Jordan. Vouddouris K, Voutsa D. Water Quality Monitoring and Assessment. Croatia: In Tech, pp 169–188

Panneerselvam B, Karuppannan S, Muniraj K (2020) Evaluation of drinking and irrigation suitability of groundwater with special emphasizing the health risk posed by nitrate contamination using nitrate pollution index (NPI) and human health risk assessment (HHRA). Hum Ecol Risk Assess Int J 27(5):1324–1348

Panneerselvam B, Muniraj K, Pande C, Ravichandran N, Thomas M, Karuppannan S (2022) Geochemical evaluation and human health risk assessment of nitrate-contaminated groundwater in an industrial area of South India. Environ Sci Pollut Res 29:86202–86219

Raju NJ, Shukla UK, Ram P (2011) Hydrogeochemistry for the assessment of groundwater quality in Varanasi: a fast-urbanizing center in Uttar Pradesh, India. Environ Monit Assess 173:279–300

Ramana RV, Rao PVN (2017) Contamination of groundwater in Jadavalli village in Guntur district of A.P. VFSTR J Stem 3(1): 2455–2062

Ramalingam S, Panneerselvam B, Kaliappan SP (2022) Effect of high nitrate contamination of groundwater on human health and water quality index in semi-arid region, South India. Arab J Geosci 15:242. https://doi.org/10.1007/s12517-022-09553-x

Rao N (2006) Nitrate pollution and its distribution in the groundwater of Srikakulam district, Andhra Pradesh, India. Environ Geol 51:631–645. https://doi.org/10.1007/s00254-006-0358-2

Reddy AGS (2013) Geochemical evaluation of nitrate and fluoride contamination in varied hydrogeological environs of Prakasam district, southern India. Environ Earth Sci 71(10):4473–4495

Reddy CKVC, Golla V, Badapalli PK, Reddy NBY (2022) Evaluation of groundwater contamination for fluoride and nitrate in Nellore urban province, Southern India: a special emphasis on human health risk assessment (HHRA). Appl Water Sci 12:32. https://doi.org/10.1007/s13201-021-01537-8

Shahraki AS, Nasab SB, Naseri AA, Mohammadi AS (2021) Evaluation of interpolation techniques for estimating groundwater level and groundwater salinity in the Salman Farsi Sugarcane Plantation. Irrig Sci Eng 44(2):67–78

Singh B, Sekhon GS (1976) Nitrate pollution of groundwater from nitrogen fertilizers and animal wastes in the Punjab, India. Agric Environ 3(1):57–67

Sudharani Y, Hema K, Mrudhula A, SambaiahA and Neelima T(2021). Assessment of quality of groundwater for irrigation in Guntur District, Andhra Pradesh. J Soil Salinity Water Qual 13(1):9–19

Sunitha V (2013) Nitrates in groundwater: health hazards and remedial measures. Indian J Adv Chem Sci 1(3):164–170

Suvarna B, Sunitha V, Reddy SY, Reddy NR (2020) Data health risk assessment of nitrate contamination in groundwater of rural region in the Yerraguntla Mandal, South India. Data Brief 30:105374. https://doi.org/10.1016/j.dib.2020.105374

USEPA (2012) Integrated risk information system. Environmental Protection Agency, United States (2012). http://cfpub.epa.gov/ncea/iris/index.cfm?fuseactioniris.showSubstanceList. Accessed 1 Feb 2023

WHO (2006) Guidelines for drinking water quality. First addendum to third edition. World Health Organization, Geneva, Switzerland

WHO (2017) Guidelines for drinking water quality, 4th edition incorporating the first addendum. World Health Organization, Geneva, Switzerland

Chapter 20
India's Current Situation with Regard to the Effects of Water Pollution on Agricultural Productivity and Public Health

Jiban Kumar Behera, Anway Kumar Jena, Manojit Bhattacharya, and Bhaskar Behera

Keywords Water pollution · Pollutant · India · Public health · Agriculture

20.1 Introduction

The most basic forms of life on earth depend on water just as much as the most sophisticated human systems. In many regions of the world, a lack of fresh water for human consumption, industrial and agricultural usage, and other crucial purposes is a limiting factor, if not the most significant one, slowing progress (Yang et al. 2020). Growing water scarcity and deteriorating water quality due to pollution in South Asia over the last several decades have raised concerns about the region's capacity to provide enough water for the needs of the twenty-first century. The growing population and shifting water consumption habits in South Asia are anticipated to lead to a precipitous drop in available water per person. While a yearly and per-capita consumption of 1000 m^3 of water is considered to be a benchmark for 'well-being' in the developed world, South Asia's annual water availability per capita is only projected to be 730 m^3 by the year 2025, and is on a declining trend in all regions of the world, including those that are thought to have ample water resources (Perveen 2023). As the climate changes, more stress will be put on South

J. K. Behera · A. K. Jena · M. Bhattacharya (✉)
Department of Zoology, Fakir Mohan University, VyasaVihar, Balasore, Odisha 756020, India
e-mail: mbhattacharya09@gmail.com

B. Behera
Department of Biosciences and Biotechnology, Fakir Mohan University, VyasaVihar, Balasore, Odisha 756020, India

P. P. Adhikary et al. (eds.), *Soil, Water Pollution and Mitigation Strategies*, Environmental Science and Engineering, https://doi.org/10.1007/978-3-031-63296-9_17

Asia's water supply. Demand, availability, and water quality will all be impacted by climate change. When there is a scarcity of water, economic, social, and environmental uses all become more competitive in water-stressed areas. With a larger population comes a greater need for irrigation and maybe industrialization, both of which might come at the price of fresh water for human use. When the environment is deteriorated by pollution and climate change, conflicts over water supplies may become a major societal consequence (Celik 2020). As compared to high-income nations, the levels of pollution in poor and middle-income countries are frequently many orders of magnitude worse (Fig. 20.1a) (Champion et al. 2022). If the costs to health, production, and the environment are nonlinear, with marginal costs growing with pollution levels, then they might be considerably larger than what would be predicted by simple linear extrapolation. It is difficult, expensive, and even impossible to purify water after it has been tainted. Even in the present day, almost 80% of the world's wastewater, including both human waste and extremely hazardous industrial (Fig. 20.1b) and municipality discharges (Fig. 20.1c), is discharged without being treated (Goswami et al. 2021). The safety of freshwater for human consumption, hygiene, and farming depends on the types and concentrations of contaminants present. The habitat and quality of life of fish and other species in freshwater habitats may also be negatively impacted by pollution (Malik et al. 2020). Pathogens, organic matter, chemical pollution, and salt are all potential sources of pollution in freshwater environments. Moreover, the amount to which these contaminants are present in our waters and the effects they have on aquatic ecosystems remain largely unclear. The lack of information on the health of the world's rivers, lakes, and groundwater puts almost 3 billion people in danger of illness (Panhwar et al. 2022). A recent study by the United Nations Environment Programme (UNEP) and partners found that 2.3 billion people live in countries classified as "water-stressed," including 721 million in areas where the water situation is "critical," while a fifth of the world's river basins are experiencing dramatic fluctuations in water availability (UNEP 2021). Climate change, biodiversity loss, pollution, and waste are the "three crises" that our world is now experiencing. UNEP has said that the world's oceans, rivers, seas, and lakes are suffering greatly due to these challenges (Uitto 2021). For better sustainable management of our water resources and universal access to clean water, "frequent, thorough, and up-to-date data collection is necessary. Around 40% of the over 75,000 bodies of water studied across 89 nations had very high levels of pollution, according to several study (Kanakaraju and Chandrasekaran 2023). According to data given at a high-level United Nations conference on water-related objectives of the 2030 agenda on March 18th, the world is slipping behind in its effort to deliver clean water to everyone. Sustainable Development Goal (SDG) 6 calls for "the availability and sustainable management of water and sanitation for all," but UNEP data shows that the world is not on track to achieve this goal by 2030. If this goal is to be achieved, then efforts will need to be doubled over the next nine years. The majority of studies examining the monetary implications of pollution have been conducted in wealthy nations, providing few grounds for extrapolating those results to developing world contexts. Compared to air pollution, water pollution has gotten far less attention from scientists and the general public (Plaas and Paerl 2020). Although public pressure has led to some improvements in air quality as a result of legislation in India, equivalent improvements in water quality have not been seen despite similar restrictions. New Delhi and Bangalore now see yearly blooms of toxic white foam on their waterways, leading to widespread fish mortality. It has proven difficult to put a price on the social costs of water contamination, even in nations with a high per capita income. While polls suggest that people care a lot about water quality, studies seldom find economically significant effects of water contamination. This might be due to the low expenses involved or to the fact that water contamination is hard to analyze. Some factors that have stymied research and dampened existing estimates include the poor quality and availability of pollution data, the challenge of modeling complicated geographical interactions, and the huge diversity of different chemicals (Naylor et al. 2023). It is estimated in this chapter how much of an influence water pollution has on India's economy, population health, and agricultural output. We elaborate on agriculture since evidence suggests it may be particularly vulnerable to the cumulative impacts of water contamination. Irrigation water is seldom treated before use, unlike drinking water, even though agriculture consumes 4 times much water as the rest of the economy combined. Because of its size and prevalence, the agricultural industry may be found near almost any kind of pollution (Banerjee et al. 2021). Water pollution is a concern anywhere; it is probably in India, where some of the world's largest industrial clusters are located. Indian major rivers are Ganga, Yamuna, Mahanadi and Godavari water quality change in large cities because of industrial and swage pollution (Table 20.1). Unlike air pollution, which may travel in any direction from its source, water pollution typically always travels in the same direction. A spatial discontinuity in pollution concentrations along a river is produced when wastewater resources are discharged into the river's flow. Pollution levels will be greater downstream from a strongly polluting industrial site than upstream, but other factors will generally be comparable. As industrial effluents may comprise hundreds of unique components and compounds, summarizing them is particularly challenging. Any of these might provide problems for people, crops, or ecosystems on their own. To estimate agricultural production, people are used indicators such as plant density, plant health, and metabolic activity to assess the impact of water contamination. These vegetation indicators have been shown to accurately forecast crop yields in a variety of environments. To avoid losses in production, farmers may have to spend a lot of money. Damage to ecosystems and human health are only two examples of the many possible forms of societal costs. Farmers and farm workers who are exposed to contaminated irrigation water may suffer health consequences (Alam and Khatun 2021). Heavy metals and other contaminants may be absorbed by produce, causing damage to consumers even if harvests remain unharmed. In this chapter, we provide evidence for three particular pollution costs but it begins by examining how many its costs due to water contamination. Domestic water pollution in the context of potable water is the subject of a substantial body of literature, while other publications focus on the consequences of water pollution in general or on agricultural runoff. Cancer rates in China and infant mortality rates in India are both affected by industrial water

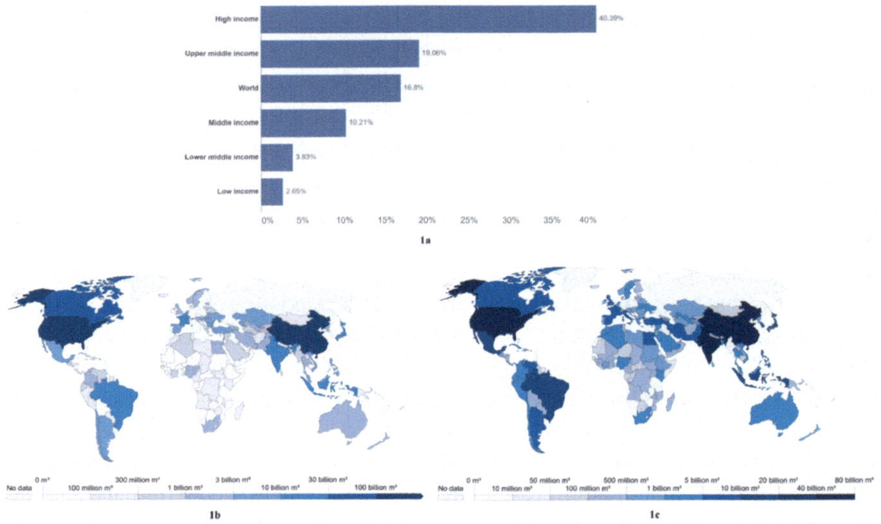

Fig. 20.1 **a** The schematic diagram represent as a percentage of industrial water used in various economic countries (source from FAO and Our World in Data). **b** Diagram presented as removal of Industrial water in worldwide (source from FAO and Our World in Data). **c** The figure indicated that removal of municipality water from various sources in different country (source from FAO and Our World in Data)

pollution (Sarkar and Devi 2022). The second focus of this chapter is on the consequences of pollution on farming. Water pollution has been studied extensively in the context of agriculture, but there are physiological indications that it may also have negative effects on crops (Bueno and Lopes 2020). Finally, this article adds to the pollution problem and understanding the connection between industrial and agricultural productivity, food production, food security, and the repercussions of incorrect water utilization in both low- and high-income nations is another area to which this study contributes. The mechanisms of maintaining public and agricultural health are also made clearer in this chapter. Lastly, this article contributes to our understanding of water contamination by expanding our toolkit of spatial computing techniques. In last, we focus the most serious issues and steps required to restore the freshwater balance to our natural systems are discussed. Moreover, it stresses the need for a global and local perspective while managing public health.

20.2 Water Pollution in India

India's present population of one billion might reach 1.7 billion by the year 2050 if the country's fertility rate of 2.1% remains unchanged (Ram and Ram 2021). India's surface and groundwater resources are vital to the country's agricultural, fishing, animal-rearing, forestry, and industrial sectors. India's agriculture and water

Table 20.1 Various water quality assessment of important river at major cities in India

State name	River	Temperature (°C)		Dissolved oxygen (mg/L)		pH		Conductivity (μmhos/cm)	
		MIN	MAX	MIN	MAX	MIN	MAX	MIN	MAX
				> 5.0		6.5–8.5			
Uttar Pradesh	Ganga	16.0	23.0	7.2	10.4	6.2	7.5	175	217
Bihar	Ganga	14.0	32.0	5.2	11.7	7.5	8.4	229	591
West Bengal	Ganga	19.0	34.0	3.5	9.9	7.3	8.4	210	443
Himachal Pradesh	Yamuna	14.0	25.0	7.4	9.4	7.3	8.4	144	385
Haryana	Yamuna	23.0	34.0	0.3	8.3	7.2	8.2	218	1576
Delhi	Yamuna	17.0	28.0	0.3	2.2	7.4	8.0	1225	1851
Uttar Pradesh	Yamuna	24.0	32.0	0.3	0.3	7.4	7.6	406	2060
Odisha	Mahanadi	24.0	29.0	7.4	7.8	7.9	8.2	214	224
Telangana	Godavari	18.0	28.0	6.0	6.8	6.8	7.5	566	780
Andhra Pradesh	Godavari	25.0	31.0	6.1	7.5	6.8	7.9	216	415
Andhra Pradesh	Godavari	24.0	27.0	6.9	8.8	6.9	8.2	110	178
Karnataka	Godavari	–	–	6.8	6.8	7.4	7.4	680	680
Maharashtra	Godavari	17.0	25.0	5.0	6.7	7.0	7.7	379	1610

(continued)

Table 20.1 (continued)

Bio-chemical oxygen demand (mg/L)		Nitrate (mg/L)		Fecal coliform (MPN/100ML)		Total coliform (MPN/100ML)		Fecal streptococci (MPN/100ML)	
MIN	MAX	MIN	MAX	MIN	MAX	MIN	MAX	MIN	MAX
< 3.0				< 2500 MPN/100ML				< 500 MPN/100ML	
1.2	2.0	0.32	0.52	200	170,000	1700	1,600,000	23	330
1.0	3.3	0.30	0.80	2700	160,000	2700	160,000	45	9200
2.6	4.7	0.32	0.97	33,000	350,000	49,000	700,000	1100	5400
1.0	2.0	0.42	1.12	8	79,000	23	110,000	49	13,000
2.0	21.0	0.30	4.32	200	280,000	900	920,000	330	94,000
38.0	83.0	0.30	1.50	70,000	22,000,000	3,300,000	22,000,000	330,000	7,000,000
8.0	127.0	0.32	4.32	17,000	2,200,000	33,000	4,900,000	2400	490,000
1.0	1.0	0.32	0.42	700	140,000	4900	2,200,000	–	–
2.1	2.1	3.10	8.16	2	6	17	31	2	4
1.8	2.8	0.57	2.47	11	20	120	210	2	2
1.0	2.2	0.30	2.09	3	4	39	75	2	2
1.8	1.8	7.12	7.12	6	6	26	26	–	–
2.4	6.0	0.30	4.10	4	17	48	140	2	2

industries both greatly changes in the timing, intensity, and path of the monsoon have caused widespread and unpredictable variations in precipitation in recent years (Table 20.1) (Kumar 2023). As agricultural output is largely controlled by the availability of water, the state water resources and the monsoon have a significant impact on the country's agrarian economy. Unsurprisingly, agriculture consumes the bulk of the country's water supply of 85.3% was used for irrigation, 1.2% for industry, 0.3% for electricity generation, 6.4% for other purposes, and 6.6% for residential usage in 1999. When there is an excessive amount of harmful substances in the water supply, we call it water pollution. Ecosystems that support human health, agricultural output, and biodiversity are weakened by polluted water. Each year, polluted water is responsible for the deaths of millions of people. As water quality declines, many businesses and sectors suffer, including agriculture, fishing, and livestock husbandry (Tubb and Seba 2021). Water pollution is a hazard to biodiversity everywhere, but it has a disproportionate impact on freshwater ecosystems. Due to water's critical role in human existence, its contamination poses a serious threat (Abell et al. 2019). Since contaminated water has disastrous impacts on plants and flora, human, and animal, preventing water pollution is crucial to ensuring that enough clean water is available to sustain the growth and development of the human and animal populations. Taking action to prevent pollution to helps guarantee that aquatic organisms have continued access to safe drinking water. Every year, India's 14 major, 55 minor, and several hundred little rivers get millions of gallons of sewage, industrial rubbish, and agricultural waste (Mishra and Dubey 2023). Water contamination in rivers is often caused by sewage from towns and industrial waste. At the moment, we only treat around 10% of wastewater before it is released into the environment. The result is contamination of both surface water and groundwater. Before it reaches our taps, drinking water is frequently already highly contaminated with microorganisms that may cause illness. Agricultural run-off, or water from fields that run into rivers carrying toxic substances like fertilizers and pesticides, is another significant cause of water pollution. Almost 80% of India's household water requirements in rural areas and 50% in urban areas are satisfied by groundwater. It's more resistant to contamination and pollution than surface water (Sathre et al. 2022). Human activities such as farming, manufacturing, mining, garbage disposal, population growth, urbanization, and climate change all have an impact on water quality. Rising salinity and nutrient enrichment are two of the most prevalent water quality challenges worldwide, and they are both largely attributable to agricultural activities.

20.3 Sources of Water Pollution

Water pollution is the term for any human-caused contaminant of water that reduces its value to both humans and other living things in the natural world. As a result of inappropriate land and water use in rural and urban areas, our watersheds' health and water quality are suffering. The contamination of our water supply threatens the

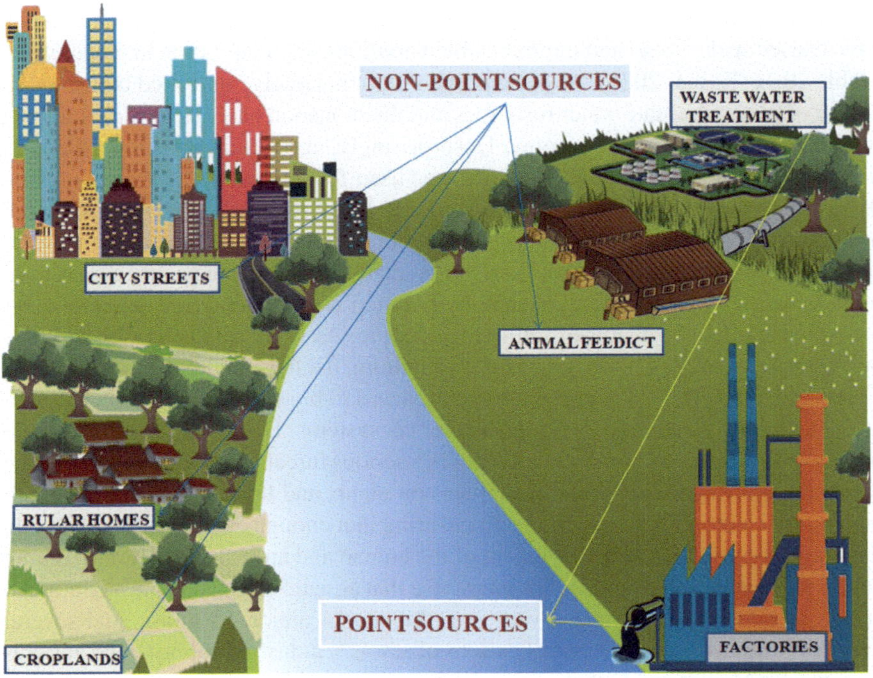

Fig. 20.2 Diagram presented as point and non-point sources of water pollution

health of people, animal, plant, and tree unless it is subjected to an expensive purification procedure (Bouwer 2002). Pesticides, herbicides, fertilizers, and hazardous chemicals are some examples of pollutants that could contaminate our water supply. The effects on the environment can be very bad when hazardous chemicals are accidentally spilled or dumped into waterways (Owa 2013). In most cases, businesses have to clean up the waterways first and get special permits before disposing of waste there. Serious fines and penalties could be incurred for non-compliance. Almost every action humans take—including those related to food production, product manufacturing, and electricity generation—raises the possibility of releasing pollution into the environment. Environmental regulatory organizations claim that there are two main categories of pollution: point-source pollution and nonpoint-source pollution (Fig. 20.2) (Ding and Liu 2019).

20.4 Point-Source

The origin of point contamination may be easily identified. As the name suggests, it has a single point of origin. The detection and prevention of pollution from nonpoint sources are more challenging. Pollution comes from a variety of different places. The

Environmental Protection Agency (EPA) of the United States defines "point source pollution" as the discharge of toxins into the environment from a single, clearly identified site. Examples include smokestacks, waste-disposal pipes, and drainage ditches. Pollution from single sources, such as industries and power plants, may have a widespread influence on the environment (Wato et al. 2020). Some of the pollutants that may be emitted into the air by smokestacks include carbon monoxide, heavy metals, sulfur dioxide, nitrogen dioxide, and "particulate matter". Water-dependent industries are allowed to discharge effluent, or wastewater containing harmful chemical pollutants, into adjacent waterways including rivers, lakes, and the ocean. That includes oil refineries, paper mills, and automotive assembly plants. Wastewater treatment plants in the community are another potential point source of pollution. Effluent dumped into rivers from wastewater treatment facilities may include nutrients and pathogenic microbes. Too many nutrients in the water might encourage algae growth. This is a good example of how a chain reaction may occur when a single event triggers another. Besides carbon monoxide and other gases, heavy metals, sulfur dioxide, nitrogen dioxide, and "particulate matter" may be released by smokestacks. Companies like are oil refineries, paper mills, and automobile manufacturers that use water in their manufacturing processes may spill effluent, or wastewater, into waterways. That wastewater may include harmful chemical pollutants. Point-source contamination is exemplified by urban wastewater treatment plants (Sayyed et al. 2021). Nutrients and infectious organisms may be released into rivers through effluent from treatment facilities. In water, algae may be fed to grow larger. When pollutants enter a water body from a specific location, such as the end of a pipe or an underground injection system, they tend to disperse outward in a plume, with the highest concentrations of the pollutant at the source and progressively lower ones as one move farther from it. Point-source pollutants in streams come from a broad range of commercial, industrial, agricultural, and urban environments. Companies in the commercial and industrial sectors often dispose of potentially dangerous byproducts from their manufacturing and maintenance activities (Syafrudin et al. 2021). Gasoline, oil, and other petroleum chemicals are examples of potential contaminants that might be present in both raw materials and garbage. Man-made chemicals such as pesticides, herbicides, and fungicides are polluting water sources and harming ecosystems. Water treatment facilities, landfills, power plants, car washes, and garages are all examples of municipal point sources. All of these scenarios have the potential for the presence of harmful substances in both the input materials used in the process and the byproducts produced by the facility. If the facility or operator does not correctly handle, store, and dispose of raw materials and garbage, it might damage the water supply. In a similar vein, the term "pipe dream" is often used to describe a situation in which an individual's expectations of an outcome differ significantly from transpired.

20.4.1 Ground Water

Volatile organic compounds, including those found in solvents, oils, paint, and fuel products, may remain in the groundwater for a very long time if not properly removed. Even brief contact with these chemicals might be worrisome (Khan et al. 2021). When it comes to trichloroethylene, a common solvent, only 4 liter of the pure material may contaminate more than 1 billion liters of water. After groundwater has been contaminated, it is difficult, costly, and often impossible to clean it.

20.4.2 Surface Water

High-temperature discharges, microbes including bacteria, viruses, and giardia, and nutrients like nitrogen and phosphorus are the most prevalent point-source contaminants in surface water. Overheated water and an influx of nutrients may spur an explosion of aquatic plant life, which in turn can deplete dissolved oxygen levels and stress or kill sensitive aquatic species (Islam et al. 2021). Both humans and aquatic species are vulnerable to the dangers posed by microorganisms. In addition to being harmful to humans and aquatic life, pesticides and other toxic compounds are also rather rare in surface water due to high dilution rates.

20.5 Non-point-Source

Pollutants from nonpoint sources are dispersed across a large region, making it the polar opposite of point-source pollution. Drops of oil from leaking automobile engines, bits of tire rubber, dog excrement, and rubbish are all washed away by rainfall as it passes over asphalt. Storm drains collect the water and the river benefits. Nonpoint-source pollution is mostly attributable to runoff (Ji et al. 2023). This is especially problematic in urban areas due to the prevalence of hard surfaces such as streets and rooftops. Some examples of nonpoint source pollution come from farms and homes that use too much fertilizer, herbicide, or pesticide. Urban runoff and energy production are major sources of oil, grease, and hazardous substances. Erosion of stream banks and sediment from unmanaged building sites, farms, and forests, Acid rain from defunct mines and salt from irrigation methods Livestock manure, pet waste, and inadequate septic systems all contribute bacteria and nutrients. Hydro modification and atmospheric deposition Although the quantity of pollution washed away from a single city block may not seem like much when multiplied over the miles and miles of pavement in a large metropolis, the issue becomes enormous. Sediment washed off the roadways in a logged-over forest tract by runoff is a common problem in rural regions. It may also be used to remove pesticides and fertilizer from agricultural areas and to transport acid from defunct mines. The waterways and bodies of water will eventually absorb all of this pollution. Acid rain is mostly caused by pollution when sulfur dioxide and nitrogen oxides react with water vapor in the atmosphere. Nonpoint-source pollution refers to acid rain's diffuse origins from several sources such as industries and power stations. The clean air act and the clean water act have both contributed to reducing pollution from both point sources and diffuse sources in the United States. Storm water, which is precipitation run-off in metropolitan areas, is a significant non-point source of pollution to our rivers and bays (Ajibade et al. 2021). In urban locations, storm water is sometimes tainted by toxicants and chemicals from automobile oil, dust, animal excrement, and the soil and sediment run-off from building projects. As a reticulated sewerage system is unavailable in certain outer-urban and urban periphery locations, sewage is instead discharged to onsite wastewater systems and septic tanks. Septic tank effluents may also contribute to non-point source contamination of streams in these locations via seepage and surface run-off. Pesticides, fertilizers, animal dung, and dirt washed into streams as a result of rainfall are all examples of non-point causes of pollution in agricultural regions. Stock that has access to stream banks may contaminate the water and hasten the stream's erosion. Soil erosion and sediment run-off may contribute to non-point source contamination of streams as a result of logging activities.

20.6 Various Causes of Water Pollution

1. **Industrial waste**

Massive amounts of industrial waste are released into the environment, polluting the air and posing health risks to humans and other animals (Fig. 20.3). Several dangerous compounds, including lead, mercury, sulfur, nitrates, and asbestos, are present in them. Without an adequate waste management system, many factories release their sewage into streams and rivers that ultimately end up in the ocean. Toxic chemicals may cause water to become a different hue, increase the concentration of minerals, cause temperature changes, and present a significant threat to aquatic life (Chowdhary et al. 2020). Water pollution is a big problem in many parts of the globe, and industrial sites and factories are important causes. Whilst many industrial sites are regulated, others still lack adequate waste management systems despite producing waste in the form of harmful chemicals and contaminants. Sometimes, factories may discharge their garbage into surrounding rivers or streams. It doesn't take much for industrial waste to poison freshwater systems if it isn't managed adequately (Ukaogo et al. 2020). Agricultural, mining and industrial facilities may all contribute to the pollution of waterways that ultimately flow into the ocean. The toxic compounds in the waste created by these companies not only have the potential to render water unfit for human consumption, but they may also cause the temperature in freshwater systems to shift, making them hazardous for many water-dwelling creatures.

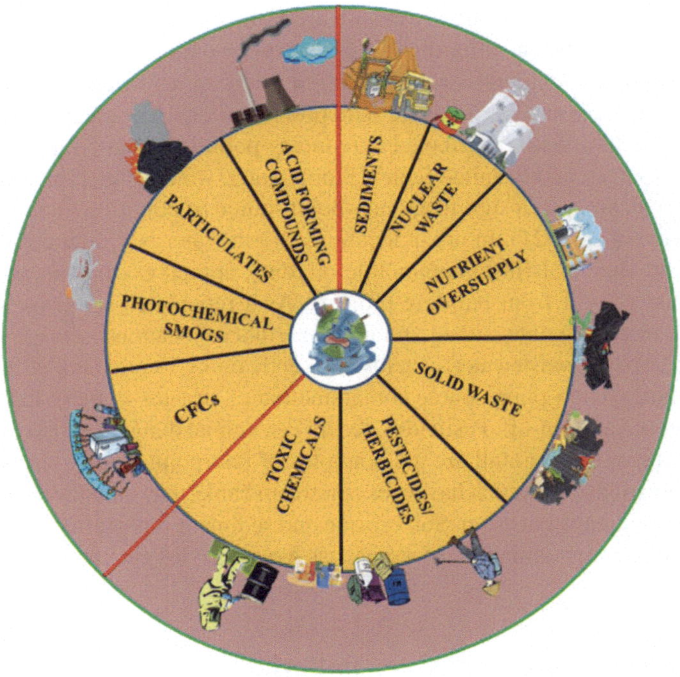

Fig. 20.3 The schematic diagram represent as various kind of water pollutant in our surrounding

2. **Sewage and wastewater**

Sewage and wastewater from individual households are chemically treated before being discharged into the ocean alongside potable water (Fig. 20.3). Sewage water is a significant health risk because it contains pathogens, a kind of water pollutant; other hazardous microorganisms; and chemical pollutants. Some of the worst illnesses are caused by microorganisms in water, which also serve as incubators for the animals that spread them (Elkhatib and Oyanedel-Craver 2020). These infected individuals come into touch with these carriers in a variety of ways. Malaria is a common illustration of this trend. Despite purification processes, sewage and wastewater still contain harmful chemicals, bacteria, and diseases. Each home's sewage and wastewater are combined with the freshwater supply and dumped into the ocean. Humans and animals alike are vulnerable to health problems because of the viruses and germs that thrive in that effluent.

3. **Mining activities**

Mining entails excavating subsurface deposits and using heavy machinery to remove valuable resources like coal (Fig. 20.3). Health issues may arise because the raw forms of these metals include dangerous compounds and may raise the concentrations of poisonous components in water (Mukherjee and Pahari 2019). Metal waste and sulfides released from rocks during mining operations are toxic to aquatic life.

4. **Marine dumping**

Household waste such as paper, plastic, food, metal, rubber, and glass are often collected and discharged into the ocean in several nations. These contaminants are harmful to marine life and contribute to water pollution when they reach the ocean (Nanda and Berruti 2021). Several nations still collect rubbish from homes and dispose of it in the ocean, which is a practice that many people find bizarre. Decomposition of these materials typically takes in between two to two hundred years.

5. **Accidental oil leakage**

When a significant volume of oil spills into the ocean and does not dissolve in the water, it poses a serious hazard to marine life. Fish, bird, and sea otter, as well as other nearby marine animals, are negatively affected and pose a serious hazard to marine life. An accident the oil released, the toxicity of the pollutants, and the vastness of the ocean, such an oil spill may harm many ocean species in differing degrees. When two substances do not mix well or at all, the idiom "like water and oil" is often used (Rudebusch et al. 2020). While often unintentional, large oil spills and leaks are a substantial contributor to water contamination. Oil drilling activities in the water or on ships that carry oil are often the source of leaks and accidents.

6. **The burning of fossil fuels**

Ash is released into the air in large quantities whenever fossil fuels are burned. As the particles combine with water vapor, they release deadly compounds that cause acid rain (Boyce et al. 2023). Moreover, burning fossil fuels releases carbon dioxide into the atmosphere, this contributes to climate change.

7. **Chemical fertilizers and pesticides**

To prevent germs and insects from damaging their crops, farmers often resort to the use of chemical pesticides and fertilizers. They contribute to the development of the plant. Nevertheless, these compounds create dangerous pollutants for plants and animals when combined with water. When it rains, the chemicals get diluted in the runoff and eventually make their way to waterways, where they may inflict considerable harm to aquatic life (Hajjar et al. 2023). Chemicals and pesticides are often used by farmers to prevent crop damage from germs and insects. Groundwater becomes contaminated when these pollutants leak into it. As it rains, the chemicals are carried away by the water and eventually end up in the rivers and streams that eventually end up in the ocean.

8. **Global warming**

As a consequence of the greenhouse effect, the average temperature of the earth is rising. Water pollution develops as a consequence of the deaths of aquatic creatures and marine species caused by the rise in water temperatures. Climate change is a big threat to water quality since it raises average world temperatures. Temperature increases in water due to global warming pose a threat to aquatic organisms (Negri et al. 2019). Water pollution is already a serious problem, and it becomes much worse when mass deaths occur. There are several simple actions you can take to help slow global warming and cut down on the water pollution it causes. Reusing materials, reducing vehicle use, and switching to energy-efficient light bulbs are all good examples.

9. **Radioactive waste**

Both fission and fusion can be used to generate nuclear energy. Uranium, an extremely hazardous substance, is employed in the generation of nuclear energy. Getting rid of the radioactive waste produced is crucial to preventing nuclear catastrophes. Inappropriate disposal of nuclear waste may result in severe ecological risks (Shayan and Ghasemzadeh 2020). Several very serious incidents have already occurred in Russia and Japan. Facilities that generate nuclear energy produce radioactive waste, which must be safely disposed of. Uranium, a key ingredient in nuclear power production, is a very dangerous substance. Sometimes, accidents do occur at these plants, leading to the discharge of harmful material into the environment.

10. **Urban development**

As the population has grown exponentially, so has the demand for housing, food, and clothing (Fig. 20.3). As more cities and towns are developed, they have resulted in the increasing use of fertilizers to produce more food, soil erosion due to deforestation, a rise in construction activities, inadequate sewer collection and treatment, landfills as more garbage is produced, and increased in chemicals from industries to produce more materials (Benton and Bailey 2019).

11. **Leakage from the landfills, underground storage and sewer lines**

Landfills are just enormous rubbish dumps that stink to high heaven and are visible from miles away (Fig. 20.3). Landfills have the potential to leak when it rains, releasing a wide range of toxins into the groundwater supply. It is commonly known that coal and other petroleum products are transported via subterranean pipelines. If a leak occurs by accident, it has the potential to harm the ecosystem and lead to soil erosion. Even a slight break in the sewage line may release harmful bacteria into the earth, making the water unusable (Obebe and Adamu 2020). Another issue that might arise if leaks aren't fixed on time is that standing water can provide a perfect environment for mosquitoes and other pests.

12. **Animal waste**

Animal feces and urine wash into the waterways during storms. Contaminated water that subsequently mixes with other hazardous substances spreads diseases like cholera, dysentery, jaundice, and typhoid (Bendigeri and Kawalekar).

13. **Eutrophication and acid rain**

Eutrophication refers to the process through which water bodies accumulate an excessive amount of nutrients. Algae blooms are reduce the quantity of oxygen in the water, which has consequences for marine life (Coffey et al. 2019). Simply, acid rain is a form of water contamination due to air pollution. Acid rain occurs when air pollution releases acidic particles into the sky, and these particles combine with water vapor in the atmosphere.

14. **Personal care, household cleaning, and pharmaceutical products**

It's important to remember that nearly all of the products we use to maintain our hygiene and clean our homes—things like laundry detergent, bleach, fabric softener, window cleaner, dusting spray, stain remover, hair dye, shampoo, conditioner, rogaine, cologne, perfume, toothpaste, mouthwash, and hand lotion—end up down the drain after use (Fig. 20.3). Similarly, when we take pharmaceuticals, ultimately excrete them, either partially metabolized or unmetabolized, into the water supply it affect up to 90% of your medication pollute water quality. Hormones and antibiotics used in livestock production also pollute our water supplies in significant amounts (Patel et al. 2019). However, most wastewater treatment plants are not able to filter out personal care goods, home products, and medications, thus a significant amount of the chemicals flow straight into the local stream that absorbs the treatment plant's ostensibly clean effluent. The hormonal effects of birth control ingredients have been studied in frogs and fish and found to be transgressive. Nonylphenol, a detergent byproduct, has been shown to interfere with fish reproduction and development (De la Parra-Guerra and Acevedo-Barrios 2023).

20.7 The Major Group of Water Pollutants

1. Biodegradable: Biodegradable water pollutants are of two types i.e. none persistent and persistent. A material is considered biodegradable if it can be broken down by bacteria or other naturally occurring creatures. The term "biodegradable waste" refers to trash that can be decomposed by microorganisms and environmental factors (Bilal et al. 2019). This kind of trash includes things like natural and organic waste. In the soil, organic matter is decomposed by a combination of microbial activity and other abiotic processes. Nature ensures that everything occurs in its own time. Hence, biodegradable waste doesn't cause significant environmental issues. Biodegradable waste, in contrast to inorganic rubbish, may be decomposed by bacteria or fungi. Biodegradable trash is often found in municipal solid waste and includes items like green rubbish, food waste, paper waste, and biodegradable plastics (Popoola 2022). Biodegradable garbage also includes human and animal feces and urine, as well as sewage and leftovers from slaughterhouses.

2. Non-biodegradable: Any substance that cannot be decomposed by naturally occurring organisms falls under the category of "non-biodegradable material" in the context of environmental contamination. Non-biodegradable garbage is far more difficult to manage than biodegradable garbage (Velvizhi et al. 2020). Non-biodegradable materials are those that cannot decompose or disappear through natural processes. It can survive thousand of years on earth without degrading; as a result the danger pose is greater. Plastic, which has applications in almost every industry, is one such example. Enhanced plastic quality is included in these materials to ensure durability. That enhanced their durability and resistance to heat, even after repeated usage. Some common examples are metal and chemical products used in agriculture and industry. These are major contributors to environmental degradation and illnesses like cancer. Non-biodegradable trash should be swapped out with something more environmentally friendly (Al-Hameedi et al. 2020). Scientists have proposed several alternatives, such as biodegradable polymers as part of the development process. It combined biodegradable materials with polymers and rendered them biodegradable in a short amount of time. For instance, some materials cannot be broken down by microorganisms or worms and are thus considered non-biodegradable trash. Unfortunately, many of these materials are also inorganic and cannot be composted. As a result, disposing of non-biodegradable trash can be rather expensive. One option is to focus on reducing the amount of non-biodegradable trash produced in the first place through measures such as using reusable containers or opting for products with minimal packaging. Additionally, it's important to properly sort recyclable and non-biodegradable wastes from non-recyclable ones to ensure that they can be repurposed efficiently (Liu et al. 2021). With a concerted effort from individuals and communities alike, we can work towards a more sustainable.

20.8 Type of Water Pollutant

1. **Pathogens**
   Infectious agents include microorganism that may cause disease in humans, including bacteria, viruses, parasites, and even certain foods. Wastewater is a common source of infectious microorganisms, including viruses like hepatitis and norwalk and yeasts like candida (Newell et al. 2010). Salmonella causes food poisoning, whereas *Vibrio cholerae* causes cholera epidemics. Diarrhea and similar gastrointestinal issues may be caused by several illnesses present in wastewater, including parasites like cryptosporidium and schistosoma (Gerba and Pepper 2019). Contaminated water sources and the subsequent spread of water-borne viruses represent a worldwide health disaster. There is a growing body of water-borne pathogen control, and unify it in a single place to present and address the problem comprehensively (Cissé 2019). Various studies provide a more complete picture of the contamination of both fresh and saltwater resources by disease-causing organisms. Water-borne pathogens in ambient water bodies

and the ensuing spread of disease are two of the most pervasive issues with water quality. It is vital to recognize and understand the degree to which pathogen contamination affects different types of ambient water bodies (Wen et al. 2020). Water storage techniques should be included in long-term plans to meet the increasing demand for water for human consumption, industrial usage, and environmental purposes. Unfortunately, these recent innovations may lower water quality and raise risks to people's health.

2. **Organic wastes**
   Components such as sulfur, hydrogen, oxygen, carbon, and nitrogen are used to create organic molecules. Pollution comes from a variety of sources, including farm and factory runoff as well as municipal and industrial effluent and sewage (Xue et al. 2022). Palmitic acid, oleic acid, docosanoic anhydride, and dodecanoyl chloride are all types of organic waste that are often seen. Detritus such as trash and yard trimmings or leaves may have a big impact on the environment. As organic materials deteriorate, they eventually wash away and contaminate nearby water supplies (Sanda et al. 2022). The overpopulation of aquatic organisms called phytoplankton is one probable source of water pollution. Nevertheless, some of the trash could be managed by breakdown mechanisms in nature. Microorganisms abound in water, and it is these organisms that do the work of breaking down the organic materials. These microbes need every last bit of oxygen in the water to survive. Water's capacity to contain liquids is proportional to its oxygen content. Water contains both oxygen from the air and oxygen produced by aquatic plants via photosynthesis. The oxygen level in the water decreases because photosynthesis stops at night, but aquatic plants still need oxygen for breathing. Yet if bacteria in water bodies consume all the oxygen in the water, the water will become oxygen-depleted. Loss of oxygen in the water causes the gradual death of organisms dependent on it. Thus, eventually, anaerobic bacteria will begin breaking down the organic waste, resulting in the release of poisonous compounds that are harmful to human health.
3. **Chemical pollutants**
   There is a serious problem in developing countries with drinking water contamination from substances like nitrite, ammonium nitrate, and heavy metals. Human activities, such as the discharge of sewage from towns and industry, lead to elevated levels of inorganic nitrogen pollutants and inorganic phosphates in river water, which in turn contribute to a wide range of health problems (Raimi et al. 2022). Increased ammonium levels and the risk of developing stomach, liver, and esophageal cancer have all been related to nitrite exposure. Waterways and fish habitats are poisoned when pesticides used on farms, roadsides, and even home lawns runoff into them. Although farms may be the first thing that comes to mind when you think of pesticide usage, the reality is that more pesticides are applied to residential lawns than to farmland (Robinson 2023). The agricultural sector is a key contributor to this problem. Groundwater is frequently used in the Midwest, and it is estimated that the annual cost of treating water for the pesticide atrazine alone is $400 million. Chemical and industrial wastes pollute almost all of the world's water sources. The United States' most hazardous liquid

waste, totaling 34 billion liters per year, is pumped directly into deep groundwater via thousands of "injection wells." This trash mostly consists of solvents, heavy metals, and radioactive substances (Perkins et al. 2022). In the late 1990s, the Central Pollution Control Board of India determined that the groundwater in all 22 of main industrial zones was unfit for human consumption (Chakraborti et al. 2011).

4. **Sediments**
   As the root systems that held the soil in place are destroyed, silt may easily wash into nearby streams. The increased amount of silt washing off the land and into surrounding bodies of water has devastating effects on fish and other aquatic life; hence, clear cutting has substantial ramifications for forest plant and animal biodiversity (Ma et al. 2022). Poor soil management practices that leave soil vulnerable to the elements are a major contributor to silt-related water pollution.
5. **Nutrients**
   Several types of pollution, including sewage, manure, and artificial fertilizers, include "nutrients," such as nitrates and phosphates. Atmospheric nitrogen deposition also contributes to the pollution of water with nutritional toxins. Overstimulation of plant and algae growth is caused by nutrients in the water (Lim et al. 2021). Although these organisms are still alive, their uncontrolled development is blocking light from reaching deeper waters and clogging our waterways. They deplete oxygen in the water as they decay, leaving behind an environment that can support just a small percentage of the marine life it formerly supported. A dead zone is a common name for these types of areas. Because of their location near the mouths of large rivers, estuaries and deltas are especially susceptible to nutrient pollution from the runoff they get from several watersheds.

Environmental effects of water pollution, both point and non-point Ecosystems suffer from the following consequences of point source pollution: Industries discharging used water or other materials into a body of water is an example of a point source of pollution. Bad influences can harm the environment and the people who live in it. Agricultural fields doused in fertilizers or pesticides are one example of a large area that contributes to pollution. Pesticide and fertilizer runoff can contaminate surface water as well as groundwater if it percolates through the soil (Wang et al. 2021). Lake pollution has far-reaching consequences, from tainting drinking water to killing fish and other wildlife to building up in the tissues of the fish we catch and eat to dulling the lake's aesthetic appeal and even harming the forest trees in the area.

Pollution from sources that aren't specifically pointed out is a far greater concern: It takes place when precipitation of any kind, including rain, snow, or irrigation water, travels over land or through the ground and picks up pollutants, and then dumps those pollutants into bodies of water. Paint, solvents, acids, and glue are all examples of potentially toxic building products that may contaminate water. Rain water in urban environments, such as roads, sidewalks, and parking lots carries contaminants into streams. Urban areas are more likely to have these impermeable surfaces. Chemicals used in lawn care and gardening, such as herbicides, insecticides, and fertilizers, have the potential to leach into groundwater or make their way into rivers (Mishra and Dubey 2023). Instead of being disposed of appropriately, hazardous liquids such as paints, oils, solvents, and cleansers are often flushed down the toilet. The salt that is used to de-ice roadways is another significant contributor to pollution. Agriculture often has a significant impact on the quality of the water. Chemicals like herbicides, insecticides, and fertilizers, in particular when applied more than what is required, have the potential to flow off into streams or seep into the groundwater. The effect of natural pollution, sometimes referred to as nonpoint source pollution, is far greater than that of manufactured pollution or point source pollution (Artiola et al. 2019). It affects water sources like surface waters and subsurface waters. The pollution of rivers occurs in three stages: the first two take place in the atmosphere, and the third stage is caused by the pollutants' deposition on the ground. Condensation, rainfall, and wastewater, sometimes known as residual water, are the three processes that make up the process of water pollution. The amount of pollution that comes from artificial sources, also known as point sources, is proportional to the degree of urban and industrial growth. Residential, industrial, and agricultural stocking residues are the primary contributors to man-made pollution. Man-made pollution may be categorized according to pollutants in nature. The processes that release pollutants may be intentional or unintentional, and divided into three categories based on the nature of the pollutants themselves: physical, chemical, and biological. It is also possible for streams to get contaminated by manure from livestock, especially in situations where the animals have direct access to the water. There can be an increase in the number of bacteria in the water if the manure is handled improperly, particularly on big animal farms. Both point and non-point sources of pollution contribute to serious environmental issues, including the following: photosynthesis in aquatic plants may be inhibited, which may affect ecosystems that are dependent on these plants; terrestrial and aquatic plants and trees may absorb pollutants from water; mud and clay from construction sites and other sources may kill plants and trees; herbicides in water may kill plants and trees; and these are the types of chemicals that are most harmful to plant and tree species (Saleh 2020). The control of pollution that is carried away by precipitation is far more difficult to circumvent than the regulation of pollution that is released from a pipe outlet that can be located and identified. However, there is a growing interest in methods for controlling non-point sources of water pollution, and this can be done through planning, cost-sharing mechanisms, voluntary best management practices, and other approaches. Although non-point sources of water pollution are typically difficult to identify and monitor, there is a growing interest in methods for controlling non-point sources of water pollution.

20.9 The Negative Effects of Pollution on People's Health

Even though there has been a strong effort to increase access to clean water for drinking, it is estimated that there are still 1.1 billion people who do not have access to an improved water source. About 10% of people are death per year due to drinking unsafe water (Fig. 20.4a, b). Each year, water-related illnesses claim the lives of over three million people, the majority of them are children (Ochilova et al. 2021). Over 2 million of these fatalities are the consequence of diarrheal illnesses, which are caused by the use of water that has been polluted by feces as well as by poor sanitation and hygiene. These disease may be prevented by maintaining acceptable levels of cleanliness and sanitation. Water supplies that are contaminated may also play a role in the spread of illnesses that are transmitted by skin contact or vectors. The respondents also had recurrent health issues or were required to take drugs regularly to keep the sickness. A significant percentage of responders had issues related to diarrhea, cholera, irritation of the skin and eyes, typhoid, and other respiratory conditions (Fig. 20.5; Table 20.2). The repercussions of poor water quality extend well beyond the realm of health (Fazal-ur-Rehman 2019). The patients, as well as society are subjected to severe social and economic expenses as a result of water-related disorders when they suffer from chronic spells of the illness. Unsafe water is one of the most significant threats to human health and the environment throughout the globe, and this is especially true for the world's most impoverished people. The global burden of disease is a large-scale worldwide research found that lead to mortality and illness due to contaminate water. One of the most significant contributors to the development of infectious illnesses, such as cholera, diarrhea, dysentery, hepatitis A, typhoid, and polio, is a lack of access to sources of clean water. In addition to this, it makes malnutrition, and particularly stunting in children, much worse (Ritchie and Roser 2021). According to the global burden of disease research, contaminated water was a contributing factor in the early deaths of 1.2 million people in 2017 (Fig. 20.4b). Unsafe water sources are responsible for 6% of all fatalities that occur in low-income nations; in 2017 individuals lost their lives as a direct consequence of hazardous water sources (Lawrencia et al. 2023). Countries with low incomes have much higher overall death rates. The death rates caused by contaminated water sources provide us with an accurate assessment of the variations in the mortality implications of these factors across nations and over time. Death rates are much higher in nations with lower per capita incomes, notably in Sub-Saharan Africa and Asia region, mortality rates often exceed 50 per 100,000 people, and in the Central African Republic and Chad were higher than 100 per 100,000 people. As compared to the mortality rates in high-income nations, such as those in Europe, where the rate is sub 0.1 fatalities per 100,000 people, it is bigger than 1000 times larger. As a result, nations with low and lower-middle incomes are the ones that are most likely to be affected by the problem of unclean sanitation.

20.10 Effects of Water Pollution on Agricultural Output

Water is essential to the processes of photosynthesis, growth, and reproduction in plants. Due to the incorporation of the water into the plant's chemical makeup and its expulsion into the air, it is impossible to recover the water that plants have absorbed. Plants must lose a significant quantity of water via transpiration to successfully

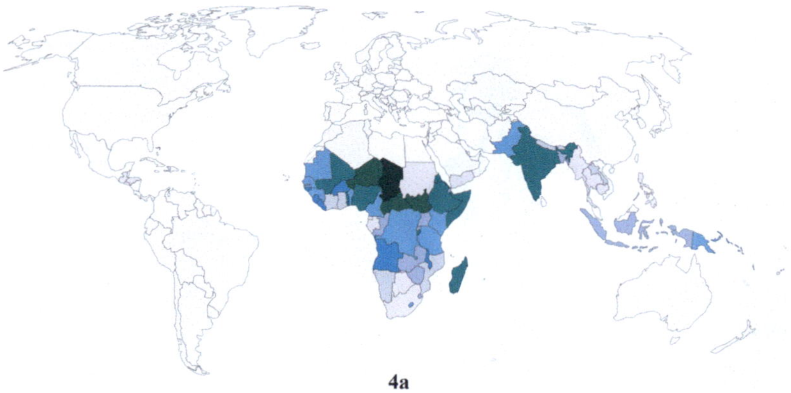

4a

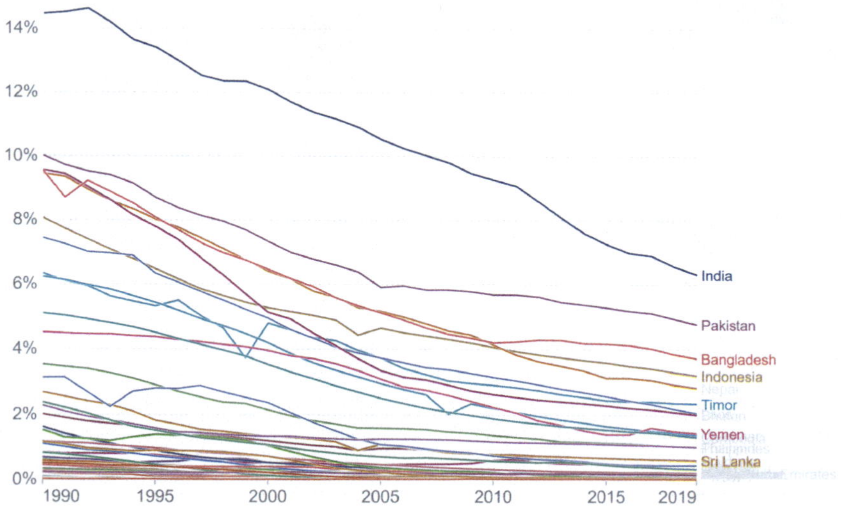

4b

Fig. 20.4 **a** The world map represented as death attributed to taking unsafe water of people (source from Global Burden Disease and Our World in Data). **b** Diagram presented as 10% of people are death by taking unsafe water per year (source from Global Burden Disease and Our World in Data)

carry out the processes of carbon dioxide fixation and temperature regulation. All over the world, agriculture uses about 70% of the freshwater extracted each year. Due to the rising demand for agricultural products, farmers are increasingly considering the use of non-conventional water sources, some of which are of substandard quality, it includes wastewater (Ma et al. 2020). Due to its high nutritional content, wastewater from households and municipalities provides an appealing alternative, particularly in situations where traditional water supplies are limited or nonexistent. The unsafe use of non-conventional sources of water, particularly wastewater, in agriculture can lead to the accumulation of microbiological and chemical pollutants

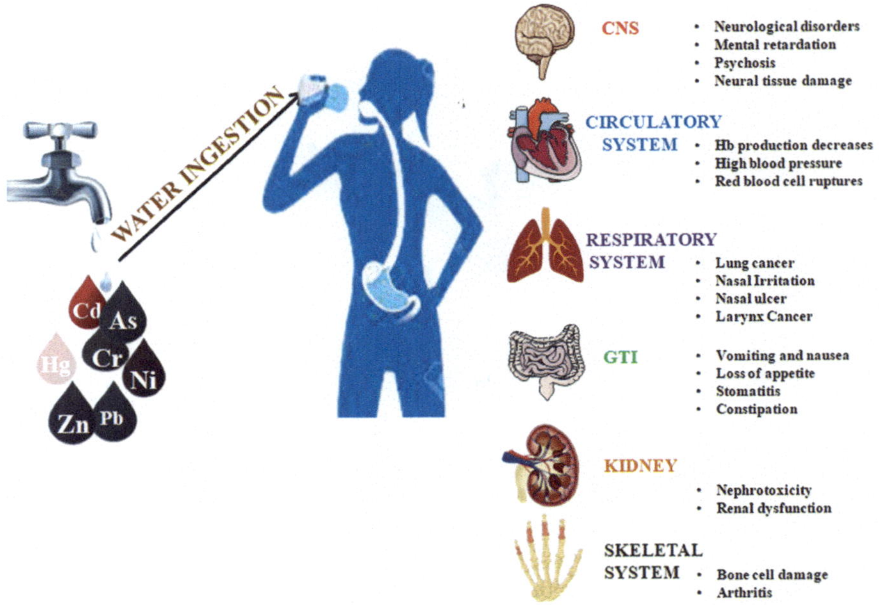

Fig. 20.5 The schematic diagram represented as effect of contaminated water in human body

in crops, livestock products, soil, and water resources, which can ultimately lead to severe health impacts among exposed plants (Fig. 20.6) and food consumers and farm workers (Fig. 20.5). It may also contribute to the development of antimicrobial resistance (de Mello et al. 2020). Moreover properly cleaned and reused in a controlled environment, wastewater has the potential to be a useful source of both water and nutrients contribute to the enhancement of food security and livelihoods. Heavy metals are thought to pose the largest threat to the quality of food in India out of all the different kinds of pollution that have been detected. Mining and smelting are two of the primary activities that contribute heavy metals to the soils of agricultural lands. Other contributors include the irrigation of sewage, the reuse of sludge, and the use of fertilizer (Evans et al. 2019). Large quantities of heavy metals can be transferred to farmland via wastewater irrigation, waste transportation, sludge application, and atmospheric deposition as a result of the extensive and nonstandard production processes utilized by some mining and smelting businesses. These has been demonstrated to be of particular significance in the northern region of India, which contains a wealth of mineral resources. The term "sewage irrigation" refers to the practice of using sewage output for irrigation purposes without first subjecting the water to any treatment or removing anything except the solids. That practice often involves the use of compounds that are poisonous or harmful. In certain regions, farmland has been sprayed directly with raw sewage that has been discharged from nearby towns that have not been cleaned. Nevertheless, the land that was irrigated with sewage was polluted with heavy metals, the most dangerous of which were mercury (Hg), lead

Table 20.2 The table represent as several types of water pollutant and its sources impact in human and plant with acceptable level

Contaminant	Sources of contaminant	Effects	Acceptable Level
Per and Polyfluoroalkyl substances (PFAS)	Meat, poultry, fish, eggs, fruits, vegetables, tubers, and roots are all good options. Utensil coated with perfluoroalkyl silanes	The immune system weakens, vaccine responses decrease, cholesterol levels rise, Coronary heart disease	8 ng/kg bw/week For PFOA, PFNA, PFHxS and PFOS
Arsenic	Groundwater, paddy rice, mushrooms, and fungus are all naturally polluted	Bladder cancer and large, ugly sores on the skin	No health-Protective value available, previous value of 2.1ug/kg bw/day withdrawn for not being health protective
Cadmium	Meat, poultry, organs, mollusks, rice, wheat, vegetables root, tuber, and leafy, and other foods	Cancer of the breast, endocrine disruption, pulmonary emphysema, Itai-Itai illness, and damage to the kidneys and renal tubules	25 μg/kg bw/ month
Phthalates	Vegetables, grains, mustard, and rapeseed	Liver cancer, testicular atrophy, and altered adipogenesis	20 μg/kg bw/day (USEPA) 50 μg/kg bw/day (EFSA)
Copper	Organ meats, fish, poultry, grains, nuts, and leafy greens all fit the bill	Liver and hepatic necrosis, methemoglobinemia, and copper toxicity in children	0.5 mg/kg bw/ day(WHO) 0.4–1.6 mg/ day(EFSA)
PCBs and other dioxin-like compounds	Soil, fish, meat, milk, and vegetables are all polluted	Elevated hepatic enzyme levels; skin, tooth, and nail abnormalities; altered growth factor signaling; cardiovascular disease, diabetes, cancer, porphyria, endometriosis, menopause, decreased testosterone and thyroid hormones, altered immunologic response, chloracne, demial hyperpigmentation and hirsutism, and altered metabolism and reproductive hormone levels	70 pg/kg bw/ month, to intake of PCDDs, PCDFs and early coplanar PCBs. Not established for other PCBs

(continued)

Table 20.2 (continued)

Contaminant	Sources of contaminant	Effects	Acceptable Level
Lead	Lean meats and poultry, chocolate and chocolate goods, fresh produce, whole grains, and cereals	Damage to children's developing brains, resulting in lower intelligence quotient(IQ) and behavioral problems, as well as anemia, hypertension, kidney injury, impairment to the genitourinary system, and impaired immunity	No health-Protective value available. Previously established provisional tolerable weekly intake of 25 ug/ kg bw withdrawn
PBDEs (flame retardants)	Everything from seafood and poultry to common foodstuffs and cooking oils	Toxicity to neurodevelopment, as well as effects on the endocrine system, the reproductive system, the metabolism, and the body weight	Not established but estimated daily intakes of 4 ng/kg bw/day for adults & up to 100 ng/kg bw/ day for breastfed infants for exposure to total PBDE congeners
Manganese	Water, and baby formula made with soy	Movement disorders like Parkinson's disease low sexual desire, impotence, and erection problems; harmful to vegetation, Discolor the fixture and ruin it	1.8–2.3 mg/day for adults (U.S.ATSDR) Not yet considered by JECFA, nor EFSA
PAHs	Leafy crops, shellfish like mussels and oysters, and preparation techniques including drying, smoking, grilling, roasting, and frying are all in this category	Alterations in neurodevelopment and behavior causes of death include cancer, dermatitis, inflammation, lowered immunity, cataracts, hepatitis, renal failure, liver failure, asthma, and haemolysis	Not established but estimated daily Intakes of 10 ng/kg bw/day

(continued)

Table 20.2 (continued)

Contaminant	Sources of contaminant	Effects	Acceptable Level
Mercury	Methyl mercury is most often found in fish caught in urban and coastal waterways, as well as in shellfish; however in mushrooms, rice, cereals, legumes, potatoes, meat, eggs, vegetables, fruits, and breast milk	Effects on kidney weight, proximal tubule damage and progressive nephropathy, and embryonic and fetal neurotoxicity	4 pg/kg bw/week (inorganic mercury) 1.61 g/kg bw/week (methyl mercury)
Beryllium	Coal, Industrial water	Toxic to plants and human Chronic Beryllium Diseas	0.2 μg/m3
Copper	Metal plating, mining, Industrial waste	Essential trace elements toxic to plants and algae at higher levels	1 mg/l

(Pb), and cadmium (Cd). Sewage irrigation is an efficient approach to reducing the strain placed on available water supplies. In addition, the findings of a recent investigation revealed that sewage irrigation regions were tainted with cadmium, arsenic (As), and polycyclic aromatic hydrocarbons (PAHs) (Gao et al. 2019). The buildup of heavy metals is accelerating, particularly in agricultural regions that make use of intensive agricultural practices and extensive irrigation networks. In the meantime, waste products from intensive livestock production include significant concentrations of As, zinc (Zn), and copper (Cu). These materials are becoming key sources of pollution as the animal husbandry sector continues to expand which are impacts on plants (Fig. 20.6).

Effects of acid deposition: Water bodies are acidified when many acid gases, aerosols, and other acidic compounds released into the atmosphere by industrial and household burning of fossil fuels make their way into water bodies through runoff precipitation from damaged soil surfaces. Acidification of lakes, rivers, and ponds due to sulfates, nitrates, and chloride has been documented in several countries (Ameen 2019).

Nutrient deficiency in aquatic ecosystem: When water becomes more acidic, fewer microorganisms, such as bacteria and fungi, can break down organic materials, slowing the recycling of nutrients. Most aquatic organisms have a pH 6.0 critical range, below this pH diversity of species decreases while acid-tolerant species proliferate. After the onset of water acidification, the fast growth of filamentous algae

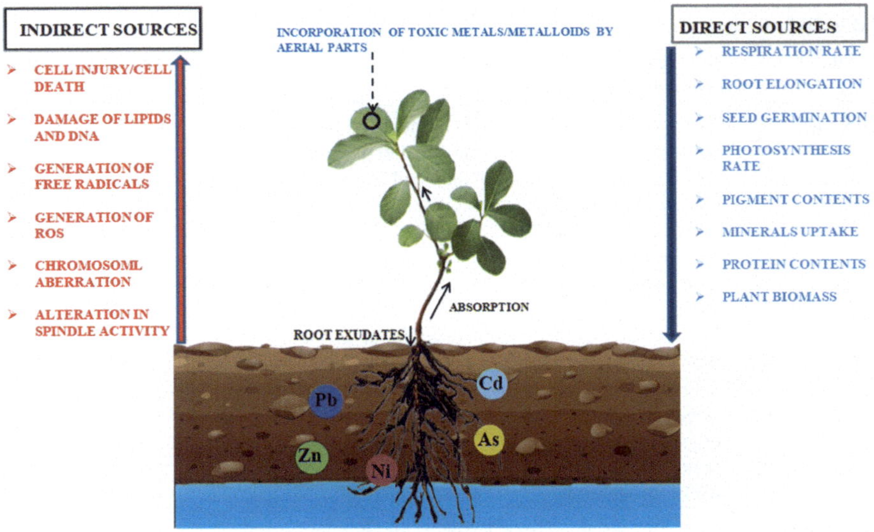

Fig. 20.6 Diagram presented as effect of contaminated water in plant through indirect and direct sources

produces a dense mat. At pH levels lower than 5.8, diatoms and green algae die off (Wurtsbaugh et al. 2019). Cladophora is a genus of algae that thrives in acidic freshwater environments. The alga Potamogeton pectoralis thrives in low-pH environments. It has been shown that plants with deep roots and rhizomes are less harmed by acidic water than those with shallow root systems.

Effects of organic matter deposition: Decomposers and microorganisms, such as aerobic and anaerobic bacteria, proliferate in water bodies when organic matter from dead and decaying plant and animal components is deposited directly from sewage discharges and washed together with rain. Due to an increase in available nutrients, planktonic green and blue-green algal blooms flourish when organic matter decomposes quickly. Increases in decomposers and microorganisms like aerobic and anaerobic bacteria are another consequence of the fast growth of macrophytes like salvinia, azolla, eicchhornia, etc., which block light from reaching the water's depths (Sonone et al. 2020). Planktonic green and blue-green algal blooms flourish when organic waste decomposes quickly because of the increased availability of nutrients in the water. In addition, the fast expansion of macrophytes results in less light reaching the water's depths, leading to the slow demise of the aquatic vegetation. Both dissolved oxygen (B.O.D) levels and the need for oxygen by living organisms drop, unlike organic matter-contaminated water, which has a B.O.D. of over 400 mg/l, unpolluted fresh water typically has a B.O.D. of less than 1 mg/l (Walker et al. 2019).

Effects of detergent deposition: When detergents used at home or in factories get into water supplies, these have a devastating impact on vegetation. Phosphate is often found in detergents, and its use enriches water supplies with the substance (Atakan et al. 2020). Phosphates are taken up by plants either through their roots or their leaves, and once inside, they inhibit enzyme activity in several metabolic pathways, including root development, root lengthening, carbon dioxide fixation, photosynthesis, cation uptake, pollen germination, and pollen tube growth, destruction of chlorophylls and cell membranes, and denaturation of proteins.

Effects of agricultural chemicals: Excessive use of chemical fertilizers, pesticides, insecticides, herbicides, etc. on crops causes runoff that the soil eventually absorbs and carries these chemicals to bodies of water. The chemical fertilizers used enrich nutrients, leading to eutrophication (Okereafor et al. 2020). Changes in water pH are another effect of pesticides, herbicides, and insecticides. The primary impact of these chemicals is a slowing of photosynthesis. Some may block the nitrate reductase enzyme or cause oxidative phosphorylation to become uncoupled. Due to their limited solubility in water, macrophytic plants have a high capacity for absorption and bioaccumulation of these chemicals.

Effects of industrial wastes: Industries discharge several kinds of biological and inorganic waste into the environment. Fly ash creates a dense, floating layer that blocks sunlight from reaching the water's depths. Fly ash increases the alkalinity of the water, preventing aquatic plants from absorbing the vital bases they require to survive (Chowdhary et al. 2020). Liquid organic effluents change the pH of the water, and different chemical compounds have various harmful effects on aquatic plants. Metals may have additive, synergistic, or antagonistic effects on plants; however, hard and buffered freshwater bodies mitigate these interactions.

Effects of silt deposition: Silt accumulates in bodies of water because of two main causes: erosion with its accompanying silt-laden water and floods. It makes the water murkier and blocks off less light in the depths, leading to less submerged vegetation. Siltation slows down the growth of aquatic vegetation (Alidadi et al. 2022). The reduced gas and nutrient exchange at the water's surface affects phytoplankton abundance. The most prevalent plants are those that are tolerant with a medium level of tolerance, and finally the least tolerant plants are polygonum, sagittaria, etc.

Effects of oil spillage: Oil tankers and storage containers contribute to water pollution, which lowers the oxygen content of the water. This happens when oil pollution blocks some of the light, slowing the development of plankton and halting photosynthesis in macrophytes (Tayeb et al. 2020).

Effects of thermal pollution: Aquatic life is negatively impacted when thermal power plants emit hot water into bodies of water. Because of the high temperature and oxygen deprivation, it decreases the activity of aerobic decomposers. A decrease in the decomposition of organic matter puts aquatic bodies' nutrient availability in danger (Adejumo and Adebiyi 2020). Aquatic plants have slower photosynthesis rates as a result of temperature-induced suppression of enzyme activity. Thermal pollution causes water bodies to become warmer, which leads to a decrease in the primary production and species diversity of aquatic plants.

Effect of nutrient enrinchement: Eutrophication occurs when water bodies are enriched with inorganic nutrients, particularly nitrogen (N) and phosphorus (P), leading to an increase in plant and algal development. Certain freshwater ecosystems may be eutrophic as part of their natural succession. Human-caused nutrient enrichment, or "culture eutrophication," is characterized by a significantly accelerated rate of enrichment. The pollution issue of eutrophication in lakes and reservoirs throughout North America was first identified around the middle of the twentieth century (Adams et al. 2020). Particularly beneficial to expansion are nitrogen and phosphorus, which both have a role in enhancing photosynthesis. Because of this, they are often found in plant foods. When algae blooms grow in nutrient-rich waters, such as those resulting from agricultural runoff, eutrophication takes place. As a consequence, fish populations collapse as a result of oxygen shortages.

Phytotoxicity effects on plants: Plants may take up chemical contaminants in the soil or water via their roots if there is an excess of such pollutants. Toxic chemical poisoning of plants is known as phytotoxicity. Lackluster development, perishing seedlings, and leaf blotches are all signs of phytotoxicity (Zhang et al. 2021). The bioaccumulation of mercury compounds in plant roots and bodies often associated with fish, may also impact aquatic plants. The more people who know about the dangers of mercury, the less likely they are to consume it.

20.11 The Repercussions of Polluted Water on the Natural World

Animal, plant, microbes, and fungi all depend on one another in some way or another for an ecosystem to be healthy. Damage to even one of these species may have a domino effect, putting whole marine ecosystems in jeopardy. As a result of the proliferation of newly supplied nutrients, plant, and algae development is stimulated, and oxygen levels in the water are reduced when water pollution generates an algal bloom in a lake or marine environment (Goudie 2018). The lack of oxygen caused by eutrophication kills plants and animals and may lead to the formation of "dead zones" in the ocean. Animals as whales and sea turtles may be harmed by the neurotoxins that can be produced by some of these toxic algal blooms. Wastewater from factories and cities also pollutes water supplies with chemicals and heavy metals. These pollutants are ingested by predators to pass them on to their prey, causing widespread harm to aquatic ecosystems. Marine debris may entangle, smother, and starve marine life, posing a serious hazard to marine ecosystems (Camargo and Alonso 2006). Plastic bag and soda cans are only two examples of the solid waste that is washed into sewers and storm drains and ultimately washed out to sea, where it combines with other rubbish to create floating garbage patches and turns the ocean into trash soup (Gall and Thompson 2015). More than 200 marine species are at risk due to trash and abandoned fishing gear. It is becoming more difficult for shellfish and coral to thrive due to ocean acidification. Despite soaking up roughly a quarter of annual carbon pollution from fossil fuel use, the seas are growing increasingly acidic. The difficulty of shell formation and possible effects on neural systems are two consequences of this process. Influence the neurological systems of sharks and clownfish, among other marine animals, while helping shellfish and other species construct shells (Di Santo 2019).

20.12 Conclusion

The use of water is proportional to the expanding human population. The prevention of pollution and the conservation of water are both very necessary if we want to guarantee that future and current generations will always have access to drinkable water. The following preventative steps that may be performed to reduce pollution: stabilization of the soil by the planting of forest trees, shrubs, and grasses with a minimum use of chemical fertilizers, herbicides, and insecticides can use effective conservation practices such as building terraces on slopes, planting grass waterways, and creating wetlands to slow down the progression of pollutants, especially in areas that are close to waterways. Moreover it avoid storing chemicals outside or provide a cover to prevent rainwater from eroding them, and they should conduct soil tests to determine the appropriate amounts of fertilizer to apply. While it may be difficult to believe, each individual is capable of contributing to the effort to maintain the world's supply of fresh water. Even though the nation's water policy is moving in the correct direction, the implementation of it has not been very successful. To curb the amount of water that is used, a combination of economic and regulatory strategies is required. It is essential to effectively execute policies, strategies, and regulatory measures to cut down on water waste. It is possible that the fight against water scarcity in the nation would be benefited by increasing awareness and teaching water users in the agriculture sector about the advantages of adopting more water-efficient production techniques. That will help the country in its fight against water limitations. In addition, the existing policymakers and planners may benefit from implementing best practices to enhance governance structures and obtain a better knowledge of crucial indicators for data-driven decision-making if they were to do so. These problems may be more effectively addressed if there were supporting policies and practices that promoted the agricultural sector to raise the efficiency with which it used water. Almost every nation is now making concerted efforts to solve environmental problems, and these efforts, together with a growing awareness of the need of incorporating environmental concerns into national development objectives, are more commonplace. To name just a few of the most significant environmental challenges facing India as a developing country, some of these challenges include industrial pollution, car emissions, hospital waste, household sewage disposal, and so on. Industrial pollution in the form of air, water, solids, thermal pollution, and other forms reaches new heights with each passing day, needing prompt attention and appropriate cures. There is a wide variety of commercial activity that contributes to the formation of an economy. Because of the inextricable connections that exist between these domains, it is impossible to treat them as separate entities. As a consequence of that, there is a general agreement among most people that the output of one industry may fulfill the requirements of another. When the idea of interconnectedness across economic sectors is considered within the context of Leontief's input–output approach, it becomes clear that certain economic sectors may not directly cause pollution, but they do so indirectly in a very substantial manner. There have been a lot of investigations carried out. While there aren't many quantitative studies done on this subject, there is an undeniable connection between water pollution and economic activity. The availability of up-to-date and accurate information has made it feasible to conduct an in-depth quantitative study linking many aspects of the Indian economy to water pollution. The analysis will look at the economy is related to water pollution. It has been determined that India's biodiversity is threatened by the amount of water pollution both an indirect and direct influence. Our synthesis will improve our knowledge of the environmental implications of contaminated water, enhance public awareness of the issue, and determine the paths of future study.

References

Abell R, Vigerstol K et al (2019) Freshwater biodiversity conservation through source water protection: quantifying the potential and addressing the challenges. Aquat Conserv Mar Freshwat Ecosyst 29(7):1022–1038

Adams J, Taljaard S et al (2020) Nutrient enrichment as a threat to the ecological resilience and health of South African microtidal estuaries. Afr J Aquat Sci 45(1–2):23–40

Adejumo IO, Adebiyi OA (2020) Agricultural solid wastes: causes, effects, and effective management. Strat Sustain Solid Waste Manag 8

Ajibade FO, Adelodun B et al (2021) Environmental pollution and their socioeconomic impacts. Elsevier, Microbe mediated remediation of environmental contaminants, pp 321–354

Al-Hameedi ATT, Alkinani HH et al (2020) Experimental investigation of bio-enhancer drilling fluid additive: can palm tree leaves be utilized as a supportive eco-friendly additive in water-based drilling fluid system? J Petrol Explor Prod Technol 10:595–603

Alam GM, Khatun MN (2021) Impact of COVID-19 on vegetable supply chain and food security: empirical evidence from Bangladesh. PLoS ONE 16(3):e0248120

Alidadi F, Hojjati M et al (2022) The effect of different types of land use on the physical, chemical and carbon deposition characteristics of the soil along the Karkhe River. J Wood Forest Sci Technol 29(4):1–21

Ameen HA (2019) Spring water quality assessment using water quality index in villages of Barwari Bala, Duhok, Kurdistan Region, Iraq. Appl Water Sci 9(8):176

Artiola J, Walworth J et al (2019) Soil and land pollution. Environmental and pollution science. Elsevier, pp 219–235

Atakan R, Tufan HA et al (2020) Protocol to assess the quality of transmission lines within smart textile structures. Measurement 152:107194

Banerjee S, Hazra S et al (2021) Investigating India's pollution-intensive 'dirty'trade specialisation: analysis with 'revealed symmetric comparative advantage'index. Environ Sci Pollut Res 28:30153–30167

Bendigeri P, Kawalekar JS, Health hazards and waste management 64

Benton TG, Bailey R (2019) The paradox of productivity: agricultural productivity promotes food system inefficiency. Glob Sustain 2:e6

Bilal M, Adeel M et al (2019) Emerging contaminants of high concern and their enzyme-assisted biodegradation–a review. Environ Int 124:336–353

Bouwer H (2002) Integrated water management for the 21st century: problems and solutions. J Irrig Drain Eng 128(4):193–202

Boyce JK, Ash M et al (2023) Environmental justice and carbon pricing: can they be reconciled? Glob Challenges 2200204

Bueno PC, Lopes NP (2020) Metabolomics to characterize adaptive and signaling responses in legume crops under abiotic stresses. ACS Omega 5(4):1752–1763

Camargo JA, Alonso Á (2006) Ecological and toxicological effects of inorganic nitrogen pollution in aquatic ecosystems: a global assessment. Environ Int 32(6):831–849

Celik S (2020) The effects of climate change on human behaviors. Environ Clim Plant Veg Growth 577–589

Chakraborti D, Das B et al (2011) Examining India's groundwater quality management, ACS Publications

Champion WM, Khaliq M et al (2022) Advancing knowledge to reduce lead exposure of children in data-poor low-and middle-income countries. Environ Sci Technol Lett 9(11):879–888

Chowdhary P, Bharagava RN et al (2020) Role of industries in water scarcity and its adverse effects on environment and human health. Environ Concerns Sustain Develop Volume 1 Air Water Energy Resour 235–256

Cissé G (2019) Food-borne and water-borne diseases under climate change in low-and middle-income countries: further efforts needed for reducing environmental health exposure risks. Acta Trop 194:181–188

Coffey R, Paul MJ et al (2019) A review of water quality responses to air temperature and precipitation changes 2: nutrients, algal blooms, sediment, pathogens. JAWRA J Am Water Resour Assoc 55(4):844–868

De la Parra-Guerra, A. C. and R. Acevedo-Barrios (2023) Studies of endocrine disruptors: nonylphenol and isomers in biological models. Environ Toxicol Chem

de Mello K, Taniwaki RH et al (2020) Multiscale land use impacts on water quality: assessment, planning, and future perspectives in Brazil. J Environ Manage 270:110879

Di Santo V (2019) Ocean acidification and warming affect skeletal mineralization in a marine fish. Proc R Soc B 286(1894):20182187

Ding X, Liu L (2019) Long-term effects of anthropogenic factors on nonpoint source pollution in the upper reaches of the Yangtze river. Sustainability 11(8):2246

Elkhatib D, Oyanedel-Craver V (2020) A critical review of extraction and identification methods of microplastics in wastewater and drinking water. Environ Sci Technol 54(12):7037–7049

Evans AE, Mateo-Sagasta J et al (2019) Agricultural water pollution: key knowledge gaps and research needs. Curr Opin Environ Sustain 36:20–27

Fazal-ur-Rehman M (2019) Polluted water borne diseases: symptoms, causes, treatment and prevention. J Med Chem Sci 2(1):21–26

Gall SC, Thompson RC (2015) The impact of debris on marine life. Mar Pollut Bull 92(1–2):170–179

Gao H, Yan C et al (2019) Effects of plastic mulching and plastic residue on agricultural production: a meta-analysis. Sci Total Environ 651:484–492

Gerba C, Pepper I (2019) Microbial contaminants. Environ Pollut Sci 191–217

Goswami RK, Mehariya S et al (2021) Microalgae-based biorefineries for sustainable resource recovery from wastewater. J Water Process Eng 40:101747

Goudie AS (2018) Human impact on the natural environment. Wiley

Hajjar MJ, Ahmed N et al (2023) Integrated insect pest management techniques for rice. Sustainability 15(5):4499

Islam MM, Iqbal MS et al (2021) A review on present and future microbial surface water quality worldwide. Environ Nanotechnol Monit Manag 16:100523

Ji Y, Gao D et al (2023) An integrated framework for agricultural non-point source pollution control technology evaluation: application of an improved multiple attribute decision making method. Expert Syst Appl 120319

Kanakaraju D, Chandrasekaran A (2023) Recent advances in TiO2/ZnS-based binary and ternary photocatalysts for the degradation of organic pollutants. Sci Total Environ 868:161525

Khan A, Kanwal H et al (2021) Volatile organic compounds and neurological disorders: from exposure to preventive interventions. Environmental contaminants and neurological disorders. Springer, pp 201–230

Kumar D (2023) Climate change, a strong threat to food security in India: with special reference to Gujarat. Climate change impacts on natural resources, ecosystems and agricultural systems. Springer, pp 153–173

Lawrencia D, Maniam G et al (2023) Critical review of household water treatment in Southeast Asian countries. Wiley Interdisciplinary Reviews: Water: e1640

Lim HR, Khoo KS et al (2021) Perspective of Spirulina culture with wastewater into a sustainable circular bioeconomy. Environ Pollut 284:117492

Liu J, Zhang T et al (2021) Microplastic pollution in China, an invisible threat exacerbated by food delivery services. Bull Environ Contam Toxicol 107:778–785

Ma Y, Shen Y et al (2020) Food waste to biofertilizer: a potential game changer of global circular agricultural economy. ACS Publications

Ma J, Li Z et al (2022) Modeling soil detachment capacity by rill flow under the effect of freeze–thaw and the root system. Nat Hazards 112(1):207–230

Malik D, Sharma AK et al (2020) A review on impact of water pollution on freshwater fish species and their aquatic environment. Adv Environ Pollut Manag Wastewater Impacts Treat Technol 1:10–28

Mishra R, Dubey S (2023) Fresh water availability and it's global challenge. J Mar Sci Res 2(1)

Mukherjee S, Pahari DP (2019) Underground and opencast coal mining methods in India: a comparative assessment. Space Cult India 7(1):39–55

Nanda S, Berruti F (2021) Municipal solid waste management and landfilling technologies: a review. Environ Chem Lett 19:1433–1456

Naylor R, Fang S et al (2023) A global view of aquaculture policy. Food Policy 116:102422

Negri AP, Smith RA et al (2019) Adjusting tropical marine water quality guideline values for elevated ocean temperatures. Environ Sci Technol 54(2):1102–1110

Newell DG, Koopmans M et al (2010) Food-borne diseases—the challenges of 20 years ago still persist while new ones continue to emerge. Int J Food Microbiol 139:S3–S15

Obebe S, Adamu A (2020) Plastic pollution: causes, effects and preventions. Int J Eng Appl Sci Technol 4(12):85–95

Ochilova NR, Muratova GS et al (2021) The importance of water quality and quantity in strengthening the health and living conditions of the population. Central Asian J Med Nat Sci 2(5):399–402

Okereafor U, Makhatha M et al (2020) Toxic metal implications on agricultural soils, plants, animals, aquatic life and human health. Int J Environ Res Public Health 17(7):2204

Owa F (2013) Water pollution: sources, effects, control and management. Mediterr J Soc Sci 4(8):65

Panhwar A, Abro R et al (2022) Global water mapping, requirements, and concerns over water quality shortages

Patel M, Kumar R et al (2019) Pharmaceuticals of emerging concern in aquatic systems: chemistry, occurrence, effects, and removal methods. Chem Rev 119(6):3510–3673

Perkins DB, Stone Z et al (2022) Development of a US national-scale, mixed-source, pesticide, rural well database for use in drinking water risk assessment: an atrazine case study. Environ Monit Assess 194(8):578

Perveen S (2023) Drinking water quality monitoring, assessment and management in Pakistan: a review. Heliyon

Plaas HE, Paerl HW (2020) Toxic cyanobacteria: a growing threat to water and air quality. Environ Sci Technol 55(1):44–64

Popoola BM (2022) Biodegradable waste. Recycling-recent advances. IntechOpen
Raimi MO, Sawyerr HO, Ezekwe IC, Gabriel S (2022) In: Saleh PH, Hassan PAI (eds) Toxicants in water: hydrochemical appraisal of toxic metals concentration and seasonal variation in drinking water quality in oil and gas field area of Rivers State, Nigeria. Heavy Metals-New Ins.
Ram U, Ram F (2021) Demographic transition in India: insights into population growth, composition, and its major drivers. Oxford Research Encyclopedia of Global Public Health
Ritchie H, Roser M (2021) Clean water and sanitation. Our world in data
Robinson RM (2023) Habitat restoration: an introduction. Environmental advocacy and local restorations. Springer, pp 1–42
Rudebusch J, Hughes BB et al (2020) Assessing anthropogenic risk to sea otters (Enhydra lutris nereis) for reintroduction into San Francisco Bay. PeerJ 8:e10241
Saleh TA (2020) Trends in the sample preparation and analysis of nanomaterials as environmental contaminants. Trends Environ Anal Chem 28:e00101
Sanda B, Bah A et al (2022) Involvement of the general public in the management of water pollution in Johor Bahru Malaysia. Arid Zone J Eng Technol Environ 18(1):79–96
Sarkar A, Devi S (2022) Health hazards of water contamination: an updated review among the COVID-19 pandemic. J Datta Meghe Inst Med Sci Univ 17(4):996–1004
Sathre R, Antharam SM et al (2022) Water security in South Asian cities: a review of challenges and opportunities. CivilEng 3(4):873–894
Sayyed AJ, Pinjari DV et al (2021) Cellulose-based nanomaterials for water and wastewater treatments: a review. J Environ Chem Eng 9(6):106626
Shayan ME, Ghasemzadeh F (2020) Nuclear power plant or solar power plant. Nuclear Power Plants-The Processes from the Cradle to the Grave, IntechOpen
Sonone SS, Jadhav S et al (2020) Water contamination by heavy metals and their toxic effect on aquaculture and human health through food Chain. Lett Appl Nanobiosci 10(2):2148–2166
Syafrudin M, Kristanti RA et al (2021) Pesticides in drinking water—a review. Int J Environ Res Public Health 18(2):468
Tayeb AM, Farouq R et al (2020) Oil spill clean-up using combined sorbents: a comparative investigation and design aspects. Int J Environ Anal Chem 100(3):311–323
Tubb C, Seba T (2021) Rethinking food and agriculture 2020–2030: The second domestication of plants and animals, the disruption of the cow, and the collapse of industrial livestock farming. Ind Biotechnol 17(2):57–72
Uitto JI (2021) Surviving the anthropocene: how evaluation can contribute to knowledge and better policymaking. Evaluation 27(4):436–452
Ukaogo PO, Ewuzie U et al (2020) Environmental pollution: causes, effects, and the remedies. Microorganisms for sustainable environment and health. Elsevier, pp 419–429
UNEP D (2021) Partnership and United Nations Environment Programme (2021). Reducing consumer food waste using green and digital technologies
Velvizhi G, Shanthakumar S et al (2020) Biodegradable and non-biodegradable fraction of municipal solid waste for multifaceted applications through a closed loop integrated refinery platform: paving a path towards circular economy. Sci Total Environ 731:138049
Walker D, Baumgartner D et al (2019) Surface water pollution. Elsevier, Environmental and pollution science, pp 261–292
Wang R, Wang Q et al (2021) Cleaner agricultural production in drinking-water source areas for the control of non-point source pollution in China. J Environ Manage 285:112096
Wato T, Amare M et al (2020) The agricultural water pollution and its minimization strategies–a review. J Resour Dev Manag 64:10–22
Wen X, Chen F et al (2020) Microbial indicators and their use for monitoring drinking water quality—a review. Sustainability 12(6):2249
Wurtsbaugh WA, Paerl HW et al (2019) Nutrients, eutrophication and harmful algal blooms along the freshwater to marine continuum. Wiley Interdiscip Rev Water 6(5):e1373

Xue S, Ding W et al (2022) Emission, dispersion, and potential risk of volatile organic and odorous compounds in the exhaust gas from two sludge thermal drying processes. Waste Manage 138:116–124

Yang H, Ye S et al (2020) Utilization of biochar for resource recovery from water: a review. Chem Eng J 397:125502

Zhang Y, Whalen JK et al (2021) Phytotoxicity and bioconcentration of microcystins in agricultural plants: meta-analysis and risk assessment. Environ Pollut 272:115966

Chapter 21
Role of Global Climate Change in Crop Yield Reductions

Priyanka Devi, Prasann Kumar, and Joginder Singh

Keywords No poverty · Climate change · Crop production · Greenhouse gas · Global warming · Zero hunger · Sustainability

21.1 Introduction

Evidence from around the world shows that climate change will affect food production and threaten food security. Agricultural systems must adapt to climate change to increase food production to meet the needs of a growing global population with shifting dietary preferences and to reduce their negative effects on the planet's ecosystems. The most important tools for assessing the potential impacts of climate change on crop yield and practical adaptation strategies for agricultural risk management are crop simulation models. The importance of crop simulation models to ensure a safe food supply is one of the main concerns. Over the past fifty years, much progress has been made in determining how environmental factors, such as temperature and precipitation, may affect agricultural output. The effects of climate change and worldwide warming are the results of unchecked emissions of greenhouse gases (GHGs). In the future, it is going to change at an alarming rate. Changes in weather patterns, such as higher or lower temperatures, more or less precipitation, and droughts or heat waves, can be traced back to global warming's worsening impact. Heat shock injuries and other biochemical and physiological changes will result from the higher temperature. Grain output and yield were thus diminished, although the extent of the decline

P. Devi · P. Kumar (✉)
Department of Agronomy, School of Agriculture, Lovely Professional University, Phagwara, Punjab 144411, India
e-mail: prasann0659@gmail.com

J. Singh
Department of Botany, Nagaland University, Hqrs. Lumami Zunheboto, Nagaland, India

P. P. Adhikary et al. (eds.), *Soil, Water Pollution and Mitigation Strategies*, Environmental Science and Engineering, https://doi.org/10.1007/978-3-031-63296-9_18

varied by the location also. The balance between C3 photosynthetic carbon assimilation and photorespiration is influenced and sustained by the increased temperature. In the case of C3, plants will be preferred under raised CO_2, and C4 plants will be preferred under higher temperatures due to the interaction between CO_2 and temperature. Increased CO_2 effects on plant development, warmer and drier weather, changes in wind speed, pest and disease pressure, and many other changes resulting from changed interactions among components of crop agroecosystems will all influence agriculture. As global warming continues to worsen, experts expect global food production to grow overall. New farming techniques and crops that lower net GHG output by generating less nitrous oxide, increasing soil organic matter content, and permitting the production of bio-products like biofuel can help in mitigating greenhouse gas impacts. Sustainable development necessitates the use of mitigation and adaptation measures to boost crop yields while reducing production risks in agriculture. Multimodal approaches that combine crop modelling with plant physiology, biochemistry, and gene-based modelling can help minimize prediction uncertainty. Creating and collecting high-quality field data for testing models is essential for new model development. Therefore, it is important to analyze agricultural output regularly to make sure that food will be available for future generations. This can be done by keeping a body of knowledge and resources that can be used to lessen the effects of climate change and improve farming methods. Increased emissions of GHGs, including CO_2, CH_4, and N_2O, from anthropogenic sources (fossil fuel burning) are driving rapid changes in the global climate (IPCC 2014). CO_2 accounts for 76% of all anthropogenic GHG concentrations, 16% originating from CH4, 6.2% originating from N_2O, and 2.0% originating from CFC gases (IPCC 2014). As more and more of these gases are released into the atmosphere, their concentrations change. The world's climate is changing as a result of human-caused global warming. This includes higher average temperatures, an increase in the frequency of extreme weather events like floods and droughts, and a shift toward wetter and more extreme higher average global sea levels). This is because it raises infrared absorption radiation, which is reflected away from Earth's surface and interferes with our bodies' ability to produce and use energy (Venkataramanan 2011; IPCC 2013). Coal combustion (45%), oil combustion (35%), and natural gas combustion (20%) were each cited as contributing to the world's CO_2 emissions. The most significant contributions to global CO_2 emissions were made in 2017 by China (27%), the United States (15%), the European Union (10%), and India (7%). Together, these five countries are responsible for approximately 59% of the world's CO_2 emissions, while the remaining nations contributed 42%. However, carbon dioxide (CO_2), the most important governing component that is responsible for warming the planet, is frequently ignored. Since 1950, the concentration of carbon dioxide in the atmosphere has increased by thirty percent, which is a significant rise following the advent of industrialization. Before the beginning of the industrial revolution in 1750 A.D., CO_2 levels in the atmosphere were at 280 parts per million. Under current conditions, the average daily concentration of carbon dioxide at Hawaii's Mauna Loa Observatory was 412 parts per million in April 2019. By the end of 2100, projections put it between 421 and 936 ppm, up from a current estimate of 443 to 541 ppm for 2050. Climate change contributes to a

0.84 °C increase in the world's average temperature (IPCC 2014; Meinshausen et al. 2011; Hartmann et al. 2013). Therefore, there is a sudden and strategic necessity for collaborative action efforts from every region of the world to reduce emissions in order to keep atmospheric CO_2 levels at the bottom of that range (about 421 ppm). Since 1870, multiple sources and sinks have been contributing to the global carbon budget, which is an accumulation of their total contributions. In this area, 202 parts per million of $CO_2$are released due to fossil fuel combustion and cement production. These emissions increased by 2.7% per year, or 63%, compared to 1990 levels. Consistent with current trends, CO_2 emissions might reach 100 $GtCO_2$. There will be an average annual emission of 36 $GtCO_2$/year by the end of the twenty-first century will have approximately tripled, and the concentration will have finally surpassed the barrier of 1000 PPM (Fuss et al. 2014).

Primary crops include cereal grains like wheat, maize, and rice because they significantly affect a person's daily caloric intake. To feed the estimated 9.8 billion people on the planet by 2050, the cereal food supply must increase by 70–100% (Godfray et al. 2010). It is widely agreed that production rates need to be increased to meet the ketorising demand, yet statistics reveal that present production levels are well below what is needed to achieve the goals set (Ray et al. 2013). Current agricultural techniques are also contributing to a dramatic decline in the quantity accessible area that is fertile and suitable for agriculture in order to develop these goods, a trend that is only expected to intensify in the foreseeable future (Hawkesford et al. 2013). Researchers believe that addressing global climate change, which can affect crop production yields, is a problem that must be tackled to achieve food security (Tripathi et al. 2016; Howden et al. 2007). Over the next 30 years, the average surface temperature will rise by 0.2 °C every decade, according to research. Estimates also predict that by the end,increasing quantities of greenhouse gases (GHGs), including carbon dioxide (CO_2) and methane (CH_4) in the twenty-first century would cause an increase in world average temperatures of between 2.5 and 4.5 °C (Solomon 2007; Bernstein et al. 2008). By raising plant respiration rates, global warming can lessen the net carbon gain, which in turn reduces food production yields and may even lead to the introduction of weeds, diseases, and pests (Asseng et al. 2011; Hogy and Fangmeier 2008). When growing wheat, for instance, a rise in temperature of just 1 °C can affect yields by 3–10% (You et al. 2009). Reduced The agricultural yields of key cereal crops can make it far more difficult to provide food security in a world where population growth is growing and nearing levels that are unsustainable. For this reason, scientists need to identify suitable crops that can be grown despite climate change. At the same time, these plants should be able to produce fewer greenhouse emissions, utilise fewer resources, and provide plenty of essential micronutrients.

Millets, a kind of cereal crop, provide several of the aforementioned benefits in comparison to other economically significant crops (such as wheat along with paddy and maize) (Jain et al. 2016). Millets are a type of grain crop that can withstand harsh conditions and have a relatively quick growth cycle, anywhere from 60 to 100 days. Millets are crops that do well in a variety of climates and situations, including drought. In addition, studies have indicated that millets produce fewer greenhouse emissions, which might help the agriculture and food industries contribute less to global warming (Jain and Bal 1997). Millets have the potential to become a significant food source by helping us reach our food and nutritional security ambitions in a world where both the water crisis and the food crisis are at the forefront. In addition to being a more sustainable option, millets are preferred as a substitute since they require less water, fertilizer, and good soil than crops like maize and rice. Paddy, for instance, needs a lot of water throughout its growth, and a lack of it might have serious consequences for the output (Yao et al. 2012). Additional methods to boost cereal crop output include breeding, technological advancement, and enhanced fertilizer efficiency. Heat and water stressors are two primary ways in which climate change may reduce grain productivity. Indirect effects on fertilizer availability, disease vectors, and insect populations are also possible (Porter et al. 2014). To minimize and repair the environmental harm caused by global warming, the majority of academics feel that immediate action is futile and that long-term policy changes and sustainable agriculture methods are necessary. The primary goal of this analysis is to determine how climate change will affect future crop yields for the most important cereal crops. In addition to advocating for the cultivation of millets as a means of crop diversification, several potential remedies for the impending problem of food insecurity have been proposed. Still, especially in areas with poor soil and water conditions, it's crucial to remember that yield projections are subject to a wide range of uncertainties due to a variety of causes.

One of the most pressing problems of the twenty-first century is how to feed the world's growing population without further straining an already fragile ecosystem as a result of climate change (Lal 2005). The impacts of climate change on water supplies, food security, hydropower, and human health have already been observed around the globe, most notably in Africa (Magadza, 2000). Having an effect on agricultural production (Howden and O'Leary 1997), water reserves in river-basin catchments, woods, factories, and the overall natural environment, to name a few, are driving the need for research into climate change and adaptation measures (Dockerty et al. 2005). Using characteristics from several climate models, crop growth models have been used to investigate crop yield and soil water balance. Still, even in high-yield and high-technology agricultural regions, climatic unpredictability is one of the most significant factors impacting annual crop productivity.

Over the course of the last few decades, there has been considerable climate change worldwide, including in many agricultural regions across the world. There have been reports of increases not only in atmospheric carbon dioxide (CO_2) and in atmospheric carbon dioxide (CO_2) but also in atmospheric ozone (O_3). The question of whether or not the very probable path that climate and CO_2 will continue to pursue in the coming decades will have an effect on the overall productivity of agricultural land throughout the world is one of the most significant concerns associated with food security for the future. We present estimates of historical and predicted consequences of climate and CO_2 trends and illustrate the processes by which these changes affect agricultural production. Despite the various different levels and results of relevance to food security, this assessment zeroes in only on world grain productivity. Global yields are expected to rise by around 1.8% per decade over the next three decades due to CO_2 trends. Warming trends will probably lower world yields by anywhere between 0 and 4% each decade in the absence of a significant response. This range's high end is a little fraction of the 8% rate of increase forecasted over the next few decades because of advances in technology and management. Higher O_3 and increased rainfall intensity are two examples of elements of global change that will certainly threaten yields but are not taken into account in most present evaluations. Over the next few decades, global food security will be shaped by several factors, including shifts in the expansion of the population, the rise and distribution of income, and dietary preferences, disease prevalence, rising rates of improvement in agricultural output, and growing demand on land and water resources for other purposes (such as the production of bioenergy, the storage of carbon, and the growth of urban areas) are some of the factors to consider. As scientists, the plant science community places a premium on this last component, which we'll refer to as "crop yield," as farmers try to sustain the substantial historical increases associated with enhanced genetics and agronomic management of key food crops (i.e., metric tonnes of grain production per hectare of land). The plant science community calls this factor "crop yield." The quality of the soil, factors such as variations in temperature (T) and precipitation (P) as well as the cost and accessibility of mineral fertilizers all have a role (P). Conditions are only a few of the many factors that contribute to agricultural productivity increases. In this update, we look at how climate change, increased CO_2, and reduced O_3 in agricultural regions have affected as well as continue to affect crop yields. By doing so, we acknowledge this is simply a little piece of the larger story about agricultural production, which is only a small piece of the larger story regarding the future of food. Food security can be affected by global change in many ways outside agricultural output, such as through effects on human illness rates and income levels; however, these are not included in this update (Solomon et al. 2007).

21.2 Various Climate-Related Models and Their Projections

Climate scenarios are plausible projections of future weather patterns derived from a variety of climatological connections and assumptions about the role of nuclear forcing. Complex three-dimensional mathematical representations are known as climate change. Climate change is characterised by processes of interaction between the atmosphere, land surface, seas, and sea ice. Change may be shown using global climate models (GCMs) and regional climate models (RCMs) (Mearns 2000). Instead of a method of prediction, climate projections should be seen as effective ways of determining potential futures under certain emission scenarios (Suppiah et al. 2007). Though they still have major inaccuracies, GCMs are valuable instruments for simulating the key features of present and future climates (IPCC 2007). The GCMs that have been used to estimate future climates are summarised by year in Table 21.1. Better regional climate models derived from GCMs with finer grid spacing enable climate scientists to better predict how global warming will affect agriculture on a local scale (Mearns 2000). When trying to mimic the known length of low-sensitivity monthly rainfall quantities, RCMs have some problems, according to Blenkinsop and Fowler (2007). Predicting climate vulnerability and climatic factors like rainfall and temperature requires integrating climate models with other modelling methodologies. Using the general circulation model, HadCM3 (Hadley Center Climate Model 3), Metzger et al. (2005) established an ATEAM (advanced terrestrial ecosystem analysis and modelling) methodology to quantify the vulnerability of climate change in 2080 predictions. Using a statistical approach, Suppiah et al. (2007) chose 15 of the best models from which they extrapolated Estimates of yearly annual, seasonal, and statewide averages of Australia's precipitation and temperature.With the use of climatic data from the past and recommended policy options, Khan (2008) explored climate variability and droughts in Australia. These included adjusting model-based surface water, and groundwater allocations, improvements in agricultural water usage efficiency, and the creation of a national legislative framework for the management of water resources in light of projected effects of climate change on water supplies are all necessary. The new analytic approach to decision-making based on robust decision-making was introduced by Groves and Lempert (2007), making it possible to put a numerical value on the Special Report on Emissions Scenarios (SRES) scenarios that had been generated with the scenario-axes technique. Using a stochastic simulation approach, Fowler et al. (2005) analysed the NSRP (Neyman-Scott Rectangular Pulse) rainfall scenario-generating model to extract climatic data on precipitation. The PRECIS climate model employed by Xu et al. (2006) shows a rise inUnder B2 projections. The 2080s will see a steady increase in both the average temperature and precipitation in China. With the downscaling method, climatic scenarios and repercussions have been predicted using GCMs in many different situations (Thomas 2008). The geographical and temporal resolution of GCM data is frequently insufficient for thorough regional study, and in many cases, GCMs have flaws that make it impossible to model even the current climate (Bonan et al. 2002). While GCMs may give acceptable accuracy regarding large-scale characteristics and other fluctuations owing to climate forcing, they have uncertainty in projecting future climate data (Salinger et al. 2000). Because every climate model has its degree of unpredictability, it is best to use many models to tackle the issue of making reliable projections (Cculeanu et al. 2002).

21.3 Conditions Necessary for the Growth of Various Plants

21.3.1 Rice

Paddy, also known as *Oryza sativa*, is farmed in damaged crops in over 95 nations and is an important staple food for many people throughout the world, notably in India and China. Growing rice requires a lot of water—roughly 500–600 mm, depending

Table 21.1 Effect of high temperature and elevated carbon dioxide on crop growth and quality

S.No.	Experimental plant	Parameter	Effect	Reference
A. Crop quality is affected by high temperatures				
1	Maize and soybean	Fe and P	Increase	Qiao et al. (2019)
2	Sweet corn, rice, wheat, pea, tomato, cabbage	Sugar content, protein, and starch	Decrease	Myers et al. (2014) Gruda and Tanny (2014)
3	Tomato	Lycopene Y, ascorbic acid	Decrease	Rosales et al. (2010)
4	Carrot, tomato, and lettuce	Carotene	Decrease	Wang and Frei (2011)
5	Broccoli, lettuce	Loosehead	Increase	Kałuzewicz et al. (2009)
B. Different effect of elevated CO_2 on crop quality				
1	Maize	Protein	500 ± 20 ppm, decrease	Abebe et al. (2016)
2	Chinese cabbage red leaf lettuce	Sugars	800–1000 ppm, increase	Becker and Klaring (2016)
3	Spinach and lettuce	N, P, K, S	700 ppm, decrease	Giri et al. (2016)
4	Rice	Ca	700 ppm, increase	Seneweera (2011)
5	Spinach and lettuce	Zn and Cu	700 ppm, decrease	Giri et al. (2016)

on the weather and the type of rice that is grown. It has been observed that 22–31 °C and 5–6 h of daily sunlight are ideal for development (Yao et al. 2012). Numerous scientific investigations have established that weather variables (rainfall, irrigation, and temperature) may have a major impact on harvest success. When using the "continuous flooding approach," paddy yielded 8.23 t ha^1 in a recent study. This is more than the 7.98 t ha^1 achieved. In the days of traditional paddy cultivation, alternative "wetting and drying irrigation technology," which was implemented in areas with a scarcity of water. A total of 296 mm of irrigation water was applied in a continuous flooding system, while an estimated 533 mm of rain fell on the field during the course of the growing season. The "wetting and drying technique" maintained the same level of precipitation while decreasing irrigation water to 224 mm.

21.3.2 Millets

For hundreds of years, people in India, China, and Africa have relied on millet for human nutrition because of its strong nutritive properties (high protein content; 11 g protein per 100 g) (Wen et al. 2014). Commonly farmed millet species include proso, pearl, finger, Kodo, and foxtail. Millet is a grain crop that can withstand dry conditions and has a short growing period of 60–100 days (Habiyaremye et al. 2017). Millet crops may grow anywhere heights between 30 and 100 cm, a substantial root system, and often do not need any fertilizer. Because of its low transpiration rate, millet can fix carbon even in extreme heat and with inadequate nitrogen (Habiyaremye et al. 2017; Goron et al. 2015). The ideal conditions for producing millets are stated: 20–35 °C with 4–6 h of sunlight each day. Millet has a broad optimum pH range, from 4.5 to 8.0, making it suitable for cultivation in a wide variety of soil types (including sandy, salty, alkaline, and acidic). Furthermore, millet types may be cultivated on low-fertility soils (sandy loam, slightly acidic soils) because of their low fertilizer demand (Changmei and Dorothy 2014). As a result of their inherent resilience to illness, most millets don't attract pests, which might be a major step toward lowering pesticide use and the pollution it causes.

21.3.3 Maize

A member of the grass family, maize (*Zea mays* L.) is also known as corn and Indian corn. The Mexicans domesticated it around 7000 years ago, and since then it has spread throughout the Americas since become an important staple crop throughout most of North and South America (Piperno and Flannery 2001). When compared to wheat and paddy, maize production ranks third (Zamir et al. 2013). It is a crucial crop for both humans and animals because of its high carbohydrate (76–888%), protein (6–166%), lipid (4–5.77%), and mineral content (1.3%) use in both developed and developing nations (Woldesenbet and Haileyesus 2016). Based on the data, we know that maize thrives best in warm climates with temperatures between 11 and 30 °C and between 5 and 6 h of daily sunlight. The ideal growing conditions for this plant are nutrient-rich, wet soil with a pH of 5.8–7.0. A total of 90–1100 days are needed for growth, and 200–4500 mm of rain is needed throughout the growing season (Yin et al. 2014). In general, it is advised to apply fertilizer containing between 125, and there should be a total of 160 kg N, 55 and 80 kg P_2O_5, and 85 and 110 kg K_2O applied per hectare. The C4 plant type, shared by maize and millets, allows for far greater water efficiency when compared to the C3 plant type, which is why C4 plants like maize are so successful insilty loamy soil soybeans and wheat). With this trait, maize can withstand higher temperatures and repair more carbon dioxide (Anon 2018).

21.3.4 Wheat

In many regions of the world, including the United States, wheat (*Triticum* spp.) is a staple food. When it is harvested, wheat is divided into two categories: winter and spring. Winter wheat germinates just before the earth freezes and then lies dormant until spring. To bloom, it has to be exposed to the cold for a few weeks. It is typically sown in the fall months of October and November and harvested in the summer months of June and July. Wheat is typically sown in the spring, between the months of March and May, and then harvested in the summer, between the months of July and September. The four-month growth period of spring wheat is significantly shorter than that of winter wheat. Wheat requires a growth temperature of 3–23 °C, a 120–180-day growing cycle, and 4–6 h of sunlight each day.

21.4 Changes in Crop Yield Due to Climate Change

Numerous scientific Environmental variables (such as irrigation, temperature, and fertilizer supply) have been demonstrated in research to play a substantial role in paddy output. Paddy yields were found to be considerably lower when employing an alternating wetting and drying irrigation scheme compared to yields obtained with continuous flooding irrigation (Ye et al. 2013). This research also assessed the efficacy of AWD irrigation in conjunction with the controlled delivery of nitrogen fertilizer. In the two years of testing, the wetting and drying approach 28 and 42% decrease in irrigation water supply compared to the flooding method. Increases in paddy production and other forms of plant biomass (roots, shoots, and panicles) were achieved with this strategy despite the decreased energy input. The grain output was enhanced by 5.76–6.76% throughout the course of the two years of the experiment when AWD irrigation was used. Paddy yields rose by 62–72%, 56–67%, and 80–93% traditional urea 240 kg N ha^{1} fertilised fields, controlled-release bulk blending fertilizer at 240 kg N ha^{1}, and polymer-coated urea at 240 kg N ha^{1}. The AWD irrigation system has been tried in nations like India and the Philippines, both of which grow paddy. The results indicate its potential for water saving without reducing crop yields. Recent research by Carrijo et al. (2017) shows that mildAWD can reduce water use by 23.4% compared to continuous flooding approaches. However, compared to the continual flooding approach, the ultimate yield drops by more than 20% of an acute water deficit (soil has dried up) has severely impacted the rice harvest substantially) (Carrijo et al. 2017). Similarly, research indicated that, over the course of two years of trial, paddy output increased from 6532 to 7868 kg ha^{1} under controlled irrigation treatments to 8516–8346 kg ha^{1} under alternating wetting and drying irrigation (Shao et al. 2014). Furthermore, several studies have demonstrated that global warming can reduce worldwide paddy yields. The world's average temperature has risen by 0.5–0.6 °C during the past century (Hansen et al. 2010). As temperatures rise, plants breathe more heavily, speeding up their carbon metabolism while reducing their paddy harvest (Zhao and Fitzgerald 2013). The paddy's blossoms may become infertile as a result of the increased temperatures, which would have a negative impact on the crop's ability to reproduce. According to a recent analysis by the International Food Policy Research Institute (IFPRI), paddy yields might drop by 10–15% due to climate change, leading to a 32–37% increase in market price (Zhao and Fitzgerald 2013; Li 2018). Based on data collected between 1999 and 2007, the ORYZA 2000 model projected that a 2 C increase in temperature might lower paddy yields in Malaysia by 0.36 t ha^1, leading to a substantial economic loss. In a study (Vaghefi et al. 2011), Researchers Matthews et al. discovered that average yields for rice across Asia will drop by 4% as a result of rising atmospheric carbon dioxide (Matthews et al. 1997). Increased worldwide temperatures (particularly in places with colder climates) might enhance yields in some locations, allowing for double-cropping. Unfortunately, the great majority of the world's rice is grown in Asia, and this would not be enough to make up for the total decrease in yields throughout the region grown (Matthews et al. 1997). Researchers concluded that shorter ripening-period cultivars should be used if yields are to be sustained. Numerous studies show that global warming will increase rice yields by anywhere from 2.9 to 34% in the most optimistic scenarios, primarily due to nighttime temperature increases and the relocation of rice production to currently cooler regions that will warm in the 2050s and 2080s, as a result of climate change (Aggarwal and Mall 2002). However, the socioeconomic effects in rice-growing areas and the options for farmers to adapt to the loss of rice production are often overlooked in this research. Furthermore, despite a marginal improvement in output, numerous analyses concur that at the turn of the century, when the world's average temperature rises, yields will plummet dramatically (Lobell et al. 2011a, b; Piao et al. 2010). Future generations will face food poverty if effective solutions to combat lower yields are not introduced soon due to global warming.

21.4.1 Millet

Millet plants have a strong root system that helps them take advantage of any lingering nitrogen, phosphate, or potassium in the soil. Because of its extensive root structure, this plant needs less fertilizer than most others. More than half of all millet grown in pearl millet (*Pennisetum glaucum*) (Singh et al. 2017). Pearl millet's physiological traits make it able to thrive in less-than-ideal situations (such as a lack of water and fertilizer) and use fewer resources than conventional cereal crops (Adhikari et al. 2018). The effects of 0, 40, 80, and 120 kg ha^1 of nitrogen on the growth of four distinct there were a few different varieties of pearl millet germplasm that wereexamined in a single field experiment (Obeng et al. 2012). In this study, we found that different N fertilizer applications had no appreciable influence on pearl millet production, plant dry weight, or head length. Dry fodder pearl millet harvests subjected to varied levels of nitrogen fertilisation (0–300 kg ha^1) were reported to differ significantly, however, in another study (Piri and Tavassoli 2012). Given the inconsistencies in the data, it's important to learn more about how N fertilizers will affect millet harvests. As a result of its deep taproot, millet is also very drought-resistant and may thrive in regions that endure extended dry spells (Hadebe et al. 2017). Multiple studies demonstrated that an increase in temperature results in a greater harvest of millets in arid regions. Millet yields increased by 30–121 kg ha^{1} on average annually when temperatures soared in the study's three Chinese towns (Xifeng, Anding, and Ganzhou). For this reason, they suggest increasing millet crops throughout the province as temperatures increase (Cao et al. 2010). Another thing is that we tweaked the CSM-CERES-Pearl millet simulation modelwas used to evaluate the effect of temperature change on millet production. The study's findings indicated that during a dry spell (when soil water was scarce) and a simulation of high temperatures (raised from 27 C to 29 °C), millet production rose by 6% and 8%, respectively. Millet output increased by 14% due to drought and heat tolerance in the face of climate change. In addition, unlike other grains, millets can thrive in the challenging conditions of hilly and mountainous places (Rasul et al. 2018). This demonstrates that millets have the potential to become an important crop that can flourish in drought, high temperatures, and hilly terrain with minimal nitrogen input. With the lowest global warming potential and resilience to global warming consequences like frequent droughts and higher average temperatures, this crop stands out as a promising option for lowering agricultural emissions.

21.4.2 Maize

Estimates of future crop yields show that climate change would reduce Malawi's maize harvest by 14% by mid-century and by 33% by the end of the century (Msowoya et al. 2016). Similarly, global warming projections for the northeastern part of China found that the region's maize output will drop by 35% in 2030 compared to 2008 levels due to the region's high-temperature increase (where temperatures rise by 1.32 °C) (Li et al. 2014) Corn yields in the United States are expected to drop by 20–50% by the 2050s, according to precipitation modelling, due to global warming. This decline is based on the present emission scenarios. The average loss in yield from 1970 to 1999 was 2.5% (Leng and Huang 2017). It's also important to have a steady supply of water when growing corn. The average yield of irrigated maize during that time period (12.44 t ha^{1}) was 16.5% higher than that of non-irrigated maize, according to the research done between 1999 and 2002 (10.68 t ha^{1}) (Nagy 2003). There was a comparison made between the harvests of corn planted using raised bed irrigation and those cultivated with drip irrigation (Amin et al. 2015). The results demonstrated that the raised bed irrigation system was superior to the drip irrigation method in terms of boosting plant height (by 1%), biological yield (by 5%) and grain production (by 21%). The effects of 2 °C warmer and 20 percent drier on maize yields in Africa have also been studied. They stated that under these conditions, yields dropped by around 10%. To learn more about how moisture affects final maize yields, Lobell and Burke (2010) examined the archival data. They found that in rain-fed circumstances, maize yields dropped by 1% per day and 1.7% per day under drought conditions whenever the crop was exposed to temperatures over 30 °C. They demonstrated the critical role that water plays in maize's resistance to the high temperatures that would result from climate change (Lobell et al. 2011a, b). In 2014, researchers looked at the sensitivity of maize yields to changes in temperature and atmospheric CO_2 and published their findings. Twenty-three models were put through their paces in four distinct countries on four separate continents (France, the United States, Brazil, and Tanzania). According to the findings, there is a direct relationship between temperature and yields, with a decrease in yields of around 0.5 t ha^{1} for every 1 °C increase in temperature. Even if expected increases in yields owing to elevated CO_2 levels materialise, they will not be enough to offset the losses we'll experience (Bassu et al. 2014). Plant growth, harvest, and kernel quality are all directly affected by the availability of nitrogen during the growing season of maize. Applying nitrogen at 0, 23, 46, 69, as well as 92 kg N ha^{l} to a high-yielding maize variety (BH 661). Accumulated grain yield, leaf area, plant height, ear length, kernel quality, ear row spacing, and ear length all increased production were seen alongside increases in nitrogen fertilizer rates. When the number of kernels per ear and ear length were both maximised with 92 kg N ha^{1}, outperforming the control group and other treatments. The BH-660 maize variety was subjected to a similar experiment grown with varying amounts of nitrogen (at 0, 46, 92, 138, and 184 kg ha^{1}) and phosphorus (at 0, 23, 46, 69, and 92 kg ha^{1}) (Tesfaye et al. 2015). The findings demonstrated that the maize grain yield and biomass yield were substantially improved when Nitrogen and Phosphorus were applied together.

21.4.3 Wheat

Wheat is one of the most extensively consumed grains, but the effects of global warming and other weather extremes on productivity will be devastating. We used a modelling technique that relied on observed yield data and gridded climate records from 1950 through 2015 to arrive at our conclusions, the French winter wheat harvest was analysed. This model anticipated that winter wheat output would fall by 14.6–2.4% by 2100, while wheat yields would drop by 3.5–12% between 2037 and 2065 owing to climate change (Gammans et al. 2017). Based on data collected between 1970 and 2000, researchers in China estimated that if the temperature rises by 1 C during the growth season, it will affect wheat output rates by 3–10%. The same study found that without increased usage of resources (such as irrigation, fertilizers, etc.), yields would have dropped by 4.5% due to the rise in temperature during the previous two decades. Wheat yields varied greatly throughout China's climate zones, as observed experimentally between 1981 and 2009. Northern China had a 1–13% increase in wheat yields, while southern China saw a 1–10% decrease (Tao et al. 2014). Using regional panel data (NUTS2), a wheat yield prediction model was developed in Turkey, and the findings indicated that wheat output will drop by 8–23% by the end of 2100. These models projected that as global temperatures rose, soil moisture levels would decrease, resulting in poorer crop yields. Wheat is a key component in many foods throughout the world. Therefore, this might raise its market price. One of the most crucial requirements for healthy wheat growth, uniform maturity, and abundant harvests is access to clean water. Three alternative irrigation durations and schedules were tested on the winter wheat cultivars Baviaans and 14SAWYT306, with no irrigation acting as the control condition. There were three potential irrigation schedules considered: one for the period before stem extension development, one for the period of one for the duration of the cropand from the time the stem extension is created until it reaches physiological maturitygrowth. Grain yield was shown to be considerably higher for both cultivars after receiving irrigation treatments compared to receiving no irrigation treatments. Irrigation during the growth period resulted in a 20.58 percent increase in tillers, a 26.07% rise in grains per spike, a 42.72% increase in grain yield, a 16.71% increase in the harvest index, and a 3.31% increase in grain protein compared to no irrigation.

21.5 Impact of Climate Change on Crops

Producing enough food to fulfil the needs of the expanding global population is one of the most pressing problems of our day. Climate warming complicates an already difficult problem: ensuring a steady supply of food (Cavagnaro et al. 2011). Depending on whether they are C3 or C4 plants, the growth time, development, and yield of crops will be accelerated or retarded by the increased concentration of atmospheric CO_2 near the century's end (Leakey et al. 2009). Plants are divided into C3, C4, and CAM categories according to which pathway they employ to convert carbon dioxide to carbohydrates. Crop expansion can only be encouraged through photosynthesis, respiration, and stomatal conductance. The plant's appearance changed as a result, with leaves getting bigger, the shoot–root ratio being tweaked, flowers opening, grains getting bigger, and harvests increasing (Seneweera and Conroy 2005).

21.5.1 Mechanisms

Rising temperatures, an acceleratedthe four most influential factors that have impacted and will influence agricultural output over the next few decades are the hydrological cycle, increasing CO_2, and improved tropospheric O_3, and they are the subject of this update. How these factors influence crop physiology is briefly discussed below. There are five basic channels via which T influences crop production. To begin, more T has been shown to reduce agricultural yields since it hastens crop growth, resulting in a shorter crop period (Stone 2001). In addition, Temperature, or T, affects grain filling, photosynthesis, and respiration rates. Despite the fact that the optimal temperature (T) for photosynthesis in C4 photosynthesis-based plants (such as maize and sugarcane *Saccharum officinarum* is greater than in C3 photosynthesis-based plants (like rice and wheat), photosynthesis still diminishes at high T in all plants. Based on (Crafts-Brandner and Salvucci 2002). Net photosynthesis (photosynthesis excluding respiration) can go up or down when temperatures are above or below the optimal range during the day, while rising expenses of respiration go up at night, while photosynthesis gains nothing from the extra darkness. Third, when the temperature rises, the saturation vapour pressure of the air increases exponentially. The difference between the saturation vapour pressure and the actual vapour pressure of the air is called the vapour pressure deficit (VPD), and it increases as the temperature difference between the air and the leaf increases, assuming a constant relative humidity. On a global basis, relative humidity has been rather stable over the past few decades (Willett et al. 2007), and it is expected to continue this trend into the foreseeable future. As VPD rises, plants waste more water for the same amount of carbon uptake, leading to less efficient water usage (Ray et al. 2002). Plants limit photosynthetic rates and raise canopy temperature in response to very high VPD, which may have further heat-related consequences. Fourth, plant cells can be directly damaged by high or low temperatures. The probability distribution of T changes as the temperature rises, making hotter temperatures more likely and colder temperatures less likely. In numerous will be an increase in the amount of time that can be spent cultivating crops without fear of frost in the spring and autumn. Frost risks are mitigated in temperate and boreal regions. The growing season is expected to be two weeks longer in 2030 in Scandinavia compared to the late twentieth century (Trnka et al. 2011). Large increases are forecasted for the time of year when temperatures in northern China, Russia, and Canada as well (Ramankutty et al. 2002). On the other hand, heat stress during the reproductive phase is more likely due to global warming, which can cause infertility, poorer decreases in the agricultural success or even yields (Teixeira et al. 2012). Finally, several crop-specific pests and illnesses may benefit from rising temperatures and increased atmospheric CO_2.

21.6 Impacts of Increasing Temperature and Carbon Dioxide Concentration on Product Quality and Yield

The increased temperature brought on by the greenhouse effect caused by increased CO_2 also has other effects on and interactions with the crop's physiology. Whether these interactions are beneficial or detrimental, they affectproduction rate and final product quality through several biochemical and physiological mechanisms (Reich et al. 2015). As a result, it shortens the growth season and reduces yield due to a rise in leaf temperature that has a negative effect on photosynthetic capacity, durability, and senescence, demonstrates the morphological and biochemical alterations brought on by elevated temperatures and CO_2 levels in crop production. For instance, cabbage crops were more prone to heat-related problems such as problems with secondary metabolites, tip burn, yellowing, loose and swollen heads, and accumulation, as well as other side effects. When comparing control and experimental conditions,

Qiao et al. (2019) found that soybean oil concentration was 9% and increased by 14% under enhanced temperature (eT) and elevated CO_2 (e$TeCO_2$), whereas maize grain oil concentration was 12% and increased by 20% under the same conditions, respectively. Increases in temperature and carbon dioxide (CO_2) cause maize plants to absorb more water, produce more chlorophyll, increase their leaf area, and improve their grain output per row and overall (Abebe et al. 2016).

21.7 Impact of Rising Temperatures and Carbon Dioxide on Agricultural Productivity in Percentage

Multiple studies reveal that high temperatures, CO_2, and their combination have either good or negative effects on the growth time, product quality, and yield of corn, wheat, and rice (IPCC 2007; Pathak et al. 2012; Vanaja et al. 2015). High temperature, in conjunction with high CO_2 concentration, mitigates the negative impacts of high temperature on the plant.In contrast, Qiao et al. (2019) discovered that the average soybean yield was increased by 31% under these circumstances, but not under high temperature alone. Underdeveloped maize in open-topped chambers (ambient + 2.1 C, 700 ppm CO_2) saw a roughly 25% increase in yield from either increased temperature and CO_2 or higher temperature alone. Alternatively, Abebe et al. (2016) discovered that elevated CO_2 and ambient + 1.5 C temperature increased yield and total biomass, whereas elevated CO_2 and ambient + 3.0 C temperature decreased yield. As a result, the results demonstrated that increased levels of CO_2 are increasing agricultural yields, but increased levels of heat stunted plant growth and development.

The physiological state of crops and their yields are affected by temperature in several ways.

- As the temperature rises, the air's saturation vapour pressure rises continuously; as a result, the air-to-leaf vapour pressure deficit (VPD) rises as well (VPD represents the difference between the saturation vapour pressure as well as the actual vapour pressure of air. When VPD levels are increased, water consumption becomes less efficient. Nonetheless, plants have a higher evapotranspiration rate per gram of assimilated carbon than animals do (Willett et al. 2007). Plants respond to increased VPD by shutting stomata, slowing photosynthesis, and raising internal temperatures. Thus, such a warming effect may amplify the effects of heat on the plant itself.
- Heat stress can be brought on by extreme temperatures. Infertility, a shorter growth time, and reduced yields can result from its direct effects on plant cells and their division, blooming, and fertilization periods.
- If temperatures are also raised, this might encourage the development, survival, and spread of a wide variety of pathogens and associated illnesses that specifically affect crops, resulting in crop failure (Ziska et al. 2011).

21.8 Strategies for Avoidance and Adjustment

There is a pressing need for long-term planning to implement mitigation and adaptation strategies. This approach has the potential to be a more effective, cost-efficient, and useful method of combating climate change and global warming. Reducing anthropogenic greenhouse gas (GHG) concentrations and preparing for the impacts of climate change are the primary goals of mitigation techniques (Al-Ghussain 2018; Saxena and Sonwani 2020). In contrast to adaptation strategies, which focus on making people more resilient to the consequences of climate change, mitigation efforts are aimed at lowering emissions of greenhouse gases (abatement) (sequestration). Reducing reliance on fossil fuels, switching to green energy sources, making the most of renewable energy, electrifying industrial units, and establishing a well-organized transportation system (including electric public transportation, bicycles, carpools, etc.) are all examples of mitigation measures that could be taken to slow the rise in pollution levels. However, many developing and developed nations will need to take adaptation measures to safeguard their revenue and food security. Howden et al. (2007), however, there were many who argued for major adaptation measures in agricultural operations in response to global warming and climate change.

- To create heat- and drought- and flood- and salt-tolerant species and cultivars to combat climate change
- Reevaluating and adjusting fertilizer dosage to preserve crop production and soil quality.
- Adopting new techniques for water harvesting and irrigating, such as drip irrigation
- Organizing agricultural tasks in light of time and place
- Controlling the flow of river water to make more water available for farming and reduce
- The chances of flooding, soil erosion, and nutrient loss
- Increasing the variety of crops grown
- To encourage organic farming in some areas
- Integrated pest and pathogen management are more likely to work if some new varieties and species are resistant to pests and diseases
- A greater understanding of the value of climate modelling and forecasting can help reduce the risks connected with agricultural output
- Combining fish and rice cultivation to boost farmers' earnings
- Grassland production, modified pasture rotation, grazing time adjustments based on livestock stocking rates, grazing time adjustments based on crop adaptations, and crop adaptations based on livestock usage
- To restore the damaged ecosystem and provide a source of income for rural communities through the introduction and promotion of forest conservation, agroforestry, and forest-based ventures.

21.9 Conclusion

This Chapter provides a synopsis of how climate change is affecting the global cereal crop harvest and the specific needs of each type of cereal crop. Some remedies to these problems and difficulties brought on by global warming were proposed, including the employment of breeding and irrigation techniques. The rising average temperature caused by climate change will have a negative impact on the harvest of key grain crops. Since dry and semi-arid regions throughout the world are more vulnerable to food insecurity, we must prioritize the usage cultivation of hardy food crops to supplement or replace vulnerable primary cereals as the climate continues to shift. If plants are placed in an area with increased $CO_{2,}$ and higher temperatures, the yield and quality of the crop may increase or decrease. In plants, processes including photosynthesis, dark respiration, stomatal conductance, efficient water use, transpiration, and phenological processes are only some of the physiological functions it influences. Therefore, the future generation's food security is threatened by the loss of crop quality and yield. The effects of climate change on crop yield are typically analysed with the potential effects on water productivity and soil water balance. Because of the correlation between soil moisture and rainfall, global warming will have direct consequences for both. Factors such as soil deterioration, temperature extremes, and lack of water throughout the growing season all have a negative impact on crop productivity. Water availability and agricultural output may decline in the future due to rising temperatures and unpredictable precipitation. While increasing agricultural output in aggregate, expanding irrigated areas runs the risk of diminishing both food and environmental quality. Water consumption efficiency may decline the potential effects of climate change in the futureeffects on soil evaporation and plant transpiration. In the context of a warming planet and a fast-expanding human population, this should be part of the plan to ensure everyone has enough to eat. Also, it's important to remember that there's no silver bullet when it comes to ending food poverty. Therefore, a comprehensive strategy is necessary. So, to ensure the sustainable growth of society, we need to establish mitigation and adaptation techniques to reduce pollution. For plants to sustain and minimize crop quality and yield loss, it is now crucial to pick crops by environmental and geographical conditions (Figs. 21.1 and 21.2).

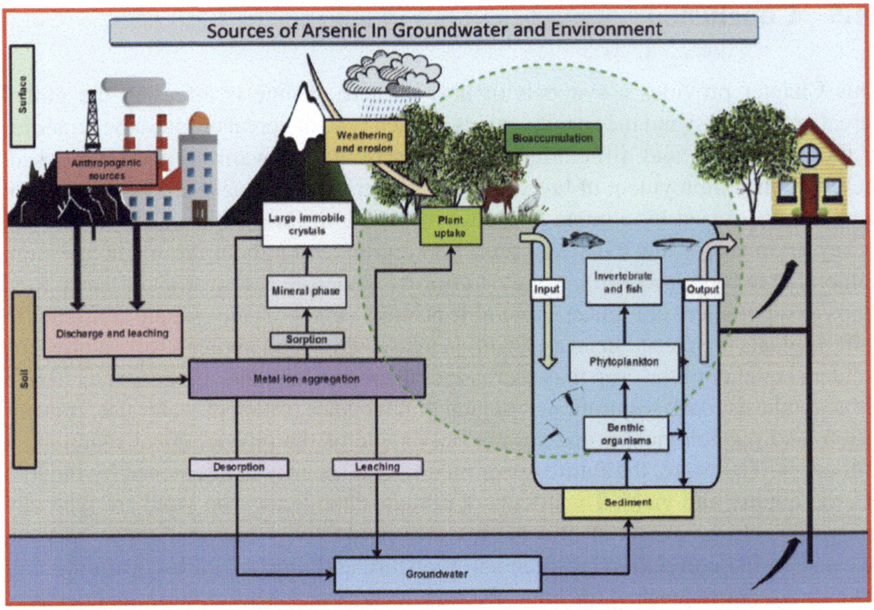

Fig. 21.1 Various models applied in the investigation of climate change's effects

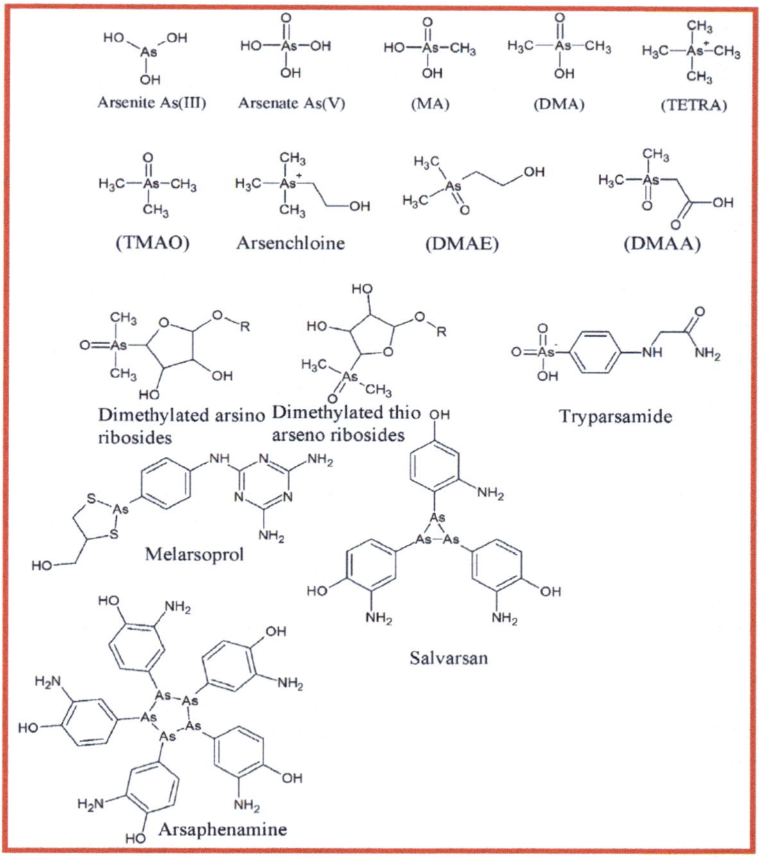

Fig. 21.2 Interaction of higher carbon dioxide in the atmosphere concentration and increased temperature on the crop plant

References

Abebe A, Pathak H, Singh SD, Bhatia A, Harit RC, Kumar V (2016) Growth, yield and quality of maize with elevated atmospheric carbon dioxide and temperature in north–west India. Agric Ecosyst Environ 218:66–72

Adhikari P, Araya H, Aruna G, Balamatti A, Banerjee S, Baskaran P, Barah BC, Behera D, Berhe T, Boruah P et al (2018) System of crop intensification for more productive, resource-conserving, climate-resilient, and sustainable agriculture: Experience with diverse crops in varying agroecologies. Int J Agric Sustain 16:1–28

Aggarwal PK, Mall R (2002) Climate change and rice yields in diverse agro-environments of India. II. Effect of uncertainties in scenarios and crop models on impact assessment. Clim Chang 52:331–343

Al-Ghussain L (2018) Global warming: review on driving forces and mitigation. Environ Prog Sustain Energy

Amin M, Anjum L, Alazba A, Rizwan M (2015) Effect of the irrigation frequency and quality on yield, growth and water productivity of maize crops. Qual Assur Saf Crops 7:721–730
Anon. Fertilizers for Corn. Available online: https://fieldcrops.cals.cornell.edu/corn/fertilizers-corn. Accessed on 24 April 2018
Asseng S, Foster I, Turner NC (2011) The impact of temperature variability on wheat yields. Glob Chang Biol 17:997–1012
Bassu S, Brisson N, Durand JL, Boote K, Lizaso J, Jones JW, Rosenzweig C, Ruane AC, Adam M, Baron C (2014) How do various maize crop models vary in their responses to climate change factors? Glob Chang Biol 20:2301–2320
Becker C, Klaring HP (2016) CO2 enrichment can produce high red leaf lettuce yield while increasing most flavonoid glycoside and some caffeic acid derivative concentrations. Food Chem 199:736–745
Bernstein L, Bosch P, Canziani O, Chen Z, Christ R, Riahi K (2008) IPCC, 2007: Climate change 2007: synthesis report. IPCC, Geneva, Switzerland
Blenkinsop S, Fowler HJ (2007) Changes in European drought characteristics projected by the PRUDENCE regional climate models. Int J Climatol 27:1595–610
Bonan GB, Oleson KW, Vertenstein M et al (2002) The land surface climatology of the community land model coupled to the NCAR community climate model. J Clim 15:3123–3149
Cao L, Wang Q, Deng Z, Guo X, Ma X, Ning H (2010) Effects of climate warming and drying on millet yield in Gansu province and related countermeasures. Ying Yong Sheng Tai Xue Bao 21:2931–2937
Carrijo DR, Lundy ME, Linquist BA (2017) Rice yields and water use under alternate wetting and drying irrigation: a meta-analysis. Field Crops Res 203:173–180
Cavagnaro RT, Gleadow R, Rebecca M (2011) Plant nutrient acquisition and utilisation in a high carbon dioxide world. Funct Plant Biol 38:87–96
Changmei S, Dorothy J (2014) Millet-the frugal grain. Int J Sci Res Rev 3:75–90
Crafts-Brandner SJ, Salvucci ME (2002) Sensitivity of photosynthesis in a C4 plant, maize, to heat stress. Plant Physiol 129:1773–1780
Cuculeanu V, Tuinea P, Balteanu D (2002) Climate change impacts in Romania: vulnerability and adaptation options. Geol J 57:203–209
Dockerty T, Lovett A, Sunnenberg G (2005) Visualising the potential impacts of climate change on rural landscapes. Comput Environ Urban Syst 29:297–332
Fowler HJ, Kilsby CG, O'Connell PE et al (2005) A weather-type conditioned multi-site stochastic rainfall model for the generation of scenarios of climatic variability and change. J Hydrol 308:50–66
Fuss S, Canadell JG, Peters GP et al (2014) Commentary: betting on negative emissions. Nat Clim Chang 4:850–853
Gammans M, Mérel P, Ortiz-Bobea A (2017) Negative impacts of climate change on cereal yields: statistical evidence from France. Environ Res Lett 12:054007
Giri A, Armstrong B, Rajashekar CB (2016) Elevated carbon dioxide level suppresses nutritional quality of lettuce and spinach. Am J Plant Sci 7(1):246
Godfray HCJ, Beddington JR, Crute IR, Haddad L, Lawrence D, Muir JF, Pretty J, Robinson S, Thomas SM, Toulmin C (2010) Food security: the challenge of feeding 9 billion people. Science 327:812–818
Goron TL, Raizada MN (2015) Genetic diversity and genomic resources available for the small millet crops to accelerate a new green revolution. Front Plant Sci 6:157
Groves D, Lempert R (2007) A new analytic method for finding policyrelevant scenarios. Global Environ Change 17:73–85
Gruda N, Tanny J (2014) Protected crops. In: Dixon GR, Aldous DE (eds) Horticulture: plants for people and places, vol 1. Springer, Netherlands, pp 327–405
Habiyaremye C, Matanguihan JB, D'AlpoimGuedes J, Ganjyal GM, Whiteman MR, Kidwell KK, Murphy KM (2017) Proso millet (*Panicummiliaceum* L.) and its potential for cultivation in the Pacific Northwest, US: a review. Front Plant Sci 7:1961

Hadebe S, Modi A, Mabhaudhi T (2017) Drought tolerance and water use of cereal crops: a focus on sorghum as a food security crop in sub-Saharan Africa. J Agron Crop Sci 203:177–191
Hansen J, Ruedy R, Sato M, Lo K (2010) Global surface temperature change. Rev Geophys 48:RG4004
Hartmann DL, Klein Tank AMG, Rusticucci M et al (2013) Observations: atmosphere and surface. In: Stocker TF, Qin D, Plattner G-K et al (eds) Climate change 2013: the physical science basis. Contribution of Working Group I to the Fifth Assessment Report of the Intergovernmental Panel on Climate Change. Cambridge University Press, Cambridge, pp 159–254
Hawkesford MJ, Araus JL, Park R, Calderini D, Miralles D, Shen T, Zhang J, Parry MA (2013) Prospects of doubling global wheat yields. Food Energy Secur. 2:34–48
Högy P, Fangmeier A (2008) Effects of elevated atmospheric CO2 on grain quality of wheat. J Cereal Sci 48:580–591
Howden SM, O'Leary GJ (1997) Evaluating options to reduce greenhouse gas emissions from an Australian temperate wheat cropping system. Environ Modell Software 12:169
Howden SM, Soussana J-F, Tubiello FN, Chhetri N, Dunlop M, Meinke H (2007) Adapting agriculture to climate change. Proc Natl Acad Sci USA 104:19691–19696
IPCC (2007) Summary for policymakers. In: Solomon S, Qin D, Manning M et al (eds) Climate change 2007: the physical science basis. Contribution of working group I to the fourth assessment report of the intergovernmental panel on climate change. Cambridge University Press, Cambridge, pp 1–18
IPCC (2013) Summary for policymakers. In: Stocker TF, Qin D, Plattner G-K, Tignor M, Allen SK, Boschung J, Nauels A, Xia Y, Bex V, Midgley PM (eds) Climate change 2013: the physical science basis. Contribution of Working Group I to the Fifth Assessment Report of the Intergovernmental Panel on Climate Change. Cambridge University Press, Cambridge
IPCC (2014) Climate change 2014: impacts, adaptation, and vulnerability. Part A: Global and sectoral aspects. In: Field CB, Barros VR, Dokken DJ, Mach KJ, Mastrandrea MD, Bilir TE, Chatterjee M, Ebi KL, Estrada YO, Genova RC, Girma B, Kissel ES, Levy AN, Mac Cracken S, Mastrandrea PR, White LL (eds) Contribution of Working Group II to the Fifth Assessment Report of the Intergovernmental Panel on Climate Change. Cambridge University Press, Cambridge, p 1132
Jain R, Bal S (1997) Properties of pearl millet. J Agric Eng Res 66:85–91
Jain N, Arora P, Tomer R, Mishra SV, Bhatia A, Pathak H, Chakraborty D, Kumar V, Dubey D, Harit R (2016) Greenhouse gases emission from soils under major crops in northwest India. Sci Total Environ 542:551–561
Kałuzewicz A, Krzesinski W, Knaflewski M (2009) Effect of temperature on the yield and quality of broccoli heads. Veg Crops Res Bull 71:51–58
Khan S (2008) Managing climate risks in Australia: options for water policy and irrigation management. Aust J Exp Agric 48:265–273
Lal R (2005) Climate change, soil carbon dynamics, and global food security. In: Lal R, Stewart B, Uphoff N et al (eds) Climate change and global food security. CRC Press, Boca Raton (FL), pp 113–143
Leakey AD et al (2009) Elevated CO_2 effects on plant carbon, nitrogen, and water relations: six important lessons from FACE. J Exp Bot 60:2859–2876
Leng G, Huang M (2017) Crop yield response to climate change varies with crop spatial distribution pattern. Sci Rep 7:1463
Li M (2018) Climate change to adversely impact grain production in China by 2030, vol 2018. IFPRI, Washington, DC, USA
Li X, Takahashi T, Suzuki N, Kaiser HM (2014) Impact of climate change on maize production in Northeast and Southwest China and risk mitigation strategies. APCBEE Proc 8:11–20
Lobell DB, Burke MB (2010) On the use of statistical models to predict crop yield responses to climate change. Agric for Meteorol 150:1443–1452
Lobell DB, Bänziger M, Magorokosho C, Vivek B (2011a) Nonlinear heat effects on African maize as evidenced by historical yield trials. Nat Clim Chang 1:42

Lobell DB, Schlenker W, Costa-Roberts J (2011b) Climate trends and global crop production since 1980. Science 333:616–620

Magadza CHD (2000) Climate change impacts and human settlements in Africa: prospects for adaptation. Environ Monit Assess 61:193–205

Matthews R, Kropff M, Horie T, Bachelet D (1997) Simulating the impact of climate change on rice production in asia and evaluating options for adaptation. Agric Syst 54:399–425

Mearns LO (2000) Climate change and variability. In: Reddy KR, Hodges HF (eds) Climate change and global crop productivity. CABI Publishing, New York, pp 7–35

Meinshausen M, Smith SJ, Calvin K et al (2011) The RCP greenhouse gas concentrations and their extensions from 1765 to 2300. Clim Chang 109:213–241

Metzger MJ, Leemans R, Schroter D (2005) A multidisciplinary multi-scale framework for assessing vulnerabilities to global change. Int J Appl Earth Observ Geoinform 7:253–267

Msowoya K, Madani K, Davtalab R, Mirchi A, Lund JR (2016) Climate change impacts on maize production in the warm heart of Africa. Water Resour Manag 30:5299–5312

Myers SS et al (2014) Increasing CO_2 threatens human nutrition. Nature 510:139–142

Nagy J (2003) Effect of irrigation on maize yield (*Zea mays* L.). ActaAgrar. Debreceniensis 1–6. Available online: http://www.date.hu/kiadvany/acta/2003-11i/nagy.pdf. Accessed on 5 May 2018

Nelson GC, Rosegrant MW, Koo J, Robertson R, Sulser T, Zhu T, Ringler C, Msangi S, Palazzo A, Batka M (2009) Climate change: impact on agriculture and costs of adaptation, vol 21. IFPRI, Washington, DC, USA

Obeng E, Cebert E, Singh BP, Ward R, Nyochembeng LM, Mays DA (2012) Growth and grain yield of pearl millet (*Pennisetum glaucum*) genotypes at different levels of nitrogen fertilization in the Southeastern United States. J Agric Sci 4:155–163

Pathak H, Aggarwal PK, Singh SD (2012) Climate change impacts, adaptations and mitigation in agriculture: methodology for assessment and application. Indian Agricultural Research Institute, New Delhi, p xix+302

Piao S, Ciais P, Huang Y, Shen Z, Peng S, Li J, Zhou L, Liu H, Ma Y, Ding Y (2010) The impacts of climate change on water resources and agriculture in china. Nature 467:43

Piperno DR, Flannery KV (2001) The earliest archaeological maize (*Zea mays* L.) from highland Mexico: new accelerator mass spectrometry dates and their implications. Proc Natl Acad Sci USA 98:2101–2103

Piri I, Tavassoli A (2012) Determining the best management of nitrogen fertilizer consumption and harvest time of forage yield of pearl millet (*Pennisetumglaucum*) in Shirvan region. Afr J Microbiol Res 6:2287–2293

Porter JR, Xie L, Challinor AJ, Cochrane K, Howden SM, Iqbal MM, Lobell DB, Travasso MI, Netra Chhetri NC, Garrett K (2014) Food security and food production systems. IPCC, Geneva, Switzerland

Qiao Y, Shujie M, Qi L, Jian J, Xiaosan L, Caixian T (2019) Elevated CO2 and temperature increase grain oil concentration but their impacts on grain yield differ between soybean and maize grown in a temperate region. Sci Total Environ 666:405–413

Ramankutty N, Foley JA, Norman J, McSweeney K (2002) The global distribution of cultivable lands: current patterns and sensitivity to possible climate change. Glob Ecol Biogeogr 11:377–392

Rasul G, Hussain A, Mahapatra B, Dangol N (2018) Food and nutrition security in the Hindu Kush Himalayan region. J Sci Food Agric 98:429–438

Ray JD, Gesch RW, Sinclair TR, Hartwell Allen L (2002) The effect of vapor pressure deficit on maize transpiration response to a drying soil. Plant Soil 239:113–121

Ray DK, Mueller ND, West PC, Foley JA (2013) Yield trends are insufficient to double global crop production by 2050. PLoS ONE 8:e66428

Reich M, Meerakker AN, Parmar S, Hawkesford MJ, De Kok LJ (2015) Temperature determines size and direction of effects of elevated CO_2 and nitrogen form on yield quantity and quality of Chinese cabbage. Plant Biol 18(S1):63–75

Rosales MA, Cervilla LM, Sanchez-Rodríguez E, Rubio-Wilhelmi MDM, Blasco B, Ríos JJ, Ruiz JM (2010) The effect of environmental conditions on nutritional quality of cherry tomato fruits: evaluation of two experimental Mediterranean greenhouses. J Sci Food Agric 91(1):152–162

Salinger MJ, Stigter CJ, Das HP (2000) Agrometeorological adaptation strategies to increasing climate variability and climate change. Agric Forest Meteorol 103:167–184

Saxena P, Sonwani S (2020) Criteria air pollutants and their impact on environmental health. Springer, Singapore

Seneweera S (2011) Effects of elevated CO_2 on plant growth and nutrient partitioning of rice (*Oryza sativa* L.) at rapid tillering and physiological maturity. J Plant Interact 6(1):35–42

Seneweera SP, Conroy JP (2005) Enhanced leaf elongation rates of wheat at elevated CO_2: is it related to carbon and nitrogen dynamics within the growing leaf blade? Environ Exp Bot 54(2):174–181

Shao G-C, Deng S, Liu N, Yu S-E, Wang M-H, She D-L (2014) Effects of controlled irrigation and drainage on growth, grain yield and water use in paddy rice. Eur J Agron 53:1–9

Singh P, Boote K, Kadiyala M, Nedumaran S, Gupta S, Srinivas K, Bantilan M (2017) An assessment of yield gains under climate change due to genetic modification of pearl millet. Sci Total Environ 601:1226–1237

Solomon S (2007) Climate Change 2007—The Physical Science Basis: Working Group I Contribution to the Fourth Assessment Report of the IPCC, vol 4. Cambridge University Press, Cambridge, UK

Solomon S, Qin D, Manning M, Chen Z, Marquis M, Averyt KB, Tignor M, Miller HL (eds) (2007) Climate change 2007: the physical science basis. Cambridge University Press, Cambridge, UK

Stone P (2001) The effects of heat stress on cereal yield and quality. In: Basra AS (ed) Crop responses and adaptations to temperature stress. Food Products Press, Binghamton, NY, pp 243–291

Suppiah R, Hennessy KJ, Whetton PH et al (2007) Australian climate change projections derived from simulations performed for the IPCC 4th assessment report. Aust Meteorol Mag 56:131–152

Tao F, Zhang Z, Xiao D, Zhang S, Rötter RP, Shi W, Liu Y, Wang M, Liu F, Zhang H (2014) Responses of wheat growth and yield to climate change in different climate zones of China, 1981–2009. Agric Meteorol 189:91–104

Teixeira EI, Fischer G, van Velthuizen H, Walter C, Ewert F (2012) Global hot-spots of heat stress on agricultural crops due to climate change. Agric Meteorol (in press)

Tesfaye K, Gbegbelegbe S, Cairns JE, Shiferaw B, Prasanna BM, Sonder K, Boote K, Makumbi D, Robertson R (2015) Maize systems under climate change in Sub-Saharan Africa: potential impacts on production and food security. Int J Clim Chang Strateg 7:247–271

Thomas A (2008) Agricultural irrigation demand under present and future climate scenarios in China. Glob Planet Change 60:306–326

Tripathi A, Tripathi DK, Chauhan D, Kumar N, Singh G (2016) Paradigms of climate change impacts on some major food sources of the world: A review on current knowledge and future prospects. Agric Ecosyst Environ 216:356–373

Trnka M, Olesen JE, Kersebaum KC, Skjelvåg AO, Eitzinger J, Seguin B, Peltonen-Sainio P, Rötter R, Iglesias A, Orlandini S (2011) Agroclimatic conditions in Europe under climate change. Glob Change Biol 17:2298–2318

Vaghefi N, Shamsudin MN, Makmom A, Bagheri M (2011) The economic impacts of climate change on the rice production in Malaysia. Int J Agric Res 6:67–74

Vanaja M, Maheswari M, Jyothi Lakshmi N, Sathish P, Yadav SK, Salini K, Vagheera P, Vijay Kumar G, Razak A (2015) Variability in growth and yield response of maize genotypes at elevated CO_2 concentration. Adv Plants Agric Res 2:42

Venkataramanan S (2011) Causes and effects of global warming. Indian J Sci Technol 4:226–229

Wang Y, Frei M (2011) Stressed food—the impact of abiotic environmental stresses on crop quality. Agric Ecosyst Environ 141(3):271–286

Wen Y, Liu J, Meng X, Zhang D, Zhao G (2014) Characterization of proso millet starches from different geographical origins of China. Food Sci Biotechnol 23:1371–1377

Willett KM, Gillett NP, Jones PD, Thorne PW (2007) Attribution of observed surface humidity changes to human influence. Nature 449:710–712

Woldesenbet M, Haileyesus A (2016) Effect of nitrogen fertilizer on growth, yield and yield components of maize (*Zea mays* L.) in Decha district, Southwestern Ethiopia. Intl J Res Granthaalayah 4:95–100

Xu Y, Zhang Y, Lin E et al (2006) Analyses on the climate change responses over China under SRES B2 scenario using PRECIS. Chin Sci Bull 51(18):2260–2267

Yao F, Huang J, Cui K, Nie L, Xiang J, Liu X, Wu W, Chen M, Peng S (2012) Agronomic performance of high-yielding rice variety grown under alternate wetting and drying irrigation. Field Crops Res 126:16–22

Ye Y, Liang X, Chen Y, Liu J, Gu J, Guo R, Li L (2013) Alternate wetting and drying irrigation and controlled-release nitrogen fertilizer in late-season rice. Effects on dry matter accumulation, yield, water and nitrogen use. Field Crops Res 144:212–224

Yin G, Gu J, Zhang F, Hao L, Cong P, Liu Z (2014) Maize yield response to water supply and fertilizer input in a semi-arid environment of northeast china. PLoS ONE 9:e86099

You L, Rosegrant MW, Wood S, Sun D (2009) Impact of growing season temperature on wheat productivity in China. Agric Meteorol 149:1009–1014

Zamir M, Yasin G, Javeed H, Ahmad A, Tanveer A, Yaseen M (2013) Effect of different sowing techniques and mulches on the growth and yield behavior of spring planted maize (*Zea mays* L.). Cercetari Agronomice in Moldova 46:77–82

Zhao X, Fitzgerald M (2013) Climate change: implications for the yield of edible rice. PLoS ONE 8:e66218

Ziska LH, Blumenthal DM, Runion GB, Hunt ER, Diaz-Soltero H (2011) Invasive species and climate change: an agronomic perspective. Clim Change 105:13–42

Index

P. P. Adhikary et al. (eds.), *Soil, Water Pollution and Mitigation Strategies*, Environmental Science and Engineering, https://doi.org/10.1007/978-3-031-63296-9

R

S

T

W

Z

The manufacturer's authorised representative in the EU is Springer Nature Customer Service Centre GmbH, Europaplatz 3, 69115 Heidelberg, Germany. If you have any concerns regarding our products, please contact ProductSafety@springernature.com

Printed and bound by CPI Group (UK) Ltd, Croydon, CR0 4YY
01/10/2026
02225910-0017